Constructions of Deviance

Social Power, Context, and Interaction

SIXTH EDITION

PATRICIA A. ADLER
University of Colorado

PETER ADLER
University of Denver

WADSWORTH
CENGAGE Learning™

Australia • Brazil • Japan • Korea • Mexico • Singapore • Spain • United Kingdom • United States

WADSWORTH
CENGAGE Learning

Constructions of Deviance: Social Power, Context, and Interaction, Sixth Edition
Patricia A. Adler, Peter Adler

Acquisitions Editor: Chris Caldeira

Assistant Editor: Tali Beesley

Technology Project Manager: Dave Lionetti

Marketing Manager: Michelle Williams

Marketing Assistant: Ileana Shevlin

Marketing Communications Manager: Linda Yip

Project Manager, Editorial Production: Cheri Palmer

Creative Director: Rob Hugel

Art Director: Caryl Gorska

Print Buyer: Barbara Britton

Permissions Editor: Roberta Broyer

Production Service/Compositor: ICC Macmillan Inc.

Copy Editor: Martha Williams

Cover Designer: Yvo Riezebos Design/ Anna Hurley

Cover Image: Jim Dandy/Images.com

For product information and technology assistance, contact us at
Cengage Learning Customer & Sales Support, 1-800-354-9706.

For permission to use material from this text or product, submit all requests online at
www.cengage.com/permissions.
Further permissions questions can be emailed to
permissionrequest@cengage.com.

Library of Congress Control Number: 2007936525

ISBN-13: 978-0-495-50429-0

ISBN-10: 0-495-50429-7

Wadsworth
10 Davis Drive
Belmont, CA 94002-3098
USA

Cengage Learning is a leading provider of customized learning solutions with office locations around the globe, including Singapore, the United Kingdom, Australia, Mexico, Brazil, and Japan. Locate your local office at: **www.cengage.com/global.**

Cengage Learning products are represented in Canada by Nelson Education, Ltd.

To learn more about Wadsworth, visit **www.cengage.com/wadsworth.**

Purchase any of our products at your local college store or at our preferred online store **www.ichapters.com.**

Printed in Canada
2 3 4 5 6 7 8 9 12 11 10 09

To Diane and Dana
Who remind us that the zest of life lies near the margins

and

To Chuck
Who brings out the deviance in everyone

and

To John
Who showed us the miracle of life and rebirth

and

To Jane
Who lives the ordinary as deviant

and

To Dubs and Linda
Who remind us what are friends are for

and

To Lois and David
Who allowed our nephews their independence to be deviant

Contents

Preface

This edition of *Constructions of Deviance,* the sixth, is being written and edited 13 years after the appearance of the first. In the Jewish religion, in a boy's or girl's thirteenth year, it is common to celebrate with a rite of passage called a *bar/bat mitzvah,* a ceremony that marks the passing of the child into young adulthood, or in modern days, more accurately into adolescence. This "coming of age" symbolizes a time when parents become less responsible for their offspring's behavior, as youngsters move toward more accountability for their own actions. In some senses, we think of this edition of the book as its *bar/bat mitzvah,* as the next step in its growth for, as we will describe below, we have taken many of the comments and feedback from previous users and continue to improve it, to react to what readers (both instructors and students) want, and to make changes that, hopefully, move it further into maturity, while maintaining its adolescent ardor. This book is a labor of love for us as we tweak each edition, remain in touch with the changing nuances in the field, and present to our audience what we feel is the most exciting research in the sociology of deviant behavior being produced today.

Born in a blizzard, this edition took birth during the now infamous holiday snowstorms that hit the Front Range region of Colorado in December and January of 2006 and 2007. Due to spend our twenty-seventh consecutive holiday season with Pete's mom in Florida, we were instead barricaded in by literally many feet of snow, impossible to get our car out of the garage, unable to fly out of Denver, and ultimately, that trip to Florida had to be canceled. Finding ourselves home when we had planned to be away, we feverishly turned to the tasks involved in constructing this volume, selecting from myriad possible readings, editing them down to a length appropriate for classroom adoption, writing the synopses that precede each selection, greatly expanding the text that comes before each major part of the book, gaining permission to reprint the articles that appear, and collecting biographical data about all contributors. The two of us, stuck in our house with the wind blaring, the snowflakes falling at record pace, and with nothing else to do, dug our teeth into what we wanted this next edition to look like, how we could capitalize on what people had reported they liked, and what we could do to improve the final product. We hope that these "trapped" days helped us create a book that represents the freedom, the liberty, and the vitality of the sociology of deviance for you.

As the shifting sands of morality in American society continue to transform our culture, deviance and its changing definitions are at the fore. Gusfield (1967) showed us 40 years ago that, with continuous social change, activities considered nondeviant a generation ago can take on deviant characteristics (the near prohibition on cigarettes in public settings is a case in point), whereas activities that were once deviant are now acknowledged as commonplace, albeit still stigmatized (the fads of tattooing and piercing, especially among teenagers). Given the Massachusetts Supreme Court ruling in 2004, perhaps one day same-sex marriage will be widely accepted, although clearly American society is not yet ready for this "revolutionary" idea. For better or worse, we are living at a time when rapid social change is occurring virtually right in front of our eyes. With this type of backdrop, there can be no better time to study the altering definitions of deviant behavior and the consequences that follow for society. It is our hope that, in this sixth edition of *Constructions of Deviance,* we are able to highlight what this field holds. We want to show students the liveliness of the debates, the graphic images that sociologists paint, and the wide array of activities that fall under the domain of deviance.

NEW TO THIS EDITION

In all the editions of this book, we have tried to keep pace with the nuances that have occurred in the research on deviant behavior. In fact, while some have decried that deviance is moribund, that the field offers little new of theoretical or empirical import, we had approximately 70 new articles from which to choose since the publication of the fifth edition. Thus, in three short years, there have been dozens of new pieces of research that we deemed as possibilities for inclusion. Our difficulty, in fact, was not in finding new pieces, but in figuring out how to winnow down all these choices to a select few. We wanted to keep the basics the same, especially because professors who assign this book have become attached to many of the pieces and continue to use them, while at the same time, the book needs an insertion of new materials to keep it fresh. We are pleased to offer you, then, many of the same selections that students have enjoyed in past editions, while presenting 18 new pieces (about 40 percent of the book) that we think will be around for the next set of students who read future editions of this book.

New to this volume are chapters on continuing debates on the definitions of deviance, the moral panic over sexual exploitation of children and the relative ease of access to pornography via the Internet, the rationalizations used by shoplifters to neutralize their deviance, the stigma associated with personal bankruptcy, the negative connotations that male cheerleaders experience, how homeless children handle their subordinate status in society while trying to maintain respectability, recreational users of the "designer drug" ecstasy, the nature of Russian organized crime, the strange bedfellows of the United States government and the Halliburton Corporation in a demonstration of how states and large conglomerates can wield their power and commit deviant acts, a new piece on sexual asphyxia that updates an older one previously published in this book, lesbian cruising, another new piece on sexual assault on campus and the nature of party rape, victims of fraud in Ponzi (pyramid) scams, the decision to commit burglary, crack use among college students, and an epilogue that tackles the issue of the current state of deviance theory and research and its relevance to sociology and to society at large.

We have continued to tweak the sections of the book that introduce concepts, that provide the overview to the various parts, and that discuss how the chapters fit into the larger dimensions of deviance research. In fact, we have added a considerable amount of new text to these sections so that, if the instructor prefers, this book can be used as a stand-alone text/reader or can be synthesized more easily with some of the standard textbooks already in the field. As this book has gone through its various transformations, it has been our intent that it be used as much as a text in its own right as an anthology of empirical works in the discipline. We recommend that you take a close look at these new sections.

As it has from the beginning, the book still leans toward the social constructionist approach, remaining a text that builds upon our own intellectual backgrounds in symbolic interactionism and ethnographic research. As such, it retains its vibrant appeal to students offering the most up-to-date empirical readings that are drawn from participant-observation studies rich in experiential descriptions of deviance from the everyday lived perspective. At the same time, the book has continued to move toward the mean of the field, incorporating classical theoretical and innovative methodological elements that satisfy the needs of professors. To that end, we have included a piece (see Chapter 4) that is intended to serve as a corrective to our general approach, in the interest of providing a more ecumenical presentation and to, hopefully, spark classroom discussion about the current relevancy of social constructionism, labeling, and social power theories of deviance. We will leave it up to you to decide on the applicability of these various theories.

We offer this collection as a testimonial to the continuing vibrancy of deviance research. Our sense is that there has never been a better time to study these phenomena and to look at the changes that they represent about society. Above all else, we hope that you find the readings enjoyable, enlightening, and thought provoking.

ACKNOWLEDGMENTS

By now literally thousands of students have been exposed to the readings we have presented and have christened in the sociology of deviance. Many people have provided critical feedback that has helped us in fashioning this latest edition.

First and foremost are our many students, particularly in Patti's class "Deviance in U.S. Society," at the University of Colorado, where more than 500 bodies cram into Chem 140 each semester, and in Peter's class "Deviance and Society," at the University of Denver. These intrepid souls continue to brave the material and exams in these courses, despite their reputation as among the toughest on each campus. Extra thanks and acknowledgments go to the valiant assistant teaching assistants (ATAs) at the University of Colorado who have dedicated two semesters of their lives to this class to personalize it for other students, to keep the exams hilarious and topical as well as challenging, and to form a cohesive working group. These students have provided us with a template of what contemporary collegians desire, do, and dream about. They remind us of the diversity of sentiments—moral and immoral, normative and deviant, radical and conservative—that exist.

There have been some special people, such as Julia Cantzler, Tim Carpenter, Katherine Coroso, Marci Eads, Marc Eaton, Abby Fagan, Molly George, Tamera Gugelmeyer, Joanna Gregson, Paul Harvey, Tom Hoffman, Katy Irwin, Jennifer Lois, Adina Nack, Patrick O'Brien, Joe Settle, Katie Sirles, Jennifer (Skadi) Snook, and Sarah Sutherland, who have not only served as teaching assistants for this course but have also provided much of the impetus for the changes and amendments we have made throughout. For this edition, we were ably assisted by Emily Tienken, a young, budding sociologist in her own right, whose work may someday grace the pages of this book. We see great things in Emily's future and credit her with the smooth running of the organizational responsibilities that allowed us to produce this book on schedule. Our friends in the discipline continue to suggest studies, supply encouragement, and lend support for our endeavors. Whether through a quick conversation in a hallway or at a convention hotel, an e-mail message, a lengthy letter, or a harangue over the telephone, they remind us that we should keep the edge and continue to search for the latest examples to hold their students' interest.

We have been extremely fortunate to get sets of constructive comments from the reviewers whom Wadsworth has commissioned over the years. For this edition, we are pleased to thank Ken Colburn, Butler University; Phil Davis, Georgia State University; James L. Williams, Texas Woman's University, Denton; Aimee Van Wagenen, Boston College; Dorothy Pawluch, McMaster University; and Lara Foley, University of Tulsa. Previous editions received helpful comments from Kim Davies, Augusta State University; Jackie Eller, Middle Tennessee State University; Kerry Ferris, University of California, Los Angeles; Kathryn Fox, University of Vermont; Scott Hunt, University of Kentucky; Phyllis Kitzerow, Westminster College, Teresa Lagrange, Cleveland State University; Matt Lee, University of Delaware; Jay Livingston, Montclair State University; Barbara Perry, Northern Arizona University; Paul Price, Pasadena City College; Angus Vail, Willamette University; Thomas C. Wilson, Florida Atlantic University; and E. Ernest Wood, Edinboro University of Pennsylvania. The original project received helpful comments from Tom Cook, Wayne State College; William M. Hall, Alfred University; John D. Hewitt, Northern Arizona University; Ruth Horowitz, University of Delaware;

Phyllis Kitzerow, Westminster College; and Michael Nusbaumer, Indiana University-Purdue University at Fort Wayne.

The stalwart staff at Wadsworth has provided unending support during the process of revisions and custom editions. We are fortunate to have worked with such diligent professionals as Matt Ballantyne, Tali Beesley, Paula Begley-Jenkens, Linda deStefano, Halee Dinsey, Jerilyn Emori, Peggy Francomb, Wendy Gordon, Jane Hetherington, Jennifer Jones, Bob Jucha, Bob Kauser, Ari Levenfeld, Kristin Marrs, Lin Marshall, Andrew Ogus, Reilly O'Neal, Michael Ryder, Erica Silverstein, Denise Simon, Steve Spangler, Jennifer Walsh, Jay Whitney, Staci Wolfram, Matthew Wright, Dee Dee Zobian, and Beth Zuber. Special commendation must go to Eve Howard, our senior editor, who has worked hand-in-hand with us since the second edition, and Serina Beauparlant, the editor who originally conceived the project. More recently, we have had the distinct pleasure to have Chris Caldeira, our acquisitions editor, as part of our team, Michelle Williams, marketing manager, and Cheri Palmer shepherding us through the entire production project from reviews to book covers. Our developing friendship with Chris and Michelle has made the work on this edition especially rewarding. Finally, we have been ably assisted by Richard Camp, Winifred Sanchez, and Jill Traut of ICC Macmillan who aided us in the copyediting and production of the book.

One of the pleasures of editing this book has been sharing it with our friends and relatives. We started a tradition with our first edition, which we have continued throughout all editions, of dedicating each volume to a different person or couple who has had a meaningful impact on our lives inside and outside the academy. We respectfully dedicated the first edition to our partners in crime, Diane Duffy and Dana Larsen, the second edition to our dear and enduring friend Chuck Gallmeier, the third edition to the implacable John Irwin, the fourth edition to our intimate friend of over 30 years, Jane Horowitz, and the fifth edition to neighbors and compadres, Linda and Dubs Jacobsen. For this sixth edition, it is our distinct joy to recognize two people who have been with us since our beginnings, who have shared life's vagaries with us, who have lent us their growing family, and who have never wavered in their belief in us, our sister and brother-in-law Lois and David Baru. We are entering a new phase in life with them, one where we look forward to spending even more time together. Finally, our children Jori and Brye keep us young with their irrepressible energy, enthusiasm, and zest for life. To all our readers of previous editions, thanks for the support; to the new readers of this sixth edition, welcome to the journey.

About the Editors

Patricia A. Adler (Ph.D., University of California, San Diego) is Professor of Sociology at the University of Colorado. She has written and taught in the area of deviance, drugs and society, and the sociology of children. A second edition of her book, *Wheeling and Dealing* (Columbia University Press), a study of upper-level drug traffickers, was published in 1993.

    Peter Adler (Ph.D., University of California, San Diego) is Professor of Sociology and Criminology at the University of Denver. His research interests include social psychology, qualitative methods, and sociology of work, sport, and leisure. His first book, *Momentum,* was published in 1981 by Sage.

    Together the Adlers served as Co-Presidents of the Midwest Sociological Society from 2006 to 2007. They have edited the *Journal of Contemporary Ethnography* and were the founding editors of *Sociological Studies of Child Development*. In 2007, the second edition of their anthology, *Sociological Odyssey,* was published by Wadsworth, and in 2001, they released *Encyclopedia of Criminology and Deviant Behavior,* Volume 1, co-edited with Jay Corzine. Among their many books are *Membership Roles in Field Research,* a treatise on qualitative methods published by Sage in 1987, *Backboards and Blackboards,* a participant-observation study of college athletes that was published by Columbia University Press in 1991, *Peer Power,* a study of the culture of elementary schoolchildren that was published by Rutgers University Press in 1998, and *Paradise Laborers,* a study of hotel workers in Hawai'i, published in 2004 by Cornell University Press. Their current project focuses on self-injurers (cutters and burners). Peter and Patti have two grown children, a son, Brye, a film developmental assistant for Plan B Productions, and a daughter, Jori, a television producer on the HBO show "In Treatment." They both graduated from Emory University and reside in Los Angeles. Peter and Patti divide their time between their homes in Boulder, Colorado, and Maui, Hawai'i.

About the Contributors

Robert Agnew, a Fellow of the American Society of Criminology, is Professor of Sociology at Emory University. His research interests focus on the causes of crime and delinquency, and he is best known for his development of general strain theory. His books include *Juvenile Delinquency: Causes and Control* (Roxbury 2005), *Why Do They Do It? A General Theory of Crime and Delinquency* (Roxbury 2005), *Criminological Theory: Past to Present* (co-edited with Francis Cullen, Roxbury 2006), and *Pressured into Crime: An Overview of General Strain Theory* (Roxbury 2006).

Elijah Anderson is currently Charles and William L. Day Professor of the Social Sciences and Professor of Sociology at the University of Pennsylvania. An expert on the sociology of Black America, he is the author of *A Place on the Corner, Streetwise,* and *Code of the Street.* He is interested in the social psychology of organizations, field methods of social research, social interaction, and social organization.

Leon Anderson is Professor of Sociology at Ohio University. He is the co-author of *Down on Their Luck: A Study of Homeless Street People* (with David Snow) and of *Analyzing Social Settings: A Guide to Qualitative Observation and Analysis* (with John Lofland, David Snow, and Lyn Lofland). He is currently completing a textbook, *Deviance: Social Constructions and Blurred Boundaries,* while also pursuing a long-term ethnographic study of recreational skydiving.

Elizabeth A. Armstrong is Associate Professor of Sociology at Indiana University. Her research interests include sociology of culture, social movements, sexuality, gender, and higher education. She is the author of *Forging Gay Identities: Organizing Sexuality in San Francisco, 1950–1994* (University of Chicago Press 2002). She is currently working on a book (with Laura Hamilton) exploring the relationship between college peer culture and social inequality.

Howard S. Becker lives and works in San Francisco. He is the author of *Outsiders, Arts Worlds, Writing for Social Scientists,* and *Tricks of the Trade.* He has taught at Northwestern University and the University of Washington. In 1998, the American Sociological Association bestowed upon him the Career of Distinguished Scholarship Award, the association's highest honor.

Michelle Bemiller received her Ph.D. in Sociology from the University of Akron in 2005. She is now an Assistant Professor of Sociology at Kansas State University. Her interests include the sociology of deviant behavior, criminology, sociology of gender, and sociology of the family. Currently, she is completing research that examines nontraditional mothers' (e.g., incarcerated mothers, non-custodial mothers) experiences with motherhood. She is also researching sexual assault and domestic violence shelter workers' experiences with occupational burnout.

Douglas J. Besharov, a lawyer, is the Joseph J. and Violet Jacobs Scholar in Social Welfare Studies at the American Enterprise Institute for Public Policy Research and Professor at the University of Maryland School of Public Affairs. He was the first director of the U.S. National Center on Child Abuse and Neglect. Among his publications is *Recognizing Child Abuse: A Guide for the Concerned* (Free Press).

Joel Best is Professor of Sociology and Criminal Justice at the University of Delaware. His books include *Threatened Children* (1990), *Random Violence* (1999), *Damned Lies and Statistics* (2001), *Deviance: Career of a Concept* (2004), and *Social Problems* (2007).

Elaine M. Blinde is Professor and Chair of the Department of Kinesiology at Southern Illinois University Carbondale. Her research relates to sociological analysis of sport, with a particular focus on gender issues. Recent work and publications relate to disability and sport, gender beliefs of young girls, and women's relationship to baseball.

Denise Bullock completed her doctoral work at the University of Missouri in Columbia and is Assistant Professor of Sociology at Indiana University East. Her primary research focus has been sexuality research. In addition to lesbian cruising, she has extensively researched sexual identity transformation and is currently examining representations of women and same-sex sexuality in 1920s to 1950s pornography.

William J. Chambliss is Professor of Sociology at George Washington University. He is the author and editor of over 20 books, including *Law, Order and Power* (with Robert Seidman); *On the Take: Petty Crooks to Presidents; Organizing Crime; Exploring Criminology; Boxman: A Professional Thief's Journey* (with Harry King); *Crime and the Legal Process; Making Law* (with Marjorie S. Zatz); and most recently an introductory text, *Sociology* (with Richard P. Applebaum). He is the Past President of the Society for the Study of Social Problems and the American Society of Criminology. He is currently writing a book on the political economy of piracy and smuggling.

Meda Chesney-Lind is Professor of Women's Studies at the University of Hawai'i at Manoa. Nationally recognized for her work on women and crime, her books include *Girls, Delinquency and Juvenile Justice, The Female Offender: Girls, Women and Crime, Female Gangs in America, Invisible Punishment,* and *Girls, Women and Crime,* published in 2004. She has just finished a book on trends in girls' violence, *Beyond Bad Girls* (Routledge).

Terry Cluse-Tolar received her M.A. and Ph.D. in Social Work from Ohio State University. She is currently Associate Professor and Chair of the Social Work Department at the University of Toledo.

Barbara J. Costello is Associate Professor of Sociology at the University of Rhode Island. She was awarded her Ph.D. in Sociology from the University of Arizona in 1994. Her primary area of interest lies in the development and testing of criminological theory, and her current research focuses on the mechanisms of peer influence on both deviant and conforming behavior.

Donald R. Cressey was Professor of Sociology at the University of California, Santa Barbara, when he died in 1987. His most well-known publications include *Principles of Criminology* (with Edwin H. Sutherland); *Other People's Money;* and *Theft of the Nation.* Cressey received many honors for his research and teaching, and he cherished none more than the Edwin H. Sutherland Award presented by the American Society of Criminology in 1967.

Paul Cromwell is Professor of Criminal Justice at Wichita State University. He received his Ph.D. in Criminology from Florida State University in 1986. His publications include 16 books and over 50 articles and book chapters. Recent publications include *Breaking and Entering: Burglars on Burglary* (with James Olson), *In Their Own Words: Criminals on Crime,* and *In Her Own Words: Women Offenders Perspectives on Crime and Victimization* (with Leanne Alarid). He has extensive experience in the criminal justice system, including service as Parole Commissioner and Chairman of the Texas Board of Pardons and Paroles.

Scott H. Decker is Professor and Director in the School of Criminology and Criminal Justice at Arizona State University. He received the B.A. in Social Justice from DePauw University, and the M.A. and Ph.D. in Criminology from Florida State University. His main research interests are in the areas of gangs, juvenile justice, criminal justice policy, and the offender's perspective. His books include *European Street Gangs and Troublesome Youth Groups* (winner of the American Society of Criminology, Division of International Criminology Outstanding Distinguished Book Award, 2006) and *The International Handbook of Juvenile Justice* (Springer-Verlag 2006). His most recent book is *Drug Smugglers on Drug Smuggling: Lessons from the Inside* (Temple University Press 2007).

Douglas Degher is Professor of Sociology at Northern Arizona University. His primary interests are in deviance, theory, and the sociology of sport. Recent interest focuses on identity change as well as work on the construction of tragedy as a means of social control. In his spare time he rides his motorcycle.

Emile Durkheim (1858–1917) was a French sociologist who is generally considered to be the father of sociology. His major works are *Suicide; The Rules of Sociological Method; The Division of Labor in Society;* and *The Elementary Forms of Religious Life.*

Kai T. Erikson has been Professor of Sociology and American Studies at Yale University for more than 30 years. He is the author of several books, including *Wayward Puritans; Everything in Its Path;* and *A New Species of Trouble.* He served as President of the American Sociological Association, 1984–1985.

Kathryn J. Fox is Associate Professor of Sociology at the University of Vermont. She teaches and studies in the area of deviance and social control.

Her ethnographic studies have included punk subcultures, an AIDS prevention project with injection drug users, and prison therapy for violent offenders. She is currently doing research on offender reentry programs.

John H. Gagnon was Professor of Sociology at the State University of New York, Stony Brook until 1998. He is the author or co-author of such books as *Sex Offenders; Sexual Conduct, Human Sexualities;* and *The Social Organization of Sexuality.* In addition, he has been the co-editor of a number of books, most recently *Conceiving Sexuality* and *Encounter with AIDS: Gay Men and Lesbians Confront the AIDS Epidemic,* as well as the author of many scientific articles.

Erich Goode is Sociology Professor Emeritus at the State University of New York, Stony Brook, where he taught for 30 years; he has also taught at New York University, the University of North Carolina, The Hebrew University of Jerusalem, and the University of Maryland. He is the author of 10 books, including *Deviant Behavior, Drugs in American Society, Deviance in Everyday Life,* and *Paranormal Beliefs,* as well as the editor of 6, including *Extreme Deviance.* In 2007, he completed a master's degree in nonfiction writing at Johns Hopkins, focusing mainly on memoir.

Michelle Gourley has been affiliated with the Aboriginal and Torres Strait Islander Health and Welfare Unit of the Australian Institute of Health and Welfare since 2004. She previously graduated with Honors in Sociology from the Australian National University in 2002.

Laura Hamilton is a doctoral student in Sociology at Indiana University Bloomington. Her research interests include gender, family, sexuality, and education. She is the author (with Simon Cheng and Brian Powell) of "Adoptive Parents, Adaptive Parents: Evaluating the Importance of Biological Ties for Parental Investment," *American Sociological Review* (2007).

Alex Heckert is Professor of Sociology at Indiana University of Pennsylvania. Most of his published research has been in the areas of family sociology, deviance, and medical sociology. Recent research in the areas of domestic violence has attempted to improve prediction of nonphysical abuse and physical reassault among batterer program participants. He has published in journals such as *Social Forces, Journal of Research in Crime and Delinquency, Journal of Marriage and the Family, Demography, Journal of Family Issues, Rural Sociology, Family Relations, Violence and Victims, Journal of Family Violence, Journal of Interpersonal Violence,* and *The Sociological Quarterly,* among others.

Druann Maria Heckert received her M.A. from the University of Delaware and her Ph.D. from the University of New Hampshire. She teaches at Fayetteville (NC) State University. Her research is in the areas of stigmatized appearance, positive deviance, and deviance theory, and her articles have appeared in journals such as *Deviant Behavior, The Sociological Quarterly, Symbolic Interaction,* and *Free Inquiry in Creative Sociology.*

Travis Hirschi received his Ph.D. in Sociology from the University of California, Berkeley. He is currently Regents Professor Emeritus at the University of Arizona. He served as President of the American Society of Criminology and has received that Society's Edwin H. Sutherland Award. His books include *Delinquency Research* (with Hanan C. Selvin); *Causes of Delinquency;*

Measuring Crime (with Michael Hindelang and Joseph Weis); and *A General Theory of Crime* (with Michael R. Gottfredson). His most recent book is a co-edited volume with Michael Gottfredson, *The Generality of Deviance.*

Malcolm D. Holmes is a Professor of Sociology at the University of Wyoming. He has published numerous articles concerning the relationships of race/ethnicity to criminal justice decisions. Recently the problem of police brutality has captured his interest. His empirical studies of the issue, which appear in *Criminology,* focus on the effects of minority threat on police brutality criminal complaints. Currently he is completing a book, provisionally entitled *Race, Threat, and Police Brutality,* which undertakes a theoretical analysis of the social psychological dynamics underlying the use of excessive force by the police.

Gerald Hughes is Professor of Sociology at the University of Northern Arizona. His primary interests are in identity theory, with the most recent work focusing on the organizational promotion of permanent deviant identities.

Curtis Jackson-Jacobs received his undergraduate degree in Sociology from the University of Wisconsin-Madison and his Ph.D. in Sociology from UCLA in 2007. Currently, he is Assistant Professor of Sociology at Northern Illinois University. He is interested in college crack users, youth fighting, and street gangs.

Peter Kaufman is Associate Professor of Sociology at the State University of New York at New Paltz. He received his B.A. from Earlham College and his Ph.D. from Stony Brook University. His teaching and research interests include education, critical pedagogy, symbolic interaction, and the sociology of sport. He is currently studying athletes who engage in social and political activism.

Edward O. Laumann is the George Herbert Mead Distinguished Service Professor in the Department of Sociology and the College at the University of Chicago. Previously he has been Editor of the *American Journal of Sociology* and Dean of the Social Sciences Division and Provost at the University of Chicago. Two volumes on sexuality were published in 1994: *The Social Organization of Sexuality* and *Sex in America,* and along with Robert Michael, he has recently published *Sex, Love, and Health: Private Choices and Public Policy* (University of Chicago Press 2001).

Lisa Laumann-Billings completed her doctorate work in developmental and clinical psychology at the University of Virginia. She has worked and published in the areas of divorce, family conflict, and child maltreatment for the past 10 years. She is currently working with Dr. David Olds at the Prevention Research Center in Denver, Colorado, on preventative strategies for reducing child maltreatment and family violence in high-risk families.

John Liederbach is Associate Professor of Criminal Justice at Bowling Green State University. His primary research interests are in the areas of white-collar and professional crime, as well as police behavior and the study of community-level influences on officer activities and citizen interactions.

Joseph Marolla is Vice Provost for Instruction, Professor of Sociology, and affiliated with the Sports Management Program at Virginia Commonwealth University. He is doing research in the sociology of sport and continues to remain interested in the social construction of deviance.

James W. Marquart is a Professor of Criminology in the School of Economic, Political & Policy Sciences at the University of Texas-Dallas as well as the Program Head of the Criminology Program. He has long-term research and teaching interests in prison organizations, capital punishment, criminal justice policy, and research methods. His current research involves an analysis of the long-term effects (i.e., prison violence, racially motivated attacks, and gang-related violence) of the in-cell racial integration policies in the California and Texas prison systems.

Daniel Martin teaches and conducts research in the areas of community, critical organizational studies, and social movements at the University of Minnesota-Duluth. He is presently working on a book, *The Politics of Sorrow: Race, Class, and Identity,* investigating the experiences and organizing efforts of parents in the aftermath of child homicide. He is the recipient of multiple teaching awards, including the Pan Hellenic Award for Outstanding Professor.

Penelope A. McLorg is Director of the Gerontology Program and Adjunct Assistant Professor of Anthropology at Indiana University–Purdue University Fort Wayne. She specializes in biological and sociocultural aspects of aging, with particular interests in health and aging. Dr. McLorg has conducted research in rural areas of Mexico and in the Midwest and has published on such topics as body composition, glucose metabolism, bone loss, eating disorders, and physical disability.

Robert T. Michael is the Eliakim Hastings Moore Distinguished Professor and Dean of the Harris Graduate School of Public Policy at the University of Chicago. From 1984 to 1989 he served as Director of the National Opinion Research Center (NORC).

Stuart Michaels served as Project Manager of the National Health and Social Life Survey (NHLS).

Jody Miller is Associate Professor of Criminology and Criminal Justice at the University of Missouri, St. Louis. Dr. Miller specializes in feminist theory and qualitative methods. Her research focuses on gender, crime and victimization, particularly in the contexts of youth gangs, urban communities, and the commercial sex industry. She has published numerous articles, as well as the monograph *One of the Guys: Girls, Gangs and Gender* (Oxford University Press 2001). Her new monograph, *Getting Played: Violence against Urban African American Girls,* is forthcoming with New York University Press.

Janet L. Mullings, Ph.D., is Associate Dean and Associate Professor in the College of Criminal Justice at Sam Houston State University in Huntsville, Texas. Her research and teaching interests include long-term consequences of victimization, child abuse and neglect, family violence, and women offenders.

Devah Pager is Assistant Professor of Sociology at Princeton University. Her research focuses on the social and economic consequences of mass incarceration, with a particular emphasis on how our crime policies contribute to enduring racial inequality. Using an experimental field methodology, Pager's research uncovers the hidden world of employment discrimination and illustrates the great barriers faced by both blacks and ex-offenders in their pursuit of economic self-sufficiency.

Lisa Pasko received her Ph.D. in Sociology at the University of Hawai'i at Manoa and her M.A. in Sociology from the University of Nevada, Reno. She is currently Assistant Professor of Sociology and Criminology at the University of Denver. In addition to gender and sex work studies, she has been involved in criminal justice research for over 10 years. As project coordinator for the University of Hawaii Youth Gang Project, she evaluated numerous prevention and intervention programs for at-risk youth. As juvenile justice research analyst for the Attorney General, she developed profiles of the serious juvenile offender and the female juvenile offender in Hawai'i. Her publications include *The Female Offender* and *Girls, Women, and Crime: Selected Readings,* which she co-authored with Meda Chesney-Lind. Additionally, Dr. Pasko has published an examination of ethnic disparities in federal drug offense sentencing and a feminist analysis of restorative justice initiatives.

Lyndy A. Potter earned her B.A. (Hons.) degree from the University of New England (Australia, 1995) and a M.S. in Clinical Psychology from Morehead State University (KY, 1998). Her specialties included forensic psychology and grief. She retired from community work in 1999 to devote her time to supporting individuals and families affected by progressive multiple sclerosis. She spends much of her time now running a website for MS-affected individuals and families and translating scientific research to that audience.

Roberto Hugh Potter received his Ph.D. in Sociology from the University of Florida (1982) and has developed a varied work history. He has been a criminal justice planner, nonprofit executive, and a Sociology/Criminology faculty member in Australia and the United States. He is currently a Health Scientist with the Centers for Disease Control and Prevention where he oversees planning activities in the area of corrections, shelters, colleges/universities, and day care. He is involved in the interface between public health and the criminal justice system, including issues such as disease prevention, testing, and linkage to services; prison rape elimination; and pandemic flu planning. Most of his writing these days is translating scientific research for practice audiences.

Douglas W. Pryor is Associate Professor in Sociology and Chair of the Department of Sociology, Anthropology & Criminal Justice at Towson University in Maryland. He is co-author of *Dual Attraction: Understanding Bisexuality* and author of *Unspeakable Acts: Why Men Sexually Abuse Children.* His current research and writing includes papers on the topics of both bisexuality and fear of rape among college women.

Richard Quinney is Professor of Sociology Emeritus at Northern Illinois University. His books include *The Social Reality of Crime; Critique of Legal Order; Class, State, and Crime; Criminology as Peacemaking* (with Harold E. Pepinsky); and *Criminal Behavior Systems* (with Marshall B. Clinard and John Wildeman). His autobiographical reflections are contained in *Journey to a Far Place* and *For the Time Being.*

Craig Reinarman is Professor of Sociology at the University of California, Santa Cruz, and Visiting Scholar at the Center for Drug Research at the University of Amsterdam. He has served on the Board of Directors of the College on Problems of Drug Dependence, as a consultant to the World Health Organization's

Programme on Substance Abuse, and a principal investigator on research grants from the National Institute on Drug Abuse. Dr. Reinarman is the author of *American States of Mind* (Yale University Press 1987), and co-author of *Cocaine Changes* (Temple University Press 1991), and co-editor of *Crack in America* (University of California Press 1997).

Anne R. Roschelle is Associate Professor and Chair of the Department of Sociology at the State University of New York at New Paltz. She earned her Ph.D. at the State University of New York at Albany and is author of numerous articles on the intersection of race, class, and gender with a focus on extended kinship networks and family poverty. Anne is the author of *No More Kin: Exploring Race, Class, and Gender in Family Networks,* which was a recipient of Choice Magazine's 1997 Outstanding Academic Book Award. She is currently writing a book about homeless families in San Francisco and a series of articles about work and family in Havana, Cuba. In addition, she is an avid hiker and plays flute in a local rock band called Questionable Authorities (with co-author Peter Kaufman).

Dawn Rothe is Assistant Professor at the University of Northern Iowa, Department of Sociology, Criminology, and Anthropology. She received her Ph.D. in Sociology from Western Michigan University in 2006. She has written, researched, and spoken on state crime, the International Criminal Court, crimes of genocide, and crimes against humanity, a subject that cuts to the heart the interrelated issues of social justice. Her current work is centered on cases relevant to the Court including Darfur, the Congo, and Uganda, analyzing the etiological and structural factors that led to the atrocities in each area. Rothe is co-author of *Symbolic Gestures and the Generation of Global Social Control: The International Criminal Court* and *Power, Bedlam, and Bloodshed: State Crime in Africa*. She has published in *Social Justice, Crime, Law and Social Change, Critical Criminology,* and *Humanity and Society.*

Robert Rush holds a Ph.D. in Sociology from the University of Delaware where he is presently a member of the supplemental faculty in the Department of Sociology and Criminal Justice. He began his career in Criminal Justice with the FBI as a Special Agent where his investigative specializations were organized crime, labor racketeering, white-collar crime, crime scene investigation, and bombing matters. For a portion of his Bureau career he was assigned to the FBI Laboratory as a Supervisory Special Agent examiner in the Instrumental Analysis Unit. He retired from the Bureau in December 1989. His current interests include criminological theory, criminal justice policy, occupational and corporate crime, organized crime, and domestic and international terrorism.

Frank R. Scarpitti received his Ph.D. from Ohio State University and is the Edward and Elizabeth Rosenberg Professor Emeritus of Sociology and Criminal Justice at the University of Delaware. He has done research and published in a number of areas of criminology and deviance, and served as President of the American Society of Criminology.

Diana Scully is Professor of Sociology and Women's Studies at Virginia Commonwealth University in Richmond, Virginia, where she is also Director of the Women's Studies Program. Her books include *Men Who Control Women's*

Health (Teachers College Press 1994) and *Understanding Sexual Violence* (Routledge 1994).

Edwin H. Sutherland (1883–1950) received his Ph.D. from the University of Chicago. His major works include his enduring textbook, *Criminology*, which was first published in 1924, and *The Professional Thief* and *White Collar Crime*. He is generally regarded as the founder of differential association theory.

Brian Sweeney received his Ph.D. from Indiana University, Bloomington. He is currently Assistant Professor of Sociology at Long Island University: CW Post Campus. His research focuses on youth and adolescence, peer cultures, and gender and sexuality.

Diane E. Taub is Professor and Chair of the Department of Sociology at Indiana University–Purdue University Fort Wayne. Her research primarily involves the sociology of deviance, with a focus on issues related to identity formation and the management of a deviant identity. Recent publications concern eating disorders and the lived experiences of women with physical disabilities.

Debb Thorne is currently Assistant Professor of Sociology at Ohio University, where she was named University Professor in 2004. She received her Ph.D., M.A., and B.A. degrees from Washington State University. As co-principal investigator of the Consumer Bankruptcy Project, she is currently conducting research that examines how the strain of personal bankruptcy affects people's relationships with their partners and children.

Quint Thurman is Professor of Criminal Justice and Department Chairperson at Texas State University-San Marcos. He received a Ph.D. in Sociology from the University of Massachusetts (Amherst) in 1987. His publications include seven books and more than 35 refereed articles that have appeared in such journals as the *American Behavioral Scientist, Crime and Delinquency, Criminology and Public Policy, Social Science Quarterly, Justice Quarterly, Police Quarterly*, and the *Journal of Quantitative Criminology*. Recently published books include *Controversies in Policing* (with co-author Andrew Giacomazzi), an anthology, *Contemporary Policing: Controversies, Challenges, and Solutions* (with Jihong Zhao), and *Police Problem Solving* (with J. D. Jamieson).

Adam Trahan is a doctoral student in the Department of Criminal Justice at Indiana University. He received his B.A. in Criminal Justice and M.A. in Criminology and Criminal Justice from Sam Houston State University. His research interests include capital punishment, white-collar crime, and criminological theory.

Justin L. Tuggle received his B.A. from Humboldt State University and his M.A. from the University of Wyoming. He teaches third grade at Grant Elementary School in Redding, California. He is married with two children.

Brent E. Turvey received a Bachelor of Science degree from Portland State University in Psychology, with an emphasis on Forensic Psychology, and an additional Bachelor of Science degree in History. He went on to receive his Masters of Science in Forensic Science after studying at the University of New Haven. Since graduating in 1996, Brent has consulted with many agencies, attorneys, and police departments in the United States, Australia, China, Canada, Barbados, and Korea on a range of rapes, homicides, and serial/multiple rape/death cases, as a forensic scientist and criminal profiler. He has also been court qualified as an expert in the areas of criminal profiling, forensic science, victimology, and crime reconstruction. He is the author of *Criminal Profiling: An Introduction to Behavioral Evidence Analysis,* 2nd Edition, and co-author of the *Rape Investigation Handbook* (2004), and *Crime Reconstruction* (2006)—all with Elsevier Science. He is currently a full partner, Forensic Scientist, Criminal Profiler, and Instructor with Forensic Solutions, LLC, and an Adjunct Professor in Criminology at Oklahoma City University.

Martin S. Weinberg received his Ph.D. at Northwestern University and is Professor of Sociology at Indiana University. He is the co-author of *Deviance: The Interactionist Perspective; Sexual Preference; Homosexualities; Male Homosexuals;* and *Dual Attraction;* and he has contributed articles to such journals as the *American Sociological Review, Social Problems, Journal of Contemporary Ethnography, Archives of Sexual Behavior,* and *Journal of Sex Research.*

Colin J. Williams is Professor of Sociology at Indiana University-Purdue University, Indianapolis. He is co-author of *Male Homosexuals; Sex and Morality in the U.S.;* and *Dual Attraction.*

Celia Williamson received her B.A. in Social Work from the University of Toledo, her M.A. in Social Work from Case Western Reserve University, and her Ph.D. in Social Work from Indiana University. She is currently Associate Professor in the Department of Social Work at the University of Toledo and Chair of the Second Chance Advisory Board that advocates for services for women in street-level prostitution and children trafficked into the sex trade.

Richard Wright is Curators' Professor and Chair of Criminology and Criminal Justice at the University of Missouri-St. Louis, where he teaches courses on qualitative research methods. He has been studying active urban street criminals, especially residential burglars, armed robbers, carjackers, and drug dealers, for over 15 years. His research has been funded by the National Institute of Justice, Harry Frank Guggenheim Foundation, Icelandic Research Council, National Consortium on Violence Research, and Irish Research Council for the Humanities and Social Sciences. His most recent book, co-authored with Bruce Jacobs, is *Street Justice: Retaliation in the Criminal Underworld* (Cambridge University Press 2006).

General Introduction

The topic of deviance has held an enduring fascination for students of sociology, gripping their interest for several reasons. Some people have career plans that include law or law enforcement and want to expand their base of practical knowledge. Others feel a special affinity for the subject of deviance based on personal experience or inclination. A third group is drawn to deviance merely because it is different, offering the promise of excitement or the exotic. Finally, some are interested in how social norms are constructed, in the ways that people and societies decide what is acceptable and what is not. The sociological study of deviance can fulfill all these goals, taking us deep into the criminal underworld, inward to the familiar, outward to the fascinating and bizarre, and finally back to the central core. In the following pages we peer into the deviant realm, looking at both deviants and those who define them as such. In so doing, we look at a range of deviant behaviors, discuss why people engage in these, and analyze how they are sociologically organized. We begin in Part I by defining deviance, in an effort to lay down the parameters of its scope.

STUDYING DEVIANCE

Reasonable theories and social policies pertaining to deviance must be based on a firm foundation of accurate knowledge. Social scientists have an array of different methodologies and sources of data at their disposal, including survey research, experimental design, historical methods, official statistics, and field research (ethnography). All these methods have obvious strengths and weaknesses and have been used by sociologists in studying deviance. While some generate statistical portraits about the extensiveness of deviant behaviors, we undertake, in this book, to offer a richer, more experiential understanding of what goes on in deviant worlds, showing *how* things happen and what they *mean* to participants. It is the reports of field researchers, individuals who immerse themselves personally in deviant settings, which yield such depth and descriptive accounts. It is also our belief that, due to the often secretive nature of deviance, methods that objectify or distance researchers from the people being researched will be less likely to accurately portray deviant worlds. Thus, the works in this book are tied together by their descriptive richness and by the belief that researchers must study deviance as it naturally occurs in the real world. The most appropriate methodology for this, field research, advocates that sociologists should get as close as possible to the people they are studying in order to understand their worlds (Adler and Adler 1987). Despite the problems that arise from the secretive and hidden nature of deviant acts, sociologists have devised techniques to penetrate secluded deviant worlds. These methodological ploys often come complete with perils, so it is wise for people who are considering studying deviance to be aware of these issues. Part III discusses field research methods and two of the other common sources of information about deviant behavior.

CONSTRUCTING DEVIANCE

In Parts I, II, and IV we delve into the origins and definitions of deviant behavior. Most sociology courses begin by examining core definitions in the field and leave these ideas behind shortly thereafter. This is not the case with deviance. Definitions of deviance pervade all aspects of the field and are therefore addressed throughout the book. Scholars, politicians, activists, moral entrepreneurs, religious zealots, journalists, and people from all walks of life frequently discuss issues of what is deviant and what is socially acceptable. Those whose definitions come to be reflected in law and social policy gain broad moral and material resources.

There are many approaches to defining and theorizing about deviance, some of which are presented in the chapters in Parts I and II. As our title suggests, we advance a social constructionist view throughout this book. We begin, here, by discussing **three perspectives on defining deviance** and locating these in relation to social constructionism.

Proponents of the **absolutist perspective** have traditionally considered defining deviance as a simple task, implying that widespread agreement exists about what is deviant and what is not. Emile Durkheim, a functionalist and one of sociology's founding fathers, represents this theoretical approach, arguing that the laws of any given society are objective facts. Laws, he believes, reflect the "collective consciousness" of each society and thereby reveal its true social nature. They exist before individuals enter the society, and they exist when individuals die; as such, laws represent a level of reality *sui generis* (unique unto themselves), transcending individual lives.

According to this position there is general agreement among citizens that there is something obvious within each deviant act, belief, or condition that makes it different from the conventional norms. At its fundamental core, each embodies the unambiguous, objective "essence" of true or real deviance. This perspective has its roots in both religious and naturalistic assumptions; these people argue that certain acts are contrary to the strictures of God or to the laws of nature. Deviance is thus immoral (possibly evil), sinful, and unnatural. Contemporary religious leaders, especially those of the charismatic and evangelical persuasions, often use these arguments in advancing their moral beliefs in written and verbal oratory. Absolutist views of deviance are eternal and global: If something is judged to have been intrinsically morally wrong in the past (for example, adultery or divorce), it should be recognized as wrong now and always in the future. Similarly, if something is considered to be morally wrong in one place, it should be judged wrong everywhere. Absolutist views on deviance flourish in homogeneous societies, where there is a high degree of universal agreement on social values.

Deviance, then, is not viewed as something that is determined by social norms, customs, or rules, but something that is intrinsic to the human condition, standing apart from and existing before the creation of these socially created codes. According to absolutists, deviant attitudes, behavior, or conditions by any name would be recognized and judged similarly. People have backed up this belief system by pointing to the existence of universal taboos surrounding such acts as murder, incest, and lying. These acts, they claim, are deviant in their very essence.

Noteworthy to this perspective is its focus on the deviance itself. Proponents believe that an absolute moral order is a necessary part of reality, enabling all people to know what is right and what is wrong. Normative behavior is inherently good and deviance is inherently bad. Violations of norms, it follows then, should be met with stern reactions. People who question the norms deserve even harsher treatment, as they challenge the moral order. At its core, then, absolutism is an objectivist approach because it relies in its definition on internal, inherent features that stand apart from subjective human judgments. We see contemporary applications of the absolutist perspective in the campaign against gay marriage, with opponents arguing that homosexuality is an abomination and a sin. The harm-based conception of deviance, presented in Costello's chapter four, offers another contemporary illustration of this viewpoint.

Functionalist theories of deviance, as represented in Erikson's Chapter 4 and Durkheim's Chapter 6, incorporate elements of the absolutist perspective by suggesting that deviance is pathological (diseased) in its substance and negative in its effect. As such, it stands apart from, and in strong contrast to, the "normal." In all ways, the absolutist perspective stands as the foil, or antithesis, to social constructionism.

A group of theories coalesce to form the *social constructionist approach,* grounded in the interactionist theory of deviance. We can draw distinctions between theories, but they all share in common a focus on the norms that bound and define deviance rather than a gaze on the deviance itself. They also share a subjectivist approach to defining deviance, guided by the belief that social meanings, values, and rules, in concrete situations, are often problematic or uncertain. Social meanings, they believe, arise in the situations where they occur, rather than being located within the essences of things, and are heavily influenced by people's perceptions and interpretations. Social constructionists study the ways that norms are created, the people who create them, the conditions under which they arise, and their consequences for different groups in society. Fundamentally, they view definitions of deviance as social products and focus on those defining deviance and their definitions, rather than on the acts that generate deviant reactions.

For example, recently there has been greater social awareness about obesity and its deleterious effects. Absolutists would say that the social definition of obesity as deviant was established by doctors as a health issue and that the level of obesity in our society can be objectively measured by a scientific instrument: the scale. They would point to the growth in portion sizes in America (supersizing), as the average dinner plate has expanded from 10 to 12 inches. Americans, they would say, are more obese because they are eating more and exercising less. On the other hand, constructionists would suggest that our collective attitudes toward obesity have changed, and that levels of weight that used to raise little attention have become less tolerated now. In addition, they would show that people react to weight differently in various cultures, as citizens in some countries prefer more "full-figured" people, and in other places the image of the skinny model is considered the ideal body shape. Clothing companies have subjectively renumbered the sizes of their garments so that size 4 is the "new" size 6 (what women should strive for), and size 8 is the "new" size 14 (you might as well forget about appearing in public). Crusaders, especially from the medical community, have made campaigns against obesity and raised social awareness about it, making it one of the chief panics in society today. The point is that, although being overweight can lead to numerous physical problems, we have shifted our definitions about how much weight is acceptable. As a result, when we look at people and try to assess if they are slightly overweight, chubby, plump, heavy, fat, or outright obese, our categories have changed. What was previously considered tolerably heavy has now been redefined as obese and labeled deviant.

Falling within social constructionism is the **relativist perspective,** articulated in Becker's Chapter 3. Spurred by the rise of subcultural studies in the 1930s, deviance theorists began to note the existence of norms that differed from, and even conflicted with, those of the larger society. This led them to consider the possibility that groups in society make up rules to fit the practical needs of their situations. The more the relativists examined norms in different places and times, the more they became convinced that definitions of deviance were not universal but varied to suit the people who hold them. This suggested that definitions of deviance derived not from absolute, unchanging universals such as God and nature, but from humans. Becker (1963, 9) articulated this position, noting:

> *Social groups create deviance by making the rules whose infraction constitutes deviance,* and by applying these rules to particular people and labeling them as outsiders. From this point of view, deviance is *not* a quality of the act the person commits, but rather a consequence of the application by others of rules and sanctions to an "offender." (*Italics in original*)

Deviance, relativists argue, is thus lodged in the eye of the beholder rather than in the act itself, and it may vary in the way it is defined by time and place. Becker (1963, 147–48) further stated,

> Formal rules such as laws require the initiative or enterprise of specific individuals and groups for their reality. They do not exist independently of the actions of moral entrepreneurs to make them real. . . . The prototype of the rule creator is the crusading reformer. He is interested in the content of the rules. The existing rules do not satisfy him because there is some evil which profoundly disturbs him. He feels that nothing can be right in the world unless rules are made to correct it. He operates within an absolute ethic; what he sees is truly and totally evil with no qualification. Any means is justified to do away with it. The crusader is fervent and righteous, often self righteous . . . [He or she] typically believe[s the] mission is a holy one.

Definitions, this position suggests, are forged by crusading reformers, reforged by them anew in different eras and locations, leading to significant constructions and reconstructions of deviance. When different social contexts frame and give meaning to the perception and interpretation of acts, the same act may be alternately perceived as deviant or normative.

This creates the possibility of multiple definitions of acts, both deviant and nondeviant, simultaneously existing among different groups. We see this clearly in cases of controversial, morally debated acts such as getting an abortion, praying in school, the use of medical marijuana, gay marriage, and illegal immigration. Major campaigns have been assembled to sway public opinion over the morality and appropriateness of these behaviors, with large segments of the country falling into oppositional camps. Groups that are fairly alike in their composition may forge shared agreements about norms that would be differently received among a broader segment of the general public. For example, more widely acceptable behavior, such as dancing, is condemned in some thoroughly conservative environments, such as Bible-Belt religious colleges. At the same time, language that is widely used throughout the country may be condemned as politically incorrect and morally offensive on extremely liberal campuses.

Even some acts that are considered universally taboo can vary in their definitions, as there are conditions under which they would be considered nondeviant. To kill somebody for personal gain, vengeance, freedom, or through negligence might be considered deviant (and, more likely, criminal), but when murder is committed by the state (for executions, war, or covert intelligence), in self-defense, such as when catching your spouse in bed with someone else (if you are a man, but not if you are a woman), under the influence of insanity, or on one's own property (in states where they have "make my day" laws), it is considered not only nondeviant, but also possibly heroic. Similarly, anthropologists have found cultures that condone certain forms of incest to quiet or soothe infants (Henry 1964; Weatherford 1986). Often, too, people differentiate between normal lies and "white" lies, accepting those designed to spare someone else's feelings as acceptable. Relativists argue that definitions of deviance are social products that are likely to be situationally invoked under certain circumstances. They represent the social constructionist approach because they focus on the circumstances under which social norms are differentially created and applied.

Paradoxically, extensions of functionalist theory bring it into the social constructionist realm. Despite their belief that deviance is pathological, functionalists overwhelmingly believe that all components of society contribute, somehow, to its existence; they all have positive functions. Since deviance is a universal feature of all societies, they admit that it must offer benefits. Four positive functions of deviance have been outlined. First, when people react against the deviance of others, they bond together into cohesion and social solidarity. We saw this after the terrorist attacks of 9/11. Second, identifying and punishing deviance redefines and reinforces the social boundaries, which are mutable and not fixed, and reinforce the dangers of transgressing those boundaries. When business executives such as Martha Stewart or Kenneth Lay and Jeffrey Skilling of Enron go to prison for insider trading, "creative accounting," and income tax evasion, it clarifies the limits of acceptable behavior. Politicians who get caught abusing their power send an impetus to others to not only avoid such behaviors, but also to pass stronger ethics rules. Third, Durkheim noted the seeds of social change in deviance, as new developments are often initially regarded skeptically or fearfully and have to go through a process of moral passage to become accepted. Without deviance, he suggested, society might stagnate. Fourth, the existence of all the occupations associated with deviance (the criminal justice system, the medical establishment, the media, the scholars, etc.) would be threatened without deviance, so that were crime or deviance to disappear entirely (a "society of saints"), new behaviors would be defined as deviant to fill this void. In sum, functionalists concluded that a certain amount of deviance is good for society. But because too much or too little is not as beneficial as just the right amount, these definitions must be continually socially constructed and adjusted to ensure the smooth functioning of society.

A third position on defining deviance, presented in Quinney's Chapter 5, is the **social power perspective.** This builds on the relativist perspective by asserting that views on crime and deviance are not arbitrarily formed by just any group of "others." Closely tied to Marx's conflict theory, the social power approach focuses on the influence that powerful groups and classes have in creating and applying laws. Rather, as Quinney (1970, p. 43) notes, these laws are a reflection of those with the greatest social power in society:

> Criminal laws support particular interests to the neglect or negation of other interests, thus representing the concerns of only some members of society. Though some criminal laws may involve a compromise of conflicting interests, more likely than not, criminal laws mark the victory of some groups over others. The notion of a compromise of conflicting interests is a myth perpetuated by the pluralistic model of politics. Some interests never find access to the lawmaking process. Other interests are overwhelmed in it, not compromised. But ultimately some interests succeed in becoming criminal law, and are able to control the conduct of other.

Thus, Quinney believes that laws reflect the interests and concerns of the dominant classes in a society. All laws reflect power relations, and their existence as well as their enforcement illustrate which groups have the power to control, and which groups are controlled.

This approach incorporates social constructionism by challenging the simplistic assumption underlying the absolutist perspective that definitions of deviance are universally shared. According to the social power viewpoint, society is characterized by conflict and struggle between groups whose interests conflict with each other, with the powerful classes dominating the subordinate groups. Members of each group pursue their own needs and desires, but because of the way society is structured, what is good for one class restricts the opportunities of others. This pits groups against each other in active struggles. Those with the greatest social power dominate both the ability to create definitions, rules, and laws, and the way these are negligently or aggressively enforced.

The conflicting interests of the dominant and subordinate groups can fall into the economic realm, as we see some political parties promoting the rights of big businesses to pay fewer taxes, less overtime compensation, and lower health care benefits, while other parties fight for the rights of workers to unionize, to gain job security, and to earn a decent living wage. Other issues along which groups in society may conflict fall into the social realm, including environmental policies, birth control, stem cell research, school prayer, gun control, health care, and drug policies. Pervading all of this, conflict theorists note a systematic differential oppression along the lines of race/ethnicity, social class, and gender.

Feminist theorists share conflict theorists' constructionist orientation, as we find in Chesney-Lind's Chapter 10. They focus on the norms, policies, and laws of the patriarchal system that uphold the social, moral, economic, and political order that fosters male privilege. Their approach questions the way men forge and maintain their dominance over women, and the role of definitions of deviance in this endeavor. They challenge the bases and benefits of patriarchy, regarding it as a form of socially constructed social control.

Deviance can be defined through the social meanings collectively applied to people's attitudes, behavior, or conditions, which are rooted in the interaction between individuals and social groups. Those who have the power to make and apply rules onto others control the normative order. The politically, socially, and economically dominant groups enforce their definitions onto the downtrodden and powerless. Deviance is thus a representation of unequal power in society.

DEVIANT IDENTITY

The second component of the social constructionist approach lies in the consequences of definitions and applications of deviance. Interactionists have argued that the two sides of constructing deviance, its articulation and its application, are both critical. Society first labels various attitudes, behaviors, and conditions as deviant, and then it labels specific individuals associated with these as deviant. Where definitions of deviance are forged, but not applied, they believe, deviance does not really exist. It's as if a tree falls in the forest, but we conclude that it makes no sound because no one is there to hear it. In effect, social constructionists are proposing that the essence of the sound does not lie in the impact of the tree on the ground or in the creation of the sound waves, but in the articulation of those waves against the hearer's eardrum. Specific individuals or groups of people must be labeled by society for deviance to be concretely envisioned.

Part V takes up this change of focus away from the construction of deviant definitions to looking at how the application of norms and laws affects people. Now, we move away from macrosocietal explanations to focus on microinteraction that occurs in everyday life. Constructionists claim that deviants are people who have undergone some sort of labeling. This section looks at how deviant labels are applied and their subsequent consequences.

Social constructionists emphasize that something profound may occur when the supposed deviants and conventional others interact. In pursuing their actions, individuals may engage in deviance but not think of themselves as deviant actors. Only when they begin to apply the deviant labels "out there" in the world to themselves do they truly become deviants. This is the process of acquiring a deviant identity. People may become dislodged from their safe identity locations within the "normal" realm through their own observations as well as the actions and remarks of others. The greater the response of others, the stronger their self-conceptions as deviant will be. This may lead to their engaging in further deviance.

Several aspects of this social psychological process are the most critical and will be addressed most vividly in this section. We begin by looking at how people acquire deviant identities. Many factors are influential, and we consider the ways people creatively use "accounts" or "motive talk" to explain, neutralize, or justify their actions to forestall being labeled as deviant. Although functionalists have suggested that labeling people as deviant has positive results for society, it has negative results for those labeled. They acquire deviant stigma, a stain or pejorative connotation associated with them or their actions. Living with known stigma makes life difficult for people, as they may be marked, disparaged, or shunned. People handle their stigma differently, and we examine these variations and their consequences in Part V.

THE SOCIAL ORGANIZATION OF DEVIANCE

We conclude this volume, in Parts VI, VII, and VIII, with a discussion of how deviants, their deviance, and their deviant careers are socially organized. Earlier sections of the book have concentrated on macro- and microlevels of addressing deviance by considering the movements and powers that shape deviant definitions at the societal level and by looking at how people's identities are shaped at the interpersonal level. Here, we take a meso (midlevel) focus by looking at how deviants organize their social organization and relationships, activities and acts, and careers in connection with others. We begin with the study of deviant organization, examining the various ways members of deviant scenes organize their relations with each other. These range from individuals acting on their own, outside of relationships with other deviants, to subcultures, to more tightly connected gangs, and to highly committed international cartels, and finally to corporations. We then consider the structure of deviant acts. Some forms of deviance involve cooperation between the participants, where people mutually exchange illicit goods or services. Others are characterized by conflict, where some parties to the act take advantage of others, often against their will. Finally, we look at the phases and contours of deviant careers, beginning with people's entry into the world of deviant behavior and associates, continuing with the way they fashion their involvement in deviance, and concluding with their often problematic, and occasionally inconclusive, retirements from the compelling world of deviance.

EPILOGUE

We finish the book by introducing readers to one of the liveliest issues currently raging in the sociology of deviance. Debate has become heated about the state of the sociology of deviance. While some have condemned research and theorizing about deviance as stale and moribund, lacking in new theoretical advances, we disagree with this assessment. We assert, rather, that the field of deviance offers unparalleled insight into society, particularly in current times. Deviance thrives in America, from the underbelly of hidden life worlds to the new frontiers of discovery and social change. Although groups resist the label of deviance and its consequences, the process of deviance making has become so important that it is understood and practiced across a broad spectrum of people. In Part IX we offer some examples of the way concepts and theories of deviance have filtered into commonsense understanding of the world and how they illuminate the stratification, dynamics, and turmoil of America.

Defining Deviance

In order to study the topic of deviance we must first clarify what we mean by the term. What behaviors or conditions fall into this category, and what is the relation between deviance and other categories such as crime? When we speak of deviance, we refer to violations of social norms. Norms are behavioral codes or prescriptions that guide people into actions and self-presentations conforming to social acceptability. Norms need not be agreed upon by every member of the group doing the defining, but a clear or vocal majority must agree.

One of the founding American sociologists, William Sumner (1906), conceptualized three types of norms: *folkways, mores,* and *laws.* He defined folkways as simple everyday norms based on custom, tradition, or etiquette. Violations of folkway norms do not generate serious outrage but might cause people to think of the violator as odd. Common folkway norms include standards of dress, demeanor, physical closeness to or distance from others, and eating behavior. People who come to class dressed in bathing suits, who never seem to be paying attention when they are spoken to, who sit or stand too close to others, or who eat with their hands instead of silverware (at least in the United States) would be violating a folkway norm. We would not arrest them nor would we impugn their moral character, but we might think that there was something peculiar about them.

Mores (pronounced "mor-ays") are norms based on broad societal morals whose infraction would generate more serious social condemnation. Interracial marriage, illegitimate childbearing, and drug addiction all constitute moral violations. Upholding these norms is seen as critical to the fabric of society, so that their violation threatens the social order. Interracial marriage threatens racial purity and the stratification hierarchy based on race; illegitimate childbearing threatens the institution of marriage and the transference of money, status, and family responsibility from one generation to the next; and drug addiction represents the triumph of hedonism over rationality, threatening the responsible behavior necessary to hold society together and to accomplish its necessary tasks. People who violate mores may be considered wicked and potentially harmful to society.

Laws are the strongest norms because they are supported by codified social sanctions. People who violate them are subject to arrest and punishment ranging from fines to imprisonment (and possibly even death). Many laws are directed toward behavior that used to be folkway or, especially, mores violations, but became encoded into laws. Others are regarded as necessary for maintaining social order. Although violating a traffic law such as speeding breaks society's rules, it will not usually brand the offender as deviant.

Following closely on Sumner's distinctions, Smith and Pollack (1976) suggested that deviance might be conceptualized as violations of the norms associated with *crime, sin,* and *poor taste.* For example, criminal acts, such as murder, rape, assault, robbery, and arson, would be violations of laws. Smith and Pollack view these as generally unacceptable to the large majority of the people in society. Acts of sin tend to be defined in relation to religious proscriptions and often involve such acts as promiscuous, lewd, extramarital, or homosexual sex, gambling, drinking, and abortion. While some of these may be subject to criminal sanction, the majority of them are not. Societal responses, rather, tend to fall into the moral category of strong disapproval. Finally, acts of poor taste, like folkways, which challenge existing standards of fashion or manners, violate social norms and are unregulated by law.

This discussion returns us to the question about the relationship between *deviance* and *crime.* Are they identical terms, is one a subset of the other, or are they overlapping categories? To answer this question, we must consider one facet of it at a time. First, do some acts fall into both categories, crime and deviance? The overlap between these two is extensive, with crimes of violence, crimes of harm, and theft of personal property considered both deviant and illegal. Second, are there types of deviance that are not crimes? Actually, much deviance, such as obesity, stuttering, physical handicaps, racial intermarriage, and unwed pregnancy, is noncriminal. In these cases, then, deviance is not a subset of crime. Finally, is there crime that is nondeviant? Although much crime is considered deviant and derives from various lesser deviant categories, some criminal violations do not violate norms or bring moral censure. Examples of this include some white-collar crimes commonly regarded as merely aggressive business practices, such as income tax evasion, and forms of civil disobedience, where people break laws to protest them. Thus, here crime would not be considered a subset of deviance. Crime and deviance, then, can best be seen as overlapping categories with independent dimensions.

People can be labeled deviant as the result of the **ABCs of deviance:** their *attitudes, behaviors,* or *conditions*. First, they may be branded deviant for alternative sets of *attitudes* or belief systems. These beliefs commonly are religiously or politically based: People who hold radical or unusual views of the supernatural (cult members, Satanists, fundamentalists) or who have extreme political attitudes (far leftists or rightists, terrorists) are considered deviant. Mental illness also falls into the deviant attitudinal category, as people with deviant worldviews ("the world is coming to an end") are often considered mentally ill and people with chemical, emotional, or psychological problems may also be considered deviant.

The *behavioral* category is the most familiar one, with people coming to be regarded as deviant for their outward actions. Deviant behaviors may be intentional or inadvertent and include such activities as violating dress or speech conventions, engaging in kinky sexual behavior, smoking marijuana, or committing murder. People cast into the deviant realm for their behaviors have an *achieved deviant status:* They have earned the deviant label through something they have done.

Other people regarded as deviant may have an *ascribed deviant status,* based on a *condition* they acquire from birth. This would include having a deviant socioeconomic status such as being either extremely poor or ultrarich; a deviant racial status such as being a person of color (in a dominantly Caucasian society); or having a congenital physical disability. Here, there is nothing that such people have done to become deviant and little or nothing they can do to repair their deviant status. Moreover, there may be nothing necessarily inherent in these statuses that make them deviant: They may become deviant through the result of a social definitional process that gives unequal weight to powerful and dominant groups in society. Conditional deviant status may be ascribed or achieved. On the one hand, people may be born with conditional deviance due to their personal and/or racial/ethnic characteristics (height, weight, color). On the other hand, a conditional deviant status can also be achieved, such as when people burn or disfigure themselves severely, when they become too fat or too thin, or when they cover their bodies with adornments such as tattoos, piercings, or scarification. Unlike ascribed statuses, some of these may also be changed, moving the deviant back within the norm.

Finally, it is interesting to consider the way deviance is assessed and categorized, as this varies between different eras, with different behaviors, and to different audiences. Deviance may be perceived and interpreted through the lens of the **three categories of Ss:** *sin, sick,* and *selected*. During the Middle Ages and many earlier times when religious paradigms about the world prevailed, deviance from the norm was attributed to religious disorders and viewed as sinful. Nonnormative attitudes, beliefs, and conditions were attributed to satanic influences, and exorcisms were performed to cure people. Religious leaders were seen as the arbiters of official morality and called upon to make judgments and administer sanctions. Even today, some religious zealots adhere to this perspective, not only in societies that are officially religious in their governance, but also within religious pockets of secular societies. The afflicted, in the eyes of these people, are strongly condemned and often viewed as contagious.

Beginning in the first half of the twentieth century, the medicalization of deviance emerged as a strong perspective in explaining its occurrence in society. Physicians and psychiatrists claimed that drug addiction, sexual misbehavior, juvenile delinquency, homosexuality, and a wide range of other behaviors were rooted in people's psychiatric problems, genetic abnormalities, inherited predispositions, and biochemical characteristics. The sick metaphor was later expanded when the recovery movement (self-help, 12-step programs, DARE programs), educators, and public health officials expanded the reach of this paradigm to include other addictions, eating disorders, self-injury, disruptive classroom actions, compulsive gambling, and violence. Psychiatrists sought to claim ownership over the diagnosis of these forms of deviance so that they could administer outpatient and inpatient therapy. Doctors and drug companies manufactured and prescribed an array of medications to "manage" or "cure" people, with the result that we are now the most legally medicated society in world history. The medicalization movement, fueled by the prestige of science and the lure of the "easy fix" of a pill, drew huge areas of attitudes, behaviors, and conditions into its sphere so that society no longer tolerates people being sad, depressed, rebellious, rambunctious, fidgety, eating too much or too little, or having too little or too much sex. The medicalization perspective had the effect of destigmatizing some deviance, since people who were seen as sick or acting on biological predispositions or inherited "differences" were not seen as making immoral choices but were doing things "beyond their control." This made it more acceptable to mainstream society, albeit not seen as "normal" or "healthy."

At the same time, some thought that the medicalization movement had become too pervasive, absorbing too great a swath of social life within its grasp. Some participants reached out to reclaim the rhetoric of selection over their attitudes, behaviors, and conditions, arguing that these were the result of voluntary choice. They, and the researchers who studied them, have found movements rejecting the medicalized interpretation of such things as homosexuality, gambling, eating disorders, drug use, and self-injury. Instead, they cast these as intentionally selected behaviors, forms of recreation, lifestyle choices, and coping strategies. This demedicalization movement has in some ways empowered and destigmatized people who participate in these forms of behavior. For them, to be seen as ill is to be derogated; to be seen as self-healing or voluntarily deviant is conventional. There is no doubt that these diverse perspectives will continue to compete for the ownership, control, and legitimacy of nonnormative forms of expression.

1

On the Sociology of Deviance

KAI T. ERIKSON

This classic selection examines the functions of deviance for society. Erikson asserts that deviance and the social reactions it evokes are key focal concerns of every community, scrutinized by the mass media, law enforcement, and ordinary citizens. Although we no longer attend public hangings or observe people held in stockades, we are aware of the way society punishes (or fails to punish) deviant acts; in so doing, we continually redraw the social boundaries of acceptability. Rather than being a fixed property, norms are subject to shifts and evolution, and the inter-actions between deviants and agents of social control locate the margins between deviance and respectability.

Erikson notes, ironically, that the very institutions and agencies mandated to manage deviance tend to reinforce it, gathering offenders together, socializing them to further deviant skills and attitudes, and hardening their identities as alienated from mainstream society. Once individuals have been identified as deviant, they undergo "commitment ceremonies" where they are negatively labeled, experiencing a status change that is hard to reverse. Society's expectations that deviants will not reform foster the "self-fulfilling prophesy" by which norm violators reproduce their deviance, living up to the negative images society holds of them. Erikson suggests several other valuable functions that deviance performs in a society: It fosters boundary maintenance so that people know what is acceptable and unacceptable; it bolsters cohesion, integration, and solidarity, thus preserving the stability of social life; and it promotes full employment, guaranteeing jobs for people working in the deviance- and crime-management sector.

Human actors are sorted into various kinds of collectivity, ranging from relatively small units such as the nuclear family to relatively large ones such as a nation or culture. One of the most stubborn difficulties in the study of deviation is that the problem is defined differently at each one of these levels: behavior that is considered unseemly within the context of a single family may be entirely acceptable to the community in general, while behavior that attracts severe censure from the members of the community may go altogether unnoticed elsewhere in the culture. People in society, then, must learn to deal separately with deviance at each

From Erikson, Kai T., *Wayward Puritans: A Study in the Sociology of Deviance*. Published by Allyn & Bacon, Boston, MA. Copyright © by Pearson Education. Reprinted by permission of the publisher.

one of these levels and to distinguish among them in his own daily activity. A man may disinherit his son for conduct that violates old family traditions or ostracize a neighbor for conduct that violates some local custom, but he is not expected to employ either of these standards when he serves as a juror in a court of law. In each of the three situations he is required to use a different set of criteria to decide whether or not the behavior in question exceeds tolerable limits.

In the next few pages we shall be talking about deviant behavior in social units called "communities," but the use of this term does not mean that the argument applies only at that level of organization. In theory, at least, the argument being made here should fit all kinds of human collectivity—families as well as whole cultures, small groups as well as nations—and the term "community" is only being used in this context because it seems particularly convenient.[1]

The people of a community spend most of their lives in close contact with one another, sharing a common sphere of experience which makes them feel that they belong to a special "kind" and live in a special "place." In the formal language of sociology, this means that communities are boundary maintaining: each has a specific territory in the world as a whole, not only in the sense that it occupies a defined region of geographical space but also in the sense that it takes over a particular niche in what might be called cultural space and develops its own "ethos" or "way" within that compass. Both of these dimensions of group space, the geographical and the cultural, set the community apart as a special place and provide an important point of reference for its members.

When one describes any system as boundary maintaining, one is saying that it controls the fluctuation of its consistent parts so that the whole retains a limited range of activity, a given pattern of constancy and stability, within the larger environment. A human community can be said to maintain boundaries, then, in the sense that its members tend to confine themselves to a particular radius of activity and to regard any conduct which drifts outside that radius as somehow inappropriate or immoral. Thus the group retains a kind of cultural integrity, a voluntary restriction on its own potential for expansion, beyond that which is strictly required for accommodation to the environment. Human behavior can vary over an enormous range, but each community draws a symbolic set of parentheses around a certain segment of that range and limits its own activities within that narrower zone. These parentheses, so to speak, are the community's boundaries.

Now people who live together in communities cannot relate to one another in any coherent way or even acquire a sense of their own stature as group members unless they learn something about the boundaries of the territory they occupy in social space, if only because they need to sense what lies beyond the margins of the group before they can appreciate the special quality of the experience which takes place within it. Yet how do people learn about the boundaries of their community? And how do they convey this information to the generations which replace them?

To begin with, the only material found in a society for marking boundaries is the behavior of its members—or rather, the networks of interaction which link these members together in regular social relations. And the interactions which do the most effective job of locating and publicizing the group's outer edges would seem to be those which take place between deviant persons on the one side and official agents of the community on the other. The deviant is a person whose activities have moved outside the margins of the group, and when the community calls him to account for that vagrancy it is making a statement about the nature and placement of its boundaries. It is declaring how much variability and diversity can be tolerated within the group before it begins to lose its distinctive shape, its unique identity. Now there may be other moments in the life of the group which perform a similar service: wars, for instance, can publicize a group's boundaries by drawing attention to the line separating the group from an adversary, and certain kinds of religious ritual, dance ceremony, and other traditional pageantry can dramatize the difference between "we" and "they" by portraying a symbolic encounter between the two. But on the whole, members of a community inform one another about the placement of their boundaries by participating in the confrontations which occur when persons who venture out to the edges of the group are met by policing agents whose special business it is to guard the cultural integrity of the community. Whether these confrontations take the form of criminal trials, excommunication hearings, courts-martial, or even psychiatric case conferences, they act as boundary-maintaining devices in the sense that they demonstrate to whatever audience is concerned where the line is drawn between behavior that belongs in the special universe of the group and behavior that does not. In general, this kind of information is not easily relayed by the straightforward use of language. Most readers of this paragraph, for instance, have a fairly clear idea of the line separating theft from more legitimate forms of commerce, but few of them have ever seen a published statute describing these differences. More likely than not, our information on the subject has been drawn from publicized instances in which the relevant laws were applied—and for that matter, the law itself is largely a collection of past cases and decisions, a synthesis of the various confrontations which have occurred in the life of the legal order.

It may be important to note in this connection that confrontations between deviant offenders and the agents of control have always attracted a good deal of public attention. In our own past, the trial and punishment of offenders were staged in the market place and afforded the crowd a chance to participate in a direct, active way. Today, of course, we no longer parade deviants in the town square or expose them to the carnival atmosphere of a Tyburn, but it is interesting that the "reform" which brought about this change in penal practice coincided almost exactly with the development of newspapers as a medium of mass information. Perhaps this is no more than an accident of history, but it is nonetheless true that newspapers (and now radio and television) offer much the same kind of entertainment as public hangings or a Sunday visit to the local gaol. A considerable portion of what we call "news" is devoted to reports about deviant behavior and its consequences, and it is no simple matter to explain why these items should be considered newsworthy or why they should command the extraordinary attention they do. Perhaps they appeal to a number of psychological perversities among the mass audience, as commentators have suggested, but at the same time they constitute one of our main sources of information about the normative outlines of society. In a figurative sense, at least, morality and immorality meet at the public scaffold, and it is during this meeting that the line between them is drawn.

Boundaries are never a fixed property of any community. They are always shifting as the people of the group find new ways to define the outer limits of their universe, new ways to position themselves on the larger cultural map. Sometimes changes occur within the structure of the group which require its members to make a new survey of their territory—a change of leadership, a shift of mood. Sometimes changes occur in the surrounding environment, altering the background against which the people of the group have measured their own uniqueness. And always, new generations are moving in to take their turn guarding old institutions and need to be informed about the contours of the world they are inheriting. Thus single encounters between the deviant and his community are only fragments of an ongoing social process. Like an article of common law, boundaries remain a meaningful point of reference only so long as they are repeatedly tested by persons on the fringes of the group and repeatedly defended by persons chosen to represent the group's inner morality. Each time the community moves to censure some act of deviation, then, and convenes a formal ceremony to deal with the responsible offender, it sharpens the authority of the violated norm and restates where the boundaries of the group are located.

For these reasons, deviant behavior is not a simple kind of leakage which occurs when the machinery of society is in poor working order, but may be, in controlled quantities, an important condition for preserving the stability of social life. Deviant forms of behavior, by marking the outer edges of group life, give the inner structure its special character and thus supply the framework within which the people of the group develop an orderly sense of their own cultural identity. Perhaps this is what Aldous Huxley had in mind when he wrote:

> Now tidiness is undeniably good—but a good of which it is easily
> possible to have too much and at too high a price. . . . The good life can
> only be lived in a society in which tidiness is preached and practised,
> but not too fanatically, and where efficiency is always haloed, as it were,
> by a tolerated margin of mess.[2]

This raises a delicate theoretical issue. If we grant that human groups often derive benefit from deviant behavior, can we then assume that they are organized in such a way as to promote this resource? Can we assume, in other words, that forces operate in the social structure to recruit offenders and to commit them to long periods of service in the deviant ranks? This is not a question which can be answered with our present store of empirical data, but one observation can be made which gives the question an interesting perspective—namely, that deviant forms of conduct often seem to derive nourishment from the very agencies devised to inhibit them. Indeed, the agencies built by society for preventing deviance are often so poorly equipped for the task that we might well ask why this is regarded as their "real" function in the first place.

It is by now a thoroughly familiar argument t
designed to discourage deviant behavior actually o
petuate it. For one thing, prisons, hospitals, and ot
and shelter to large numbers of deviant persons,
advantage in the competition for social resourc
tions gather marginal people into tightly segreg
tunity to teach one another the skills and attitudes o.
provoke them into using these skills by reinforcing their sen.
the rest of society.[3] Nor is this observation a modern one:

> The misery suffered in gaols is not half their evil; they are filled with
> every sort of corruption that poverty and wickedness can generate; with
> all the shameless and profligate enormities that can be produced by the
> impudence of ignominy, the range of want, and the malignity of dispair.
> In a prison the check of the public eye is removed; and the power of the
> law is spent. There are few fears, there are no blushes. The lewd inflame
> the more modest; the audacious harden the timid. Everyone fortifies
> himself as he can against his own remaining sensibility; endeavoring to
> practise on others the arts that are practised on himself; and to gain
> the applause of his worst associates by imitating their manners.[4]

These lines, written almost two centuries ago, are a harsh indictment of prisons, but many of the conditions they describe continue to be reported in even the most modern studies of prison life. Looking at the matter from a long-range historical perspective, it is fair to conclude that prisons have done a conspicuously poor job of reforming the convicts placed in their custody; but the very consistency of this failure may have a peculiar logic of its own. Perhaps we find it difficult to change the worst of our penal practices because we *expect* the prison to harden the inmate's commitment to deviant forms of behavior and draw him more deeply into the deviant ranks. On the whole, we are a people who do not really expect deviants to change very much as they are processed through the control agencies we provide for them, and we are often reluctant to devote much of the community's resources to the job of rehabilitation. In this sense, the prison which graduates long rows of accomplished criminals (or, for that matter, the state asylum which stores its most severe cases away in some back ward) may do serious violence to the aims of its founders; but it does very little violence to the expectations of the population it serves.

These expectations, moreover, are found in every corner of society and constitute an important part of the climate in which we deal with deviant forms of behavior.

To begin with, the community's decision to bring deviant sanctions against one of its members is not a simple act of censure. It is an intricate rite of transition, at once moving the individual out of his ordinary place in society and transferring him into a special deviant position.[5] The ceremonies which mark this change of status, generally, have a number of related phases. They supply a formal stage on which the deviant and his community can confront one another (as in

nal trial); they make an announcement about the nature of his deviancy
ct or diagnosis, for example); and they place him in a particular role which
ught to neutralize the harmful effects of his misconduct (like the role of pris-
r or patient). These commitment ceremonies tend to be occasions of wide
ublic interest and ordinarily take place in a highly dramatic setting.[6] Perhaps
the most obvious example of a commitment ceremony is the criminal trial,
with its elaborate formality and exaggerated ritual, but more modest equivalents
can be found wherever procedures are set up to judge whether or not someone is
legitimately deviant.

Now an important feature of these ceremonies in our own culture is that
they are almost irreversible. Most provisional roles conferred by society—those
of the student or conscripted soldier, for example—include some kind of terminal
ceremony to mark the individual's movement back out of the role once its temporary advantages have been exhausted. But the roles allotted the deviant seldom
make allowance for this type of passage. He is ushered into the deviant position
by a decisive and often dramatic ceremony, yet is retired from it with scarcely a
word of public notice. And as a result, the deviant often returns home with no
proper license to resume a normal life in the community. Nothing has happened
to cancel out the stigmas imposed upon him by earlier commitment ceremonies;
nothing has happened to revoke the verdict or diagnosis pronounced upon him
at that time. It should not be surprising, then, that the people of the community
are apt to greet the returning deviant with a considerable degree of apprehension
and distrust, for in a very real sense they are not at all sure who he is.

A circularity is thus set into motion which has all the earmarks of a "self-
fulfilling prophesy," to use Merton's fine phrase. On the one hand, it seems
quite obvious that the community's apprehensions help reduce whatever chances
the deviant might otherwise have had for a successful return home. Yet at the
same time, everyday experience seems to show that these suspicions are wholly
reasonable, for it is a well-known and highly publicized fact that many if not
most ex-convicts return to crime after leaving prison and that large numbers of
mental patients require further treatment after an initial hospitalization. The common feeling that deviant persons never really change, then, may derive from a
faulty premise; but the feeling is expressed so frequently and with such conviction
that it eventually creates the facts which later "prove" it to be correct. If the
returning deviant encounters this circularity often enough, it is quite understandable that he, too, may begin to wonder whether he has fully graduated from the
deviant role, and he may respond to the uncertainty by resuming some kind of
deviant activity. In many respects, this may be the only way for the individual
and his community to agree what kind of person he is.

Moreover this prophesy is found in the official policies of even the most
responsible agencies of control. Police departments could not operate with any
real effectiveness if they did not regard ex-convicts as a ready pool of suspects
to be tapped in the event of trouble, and psychiatric clinics could not do a successful job in the community if they were not always alert to the possibility of
former patients suffering relapses. Thus the prophesy gains currency at many levels within the social order, not only in the poorly informed attitudes of the

community at large, but in the best informed theories of most control agencies as well.

In one form or another this problem has been recognized in the West for many hundreds of years, and this simple fact has a curious implication. For if our culture has supported a steady flow of deviation throughout long periods of historical change, the rules which apply to any kind of evolutionary thinking would suggest that strong forces must be at work to keep the flow intact—and this because it contributes in some important way to the survival of the culture as a whole. This does not furnish us with sufficient warrant to declare that deviance is "functional" (in any of the many senses of that term), but it should certainly make us wary of the assumption so often made in sociological circles that any well-structured society is somehow designed to prevent deviant behavior from occurring.[7]

It might be then argued that we need new metaphors to carry our thinking about deviance onto a different plane. On the whole, American sociologists have devoted most of their attention to those forces in society which seem to assert a centralizing influence on human behavior, gathering people together into tight clusters called "groups" and bringing them under the jurisdiction of governing principles called "norms" or "standards." The questions which sociologists have traditionally asked of their data, then, are addressed to the uniformities rather than the divergencies of social life: how is it that people learn to think in similar ways, to accept the same group moralities, to move by the same rhythms of behavior, to see life with the same eyes? How is it, in short, that cultures accomplish the incredible alchemy of making unity out of diversity, harmony out of conflict, order out of confusion? Somehow we often act as if the differences between people can be taken for granted, being too natural to require comment, but that the symmetry which human groups manage to achieve must be explained by referring to the molding influence of the social structure.

But variety, too, is a product of the social structure. It is certainly remarkable that members of a culture come to look so much alike; but it is also remarkable that out of all this sameness a people can develop a complex division of labor, move off into diverging career lines, scatter across the surface of the territory they share in common, and create so many differences of temper, ideology, fashion, and mood. Perhaps we can conclude, then, that two separate yet often competing current sare found in any society: those forces which promote a high degree of conformity among the people of the community so that they know what to expect from one another, and those forces which encourage a certain degree of diversity so that people can be deployed across the range of group space to survey its potential, measure its capacity, and, in the case of those we call deviants, patrol its boundaries. In such a scheme, the deviant would appear as a natural product of group differentiation. He is not a bit of debris spun out by faulty social machinery, but a relevant figure in the community's overall division of labor.

NOTES

1. In fact, the first statement of the general notion presented here was concerned with the study of small groups. See Robert A. Dentler and Kai T. Erikson, "The Functions of Deviance in Groups," *Social Problems*, VII (Fall 1959), pp. 98–107.

2. Aldous Huxley, *Prisons: The "Carceri" Etchings by Piranesi* (London: The Trianon Press, 1949), p. 13.

3. For a good description of this process in the modern prison, see Gresham Sykes, *The Society of Captives* (Princeton, N.J.: Princeton University Press, 1958). For discussions of similar problems in two different kinds of mental hospital, see Erving Goffman, *Asylums* (New York: Bobbs-Merrill, 1962) and Kai T. Erikson, "Patient Role and Social Uncertainty: A Dilemma of the Mentally Ill," *Psychiatry*, XX (August 1957), pp. 263–274.

4. Written by "a celebrated" but not otherwise identified author (perhaps Henry Fielding) and quoted in John Howard, *The State of the Prisons*, London, 1777 (London: J. M. Dent and Sons, 1929), p. 10.

5. The classic description of this process as it applies to the medical patient is found in Talcott Parsons, *The Social System* (Glencoe, Ill.: The Free Press, 1951).

6. See Harold Garfinkel, "Successful Degradation Ceremonies," *American Journal of Sociology*, LXI (January 1956), pp. 420–424.

7. Albert K. Cohen, for example, speaking for a dominant strain in sociological thinking, takes the question quite for granted: "It would seem that the control of deviant behavior is, by definition, a culture goal." See "The Study of Social Disorganization and Deviant Behavior" in Merton et al., *Sociology Today* (New York: Basic Books, 1959), p. 465.

2

An Integrated Typology of Deviance Applied to Ten Middle-Class Norms

ALEX HECKERT AND DRUANN MARIA HECKERT

Deviance has traditionally been regarded as behavior that underconforms to society's norms of acceptability and is negatively received. Deviants were considered antisocial misfits either through their failure to live up to societal standards or through their intentional defiance of norms. In previous works, Heckert and Heckert challenged this view, exploring both of its dimensions: the excessive enactment of pro-social behavior, and people's positive reactions to deviance. They offered a controversial fourfold table interrelating the axis of under/nonconformity and overconformity (how the behavior corresponds to normative expectations) with the suggestion that people do not always react to deviance negatively but sometimes admire it (their social reactions and collective evaluations). By cross-tabulating the dimension of underconformity and over-conformity with that of social evaluation, positive and negative, they yielded four analytic types of deviance: negative deviance, rate busting, deviance admiration, and positive deviance.

In this chapter they further articulate their typology by applying it to a list of 10 middle-class norms developed by Tittle and Paternoster (2000). Addressing such issues as lying, disloyalty, hedonism, irresponsibility, and invasion of privacy, Heckert and Heckert evaluate these forms of "negative deviance" to suggest how such acts might translate into their deviance admiration, positive deviance, and rate-busting counterparts. This creative analysis strengthens their previous typology by fleshing it out with concrete examples that further challenge traditional thinking about definitions of deviance. We encourage readers to think about Heckert and Heckert's assertions carefully rather than just accepting them, and to forge your own decisions about whether you agree with their propositions or not.

From Alex Heckert and Druann Maria Heckert, "An Integrated Typology of Deviance Applied to Ten Middle-Class Norm," *The Sociological Quarterly.* Copyright © 2004 by Blackwell Publishers. Reprinted by permission of the publisher.

A TYPOLOGY OF DEVIANCE INTEGRATING NORMATIVE AND REACTIVIST PERSPECTIVES

An assortment of definitions have clouded the substantive area of deviance, ranging from an absolutist approach to a statistical one to the more ubiquitous normative and reactivist approaches (Goode 2001). Nevertheless, the predominant bifurcation within the discipline is between the normative and reactivist perspectives (Liska 1981). Normative, or objectivist, definitions, on the one hand, primarily focus on the violation of norms (c.f., Merton 1966); simply and succinctly, Cohen defined deviance as "behavior that violates normative rules" (1966, p. 12). On the other hand, reactivist or subjectivist definitions focus on the dynamics of the reaction of a social audience (c.f., Becker 1963; Erikson 1964; Lemert 1972). As Becker (1963, p. 11) so famously wrote, "Social groups constitute deviance by making rules whose infractions constitute deviance, and by applying those rules to particular people and labeling them as outsiders." Each definition presents a major paradigmatic understanding of deviance. These definitions clearly tend to emphasize behaviors. Most sociologists do, however, extend definitions of deviance to include what Adler and Adler (2000, p. 8) term the "ABCs of deviance: their attitudes, behaviors, or conditions." Although the examples we provide in the current article focus primarily on behaviors, attitudes (or beliefs) and conditions are assuredly essential (since normative expectations and social reactions pertain to both) with regard to deviance and can be fruitfully analyzed using our conceptual framework.

Our typology seeks to integrate the underlying phenomena addressed by each definition (Heckert and Heckert 2002). This typology recognizes the distinction as a false dichotomy and acknowledges the existence of both normative expectations and social reactions. Behavior and conditions can underconform (or nonconform) or overconform to normative expectations, as well as lead to negative or positive reactions. As such, we have cross-classified the significance of both norms and reactions, creating four possible scenarios. As Figure 2.1 shows, negative deviance, on the one hand, is underconformity or nonconformity that results in negative reactions. On the other hand, rate busting is overconformity that is negatively evaluated. Next, deviance admiration is underconformity or nonconformity that is positively appraised by the collectivity. Finally, positive deviance is overconformity that receives positive reactions. Each type of deviance will be briefly described.

The substantive area of deviance has first and foremost involved the scrutiny of negative deviance. Negative deviance refers to underconformity or nonconformity that is also negatively evaluated; metaphorically, negative deviance is the "Jeffrey Dahmer phenomenon." As Adler and Adler (2000) described this situation, attitudes, behaviors, or conditions can be deviantized. Negative deviants can range from most criminals to the mentally ill to substance abusers to those with unpopular political or religious stances. Essentially, the standard in the

Normative Expectations

Social Reactions And Collective Evaluations	Underconformity or Nonconformity	Overconformity
Negative Evaluations	Negative Deviance	Rate busting
Positive Evaluations	Deviance Admiration	Positive Deviance

FIGURE 2.1 Deviance Typology

field of deviance is the negative reaction to (and sanctioning of) underconformity or nonconformity.

Alternatively, rate busting is negatively appraised overconformity. At the core this cell epitomizes the "geek phenomenon." Various social scientists have noted that norms operate at two levels; the idealized (i.e., that which is believed sublimely better but improbable for most people) and the realistic (i.e., that which is viewed as achievable by typical people) (Homans 1950; Johnson 1978; Heckert 1998). Nevertheless, this overconformity—even if idealized or perhaps because it is idealized—is often subjected to negative evaluations. Rate busting has occurred in various realms of life. For example, Huryn (1986) determined that gifted students are often rejected by their peers; Krebs and Adinolfi (1975) concluded that the attractive are often slighted by members of their same sex; and Heckert (2003) found that blonde women (defined as over-conforming to European standards of beauty) are subjected to epic stereotyping, especially related to their intellectual capacity (or assumed lack thereof).

The third cell is deviance admiration, which focuses on underconformity or nonconformity that is favorably assessed. For all intents and purposes, this condition could be labeled the "John Gotti phenomenon." Although the attribute, behavior, or condition is normatively unacceptable to a collectivity, the response of that group is still assenting. Kooistra (1989) concluded that a fair number of criminals, even brutal ones, have been transformed—by the collective imagination—from thugs to icons. In other words, Robin Hood is not relegated to the realm of myth but has bona fide real-life counterparts, ranging from

Billy the Kid to D. B. Cooper to Butch Cassidy. As another example, in a society that castigates those individuals who do not meet culturally dominant and created aesthetic images of appearance, there are also many that admire the stigmatized attribute (e.g., people of size and redheads) and individuals with that attribute.

The final category is positive deviance, which describes overconformity that is responded to in a confirmatory fashion. Essentially, this is the "Mother Theresa phenomenon." Similar to negative deviance, positive deviance has been previously defined from various viewpoints, including from a normative perspective (c.f., Wilkins 1965; Winslow 1970), from a labeling point of view (c.f., Freedman and Doob 1968; Hawkins and Tiedeman 1975; Scarpitti and McFarlane 1975; Steffensmeier and Terry 1975), and from the perspective that positive deviance is overconformity that is positively evaluated (Dodge 1985; Heckert and Heckert 2002). Examples of positively appraised overconformity include good neighbors and saints (Sorokin 1950), Congressional Medal of Honor winners (Steffensmeier and Terry 1975), and the physically attractive (c.f., Berscheid and Walster 1964; Byrne 1971; Reis et al. 1982). Exceeding normative expectations, at times, creates a scenario whereby positive reactions and evaluations flow, often producing all kinds of added advantages; thus, the response transcends even the norm at stake.

All in all, this typology recognizes that the traditional distinction between normative expectations and societal reactions involves a false dichotomy and seeks to integrate normative definitions with reactivist definitions. Thus, attitudes, behaviors, and conditions can become conceptualized as negative deviance, rate busting, deviance admiration, and positive deviance. Clearly, as theorists in the field of deviance have long contended, contexts do have to be considered as audiences can differentially react to the same behavior or conditions. In other words, gifted and overachieving students are rate busters to their peers at the same time they are positive deviants to their teachers. Elite tattoo collectors are simultaneously positive deviants to their subculture and negative deviants to the dominant society (Irwin 2003). As well, many individuals who were admired as extraordinary in a later era were castigated in their own era because they dared to not conform to their time's antiquated or oppressive normative code (Coser 1967; Merton 1968; Dinitz, Dynes, and Clarke 1969). Examples include Galileo, Jesus (to Christians), Socrates, and the French Impressionists. Essentially, for divergent reasons, underconformers or nonconformers are often positively evaluated, and overconformers are often negatively evaluated.

This typology allows sociologists to continue to utilize a traditional notion that deviance is relative and that context is critical (Thio 1983). It is important to analyze why underconformity or nonconformity can result in positive evaluations or negative evaluations, depending on the era, place, or social group involved. The same is true for overconformity (Heckert and Heckert 2002). Power and people's self-interest are critical components. In reference to either underconformity or overconformity, negative reactions are likely to emerge when attitudes, behaviors, or conditions threaten the interests of the social group. . . .

NORMS AND PREDOMINANT MIDDLE-CLASS NORMS IN THE UNITED STATES

Integral components of any cultural system, norms guide expected behavior and are based directly on values held to be important. For example, based on fourteen common definitions of norms, Gibbs (1981, p. 7) asserted that most commonly a norm has been defined as "a belief shared to some extent by members of a social unit as to what conduct *ought to be* in particular situations or circumstances." Gusfield (1963, p. 65) further suggested that these systems produce "regularities of action."

Other sociologists have indicated that norms could be dually conceptualized between the idealized and the more realistic (Johnson 1960; Johnson 1978) or operative norms. In essence, the more idealized are really values (c.f., White 1961; Blake and Davis 1964). Williams (1965) created the seminal and most comprehensive description of the value system of the United States, outlining the following dominant values: achievement and success, individualism, activity and work, efficiency and practicality, science and technology, progress, material comfort, humanitarianism, freedom, democracy, equality, and racism and group superiority. To that list Henslin (1975) added education, religiosity, romantic love, and monogamy. Thus, the American value system has been amply addressed.

Recently, Tittle and Paternoster (2000) have expanded sociological understanding by outlining not the idealized scheme of values but the normative system itself. Their outline is restricted to middle-class values in the United States and includes the following dominant norms: group loyalty, privacy, prudence, conventionality, responsibility, participation, moderation, honesty, peacefulness, and courtesy. Loyalty describes the necessity of the group to survive over individual concerns. Privacy indicates that individuals need to be in command of certain spaces and places. Prudence refers to placing pleasure in perspective and to practicing pleasure with moderation. Conventionality is descriptive of people choosing habits and life scenarios that are similar to those of other people. Responsibility is dependability, especially when others must count on that characteristic. Participation implies personal involvement in both economic and social spheres; alienation constitutes the nonconforming counterpart. Moderation means an avoidance of the extremes in life. Honesty denotes veracity and candor. Peacefulness illustrates a sedate and calm style. Finally, courtesy revolves around refined social etiquette in human interaction. Overall, one set of consequential norms has been thoroughly delineated.

APPLYING THE INTEGRATED NORMATIVE-REACTIVIST TYPOLOGY TO MIDDLE-CLASS NORMS OF THE UNITED STATES

To verify the utility of our integrative typology, we seek to apply this typology to a normative system. In this initial foray, we apply the typology to dominant and current American middle-class norms. Tittle and Paternoster (2000) documented

TABLE 2.1   A Classification of U.S. Middle-Class Deviance

Norm	Negative Deviance	Deviance Admiration	Rate Busting	Positive Deviance
Group loyalty	Apostasy	Rebellion	Fanaticism	Altruism
Privacy	Intrusion	Investigation	Seclusion	Circumspection
Prudence	Indiscretion	Exhibitionism	Puritanism	Discretion
Conventionality	Bizarreness	Faddishness	Provincialism	Properness
Responsibility	Irresponsibility	Adventurousness	Priggishness	Hyperresponsibility
Participation	Alienation	Independence	Dependence	Cooperation
Moderation	Hedonism	Roguishness	Asceticism	Temperance
Honesty	Deceitfulness	Tactfulness	Tactlessness	Forthrightness
Peacefulness	Disruption	Revelry	Wimpishness	Pacifism
Courtesy	Uncouthness	Irreverence	Obsequiousness	Gentility

Note: Types of negative deviance taken from Tittle and Paternoster 2000, p. 35.

these norms; as well, they highlighted the deviance (or what we will subsequently refer to as, more specifically, negative deviance) forms of each norm. The deviance admiration, rate busting, and positive deviance variety of each norm is specified in Table 2.1 and will be further discussed in the following sections.

Negative Deviance

To briefly reiterate, negative deviance is underconformity or nonconformity that is negatively evaluated. Tittle and Paternoster (2000) focused on negative deviance and have outlined illuminating examples of each category which are listed in Table 2.2. Again, these examples are quintessentially representative of the traditional nucleus of the substantive area of the sociological study of deviance.

As an example of the valuable work of Tittle and Paternoster (2000), as the survival of the group is viewed as paramount, apostasy becomes deviantized. Consequently, actions and/or actors such as revolution or revolutionaries are deemed deviant. As privacy is critical, intrusion in the form of theft or rape is deviant. Since prudence is valued, indiscretion in the form of prostitution is negatively stigmatized. Because conventionality forms an important normative idea, bizarreness that can include some mentally ill behavior is deviant. For the reason that responsibility is valued, irresponsibility such as family desertion becomes unacceptable to middle-class Americans. Given that participation is normative, alienation—which can include nonparticipatory lifestyles—becomes snubbed. In view of the fact that moderation is critical to middle-class Americans, hedonism in the guise of chiseling or atheism is negatively appraised. Although Tittle and Paternoster (2000) include asceticism as a type of underconformity to the norm of moderation, utilizing this new model we prefer to conceptualize it as overconformity. Because honesty is appreciated, deceitfulness epitomized by

TABLE 2.2 A Classification of U.S. Middle-Class Deviance, Negative Deviance

Norm	Negative Deviance	Examples
Group loyalty	Apostasy	Revolution; betraying national secrets; treason; draft dodging; flag defilement; giving up citizenship; advocating contrary government philosophy
Privacy	Intrusion	Theft; burglary; rape; homicide; voyeurism; forgery; record spying
Prudence	Indiscretion	Prostitution; homosexual behavior; incest; bestiality; adultery; swinging; gambling; substance abuse
Conventionality	Bizarreness	"Mentally ill behavior" (handling excrement, nonsense talking, eating human flesh, fetishes); separatist lifestyles
Responsibility	Irresponsibility	Family desertion; reneging on debts; unprofessional conduct; improper role performance; violations of trust; pollution; fraudulent business
Participation	Alienation	Nonparticipatory lifestyles (being a hermit, street living); perpetual unemployment; receiving public assistance; suicide
Moderation	Hedonism	Chiseling; atheism; alcoholism; total deceit; wasting; ignoring children
Honesty	Deceitfulness	Selfish lying; price-fixing; exploitation of the weak and helpless; bigamy; welfare cheating
Peacefulness	Disruption	Noisy disorganizing behavior; boisterous reveling; quarreling; fighting; contentiousness
Courtesy	Uncouthness	Private behavior in public places (picking nose, burping); rudeness (smoking in prohibited places, breaking in line); uncleanliness

Note: Table from Tittle & Paternoster 2000, p. 35.

bigamy and price-fixing deviantized. Seeing that peacefulness is valued, disruption in the form of noisy disorganizing behavior or boisterous reveling is not appreciated in the context of this particular normative system. As courtesy is normalized, uncouthness as exemplified by rudeness or uncleanliness is stigmatized. All in all, Tittle and Paternoster (2000) have provided a comprehensive accounting of various deviantized behaviors within the middle class of the United States. This outline sketches negative deviance, the substance of the long-established and accepted core of the field of deviance.

Rate Busting

Rate busting refers to the negative reaction not to underconformity but to overconformity and constitutes the "geek phenomenon." In other words, adhering to

TABLE 2.3 A Classification of U.S. Middle-Class Deviance, Rate Busting

Norm	Rate Busting	Examples
Group loyalty	Fanaticism	Various religious cults; Ku Klux Klan (who, of course, engage in acts of negative deviance); Aryan Nation; NRA; religious fanatics; superpatriots; "holier than thou"
Privacy	Seclusion	Hermits; loners; Amish and other reclusive religious sects; secretive or reclusive behavior; Howard Hughes
Prudence	Puritanism	Negative attitudes & behavior toward Amish and other conservative religious sects
Conventionality	Provincialism	Stepford wives; Martha Stewart & her followers; keeping up with the Joneses in following every little rule so as to be accepted; individuals who are negatively evaluated for ritualistically following convention
Responsibility	Priggishness	Negative attitude regarding people who are self-righteous; jokes and meanness toward Martha Stewart and her followers; some workaholics; straight-A student
Participation	Dependence	Brownnoser; a person trying so hard to be accepted by every group that they lose their individuality; the concepts of codependency and enabling behavior
Moderation	Asceticism	Negative reactions to not drinking; a person so meek that they never take a stand on anything so as not to offend anyone
Honesty	Tactlessness	Being too honest with friends (e.g. about ugly clothing, unattractive haircut); a person so honest they won't tell a "white lie" or who will say something mean to a child because it is honest rather than to protect child's feelings
Peacefulness	Wimpishness	A person who never parties or "lets their hair down"; a person who will never take a stand or stand up to people; "yes man"
Courtesy	Obsequiousness	The person who is overly polite; making fun of Miss Manners and people who are overly courteous

the idealized level of the norm induces unfavorable evaluations. As Tittle and Paternoster (2000) analyzed a deviant (or negative deviant) form of each norm, we outline a rate busting form of each of the ten norms in Table 2.3. We also provide a nonexhaustive list of examples for each category of rate busting.

Regarding the norm of group loyalty, overconforming to the group can result in negative evaluations. This category is referred to as fanaticism. As examples, NRA participants and hate group members, accepted subculturally, are too responsive to their groups and are deviantized—by various parts of culture in the first case and by the dominant society in the second case—on this basis (and other important ones as well). Religious cult members are also viewed as too immersed in their groups; their individuality is perceived to be lost to the group. Individuals who are "holier than thou" within dominant religious traditions often estrange members of their own group. "Super patriots" who are too enthusiastic in their display of symbolic representations of nation and who will support government under any circumstance often alienate others. Concerning the norm of privacy, the rate busting equivalent is seclusion. Among those who are negatively reacted to in response to what is deemed their seclusive tendencies are the following: hermits, secretive or reclusive individuals such as Howard Hughes in his later years, and reclusive religious groups, such as the Amish. In reference to the normative orientation respecting prudence, Puritanism refers to those who are overly prudent and negatively stigmatized. For example, very conservative religious sects that seem to eschew what dominant Americans view as necessary pleasures—such as the Amish, Hutterites, Shakers (the very few remaining), and Jehovah's Witnesses—are shunned and/or created into exotic humans and not accepted into the dominant cultural fold. With the situation of conventionality, the rate busting scenario is provincialism. Being overly conventional or mindlessly attempting to conform to the habits of others are often negatively appraised. For example, mass media created the Stepford wives, who are fictionally representative of this category. As a flesh-and-blood counterpart, Martha Stewart and her fans have been constant grist for comedians; this stigmatization occurred even prior to her legal troubles. Pertaining to the norm of responsibility, priggishness is the rate busting form. Some workaholics—those motivated by an intense sense of responsibility—have long been subjected to negative treatment by their coworkers, even if praised by their bosses. As well, peers have long taunted overachieving students, who set a high standard in the academic realm.

Concerning participation, dependence is the rate busting counterpart. Among those who overly participate in social and/or economic life are the brownnoser, the codependent and enabling family member of a substance abuser, and the person who wants to be accepted so badly by the group that he or she sacrifices individuality. With regards to moderation, asceticism is overconformity that is negatively appraised. Individuals who drink socially and moderately are conforming; those who do not drink are overconforming and ironically—similarly to pathological drinkers—are negatively evaluated on that basis (even if the stigmatization is not the same type of stigmatization). For example, nondrinkers will often develop explanations (i.e., designated driver) so as to attempt to avoid the almost inevitable questions rooted in disbelief about their choice. In connection with honesty, tactlessness is the overconforming response that will not be positively received. For example, persons so honest that they will never tell a lie, even a socially acceptable lie, are often viewed as tactless. Thus, the following scenario could occur. If a young girl asks an adult if her dress is pretty and the adult replies that it is the ugliest dress ever created, the adult would be veracious and overconforming to honesty. Nevertheless, under middle-class normative conditions, the adult would not be positively received. In respect to peacefulness, the rate busting sort is wimpishness. A meek person who never "lets her or his hair down" or who will never stand up to anybody so as to avoid contentiousness would be looked down upon and labeled with the following appellation: "mousy" or "yes-man." Vis-à-vis courtesy, obsequiousness is the rate busting form. A person who is so very courteous in social interaction that every diminutive rule of etiquette is exactly followed is viewed as overconforming and often spurned. Consequently, Miss Manners becomes almost a comically viewed person, rather than the serious person she professes to be as the arbiter of such matters. Eddie Haskell, a fictionalized character from *Leave It To Beaver,* is also viewed similarly.

All in all, for each of the ten norms, a rate busting form of deviance is listed in Table 2.3. As well, examples readily illuminate each possibility.

Deviance Admiration

To recap, deviance admiration or the "John Gotti phenomenon" occurs in the context of underconformity that is positively evaluated. In other words, people do not view John Gotti as doing right; rather, they accept that he has violated considerably cherished norms and endorse him anyway. Potentially, deviance admiration can surface for each of these norms as shown in Table 2.4. Each of the examples of negative deviance provided by Tittle and Paternoster (2000), then, could be positively evaluated and therefore potentially could serve as examples of deviance admiration.

Rebellion is the deviance admiration form of group loyalty. Among the most historically revered figures in middle-class American life are the American revolutionaries, including George Washington, Thomas Jefferson, John Hancock, Nathan Hale, and Thomas Paine. They are now emblematic of ultimate patriotism because of their rebellion. Investigation constitutes the deviance admiration of the norm of privacy. For example, revelations on the *Jerry Springer Show* constitute a violation of the norm of privacy; though perhaps not universally admired, the enduring tenure of the show suggests that there are many fans. Exhibitionism composes the deviance admiration type of prudence, whereby pleasure is bounded. Famous strippers and entertainers, such as RuPaul, have fashioned abundant economic lives on this premise. Additionally, participants in extreme sports certainly underconform to norms of prudence and are tremendously acclaimed for this imprudence, especially by the young. Faddishness embodies the deviance admiration form of conventionality. Faddishness describes underconformity to the habits of most others in the context of an appreciation of that behavior. For example, Americans are presently charmed by the Osbournes; other than the love that permeates their family bonds, they are not conventional. As well, tattoos (and body piercing) and individuals—such as Dennis Rodman, Allen Iverson, and virtually any rock star—who acquire these accoutrements of a youthful imaginative aesthetic are often appreciated. Adventurousness makes up the deviance admiration form of responsibility. Many adventurers of unconquered terrains, including expeditionary individuals who made early conquests of the Artic, Antarctica, Mount Everest, and outer space, have created a permanent place in history for themselves.

**TABLE 2.4 A Classification of U.S. Middle-Class Deviance,
Deviance Admiration**

Norm	Deviance Admiration	Examples
Group loyalty	Rebellion	James Dean (rebel without a cause); American Revolution; Pentagon papers; admiration for men who went to Canada to avoid draft during Vietnam War
Privacy	Investigation	Investigative journalism; private investigators; revelations on *Jerry Springer Show*
Prudence	Exhibitionism	Famous strippers; gay rights movement; famous gamblers; flamboyant entertainers (e.g., Liberace, RuPaul, transvestites, female impersonators); extreme sports; reality television like *The Osbournes* & the *Anna Nicole Smith Show*
Conventionality	Faddishness	*The Osbournes;* tattoos; bizarre subcultures; body-piercing; wiccan and similar subcultures
Responsibility	Adventurousness	Extreme sports (e.g., hang gliding, mountain climbing, free diving, etc.); *Jackass the Movie* humor; explorers
Participation	Independence	Romantic loners; mysterious strangers; drifters; hoboes; explorers
Moderation	Roguishness	Charming rogue; the lovable drunk (e.g., Otis on the *Andy Griffith Show*); Bill Clinton
Honesty	Tactfulness	Honest humor on *The Man Show*; telling "tall tales"; the liar who is admired; telling "white lies"; person who lies to achieve greater goal (e.g. national security, confession from serial killer using "good cop," "bad cop" strategies, an operative who lies to be able to capture an "enemy" such as a terrorist)
Placefulness	Revelry	Mardi Gras; fraternity partying; loud entertainer at a party; Spring Break phenomenon
Courtesy	Irreverence	Class clown; humor on *The Man Show*; rude humor on shows such as *The Simpsons, South Park,* and *Beavis and Butthead*

Independence represents the deviance admiration counterpart to participation or involving oneself in the social and economic life of a community. Hoboes or other historical figures viewed to be voluntarily not attached to place or community and keenly uninvolved in community geographical boundaries were almost symbolic of freedom—rather than economic hard times—in another era. As well, those individuals who spend time in settlements in Antarctica, especially those who winter over and thus are effectively cut off from the rest of the planet, fall into this category. Roguishness is illustrative of the deviance admiration component of moderation. Assuredly, dominant middle-class culture has often preferred the enchanting and charming rogue to the earnest, but dull, person. Fictionally, Otis on the *Andy Griffith Show* is representative of this phenomenon; amongst the living, so is Bill Clinton. Tactfulness, particularly as resourcefulness (depending on the context), is the deviance admiration characterization of honesty. In other words, middle-class culture does value lying under certain conditions or for certain reasons. For example, the cop who plays loose with the truth to extract a confession from a serial killer would be positively assessed for his or her resourcefulness in underconforming to the norm of honesty, as would an operative who is dishonest in order to capture a terrorist. On the level of social interaction, those who tactfully tell "white lies" or socially accepted lies—for the purpose of protecting the emotional feelings of another—are usually positively reviewed. Revelry serves as the deviance admiration element of peacefulness. Much vacation behavior falls into this category. Among popular vacation ventures and/or venues that are often adored amongst middle-class Americans are the Mardi Gras, spring break on any warm beach, and Times Square on New Year's Eve. Finally, irreverence is the deviance admiration type of courtesy. Mischievous individuals engage in sly impudence or adopt a mildly mocking stance toward etiquette. The irreverent range from the fictional Dennis the Menace to the real life class clown to virtually any comedian.

Essentially, there is a corresponding deviance admiration form for each of the middle-class norms outlined by Tittle and Paternoster (2000). Various attitudes, behaviors, or conditions (Adler and Adler 2000) exemplify each category.

Positive Deviance

To echo the model, positive deviance denotes overconformity to the norms that is positively evaluated. The "Mother Theresa phenomenon" occurs when positive reactions or labeling follows overconformity to norms. In other words, Mother Theresa overconformed to the norm of altruism; she was almost universally praised as a saintly woman. For each of the norms suggested by Tittle and Paternoster (2000), a corresponding positive deviance category is proffered in Table 2.5.

Altruism is the positive deviance form of loyalty. In this case, the person is extraordinarily loyal to the group, overconforming to that norm, and that which emerges due to that loyalty is positively viewed. For example, historically, religious and political martyrs have fallen into this category. That is, the person is viewed as having sacrificed greatly for his or her group. Examples would be the Sullivan brothers from Iowa in World War II or Joan of Arc. National holidays commemorate this type of positive deviance. Every soldier is honored on Veteran's Day, and soldiers who have died for their society are venerated on Memorial Day. Religious martyrs include figures such as Stephen (from the Christian faith). As well, assassinated political leaders such as Abraham Lincoln, John F. Kennedy, Martin Luther King Jr., and Robert F. Kennedy are greatly admired.

Circumspection constitutes the positive deviance form of privacy. Individuals who have security clearances and honor that code are admired for their

T A B L E 2.5 A Classification of U.S. Middle-Class Deviance, Positive Deviance

Norm	Positive Deviance	Examples
Group loyalty	Altruism	Kamikaze pilots; martyrs; sharing food or other resources; daring rescues (e.g., at sea, mountains); patriot
Privacy	Circumspection	CIA operatives; FBI/Justice Department agents; loyal company employees (e.g., Oliver North)
Prudence	Discretion	The good friend who practices discretion; a person who denies themselves pleasures to achieve a goal (e.g., Olympians & other athletes)
Conventionality	Properness	Junior League members; Martha Stewart's followers; Boy Scouts, Girl Scouts, and similar groups
Responsibility	Hyperresponsibility	Overachiever; straight-A student (as viewed by parents and teachers); workaholic (as viewed by management); overzealous athlete (e.g. Tiger Woods, Michael Jordan)
Participation	Cooperation	Athletic team where individual talents are deemphasized so the team can win; employees who are positively viewed as team players
Moderation	Temperance	Monks; Nuns; Women's Temperance Movement; Mothers Against Drunk Driving
Honesty	Forthrightness	Honest Abe Lincoln; George Washington cutting down the cherry tree
Peacefulness	Pacifism	Gandhi; Martin Luther King Jr.; Jimmy Carter
Courtesy	Gentility	Miss Manners; the old-fashioned practitioners of southern hospitality who are admired; the gentleman; the "southern belle"

circumspection; as well, so are professionals—such as doctors, nurses, and clergy—who deal with confidential aspects of the lives of people and honor those confidences. Discretion composes the positive deviance aspect of prudence. In other words, someone who eschews unnecessary frivolous and pleasurable activities so as to achieve an important goal is often admired. Olympic athletes are often viewed as having to steadfastly deny themselves various pleasures so as to diligently advance toward their goal. Properness is the positive deviance type of conventionality. At times, admiration is granted toward those who over-conform to the expected habits and lifestyles of the middle class. Some admire the Junior Leaguers who engage in exceedingly conventional lifestyles. Hyperres-ponsibility is the positive deviance form of the norm of responsibility. For example, those who overconform on the norm of responsibility are often positively evaluated. Management values the overachieving worker; parents and teachers affirm the overachieving student; and almost everyone acknowledge the overze-alous athlete (e.g., Michael Jordan and Tiger Woods).

Cooperation (doing it for the team) represents the positive deviance profile of participation. At times, individuals are so involved in community life that they will diminish their own egos for the good of community involvement. As an example, when individual talents are deemphasized so that an athletic team can win, this process occurs. A long-standing joke in the sports world is that Dean Smith was the only person capable of keeping Michael Jordan under twenty points per game; nevertheless, that team's first and foremost philosophy allowed Michael Jordan and his UNC teammates to defeat Georgetown for the national championship. Temperance illustrates the positive deviance of moderation. When their purpose is a more transcendent one, ascetics or others who overconform to moderation are positively evaluated. Examples include people who choose religion for their life's calling, including priests, monks, nuns, preachers, and rabbis. Specifically, people—from many faith traditions—admire the Dalai Lama for the sublime and spiritual path of his life.

Forthrightness exemplifies positive deviance for the norm of honesty. Virtually every school child in this country is successfully socialized to esteem the importance of honesty through the story of George Washington and the cherry tree and through the honest individual as epitomized by the iconic and, perhaps, most reverentially viewed president—Abraham Lincoln. Pacifism demonstrates the deviance admiration category of peacefulness. Yearly, the Nobel Prize for Peace serves as a formalized positive reward for individuals in this category. Among individuals nearly universally admired in the context of dominant middle-class American values are individuals—such as Gandhi, Martin Luther King Jr., and Jimmy Carter—who practiced peacefulness while working toward social change. Finally, gentility is the positive deviance form of courtesy. Old-world manners (or southern hospitality) are overconforming to norms of courtesy and are often admired.

Certainly, a positive deviance form exists for each of the ten norms assembled by Tittle and Paternoster (2000). Various examples illustrate each of these categories. . . .

CONCLUSION

Long ago Durkheim (1982) pointed out the inevitability or normalcy of deviance; furthermore, Durkheim (1964; 1982) illuminated positive functions of deviance. A variety of social-psychological experiments, such as those by Sherif (1936), Asch (1952), and Schachter (1951), also suggested the fundamental existence of the processes of norms production, conformity, deviance production, and individual and social reactions to (potentially) deviant behaviors, attitudes, and conditions. The importance of power and social context in influencing norms, social reactions, and deviance outcomes cannot be ignored. The behaviors and conditions of more powerful actors are less likely to be deviantized than those of the less powerful. The reactions of powerful actors and social groups, moreover, are more important in determining who and what is successfully labeled deviant. In fact, we have argued that deviant behavior actually serves as a test of power relationships (Heckert and Heckert 2002). When a child lies, for example, she is testing the parent's ability to discern and react to the lie. When the child successfully "gets away" with the lie, a shift in the relative power of the child and parent has occurred.

The central place of norms, social reactions, efforts at social control, and deviance processes, therefore, suggest the importance of an integrated framework for understanding these factors and their interrelationships. Our integrated typology is an initial attempt to eliminate the false dichotomy imposed by previous conceptualizations of deviance.

REFERENCES

Adler, Patricia A., and Peter Adler. 2000. *Constructions of Deviance: Social Power, Context, and Interaction.* 3rd ed. Belmont, CA: Wadsworth.

Asch, Solomon. 1952. *Social Psychology.* Englewood Cliffs, NJ: Prentice Hall.

Becker, Howard. 1963. *Outsiders.* New York: The Free Press of Glencoe.

Berscheid, Ellen, and Elaine Walster. 1964. "Physical Attractiveness." Pp. 157–215 in *Advances in Experimental Social Psychology.* Vol. 7. New York: Academic Press.

Blake, Judith, and Kingsley Davis. 1964. "Norms, Values, and Sanctions." Pp. 456–484 in *Handbook of Modern Sociology,* edited by Robert E. L. Faris. Chicago: Rand McNally.

Byrne, Don. 1971. *The Attraction Paradigm.* New York: Academic Press.

Cohen, Albert K. 1966. *Deviance and Control.* Englewood Cliffs, NJ: Prentice Hall.

Coser, Lews A. 1967. *Continuities in the Study of Social Conflict.* New York: The Free Press.

Dinitz, Simon, Russell R. Dynes, and Alfred C. Clarke. 1969. *Deviance: Studies in the Process of Stigmatization and Societal Reaction.* New York: Oxford University Press.

Dodge, David L. 1985. "The Over-negativized Conceptualization of Deviance: A Programmatic Exploration." *Deviant Behavior* 6:17–37.

Durkheim, Emile. [1895] 1964. *The Division of Labor in Society.* Trans. J. Solovay and J. Mueller. New York: The Free Press.

———. [1893] 1982. *The Rules of Sociological Method.* Trans. S. Lukes. London: Macmillan.

Erikson, Kai T. 1964. "Notes on the Sociology of Deviance." Pp. 9–20 in *The Other Side,* edited by Howard Becker. New York: The Free Press.

Freedman, Jonathan L., and Anthony N. Doob. 1968. *Deviancy: The Psychology of Being Different.* New York: Academic Press.

Gibbs, Jack P. 1981. *Norms, Deviance and Social Control.* New York: Elsevier.

Goode, Erich. 2001. *Deviant Behavior.* 6th ed. Upper Saddle River, NJ: Prentice Hall.

Gusfield, Joseph R. 1963. *Symbolic Crusade.* Urbana, IL: University of Illinois Press.

Hawkins, Richard, and Gary Tiedeman. 1975. *The Creation of Deviance.* Columbus, OH: Charles E. Merrill Publishing Company.

Heckert, Alex, and Druann Maria Heckert. 2002. "A New Typology of Deviance: Integrating Normative and Reactivist Definitions of Deviance." *Deviant Behavior* 23:449–479.

Heckert, Druann. 1998. "Positive Deviance: A Classificatory Model." *Free Inquiry in Creative Sociology* 26:23–30.

———. 2003. "Mixed Blessings: Women and Blonde Hair." *Free Inquiry in Creative Sociology* 31:47–72.

Henslin, James M. 1975. *Introducing Sociology: Toward Understanding Life in Society.* New York: Free Press.

Homans, George C. 1950. *The Human Group.* New York: Harcourt, Brace and Company.

Huryn, Jean Scherz. 1986. "Giftedness as Deviance: A Test of Interaction Theories." *Deviant Behavior* 7:175–186.

Irwin, Katherine. 2003. "Saints and Sinners: Elite Tattoo Collectors and Tattooists as Positive and Negative Deviants." *Sociological Spectrum* 23:27–58.

Johnson, Elmer Hubert. 1978. *Crime, Correction, and Society.* Homewood, IL: The Dorsey Press.

Johnson, Harry M. 1960. *Sociology.* New York: Harcourt, Brace and World.

Kooistra, Paul. 1989. *Criminals as Heroes: Structure, Power and Identity.* Bowling Green: Bowling Green State University Popular Press.

Krebs, Dennis, and Allen A. Adinolfi. 1975. "Physical Attractiveness, Social Relations, and Personality Style." *Journal of Personality and Social Psychology* 31:245–253.

Lemert, Edwin M. 1972. *Human Deviance, Social Problems, and Social Control.* Englewood Cliffs, NJ: Prentice-Hall.

Liska, Allen W. 1981. *Perspectives on Deviance.* Englewood Cliffs, NJ: Prentice-Hall.

Merton, Robert K. 1966. "Social Problems and Sociological Theory." Pp. 775–827 in *Contemporary Social*

Problems, 2nd ed., edited by Robert K. Merton and Robert Nisbet. New York: Harcourt Brace.

———. 1968. *Social Theory and Social Structure*. New York: The Free Press.

Reis, Harry T., Ladd Wheeler, Nancy Spiegel, Michael H. Kerns, John Nezlik, and Michael Perry. 1982. "Physical Attractiveness as Social Interaction II: Why Does Appearance Affect Social Experience?" *Journal Personality and Social Psychology* 43:979–996.

Scarpitti, Frank R., and Paul T. McFarlane. 1975. *Deviance: Action, Reaction, Interaction*. Reading, MA: Addison-Wesley Publishing.

Schachter, Stanley. 1951. "Deviance, Rejection and Communication." *Journal of Abnormal Social Psychology* 46:190–207.

Sherif, Muzifer. 1936. *The Psychology of Social Norms*. New York: Harper.

Sorokin, Pitirim A. 1950. *Altruistic Love*. Boston: The Beacon Press.

Steffensmeier, Darrel J., and Robert M. Terry. 1975. *Examining Deviance Experimentally*. Port Washington, NY: Alfred Publishing.

Thio, Alex. 1983. *Deviant Behavior*. 2nd ed. Boston: Houghton Mifflin.

Tittle, Charles R., and Raymond Paternoster. 2000. *Social Deviance and Crime*. Los Angeles: Roxbury Publishing Company.

White, Winston. 1961. *Beyond Conformity*. Westport, CT: Greenwood Press.

Wilkins, Leslie. 1965. *Social Deviance*. Englewood Cliffs, NJ: Prentice Hall.

Williams, Robin M. Jr. 1965. *American Society*, 2nd ed. New York: Knopf.

Winslow, Robert W. 1970. *Society in Transition: A Social Approach to Deviance*. New York: The Free Press.

3

Relativism: Labeling Theory

HOWARD S. BECKER

Becker's classic statement of labeling theory advances the relativistic perspective on defining deviance. Here he argues that the essence of deviance is not contained within individuals' behaviors but in the response others have to these. Deviance, he claims, is a social construction forged by diverse audiences. Becker assesses the level of deviance attached to a behavior by the social reactions to it. He supports this assertion by pointing out that the same behaviors may be received very differently under varying conditions. He notes that variations in the degree of deviance attached to an act may arise due to the temporal or historical contexts framing acts, to the social position and power of those who commit or who have been harmed by acts, and by the consequences that arise from acts. These framing elements, which are sometimes unrelated to the behavior itself, may lead one act to be designated as heinous and relegate another, similar one, to obscurity. Becker thus locates the root of deviance in the response of people rather than the act itself and in the chain of events that is unleashed once people have labeled acts and their perpetrators as deviant.

The interactionist perspective . . . defines deviance as the infraction of some agreed-upon rule. It then goes on to ask who breaks rules, and to search for the factors in their personalities and life situations that might account for the infractions. This assumes that those who have broken a rule constitute a homogeneous category, because they have committed the same deviant act.

Such an assumption seems to me to ignore the central fact about deviance: it is created by society. I do not mean this in the way it is ordinarily understood, in which the causes of deviance are located in the social situation of the deviant or in "social factors" which prompt his action. I mean, rather, that *social groups create deviance by making the rules whose infraction constitutes deviance, and by applying* those rules to particular people and labeling them as outsiders. From this point of view, deviance is *not* a quality of the act the person commits, but rather a consequence of the application by others of rules and sanctions to an "offender." The deviant is one to whom the label has successfully been applied; deviant behavior is behavior that people so label.[1]

Since deviance is, among other things, a consequence of the responses of others to a person's act, students of deviance cannot assume that they are dealing with a homogeneous category when they study people who have been labeled deviant. That is, they cannot assume that those people have actually committed a deviant act or broken some rule, because the process of labeling may not be infallible; some people may be labeled deviant who in fact have not broken a rule. Furthermore, they cannot assume that the category of those labeled deviant will contain all those who actually have broken a rule, for many offenders may escape apprehension and thus fail to be included in the population of "deviants" they study. Insofar as the category lacks homogeneity and fails to include all the cases that belong in it, one cannot reasonably expect to find common factors of personality or life situation that will account for the supposed deviance. What, then, do people who have been labeled deviant have in common? At the least, they share the label and the experience of being labeled as outsiders. I will begin my analysis with this basic similarity and view deviance as the product of a transaction that takes place between some social group and one who is viewed by that group as a rule-breaker. I will be less concerned with the personal and social characteristics of deviants than with the process by which they come to be thought of as outsiders and their reactions to that judgement. . . .

The point is that the response of other people has to be regarded as problematic. Just because one has committed an infraction of a rule does not mean that others will respond as though this had happened. (Conversely, just because one has not violated a rule does not mean that he may not be treated, in some circumstances, as though he had.)

The degree to which other people will respond to a given act as deviant varies greatly. Several kinds of variation seem worth noting. First of all, there is variation over time. A person believed to have committed a given "deviant" act may at one time be responded to much more leniently than he would be at some other time. The occurrence of "drives" against various kinds of deviance illustrates this clearly. At various times, enforcement officials may decide to make an all-out attack on some particular kind of deviance, such as gambling, drug addiction, or homosexuality. It is obviously much more dangerous to engage in one of these activities when a drive is on than at any other time. (In a very interesting study of crime news in Colorado newspapers, Davis found that the amount of crime reported in Colorado newspapers showed very little association with actual changes in the amount of crime taking place in Colorado. And, further, that people's estimate of how much increase there had been in crime in Colorado was associated with the increase in the amount of crime news but not with any increase in the amount of crime.)[2]

The degree to which an act will be treated as deviant depends also on who commits the act and who feels he has been harmed by it. Rules tend to be applied more to some persons than others. Studies of juvenile delinquency make the point clearly. Boys from middle-class areas do not get as far in the legal process when they are apprehended as do boys from slum areas. The middle-class boy is less likely, when picked up by the police, to be taken to the station; less likely when taken to the station to be booked; and it is extremely unlikely that he will be convicted and sentenced.[3] This variation occurs even though the original infraction of the rule is the same in the two cases. Similarly, the law is differentially applied to Negroes and whites. It is well known that a Negro believed to have attacked a white woman is much more likely to be punished than a white man who commits the same offense; it is only slightly less well known that a Negro who murders another Negro is much less likely to be punished than a white man who commits murder.[4] This, of course, is one of the main points of Sutherland's analysis of white-collar crime: crimes committed by corporations are almost always prosecuted as civil cases, but the same crime committed by an individual is ordinarily treated as a criminal offense.[5]

Some rules are enforced only when they result in certain consequences. The unmarried mother furnishes a clear example. Vincent[6] points out that illicit sexual relations seldom result in severe punishment or social censure for the offenders. If, however, a girl becomes pregnant as a result of such activities, the reaction of others is likely to be severe. (The illicit pregnancy is also an interesting example of the differential enforcement of rules on different categories of people. Vincent notes that unmarried fathers escape the severe censure visited on the mother.)

Why repeat these commonplace observations? Because, taken together, they support the proposition that deviance is not a simple quality, present in some kinds of behavior and absent in others. Rather, it is the product of a process which involves responses of other people to the behavior. The same behavior may be an infraction of the rules at one time and not at another or may be an infraction when committed by one person, but not when committed by another; some rules are broken with impunity, others are not. In short, whether a given act is deviant or not depends in part on the nature of the act (that is, whether or not it violates some rule) and in part on what other people do about it.

Some people may object that this is merely a terminological quibble, that one can, after all, define terms any way he wants to and that if some people want to speak of rule-breaking behavior as deviant without reference to the reactions of others they are free to do so. This, of course, is true. Yet it might be worthwhile to refer to such behavior as *rule-breaking behavior* and reserve the term *deviant* for those labeled as deviant by some segment of society. I do not insist that this usage be followed. But it should be clear that insofar as a scientist uses "deviant" to refer to any rule-breaking behavior and takes as his subject of study only those who have been *labeled* deviant, he will be hampered by the disparities between the two categories.

If we take as the object of our attention behavior which comes to be labeled as deviant, we must recognize that we cannot know whether a given act will be categorized as deviant until the response of others has occurred. Deviance is not a quality that lies in behavior itself, but in the interaction between the person who commits an act and those who respond to it. . . .

In any case, being branded as deviant has important consequences for one's further social participation and self-image. The most important consequence is a drastic change in the individual's public identity. Committing the improper act and being publicly caught at it place him in a new status. He has been revealed as a different kind of person from the kind he was supposed to be. He is labeled a "fairy," "dope fiend," "nut," or "lunatic," and treated accordingly.

In analyzing the consequences of assuming a deviant identity let us make use of Hughes' distinction between master and auxiliary status traits.[7] Hughes notes that most statuses have one key trait which serves to distinguish those who belong from those who do not. Thus the doctor, whatever else he may be, is a person who has a certificate stating that he has fulfilled certain requirements and is licensed to practice medicine; this is the master trait. As Hughes points out, in our society a doctor is also informally expected to have a number of auxiliary traits: most people expect him to be upper middle-class, white, male, and Protestant. When he is not, there is a sense that he has in some way failed to fill the bill. Similarly, though skin color is the master status trait determining who is Negro and who is white, Negroes are informally expected to have certain status traits and not to have others; people are surprised and find it anomalous if a Negro turns out to be a doctor or a college professor. People often have the master status trait but lack some of the auxiliary, informally expected characteristics; for example, one may be a doctor but be a female or a Negro.

Hughes deals with this phenomenon in regard to statuses that are well thought of, desired, and desirable (noting that one may have the formal qualifications for entry into a status but be denied full entry because of lack of the proper auxiliary traits), but the same process occurs in the case of deviant statuses. Possession of one deviant trait may have a generalized symbolic value, so that people automatically assume that its bearer possesses other undesirable traits allegedly associated with it.

To be labeled a criminal one need only commit a single criminal offense, and this is all the term formally refers to. Yet the word carries a number of connotations specifying auxiliary traits characteristic of anyone bearing the label. A man who has been convicted of housebreaking and thereby labeled criminal is presumed to be a person likely to break into other houses; the police, in rounding up known offenders for investigation after a crime has been committed, operate on this premise. Further, he is considered likely to commit other kinds of crimes as well, because he has shown himself to be a person without "respect for the law." Thus, apprehension for one deviant act exposes a person to the likelihood that he will be regarded as deviant or undesirable in other respects.

There is one other element in Hughes' analysis we can borrow with profit: the distinction between master and subordinate statuses.[8] Some statuses, in our society as in others, override all other statuses and have a certain priority. Race is one of these. Membership in the Negro race, as socially defined, will override most other status considerations in most other situations; the fact that one is a physician or middle-class or female will not protect one from being treated as a Negro first and any of these other things second. The status of deviant (depending on the kind of deviance) is this kind of master status. One receives the status as a result of breaking a rule, and the identification proves to be more important than most others. One will be identified as a deviant first, before other identifications are made. . . .

NOTES

1. The most important earlier statements of this view can be found in Frank Tannenbaum, *Crime and the Community* (New York: Columbia University Press, 1938), and E. M. Lemert, *Social Pathology* (New York: McGraw-Hill Book Co., 1951). A recent article stating a position very similar to mine is John Kitsuse, "Societal Reaction to Deviance: Problems of Theory and Method," *Social Problems* 9 (Winter 1962): 247–256.

2. F. James Davis, "Crime News in Colorado Newspapers," *American Journal of Sociology* LVII (January 1952): 325–330.

3. See Albert K. Cohen and James F. Short, Jr., " Juvenile Delinquency," p. 87 in Robert K. Merton and Robert A. Nisbet, eds., *Contemporary Social Problems* (New York: Harcourt, Brace and World, 1961).

4. See Harold Garfinkel, "Research Notes on Inter- and Intra-Racial Homicides," *Social Forces* 27 (May 1949): 369–381.

5. Edwin Sutherland, "White Collar Criminality," *American Sociological Review* V (February 1940): 1–12.

6. Clark Vincent, *Unmarried Mothers* (New York: The Free Press of Glencoe, 1961), pp. 3–5.

7. Everett C. Hughes, "Dilemmas and Contradictions of Status," *American Journal of Sociology* L (March 1945): 353–359.

8. *Ibid.*

4

Against Relativism: A Harm-Based Conception of Deviance

BARBARA J. COSTELLO

Costello reacts against Becker's relativist position by advancing an absolutist perspective on defining deviance. She suggests that cultural relativity can be overstated and that while reactions to deviance are not always universal, examples of the relativity of deviance are based on anecdotes rather than scientifically tested. She accuses both right- and left-wing approaches to deviance of guilt in making this relativist assumption and dismisses them both. In their place she offers her own view: Acts are considered deviant on the basis of the harm they cause to others. Thus, she argues that while homosexuality may run counter to some individuals' moral preferences, most people would agree that voluntary same-sex liaisons cause less harm than extramarital affairs, which hurt the uninvolved spouse through their betrayal. Costello's harm-based approach to defining deviance is inherently absolutist because it advances a situationally and temporally consistent universal yardstick for assessing the meaning of behavior that is lodged in the essence of the behavior itself.

The major problem with the sociology of deviant behavior is that it rests too heavily on an assumption—the assumption of cultural relativism, or that the values of social groups vary substantially both within and between cultures. There is ample anecdotal evidence that group values vary to some extent and it would be foolish to deny this. However, in order for the study of "deviant" behavior to make a contribution to sociology it requires a greater degree of cultural variation in values than actually exists. The arguments of scholars with a liberal or radical political bias require that those in power are free to make any rules that suit their interests and that they can convince the rest of society that those rules are right and just.[1] The arguments of scholars with a more conservative bias require that any behavior can be valued or at least allowed because for them individual values are the only cause of "deviant" behavior. Neither of these positions is correct because they exaggerate both the extent and importance of variation in group values.

I argue that deviant behavior has always been evaluated in terms of the harm it causes to others, the individual actor him- or herself, and the social order. Those behaviors that are viewed as deviant because they violate some perceived "natural order of things," but which do not cause harm to others, such as body piercing or homosexuality, have never been considered to be as serious or as worthy of sanction as those that more objectively cause pain and suffering for others.[2] Deviance scholars have failed to acknowledge the full significance and predictability of the variation in social reaction to various behaviors. In short, not all social reactions are the same, and they vary both quantitatively and qualitatively depending on the harmfulness of the behavior in question.

CULTURAL RELATIVISM IN THE STUDY OF DEVIANCE

Some of the best known statements of the sociological study of deviance provide the clearest examples of cultural relativism, or the notion that group norms and values vary widely by culture, subculture, or historical period. As Becker put it, "social groups create deviance by making the rules whose infraction constitutes deviance . . . [D]eviance is not a quality of the act the person commits, but rather a consequence of the application by others of rules and sanctions to an 'offender'" (1963, p. 9). Later, Becker continues, "Deviance is not a simple quality, present in some kinds of behavior and absent in others. Rather, it is the product of a process which involves responses of other people to the behavior (1963, p. 14). Taken at face value, these statements hold that there are no objective means to determine whether a behavior is likely to be considered unacceptable and worthy of sanction, and that social groups are completely free to define any behavior as either acceptable or unacceptable.[3] Although this is a fairly extreme position, it is repeated in more modern work in the field. For example, Goode's explanation of the concept relativity is "the empirical fact that at certain times and in certain places, acts that we may regard as wrong were or are not so regarded by persons living at those times and in those places" (2003, p. 527). Goode further argues that if we use the standard of harm to predict which acts might be seen as deviant, "conventional morality makes no sense whatsoever" (2003, p. 529). Ben-Yehuda (2006) claims that cultural relativity in normative systems is "no great revelation," and that "as cultures change, views of what is and what is not deviance change as well . . . [D]eviance is a relative phenomenon, among cultures and within cultures. Researchers of deviance have known, discussed, and written about this issue for dozens of years" (p. 561).

A consequence of the lack of recognition of cultural relativism as an empirical question is that scholars with competing political views are free to assert their own position on what is or should be considered truly deviant as if it is the obvious truth. A second consequence of the lack of recognition of cultural relativism as an empirical question is that those arguing from conservative political positions and those arguing from liberal or radical positions fail to recognize that they're making arguments that are exactly the same in form but different in content. They argue that the study of deviance should be the study of behavior that is truly socially harmful and they argue that their position is the correct one for recognizing what is socially harmful. I agree with the argument that we should study and try to reduce truly harmful behavior, but political ideologies are not the correct starting place for recognizing what is truly harmful.

THE ARGUMENTS OF THE RIGHT AND THE LEFT

As Best (2004) outlines in his book tracing the history of the concept of deviance, labeling theory became very influential in the 1960s. Its appeal was in part its tendency to question authority that fit so well with the spirit of the times. Best (2004) argues that labeling theory's tendency to romanticize the deviant and vilify the agents of social control "turned traditional criminology upside down and made it seem stodgy" (p. 31). The virtual institutionalization of this view led Hirschi to argue that to "appreciate deviance" and "not assume the values of the group making the rules" had become "procedural rules" for the study of deviance ([1973] 2002, pp. 31–32).

One of the best-known statements of these views is Liazos's (1972) critique of the study of deviance. The heart of his critique was that the field ignored harmful behaviors engaged in by those in power and ignored the social conditions that led to the "deviance" and suffering of the poor and powerless. Liazos argued that the rules of society are rules created and enforced by those in power to serve their own interests and, as a consequence, much truly harmful behavior is legal or at least unpunished. Social conditions created by those in power are the true cause of the "deviance" that sociologists are so distracted by, the "nuts, sluts, and perverts" that filled the deviance texts of the time (Liazos 1972, p. 118). Liazos goes so far as that deviants are actually "victims" of their society (1972, p. 119), and that "most prisoners are political prisoners" whose actions result largely from the social conditions of the time (1972, p. 108).

On the face of it, Liazos's argument is not a relativistic one—he recognizes that crime is real and has real, identifiable causes. However, at root his argument is that the crimes committed by the powerful are left unchecked because we as a society do not recognize their wrongness. Those in power are free to create any rules they like to serve their own interests and the rest of society (except Liazos, interestingly) is blinded to the wrongness of the actions of the powerful, in a state of false consciousness. We are also apparently unable to recognize the harm caused by those in power because we're so distracted by the nuts and sluts who are so interesting to us. Liazos states that "actions of violence are committed daily by the government and corporations; but in these days of misplaced emphasis, ignorance, and manipulation we do not see the destruction inherent in these actions" (1972, p. 112). Liazos also implies that a lack of perceived seriousness of white-collar crime and lenient penalties once it is detected are to blame for its prevalence. Thus, the cause of white-collar crime is the failure to strongly condemn the offense, a classic cultural relativist argument.[4]

On the other side of the political spectrum, the argument is the same in form but very different in content. Moynihan argued that "defining deviancy down" was leading to the downfall of our society (1993), and more recently Hendershott (2002) echoes his concerns. She writes,

> Moynihan had captured the essence of a disturbing trend in the United States: the decline of our quality of life through our unqualified acceptance of too many activities formerly considered unacceptable. Out-of-wedlock births, teenage pregnancy and promiscuity, drug abuse, welfare dependency, and homelessness all seemed to be increasing . . . Worse, these behaviors appeared to be morally condoned. (2002, p. 9)

Hendershott devotes most of the book to a presentation of carefully selected anecdotal evidence in support of her hypothesis that changes in public opinion toward deviant behavior cause changes in rates of the behavior. For example, she argues that teenage promiscuity has increased to alarming proportions because it has become so acceptable (except to her, interestingly). She advocates a restigmatization of such behaviors as drug abuse and mental illness to help reduce the rates of these social problems.

Like Liazos, then, Hendershott argues that the lack of definition of certain behaviors as deviant, criminal, or wrong is the major (if not the only) cause of these behaviors. Both also argue that there are truly harmful behaviors that are not recognized or defined as harmful, but neither provides a clear means of determining which acts are harmful and which are not, aside from those they specifically mention. Liazos at least gives a few hints such as his repeated use of the term "violence," implying that actions that kill or injure others should be considered deviant. Hendershott, on the other hand, only tells us that we need to draw on "nature, reason, and common sense" to define deviance, apparently implying that we should rely on her reasoning and common sense (2002, p. 11). The problem, then, is that both sides of the political debate are arguing that socially harmful behaviors exist because we fail to recognize them as harmful and punish them, but they're pointing to two different sets of behaviors as examples of harmful behavior that is considered acceptable (Goode 2003). This leaves us with no means to resolve the argument.

A HARM-BASED CONCEPTION OF DEVIANCE

Both the right and left are arguing that we need to recognize and stigmatize truly harmful behaviors. Ironically, although both sides of the debate adopt a position of cultural relativism, neither one maintains fidelity to this position because they both acknowledge that some behaviors are objectively harmful.

I argue that Liazos's and Hendershott's reliance on the concept "harm" to define the wrongfulness of certain actions is no coincidence, that all cultures' moral codes rely on the same concept, and that it is not possible to argue that a behavior is morally wrong if the behavior is seen as harmless to the individual engaging in it, harmless to others, and harmless to the social order. While it is certainly possible to manipulate public opinion on the harmfulness of certain behaviors, ultimately the harmfulness or harmlessness of a behavior is an empirical question, so the ability to manipulate opinion is limited by people's ability to observe the world around them and by their access to disconfirming information.

One example of the attempt to use harm as a way to sway public opinion against a certain behavior is found in arguments against gay marriage. These point to the harm gay marriage would bring to the individuals themselves, their children, or the community in general. The Illinois Family Institute's website (2006) claims that legalization of same-sex marriage in Canada "has sparked a devastating social revolution and has led to the persecution of Christians." The Family Research Institute website (2006) argues that same-sex marriage "would harm society in general and homosexuals in particular" and that it "poses a clear and present danger to the health of the community." Although these sources argue that gay marriage should not be legal because of the harm this would bring to many people, these arguments are not particularly successful. The problem is that they are empirically falsifiable; anyone reading such claims with an open mind and the ability to think critically would recognize that they are empirically false—there is no devastating social revolution in Canada and the communities in Massachusetts are doing just fine.

Another example of the reliance on the concept of harm to demonstrate the wrongfulness of a behavior or the seriousness of a social problem can be found in the dramatization of statistics on the frequency with which certain behaviors occur. As Best (2001) notes, inflated estimates of the prevalence of social problems such as the abduction of children by strangers are often used to emphasize the harm caused by the problem and to drum up support for efforts to reduce the problem. Once again, however, the ability of advocates to manipulate public opinion is limited by empirical reality and people's ability to observe it; the claim that 50,000 children are abducted every year in the U.S. didn't survive very long (Best 2001). Similarly, concern over crimes committed by satanic cults quickly diminished once it became clear that there was little evidence for their occurrence (Best 1999).

Other examples of harm-based arguments can be found in the movements against drunk driving and cigarette smoking. MADD, Mothers Against Drunk Driving, argues that drunk driving should be considered a "violent crime" (MADD website 2006) and has effectively publicized the obvious negative consequences of accidents caused by drunk driving. There is evidence that attitudes toward cigarette smoking and laws regulating smoking in public places were affected by the awareness and publicizing of the dangers of second-hand smoke to others, emphasizing the potential harm caused by smoking even to those who do not smoke.

The reason that those promoting causes or publicizing social problems draw on the concept of harm is quite simply because it works. People are concerned about behavior that causes harm to individuals, to others, and to communities. There is evidence that this is the case from studies of crime seriousness. Warr (1989) found that judgments of harmfulness of crimes explained 85% of the variance in judgments of wrongfulness of crimes. A substantial minority of his sample of Dallas, Texas, residents did not distinguish between crimes on the basis of their moral wrongfulness, but did show variation in their judgments of harmfulness. For this, group judgments of harmfulness explained 94% of the variance in judgments of the seriousness of these crimes. Although Warr's study did not include behaviors that might be considered deviant but noncriminal, this study provides evidence that people evaluate behaviors in predictable ways, and that there tends to be a great deal of agreement regarding which behaviors are least and most serious.

Other studies of crime seriousness also yield no real surprises about how people evaluate different behaviors. In general, as we would expect, behavior that causes death or physical or psychological injury to others is seen as being the most serious, behaviors that involve theft or destruction of others' property are somewhat less serious, behaviors that don't have clear victims but which can lead to negative consequences to the individuals engaging in them (especially when they're children) are somewhat less serious, and behaviors that are simply annoying or unpleasant to see (such as public drunkenness or body piercing) are seen as the least serious. Even within offense types such as white-collar crimes, people distinguish between violent offenses (those involving injury to others) and nonviolent ones, ranking violent offenses as more serious. . . .

Of course, there are criticisms of the view that people evaluate behavior based on the harm it causes to others. Obviously, people make distinctions between types of behaviors depending on the circumstances and the people involved. And there are clear examples of behaviors being seen as very wrong even when they cause only psychological harm to others, such as flag burning or desecrating religious symbols. Further, as noted previously, perceptions of the harmfulness of behaviors can certainly be affected by misinformation or direct attempts at manipulation. However, I argue that public opinion is most easily manipulated when the behavior in question is the least objectively harmful because those are behaviors over which there is less consensus to begin with. Although some people do strongly condemn some victimless crimes such as flag burning, this does not prove the point of the cultural relativists—it merely proves that there is less-than-perfect consensus within and between societies. . . .

CONCLUSIONS

I believe that deviance as a field of inquiry has suffered from its reliance on cultural relativism to understand the causes or consequences of deviance and stigmatization. If we remove from the field of deviance the ideas that any behavior can be valued and that any behavior can be stigmatized, it can't make much of a contribution to sociology. As Goode notes, "the early sociology of deviance was above all a theory of the social construction of phenomena as deviant"(2006, p. 552). If my claims are correct and there are definite limits to what can be socially constructed as deviant, there isn't much left of the field's dependent variable. If I am correct in my assertion that there is more variation in attitudes toward less harmful behavior, this implies that the field of deviance is a field centered around the study of attitudes toward trivial behavior.

NOTES

1. Of course, it is not necessary to convince people that rules are right and just if those in power can control the powerless through sheer force or threat of force.

2. This is not to deny that people have been murdered for being homosexual. However, I would argue that killing someone for being homosexual is generally considered to be a crime and that most people in most cultures would not think it appropriate to administer the death penalty for homosexuality.

3. Becker acknowledges the argument that "there are some rules that are very generally agreed to by everyone . . . ." He notes that this is an empirical question, but also states, "I doubt there are many such areas of consensus. . . ." (1963, p. 8).

4. As Cullen, Link, and Polanzi (1982) note, this claim was not a new one at the time Liazos made it, and they trace it back to work published in the early 1900s.

REFERENCES

Becker, Howard S. 1963. *Outsiders: Studies in the Sociology of Deviance*. New York: Free Press.

Ben-Yehuda, Nachman. 2006. "Contextualizing Deviance Within Social Change and Stability, Morality, and Power." *Sociological Spectrum* 26:559–580.

Best, Joel. 1999. *Random Violence: How We Talk About New Crimes and New Victims*. Berkeley: University of California Press.

———. 2001. *Damned Lies and Statistics: Untangling Numbers from the Media, Politicians, and Activists*. Berkeley: University of California Press.

———. 2004. *Deviance: Career of a Concept*. Belmont, CA: Thomson/Wadsworth.

Cullen, Francis T., Bruce G. Link, and Craig W. Polanzi. 1982. "The Seriousness of Crimes Revisited: Have Attitudes Toward White-Collar Crime Changed?" *Criminology* 20:83–102.

Family Research Institute. 2006. http://www. familyresearchinst.org/FRI_Edu Pamphlet7.html.

Goode, Erich. 2003. "The MacGuffin That Refuses to Die: An Investigation Into the Condition of the Sociology of Deviance." *Deviant Behavior* 24: 507–533.

———. 2006. "Is the Deviance Concept Still Relevant to Sociology?" *Sociological Spectrum* 26:547–558.

Hendershott, Anne. 2002. *The Politics of Deviance*. San Francisco: Encounter Books.

Hirschi, Travis. [1973] 2002. "Procedural Rules for the Study of Deviant Behavior." Pp. 31–44 in *The Craft of Criminology: Selected Papers*, edited by J. Laub. New Brunswick, NJ: Transaction.

Illinois Family Institute, 2006. http:// www.illinoisfamily.org/news/contentview. asp?c = 26914.

Liazos, Alexander. 1972. "The Poverty of the Sociology of Deviance: Nuts, Sluts, and Perverts." *Social Problems* 20:103–120.

Mothers Against Drunk Driving. 2006. http:// www.madd.org/aboutus/1094.

Moynihan, Daniel P. 1993. "Defining Deviancy Down." *American Scholar* 62:17–31.

Warr, Mark. 1989. "What is the Perceived Seriousness of Crimes?" *Criminology* 27:795–821.

5

Social Power: Conflict Theory of Crime

RICHARD QUINNEY

Quinney's conflict theory of crime represents the social power perspective on defining deviance. He builds on Becker's relativist approach by asserting that definitions of deviance are social constructions and not absolute "givens." He also rejects the essentialist view of deviance as inherent in specific acts. But while Becker is not specific about the identity of those who formulate the definitions of deviance, Quinney locates the decision makers in the dominant class. Definitions of deviance, then, stem from the views of those who have the power to make and enforce them. Members of this group formulate the definitions of deviance with the express purpose of advancing themselves by labeling behaviors that threaten their class interests as criminal. They then enforce the definitions unequally: more harshly against their opponents and more leniently against members of their own group. Yet they disguise the self-serving basis of their rule, creating and enforcing through justifications that legitimate their actions and cast these as rational or beneficial to others. In turn, those defined as criminal become more likely to engage in future behavior that will be defined as criminal.

Quinney thus goes beyond Becker to offer a view of society that radically departs from previous perspectives. While Erikson proposed a fairly unified view of society, assuming a broad social consensus about boundaries and the appropriate means of enforcing them, Quinney offers a more fragmented view of society. The conflict perspective envisions two groups in society: the dominant class and those they dominate. Criminals are conceptualized as powerless and oppressed people who threaten the interests of the ruling class. Defining and enforcing crime becomes a means of reproducing the power and socioeconomic inequalities between these groups. Quinney's conflict theory suggests that definitions of deviance represent one of the coercive means through which the elite maintain their dominance over the masses.

A theory that helps us begin to examine the legal order critically is the one I call the *social reality of crime.* Applying this theory, we think of crime as it is affected by the dynamics that mold the society's social, economic, and political structure. First, we recognize how criminal law fits into capitalist society. The legal order gives reality to the crime problem in the United States. Everything that makes up crime's social reality, including the application of criminal law, the behavior

From Richard Quinney, *Criminology* (Boston: Little, Brown, 1975), pp. 37–41. Reprinted with permission of the author.

patterns of those who are defined as criminal, and the construction of an ideology of crime, is related to the established legal order. The social reality of crime is constructed on conflict in our society.

The theory of the social reality of crime is formulated as follows.

I. **The Official Definition of Crime:** Crime as a legal definition of human conduct is created by agents of the dominant class in a politically organized society.

The essential starting point is a definition of crime that itself is based on the legal definition. Crime, as *officially* determined, is a *definition* of behavior that is conferred on some people by those in power. Agents of the law (such as legislators, police, prosecutors, and judges) are responsible for formulating and administering criminal law. Upon *formulation* and *application* of these definitions of crime, persons and behaviors become criminal.

Crime, according to this first proposition, is not inherent in behavior, but is a judgment made by some about the actions and characteristics of others. This proposition allows us to focus on the formulation and administration of the criminal law as it applies to the behaviors that become defined as criminal. Crime is seen as a result of the class-dynamic process that culminates in defining persons and behaviors as criminal. It follows, then, that the greater the number of definitions of crime that are formulated and applied, the greater the amount of crime.

II. **Formulating Definitions of Crime:** Definitions of crime are composed of behaviors that conflict with the interests of the dominant class.

Definitions of crime are formulated according to the interests of those who have the power to translate their interests into public policy. Those definitions are ultimately incorporated into the criminal law. Furthermore, definitions of crime in a society change as the interests of the dominant class change. In other words, those who are able to have their interests represented in public policy regulate the formulation of definitions of crime.

The powerful interests are reflected not only in the definitions of crime and the kinds of penal sanctions attached to them, but also in the *legal policies* on handling those defined as criminals. Procedural rules are created for enforcing and administering the criminal law. Policies are also established on programs for treating and punishing the criminally defined and programs for controlling and preventing crime. From the initial definitions of crime to the subsequent procedures, correctional and penal programs, and policies for controlling and preventing crime, those who have the power regulate the behavior of those without power.

III. **Applying Definitions of Crime:** Definitions of crime are applied by the class that has the power to shape the enforcement and administration of criminal law.

The dominant interests intervene in all the stages at which definitions of crime are created. Because class interests cannot be effectively protected merely by formulating criminal law, the law must be enforced and administered. The interests of the powerful, therefore, also operate where the definitions of crime reach the *application* stage. As Vold has argued, crime is "political behavior and the criminal becomes in fact a member of a 'minority group' without sufficient public support to dominate the control of the police power of the state." Those whose interests conflict with the ones represented in the law must either change their behavior or possibly find it defined as criminal.

The probability that definitions of crime will be applied varies according to how much the behaviors of the powerless conflict with the interests of those in power. Law enforcement efforts and judicial activity are likely to increase when the interests of the dominant class are threatened. Fluctuations and variations in applying definitions of crime reflect shifts in class relations.

Obviously, the criminal law is not applied directly by those in power; its enforcement and administration are delegated to authorized *legal agents*. Because the groups responsible for creating the definitions of crime are physically separated from the groups that have the authority to enforce and administer law, local conditions determine how the definitions will be applied. In particular, communities vary in their expectations of law enforcement and the administration of justice. The application of definitions is also influenced by the visibility of offenses in a community and by the public's norms about reporting possible violations. And especially important in enforcing and administering the criminal law are the legal agents' occupational organization and ideology.

The probability that these definitions will be applied depends on the actions of the legal agents who have the authority to enforce and administer the law. A definition of crime is applied depending on their evaluation. Turk has argued that during "criminalization," a criminal label may be affixed to people because of real or fancied attributes: "Indeed, a person is evaluated, either favorably or unfavorably, not because he *does* something, or even because he *is* something, but because others react to their perceptions of him as offensive or inoffensive." Evaluation by the definers is affected by the way in which the suspect handles the situation, but ultimately the legal agents' evaluations and subsequent decisions are the crucial factors in determining the criminality of human acts. As legal agents evaluate more behaviors and persons as worthy of being defined as crimes, the probability that definitions of crime will be applied grows.

IV. **How Behavior Patterns Develop in Relation to Definitions of Crime:** Behavior patterns are structured in relation to definitions of crime, and within this context people engage in actions that have relative probabilities of being defined as criminal.

Although behavior varies, all behaviors are similar in that they represent patterns within society. All persons—whether they create definitions of crime or are the objects of these definitions—act in reference to *normative systems* learned in relative social and cultural settings. Because it is not the quality of the behavior but the action taken against the behavior that gives it the character of criminality, that which is defined as criminal is relative to the behavior patterns of the class that formulates and applies definitions. Consequently, people whose behavior patterns are not represented when the definitions of crime are formulated and applied are more likely to act in ways that will be defined as criminal than those who formulate and apply the definitions.

Once behavior patterns become established with some regularity within the segments of society, individuals have a framework for creating *personal action patterns*. These continually develop for each person as he moves from one experience to another. Specific action patterns give behavior an individual substance in relation to the definitions of crime.

People construct their own patterns of action in participating with others. It follows, then, that the probability that persons will develop action patterns with a high potential for being defined as criminal depends on (1) structured opportunities, (2) learning experiences, (3) interpersonal associations and identifications, and (4) self-conceptions. Throughout the experiences, each person creates a conception of self as a human social being. Thus prepared, he behaves according to the anticipated consequences of his actions.

In the experiences shared by the definers of crime and the criminally defined, personal-action patterns develop among the latter because they are so defined. After they have had continued experience in being defined as criminal, they learn to manipulate the application of criminal definitions.

Furthermore, those who have been defined as criminal begin to conceive of themselves as criminal. As they adjust to the definitions imposed on them, they learn to play the criminal role. As a result of others' reactions, therefore, people may develop personal-action patterns that increase the likelihood of their being defined as criminal in the future. That is, increased experience with definitions of crime increases the probability of their developing actions that may be subsequently defined as criminal.

Thus, both the definers of crime and the criminally defined are involved in reciprocal action patterns. The personal-action patterns of both the definers and the defined are shaped by their common, continued, and related experiences. The fate of each is bound to that of the other.

V.   **Constructing an Ideology of Crime:** An ideology of crime is constructed and diffused by the dominant class to secure its hegemony.

This ideology is created in the kinds of ideas people are exposed to, the manner in which they select information to fit the world they are shaping, and their way of interpreting this information. People behave in reference to the *social meanings* they attach to their experiences.

Among the conceptions that develop in a society are those relating to what people regard as crime. The concept of crime must of course be accompanied by ideas about the nature of crime. Images develop about the relevance of crime, the offender's characteristics, the appropriate reaction to crime, and the relation of crime to the social order. These conceptions are constructed by communication, and, in fact, an ideology of crime depends on the portrayal of crime in all personal and mass communication. This ideology is thus diffused throughout the society.

One of the most concrete ways by which an ideology of crime is formed and transmitted is the official investigation of crime. The President's Commission on Law Enforcement and Administration of Justice is the best contemporary example of the state's role in shaping an ideology of crime. Not only are we as citizens more aware of crime today because of the President's Commission, but official policy on crime has been established in a crime bill, the Omnibus Crime Control and Safe Streets Act of 1968. The crime bill, itself a reaction to the growing fears of class conflict in American society, creates an image of a severe crime problem and, in so doing, threatens to negate some of our basic constitutional guarantees in the name of controlling crime.

Consequently, the conceptions that are most critical in actually formulating and applying the definitions of crime are those held by the dominant class. These conceptions are certain to be incorporated into the social reality of crime. The more the government acts in reference to crime, the more probable it is that definitions of crime will be created and that behavior patterns will develop in opposition to those definitions. The formulation of definitions of crime, their application, and the development of behavior patterns in relation to the definitions are thus joined in full circle by the construction of an ideological hegemony toward crime.

VI. **Constructing the Social Reality of Crime:** The social reality of crime is constructed by the formulation and application of definitions of crime, the development of behavior patterns in relation to these definitions, and the construction of an ideology of crime.

The first five propositions are collected here into a final composition proposition. The theory of the social reality of crime, accordingly, postulates creating a series of phenomena that increase the probability of crime. The result, holistically, is the social reality of crime.

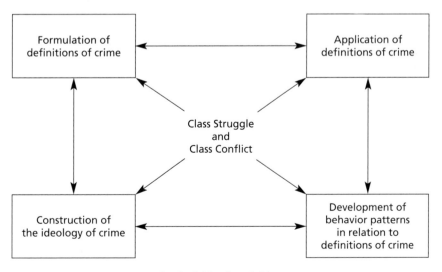

The Social Reality of Crime

Because the first proposition of the theory is a definition and the sixth is a composite, the body of the theory consists of the four middle propositions. These form a model of crime's social reality. The model, as diagrammed, relates the proposition units into a theoretical system (see figure on p. 57). Each unit is related to the others. The theory is thus a system of interacting developmental propositions. The phenomena denoted in the propositions and their relationships culminate in what is regarded as the amount and character of crime at any time— that is, in the social reality of crime.

The theory of the social reality of crime as I have formulated it is inspired by a change that is occurring in our view of the world. This change, pervading all levels of society, pertains to the world that we all construct and from which, at the same time, we pretend to separate ourselves in our human experiences. For the study of crime, a revision in thought has directed attention to the criminal process: All relevant phenomena contribute to creating definitions of crime, development of behaviors by those involved in criminal-defining situations, and constructing an ideology of crime. The result is the social reality of crime that is constantly being constructed in society.

PART II

Theories of Deviance

Deviance holds a special intrigue for scholars of theory. Given its pervasive nature in society, its enigmatic conditions, and its generic appeal, even the earliest sociologists attempted to explain how and why deviance occurs. Especially considering people's inclination to conformity, the pressing question for scholars has dealt with *why* individuals engage in norm-violating behavior. Explanations for deviant behavior are as divergent as the acts they explain, ranging from acts of delinquency to professional theft, acts of integrity to a search for kicks, acts of desperation to those of bravado and daring. We next outline some of the major attempts at understanding deviant behavior.

THE STRUCTURAL PERSPECTIVE

The dominant theory in sociology for the first half of the twentieth century, **structural functionalism,** also commanded the greatest amount of sway in explaining deviant behavior. Durkheim advanced the theory that society is a moral phenomenon. He believed that at its root, the morals (norms, values, and laws) that individuals are taught constrain their behavior. Youngsters are taught the "rights" and "wrongs" of society early in life, with most people conforming to these expectations throughout adulthood. These moral beliefs, in large measure, determine how people behave, what they want, and who they are. Durkheim suggested that societies with high degrees of social integration (bonding, cohesion, community involvement) would increase the conformity of its members. However, and this is what concerned Durkheim, in the modern French society in which he was living, more and more people were becoming distanced from each other, people were partially losing their sense of belonging to their communities, and the norms and expectations of their groups were becoming less clearly defined. He believed that this condition, which he referred

to as *anomie,* was producing a concomitant amount of social disintegration, leading to greater degrees of deviance. Thus, for Durkheim, while norms still existed
on the societal level, the lack of social integration created a situation where they
were no longer becoming as significant a part of each individual.

Despite his concerns about the increasing rates of deviance that society would produce, Durkheim also subscribed to the idea that deviance was functional for society. As
we noted earlier, despite its obvious negative effects, Durkheim felt that deviance produces some positive benefits as well. At a time when people are worrying about the
moral breakdown and social disintegration of society, deviance serves to remind us
of the moral boundaries in society. Each time a deviant act is committed and publicly
announced, society is united in indignation against the perpetrator. This serves to bring
people together, rather than tearing them apart. At the same time, society is reminded
about what is "right" and "wrong" and, for those who conform, greater social integration ensues. These ideas were perhaps best illustrated by Yale sociologist Kai Erikson
who, in his 1966 book *Wayward Puritans* (excerpted in Chapter 1), demonstrated the
role of deviance in defining morality and bringing people together. Erikson examined
Puritan patterns of isolating and treating offenders. He believed that deviance serves as
a means to promote a contrast with the rest of the community, thus giving members of
the larger society more strength in their moral convictions. Erikson's analysis focused
on the transformation of the seventeenth-century Bay Colony, as a group of revolutionaries tried to establish a new community in New England. These deviants, the revolutionaries, played an important role in the transformation of norms and values—
their behavior elicited societal reaction, which served to clearly define the new community's norms and values. In addition, punishing some people for norm violations
reminded others of the rewards for conformity.

Chapter 6, "The Normal and the Pathological," by Durkheim, lays out his
theory of the inevitability of deviance in all societies. In his ironic twist, Durkheim
argues that deviance is normal rather than pathological, serving a positive function
in society. To achieve the maximum benefit, however, a society needs a manageable
amount of deviance. When the numbers of people declared deviant by current moral
standards rises or falls too much, society alters its moral criteria to maintain the level
of deviance in the optimal range. At different times it may "define deviancy down," as
Moynihan (1993) has suggested in looking at the way the bar defining acceptable
behavior has been lowered. When society lowers the bar of acceptability, fewer
acts are viewed as deviant and more become recast as tolerable. Over the years we
have normalized violence, divorce, premarital sex, tattooing and piercing, and
unwed pregnancy. Or society may "define deviancy up," as Krauthammer (1993)
has suggested, raising the bar defining normality. When the bar is raised, behavior
formerly considered acceptable becomes redefined as deviant. Things now considered deviant (or even criminal) that used to be regarded as tolerable, even if

not exactly embraced, include spanking, date rape, sexual harassment, panhandling, talking in movie theaters, and hate speech. What deviance does for society is to define the moral boundaries for everyone. Violation of norms serves to remind the masses what is acceptable and what is not; in Durkheim's words, it enforces the "collective conscience" of the group. We saw this in the public outcry over the inadequate governmental response to the Hurricane Katrina victims. Perhaps it is difficult to imagine that behavior that disgusts, reviles, or even nauseates you are not the acts of immoral, sick, or evil people but are typical and even beneficial parts of all societies. Durkheim's theory suggests that structural needs of the society as a whole, beyond the scope of its individual members, foster the continuing recurrence of deviance.

The structural perspective locates the root cause of crime and deviance outside of individuals, in the invisible social structures that make up any society. Structural explanations for deviance look at features of society that seem to generate higher rates of crime or deviance among some societies or groups within them. In looking for explanations for why some societies are likely to have higher crime rates than others, sociologists have suggested that those with greater degrees of inequality are likely to show more crime than those where people have roughly similar amounts of what the society values. Looking within each society, structuralists locate the cause of crime in two main factors: the differential opportunity structure, and prejudice and discrimination toward certain groups. In a society with inequality, some groups will clearly have greater structural access to certain opportunities than others. Groups with access to greater power and political and economic opportunity may use these to define their acts as legitimate and the acts of others as deviant, at the same time as they corruptly use their power to their own advantage. Not everyone has equal opportunity to dispense political favors, to manipulate stock prices, or to conduct covert operations. At the same time, groups with less access to the legitimate opportunity structure due to reduced educational opportunity, diminished access to healthcare, lower-class background, and disadvantaged legitimate networks and connections do not have the same opportunity to succeed normatively. Members of these groups may be propelled by their position in the social structure into alternate pathways.

It was Robert Merton, a mid-twentieth-century sociologist from Columbia University, who actually extended Durkheim's ideas and built them into a specific structural **strain theory** of deviant behavior. He claimed that contradictions are implicit in a stratified system in which the culture dictates success goals for all citizens, while institutional access is limited to just the middle and upper strata. In other words, despite the American Dream of rags-to-riches, some people, most often lower-class individuals, are systematically excluded from the competition. Instead of merely going through the motions while knowing that their legitimate path to success (measured in American society by financial wealth) is blocked, some members of the lower class retaliate by choosing a deviant alternative. Merton believed that these people have accepted society's goals (to be comfortable, to get rich), but they have insufficient access to the approved means of attaining these goals (deferred gratification, education, hard work). The problem lies in the social structure in society, where even if people follow the approved means, there are "roadblocks" prohibiting them from rising through the stratification system. Deviant behavior occurs when socially sanctioned means are not available for the realization of highly desirable goals. The only way to achieve these goals is to "detour" around them, to bypass the approved means in order to get at the approved goals. For example, for young men raised in urban ghettos with poor housing facilities, dilapidated schools, and inconsistent family lives, their road to "success" is more likely to be through dealing drugs, pimping, or robbing than it is through the normative route of school and hard work. According to Merton, then, *anomie* results from the lack of access to culturally prescribed goals and the lack of availability of legitimate means for attaining those goals. Deviance (or, more specifically, crime) is the obvious alternative. Once again, structural opportunities, or lack thereof, are seen as the root cause of deviance, rather than some psychological, or individual pathology. Agnew offers a contemporary articulation of strain theory in Chapter 8.

Richard Cloward and Lloyd Ohlin, in *Delinquency and Opportunity* (1960), thought that Merton was correct in directing us toward the notion that members of disadvantaged socioeconomic groups have less opportunity for achieving success in a legitimate manner, but they thought that Merton wrongly assumed that those groups, when confronted with the problem of differential opportunity, could automatically choose deviance and crime. In their **differential opportunity theory,** Cloward and Ohlin suggested that while all disadvantaged people have some lack of opportunity for legitimate pursuits, they do not have the same opportunity for participating in illegitimate practices. What Cloward and Ohlin believed was that deviant behavior depends on people's access to illegitimate opportunities. They found that three types of deviant opportunities are present:

> Criminal: Similar to the type Merton described, these arise from access to deviant subcultures, though not all disadvantaged youth enjoy these avenues.
> Conflict: These attract people who have a propensity for violence and fighting.
> Retreatist: These attract people, such as drug users, who are not inclined toward illegitimate means or violent actions but who want to withdraw from society.

According to Cloward and Ohlin, groups of people may have greater or lesser opportunity to climb the illicit opportunity ladder by virtue of several factors: (1) Some neighborhoods are rife with more criminal opportunities, networks, and enterprises than others, and people reared in these grow up amid these better opportunities; (2) some forms of illicit enterprise are dominated by people of particular racial or ethnic groups, so that members of these groups have an easier time rising to the top of these businesses or organizations; (3) the upper echelons of crime display a distinct glass ceiling for women, with men dominating the positions of decision making, earning, and power. Thus, Cloward and Ohlin extended Merton's theory by specifying the existence of differential illegitimate opportunities available to members of disadvantaged groups. It is in this illegitimate opportunity structure, rather than individual motivation, they argue, that the explanation for deviance can be found.

Chapter 5, "Conflict Theory of Crime," by Richard Quinney (discussed in connection to the social power perspective in the General Introduction) offers a **conflict theory** explanation that is not functionalist but is still structural. Conflict theorists see society differently from functionalists in their view of society as pluralistic, heterogeneous, and conflictual rather than unified and consensual. Social conflict arises out of the incompatible interests of diverse groups in society, such as businesses versus their workers, conservatives versus liberals, whites versus people of color, and the rich versus the poor. These groups have a structural conflict of interest with each other that stands above and beyond the individual members, framing the way they come to recognize their interests and act in the world. Not only is conflict a natural outcome of this arrangement, but crime is as well. In a succinct summation of conflict theory's major tenets, Quinney describes how crime exists because some behaviors conflict with the interests of the dominant class. These powerful members of society create legal definitions of human conduct, casting those behaviors that threaten its interests as criminal. Then, the dominant class enforces those laws onto the less powerful groups in society, through the police and the legal and the criminal justice systems, ensuring that their interests are protected. Members of subordinate classes are compelled to commit those actions that have been defined as crimes because their poverty presses them to do so. The dominant group can then create and disseminate their ideology of crime, which is that the most dangerous criminal elements in society can be found in the subordinate classes, and that these groups deserve arrest, prosecution, and imprisonment. Through class struggle and class conflict, crime is constructed, formulated, and applied so that less powerful groups are subdued and more powerful groups are strengthened. These processes are illustrated by the diagram in Quinney's chapter. This approach shows how larger social forces, such as group and class interests, shape the behavior of individual members, leading some to use their advantage to dominate over others, while the others react to their structural subordination by engaging in those behaviors already defined as deviant and deserving of punishment. All these structural theories place the cause for the incidence of deviance on the structures of society, rather than on individuals and their problems.

Feminist theory, the subject of Chapter 10, takes a structural approach as well, locating the pervasive discrimination and oppression of women in society in the overarching patriarchal system. Through the intertwined effects of major institutions and social structures such as our legal codes, the economy, our political system, social and cultural practices, religion, the family, the educational system, and the media, women are systematically disadvantaged. Women, they argue, are unprotected against verbal, physical, and sexual abuse, and their individual attempts to rise up and protect themselves often subject them to being labeled as offenders. When they flee abusive situations, the patriarchy of the streets oppresses them further, funneling them into acts of survival defined as deviant by the male hegemony. Feminists maintain that theories of deviance are male-centered when they impose stereotyped gender role requirements onto teenage girls and when they problematize women's attempts to survive under oppressive conditions in a system that systematically deprives them of resources.

THE CULTURAL PERSPECTIVE

Whereas structuralist theories had enormous impact on sociologists' thinking about deviance, other authors arose who felt that these were not all-encompassing explanations. These theorists believed that deviance was a collective act, driven and carried out by groups of people. Building on conflict theory's view that multiple groups with different interests exist in society, the subcultural theorists examined the implications of membership in these groups. Groups with conflicting interests include not only dominant and subordinate groups, but also a variety of social, religious, political, ethnic, and economic factions. Membership in each of these groups places people in distinct subcultures, each of which contains their own set of distinct norms and values. A pluralistic nation that was once thought of as the world's "melting pot," we have become, in part, a nation of many different groups, each with its own distinct subculture.

Thorsten Sellin, writing about "The Conflict of Conduct Norms" (1938), suggested in his **culture conflict theory** that while the norms and values of these subcultures to some extent incorporated and meshed with the norms and values of the overarching American culture, they were also different and in conflict. The disparities and different cultural codes between subcultural groups may become apparent in three situations. First, when people from one culture "migrate," or cross over into the territory of another culture, such as when people from a rural area move to the city, they find that their country ways do not mesh with modern urbanity. In these cases, they may find themselves subject to the urban norms and values. Second, cultural conflict may occur during a "takeover" situation, when the laws of one cultural group are extended to apply to another, such as when one group moves into and takes over the territory of another. This may happen when middle-class people regentrify a run-down neighborhood, and the latitude that used to be enjoyed by the former occupants to congregate outdoors, be homeless, do or deal drugs, or solicit prostitution becomes lost. In these cases, the norms of the group that has taken over may apply. Third, cultural codes may clash on the "border" of contiguous cultural areas, such as when people from different cultural groups find themselves in contact. They may be on a national border, a neighborhood border, or they may simply be people encountering members of another subculture. In these cases, no clear set of norms and values necessarily dominates, but individuals have to negotiate their cultural understandings delicately, trying to understand each other's norms and values. This could happen when new first-year students find themselves rooming with someone from a distinctly different background, where neither of their ways necessarily predominates. Sometimes this even happens when college students go home for the holidays, only to find that the norms under which they were living at college are rather different from those in their parents' houses.

In each of these three cases, people may find themselves torn between the norms and values of different group memberships. Following the norms and values of their subcultural group may produce behavior that becomes defined as deviant by the standards of the broader culture. Yet from their subcultural perspective, their behavior may be viewed as representing the acts of good people working to uphold the behaviors they honor. In his writing, Sellin was particularly thinking about the deviance of children belonging to immigrant ethnic or racial groups moving into the United States, caught in the struggle between two cultures, but his theory applies equally well to the large number of diverse subcultural groups in our country. He extended his model to apply to all conflicts between cultural groups that share a close geographic area, especially where there is normative and value domination by one culture over another.

Building on this idea, Albert Cohen, in *Delinquent Boys* (1955), posited a **reaction theory,** where working-class adolescent males develop a subculture with a different value system from the dominant American culture. These boys, Cohen asserted, have the greatest degree of difficulty in achieving success, since the establishment's standards are so different from their own. They try, at first, to fit in with the cultural expectations but find that they are unsuccessful. Exposed to middle-class aspirations and judgments they cannot reasonably fulfill, they develop a blockage (or strain) that leads them to experience "status frustration." What results from this frustration is the formation of an oppositionally reactive subculture that allows them to achieve status based on nonutilitarian, malicious, and negativistic behavior. These boys, in reaction against society's perceived unfairness toward them, substitute norms that reverse those of the larger society. Cohen claimed that delinquent boys, due to their societal rejection, turn the society's norms "upside down," rejecting middle-class standards and adopting values in direct opposition to those of the majority.

Walter Miller (1958), writing just after Cohen, further delineated the importance of subcultural values for the development of deviant behavior. He believed that the values of the lower-class culture produce deviance for its members because these are "naturally" in discord with middle-class values. Young people who conform to the lower-class culture in which they were born almost automatically become deviant. The culture of the lower class, by which members attain status in the eyes of their peers, is characterized by several "focal concerns": getting into trouble, showing toughness, maintaining autonomy, demonstrating street smarts, searching for excitement, and being tied in their lot to the capricious whims of fate. **Lower-class culture theory** suggests that when these individuals follow the norms of their subculture, they become deviant according to the predominantly middle-class societal norms and values.

The lasting impact of subcultural theories has been to suggest that conflicting values may exist in society. When one part of society can impose its definitions on other parts, the dominant group has the ability to label the minority group's norms and values, and the behavior that results from these, as deviant. Thus, any act can be considered deviant if it is so defined. These theories are suited to illustrating the motivations of people from minority, youth, alternative, or disadvantaged subcultures that are not well aligned with the dominant culture. They locate the explanation for deviance not in the structures that shape society, but in the flesh of the norms and values that compose different subcultural groups. Through cultural transmission, groups pass their norms and values down from one generation to the next, ensuring their survival and social placement as well as the continuance of cultural conflict.

THE INTERACTIONIST PERSPECTIVE

Although these previous theories produced insight into some explanations for deviance, there are interactional forces that inevitably intervene between the larger causes these sociologists propose and the way deviant behavior takes shape. Many people are exposed to the same structural conditions and the same cultural conflicts and pressures that have been theorized as accounting for deviance but still resist engaging in deviant behaviors. Left unaddressed is how people from the same structural groups and same subcultures can turn out so differently, how members of some families turn to deviance while others do not, and how members of the same family turn out so differently from each other. Interactionist theories fill this void by looking in a more microfashion at people's everyday behavior to try to understand why some people engage in deviance and become so labeled, while others do not. Interactionist theories deal with real flesh-and-blood people in specific times and places. They look at how people actually encounter specific others, and they look at the influence of these others. They seek to understand not only why deviance occurs, but also how it happens. Many of these theories look at specific social-psychological and interactional dynamics, such as family dynamics, the influence of role models, and the role of peer groups. When people confront the problems, pressures, excitements, and allures of the world, they most often do so in conjunction with their peer groups. It is within peer groups that people make decisions about what they will do and how they will do it. Their core feelings about themselves develop and become rooted in such groups. People's actions and reactions are thus guided by the collective perceptions, interpretations, and actions of their peer groups.

Edwin Sutherland and Donald Cressey recognized this point when they proposed their **differential association theory** of deviance, the subject of Chapter 8. The key feature of this view is the belief that deviant behavior is socially learned, and not from just anyone, but from people's most intimate friends and family members. People may be exposed to a variety of deviant and nondeviant ideas and contacts without that necessarily leading them to engage in deviance. But as their circle of contacts shifts from being primarily composed of people who hold nondeviant ideas to those holding deviant ideas and favorable definitions of deviant acts, they become more likely to engage in deviance. The more their friends hold deviant attitudes and engage in deviant behavior, the more likely they are to follow suit.

Sutherland and Cressey further suggested that people learn a variety of elements critical to deviance from their associates: the norms and values of the deviant subculture, the rationalizations for legitimizing deviant behavior, the techniques necessary to commit the deviant acts, and the status system of the subculture, by which members evaluate themselves and others. People thus do not decide, at a fixed point in time, to become deviant, but move toward these attitudes and behavior as they shift their circle of associates from more normative friends to more deviant friends. Sutherland argued that people rarely stumble onto deviance through their own devising or by seeing acts of deviance in the mass media (as many would suggest), but rather by having the knowledge, skills, attitudes, values, traditions, and motives passed down to them through interpersonal (not impersonal) means. Influencing this is the age at which they encounter this deviance (earlier in life is likely to be more significant) and the intimacy of the deviant relationships (closer friends and relatives will have greater sway).

Also looking at the interactional level, David Matza (1964) proposed **drift theory,** noting that this movement into deviant subcultures occurs through a process of drift, as people gradually leave their old crowd and become enmeshed in a circle of deviant associates. In so doing, Matza suggested that rather than just jumping immediately into deviance, they may drift between deviance and legitimacy, keeping one foot in each world. By simultaneously participating in both deviant and legitimate worlds, people can learn about and experience the nuances of the deviant world without having to abandon the advantages of their status within the legitimate world. They may drift indefinitely, without having to make a commitment to either for quite some time.

For example, college students may experiment with a different sexual orientation without revealing this to everyone, and without necessarily giving up their claim to heterosexuality. At some point they may decide to align more firmly with one side, or this may be forced on them by outside events (getting caught, moving away, becoming sick). Being confronted by someone who discovers the deviance may force people to make a choice and get off the fence, or perhaps just leaving college and having to choose which lifestyle and social network to align themselves with may force a choice. An alternate is that individuals choose one path, after a time, and decide to follow it. But Matza suggests that the dual membership condition may precede such a decision, where individuals try out both alternatives for a time without making a commitment. Thus, Matza suggests that it is rare for people to turn to deviance overnight, but more commonly, they take smaller steps, gradually moving to making deviant acquaintances, becoming familiar with deviant ideology, thinking about engaging in deviant acts, trying some out, and then expanding their frequency and range of deviance. Quitting deviance may be a similarly gradual and difficult process, requiring the abandonment of the group of deviant friends and reintegration in conventional circles, perhaps one foot in both worlds again, before normative behavior becomes the mode that is finally chosen.

A third theory under the interactionist perspective is **labeling theory,** the subject of Chapter 3. This approach suggests that many people dabble to greater or lesser degrees in various forms of deviance. Studies of juvenile delinquency suggest that rates of youthful participation are extremely widespread, nearly universal. How many people can claim to have reached adulthood without experimenting in illicit drinking, drug use, cheating, stealing, or vandalism? Yet do all of these people consider themselves deviants? Most do not. Many people retire out of deviance as they mature and avoid developing the deviant identity altogether. Others go on to engage in what Becker (1963) has called "secret deviance," conducting their acts of norm violation without ever seriously encountering the deviant label. Yet others, many of them no more experienced in the ways of deviance than the youthful delinquent or the secret deviant, become identified and identify themselves as deviants. What causes this difference? One critical difference, labeling theory suggests, lies in who gets caught. Getting caught sets off a chain reaction of events that leads to profound social and self-conceptual consequences. Frank Tannenbaum (1938) has described how individuals are publicly identified as norm violators and branded with that tag. They may go through official or unofficial social sanctioning where people identify and treat them as deviant. Becker noted that "the deviant is one to whom that label has successfully been applied; deviant behavior is behavior that people so label" (1963, p. 9). Deviance exists at the macrosocietal level of social norms and definitions through the collective attitudes we assign to certain acts and conditions. But it also comes into being at the everyday microlife level when the deviant label is applied to someone. The thrust of labeling theory is twofold, focusing on diverse levels and forces. As Edwin Schur (1979, p. 160) summarized its complexity:

> The twin emphases in such an approach are on *definition* and *process* at all the levels that are involved in the production of deviant situations and outcomes. Thus, the perspective is concerned not only with what happens to specific individuals when they are branded with deviantness ("labeling," in the narrow sense) but also with the wider domains and processes of social definitions and collective rule-making that frequently lie behind such concrete applications of negative labels. (*italics in original*)

In the chapter on labeling theory, Becker emphasizes that deviance lies in the eye of the beholder. There is nothing inherently deviant in any particular act, he claims, until some powerful group defines the act as deviant. Taking the onus off of the individual, Becker emphasizes the importance of looking at the process by which people are labeled deviant, and understanding that deviance is a consequence of others' reactions. This approach forces us to look, at how people are defined as deviant, why some acts are labeled and others ignored, and the circumstances that surround the commission of the act. Thus, deviance only exists when it is created by society. The key emphasis of the labeling theory approach to deviance, then, lies in the importance of human peer interaction in understanding the cause of human behavior.

Rooted at the microlevel, but looking less at the specific dynamics of interaction and more at the relationship between individuals and society, is Travis Hirschi's **control theory of delinquency,** the subject of Chapter 9. Like labeling theory, which took as given that people readily engage in acts of deviance but focused its explanation on the process of identity change that occurs when individuals are caught and labeled, control theory finds it unnecessary to look for the causes of deviant behavior. These are obvious, Hirschi asserts, as deviance and crime may not only be fun, but they also offer shortcuts and yield immediate, tangible benefits (albeit, with a risk). What we should be seeking to understand, instead, is what holds people back from committing these acts, what forces constrain and control deviance. Hirschi's answer is that social control lies in the extent to which people develop a stake in conformity, a *bond to society.* People who have a greater investment in society will be less likely to risk losing this through norm or law violations and will follow the rules more willingly. They may have such a stake through their job, relationships to friends and family, or their reputation in the community. Their stake may be fostered by any of the four components discussed in this chapter: attachment to conventional others, commitment to conventional institutions, involvement in conventional activities, and deep beliefs in conventional norms. The extent to which a society is able to foster greater bonds between itself and its potentially deviant members, by giving them a greater stake in achieving success, will affect the constraint or spread of its deviance, particularly, Hirschi notes, of the delinquent variety. It is these ties, interactionally forged and maintained, that influence individuals in their choice between deviant or nondeviant pathways.

Although these overarching perspectives and the theories nested within them differ in the level at which they place their explanations, they all locate them squarely in the social domain. In this they renounce the prevailing tendency toward unidimensional psychological explanation that lodges causation in pathology, compulsion, neurosis, or maladjustment, explanations whose oversimplicity and inadequacy in a modern, complex world cannot be overstated.

The most recent perspective on deviance theory has been advanced by social constructionists, and Joel Best, one of its leading proponents, describes its history and views in Chapter 11, "The Constructionist Stance." Initially coming from a microinteractionist approach, constructionist theorists sought to revitalize labeling theory by bridging the gap between the way labels are applied to individuals who then internalize them in their everyday-life context, and a larger, more macro awareness of the power structure in society that influences the way these labels are defined and enforced. When some citizens have access to greater degrees of social power that enables the dominant groups to rationalize that their ideologies and behavior may be legitimate, they are simultaneously defining the actions of less powerful groups as deviant. In so doing, they use the vehicle of deviance and its enforcement to boost their own power while disempowering those they construct as illegitimate. The definitions they forge are then applied to individuals who lack the power to repel them, and the agents of social control act to carry out their moral edicts. The powerless segments of society become the recipients of the activities of moral entrepreneurs (rule creators and rule enforcers) who lay claims that certain behaviors are menacing and dangerous. Individuals connected to these activities may be defined as deviant, isolated, sought out, labeled, and stigmatized. Social constructionism thus builds on the basic labeling theory foundation of identity construction by integrating conflict theory's sensitivity to inequality, looking at how the power struggle between dominant and subordinate groups is directly tied to interactional and identity consequences.

6

Functionalism: The Normal
and the Pathological

EMILE DURKHEIM

*One of the founders of the functionalist approach to sociology, Durkheim inte-
grates deviance into his overarching view of society. If, as he believes, society can be
compared to a living organism, then all of its social institutions, like the parts of a
human body, must contribute to its continuing existence. Under this view,
deviance, because of its pervasive presence cross-culturally and over the entire course
of history, is not an illness of pathology of the system, but rather something
that contributes to society's positive functioning. It fact, like Erikson, he notes
several positive functions that it provides, including serving as a means of introducing social change, as new behaviors move through criminal and deviant status
into respectability. Durkheim argues that deviance is so critical to society that if
people stopped engaging in it immediately (if we lived in a "society of saints"), we
would have to redefine acts now considered acceptable as deviant. Punishing and
curing criminals cannot simply, then, be regarded as the objective of society.*

Crime is present not only in the majority of societies of one particular species
but in all societies of all types. There is no society that is not confronted
with the problem of criminality. Its form changes; the acts thus characterized are
not the same everywhere; but, everywhere and always, there have been men
who have behaved in such a way as to draw upon themselves penal repression.
If, in proportion as societies pass from the lower to the higher types, the rate of
criminality, i.e., the relation between the yearly number of crimes and the population, tended to decline, it might be believed that crime, while still normal, is
tending to lose this character of normality. But we have no reason to believe
that such a regression is substantiated. Many facts would seem rather to indicate
a movement in the opposite direction. From the beginning of the [nineteenth]
century, statistics enable us to follow the course of criminality. It has everywhere
increased. In France the increase is nearly 300 percent. There is, then, no

phenomenon that presents more indisputably all the symptoms of normality, since it appears closely connected with the conditions of all collective life. To make of crime a form of social morbidity would be to admit that morbidity is not something accidental, but, on the contrary, that in certain cases it grows out of the fundamental constitution of the living organism; it would result in wiping out all distinction between the physiological and the pathological. No doubt it is possible that crime itself will have abnormal forms, as, for example, when its rate is unusually high. This excess is, indeed, undoubtedly morbid in nature. What is normal, simply, is the existence of criminality, provided that it attains and does not exceed, for each social type, a certain level, which it is perhaps not impossible to fix in conformity with the preceding rules.[1]

Here we are, then, in the presence of a conclusion in appearance quite paradoxical. Let us make no mistake. To classify crime among the phenomena of normal sociology is not to say merely that it is an inevitable, although regrettable phenomenon, due to the incorrigible wickedness of men; it is to affirm that it is a factor in public health, an integral part of all healthy societies. This result is, at first glance, surprising enough to have puzzled even ourselves for a long time. Once this first surprise has been overcome, however, it is not difficult to find reasons explaining this normality and at the same time confirming it.

In the first place crime is normal because a society exempt from it is utterly impossible. Crime, we have shown elsewhere, consists of an act that offends certain very strong collective sentiments. In a society in which criminal acts are no longer committed, the sentiments they offend would have to be found without exception in all individual consciousnesses, and they must be found to exist with the same degree as sentiments contrary to them. Assuming that this condition could actually be realized, crime would not thereby disappear; it would only change its form, for the very cause which would thus dry up the sources of criminality would immediately open up new ones.

Indeed, for the collective sentiments which are protected by the penal law of a people at a specified moment of its history to take possession of the public conscience or for them to acquire a stronger hold where they have an insufficient grip, they must acquire an intensity greater than that which they had hitherto had. The community as a whole must experience them more vividly, for it can acquire from no other source the greater force necessary to control these individuals who formerly were the most refractory. For murderers to disappear, the horror of bloodshed must become greater in those social strata from which murderers are recruited; but, first it must become greater throughout the entire society. Moreover, the very absence of crime would directly contribute to produce this horror; because any sentiment seems much more respectable when it is always and uniformly respected.

One easily overlooks the consideration that these strong states of the common consciousness cannot be thus reinforced without reinforcing at the same time the more feeble states, whose violation previously gave birth to mere infraction of convention—since the weaker ones are only the prolongation, the attenuated form, of the stronger. Thus robbery and simple bad taste injure the same single altruistic sentiment, the respect for that which is another's. However, this same sentiment is less grievously offended by bad taste than by robbery; and since, in addition, the average consciousness has not sufficient intensity to react keenly to the bad taste, it is treated with greater tolerance. That is why the person guilty of bad taste is merely blamed, whereas the thief is punished. But, if this sentiment grows stronger, to the point of silencing in all consciousnesses the inclination which disposes man to steal, he will become more sensitive to the offenses which, until then, touched him but lightly. He will react against them, then, with more energy; they will be the object of greater opprobrium, which will transform certain of them from the simple moral faults that they were and give them the quality of crimes. For example, improper contracts, or contracts improperly executed, which only incur public blame or civil damages, will become offenses in law.

Imagine a society of saints, a perfect cloister of exemplary individuals. Crimes, properly so called, will there be unknown; but faults which appear venial to the layman will create there the same scandal that the ordinary offense does in ordinary consciousness. If, then, this society has the power to judge and punish, it will define these acts as criminal and will treat them as such. For the same reason, the perfect and upright man judges his smallest failings with a severity that the majority reserve for acts more truly in the nature of an offense. Formerly, acts of violence against persons were more frequent than they are today, because respect for individual dignity was less strong. As this has increased, these crimes have become more rare; and also, many acts violating this sentiment have been introduced into the penal law which were not included there in primitive times.[2]

In order to exhaust all the hypotheses logically possible, it will perhaps be asked why this unanimity does not extend to all collective sentiments without exception. Why should not even the most feeble sentiment gather enough energy to prevent all dissent? The moral consciousness of the society would be present in its entirety in all the individuals, with a vitality sufficient to prevent all acts offending it—the purely conventional faults as well as the crimes. But a uniformity so universal and absolute is utterly impossible; for the immediate physical milieu in which each one of us is placed, the hereditary antecedents, and the social influences vary from one individual to the next, and consequently diversify consciousnesses. It is impossible for all to be alike, if only because each one has his own organism and that these organisms occupy different areas in space. That is why, even among the lower peoples, where individual originality is very little developed, it nevertheless does exist.

Thus, since there cannot be a society in which the individuals do not differ more or less from the collective type, it is also inevitable that, among these divergences, there are some with a criminal character. What confers this character upon them is not the intrinsic quality of a given act but that definition which the collective conscience lends them. If the collective conscience is stronger, if it has enough authority practically to suppress these divergences, it will also be more sensitive, more exacting; and, reacting against the slightest deviations with the energy it otherwise displays only against more considerable infractions, it will attribute to them the same gravity as formerly to crimes. In other words, it will designate them as criminal.

Crime is, then, necessary; it is bound up with fundamental conditions of all social life, and by that very fact it is useful, because these conditions of which it is part are themselves indispensable to the normal evolution of morality and law.

Indeed, it is no longer possible today to dispute the fact that law and morality vary from one social type to the next, nor that they change within the same type if the conditions of life are modified. But, in order that these transformations may be possible, the collective sentiments at the basis of morality must not be hostile to change, and consequently must have but moderate energy. If they were too strong, they would no longer be plastic. Every pattern is an obstacle to new patterns, to the extent that the first pattern is inflexible. The better a structure is articulated, the more it offers a healthy resistance to all modification; and this is equally true of functional, as of anatomical, organization. If there were no crimes, this condition could not have been fulfilled; for such a hypothesis presupposes that collective sentiments have arrived at a degree of intensity unexampled in history. Nothing is good indefinitely and to an unlimited extent. The authority which the moral conscience enjoys must not be excessive; otherwise no one would dare criticize it, and it would too easily congeal into an immutable form. To make progress, individual originality must be able to express itself. In order that the originality of the idealist whose dreams transcend his century may find expression, it is necessary that the originality of the criminal, who is below the level of his time, shall also be possible. One does not occur without the other.

Nor is this all. Aside from this indirect utility, it happens that crime itself plays a useful role in this evolution. Crime implies not only that the way remains open to necessary changes but that in certain cases it directly prepares these changes. Where crime exists, collective sentiments are sufficiently flexible to take on a new form, and crime sometimes helps to determine the form they will take. How many times, indeed, it is only an anticipation of future morality—a step toward what will be! According to Athenian law, Socrates was a criminal, and his condemnation was no more than just. However, his crime, namely, the independence of his thought, rendered a service not only to humanity but to his country. It served to prepare a new morality and faith which the Athenians needed, since the traditions by which they had lived until then were no longer in harmony with the current conditions of life. Nor is the case of Socrates unique; it is reproduced periodically in history. It would never have been possible to establish the freedom of thought we now enjoy if the regulations prohibiting it had not been violated before being solemnly abrogated. At that time, however, the violation was a crime, since it was an offense against sentiments still very keen in the average conscience. And yet this crime was useful as a prelude to reforms which daily became more necessary. Liberal philosophy had as its precursors the heretics of all kinds who were justly punished by secular authorities during the entire course of the Middle Ages and until the eve of modern times.

From this point of view the fundamental facts of criminality present themselves to us in an entirely new light. Contrary to current ideas, the criminal no longer seems a totally unsociable being, a sort of parasitic element, a strange and unassimilable body, introduced into the midst of society.[3] On the contrary, he plays a definite role in social life. Crime, for its part, must no longer be conceived as an evil that cannot be too much suppressed. There is no occasion for self-congratulation when the crime rate drops noticeably below the average level, for we may be certain that this apparent progress is associated with some social disorder. Thus, the number of assault cases never falls so low as in times of want.[4] With the drop in the crime rate, and as a reaction to it, comes a revision, or the need of a revision in the theory of punishment. If, indeed, crime is a disease, its punishment is its remedy and cannot be otherwise conceived; thus, all the discussions it arouses bear on the point of determining what the punishment must be in order to fulfill this role of remedy. If crime is not pathological at all, the object of punishment cannot be to cure it, and its true function must be sought elsewhere.

NOTES

1. From the fact that crime is a phenomenon of normal sociology, it does not follow that the criminal is an individual normally constituted from the biological and psychological points of view. The two questions are independent of each other. This independence will be better understood when we have shown, later on, the difference between psychological and sociological facts.

2. Calumny, insults, slander, fraud, etc.

3. We have ourselves committed the error of speaking thus of the criminal, because of a failure to apply our rule (*Division du travail social*, pp. 395–96).

4. Although crime is a fact of normal sociology, it does not follow that we must not abhor it. Pain itself has nothing desirable about it; the individual dislikes it as a society does crime, and yet it is a function of normal physiology. Not only is it necessarily derived from the very constitution of every living organism, but it plays a useful role in life, for which reason it cannot be replaced. It would, then, be a singular distortion of our thought to present it as an apology for crime. We would not even think of protesting against such an interpretation, did we not know to what strange accusations and misunderstandings one exposes oneself when one undertakes to study moral facts objectively and to speak of them in a different language from that of the layman.

7

Strain Theory

ROBERT AGNEW

Agnew's strain theory builds on the work of Merton, a structural theorist, who suggested that broader patterns of crime characterize certain groups of people because of the social "strain" they face between the good things society offers and their inability to legitimately attain them. Merton identifies a disjuncture for people between what he calls cultural goals, or the ends toward which we are all socialized to strive (largely conceived by him in a material sense) and institutional norms, or the acceptable means that we use to reach these goals. He refers to this condition, where people cannot attain their goals legitimately, as a "blocked opportunity structure." Societies where goals are more strongly emphasized than the legitimate means of achieving them are likely to see people use deviant innovations, as when schools make grades more critically important than the learning required to attain them, or when some groups lack the resources to legitimately earn good grades. Merton sketches other ways that groups adapt to their balance of means and ends (conformity, ritualism, retreatism, rebellion) as alternate adaptations to their structural conditions in society.

Agnew's version of strain theory is newer and much broader, giving us an enhanced definition of social strain that includes not only the failure to achieve positively valued social goals, but also the loss of valued possessions and negative treatment by others. He offers an insightful distinction between objective and subjective strains, elaborating as well on the difference between experienced, vicarious, and anticipated strains. Some of these strains, he notes, are more likely to lead to deviance than others, an argument that he fleshes out more fully in the second part of this chapter. He ties the role of emotions into the gap between structural conditions and deviance by examining the role of various feelings in leading people to respond to strains with crime. This brings Merton's structural approach closer to an individual level of explanation.

According to general strain theory (GST), people engage in crime because they experience strains or stressors. For example, they are in desperate need of money or they believe they are being mistreated by their family members. They become upset, experiencing a range of negative emotions, including anger, frustration, and depression. And they may cope with their strains and negative emotions through crime. Crime is a way to reduce or escape from strains. For example, individuals engage in theft to obtain the money they desperately need or they run away from home to escape their abusive parents. Crime is a way for individuals to seek revenge against those who have wronged them. For example, individuals assault those who have mistreated them. And crime is a way to alleviate the negative emotions that result from strains. For example, individuals use illegal drugs to make themselves feel better.

Not all individuals, however, respond to strains with crime. If someone steps on your foot, for example, you are probably unlikely to respond by punching the person. Some people are more likely than others to cope with strains through crime. Criminal coping is more likely when people lack the ability to cope in a legal manner. For example, crime is more likely when people do not have the verbal skills to negotiate with those who mistreat them or do not have others they can turn to for help. Criminal coping is more likely when the costs of crime are low. For example, crime is more likely when people are in environments where the likelihood of being sanctioned for crime is low. And criminal coping is more likely when people are disposed to crime. For example, assault is more likely when people believe that violence is an appropriate response to being treated in a disrespectful manner.

I briefly elaborate on these arguments below. First, I define strains and describe the types of strains most likely to lead to crime. Next, I discuss why strains increase the likelihood of crime. Finally, I examine why some people are more likely than others to respond to strains with crime.

WHAT ARE STRAINS?

Strains refer to events or conditions that are disliked by the individual. There are three major types of strains. Individuals may lose something they value (lose something good). Perhaps their money or property is stolen, a close friend or family member dies, or a romantic partner breaks up with them. Individuals may be treated in an aversive or negative manner by others (receive something bad). Perhaps they are sexually or physically abused by a family member, their peers insult or ridicule them, or their employer treats them in a disrespectful manner. And individuals may be unable to achieve their goals (fail to get something they want). Perhaps they have less money, status, or autonomy than they want.

Objective and Subjective Strains Some events and conditions are disliked by most people or at least by most people in a given group. For example, most people dislike being physically assaulted or deprived of adequate food and shelter. And it has been argued that most males dislike having their masculine status called into question. I refer to these events and conditions as objective strains, because they are generally disliked. It is possible to determine the **objective strains** for a group by interviewing a carefully selected sample of group members or people familiar with the group being examined. We can ask these people how much they (or the group members) would dislike a range of events and conditions.

It is important to keep in mind, however that people sometimes differ in their subjective evaluation of the same events and conditions—even those events and conditions classified as objective strains. So a given objective strain, like a death in the family, may be strongly disliked by one person but only mildly disliked by another. This is because the subjective evaluation of objective strains is influenced by a range of factors, including peoples' personality traits, goals and values, and prior experiences. Wheaton (1990), for example, found that there was some variation in how people evaluated their divorce. Among other things, the quality of their prior marriage strongly influenced their evaluation, with people in bad marriages evaluating their divorce in positive terms. I therefore make a distinction between objective and **subjective strains.** While an objective strain refers to an event or condition that is disliked by most people or most people in a given group, a subjective strain refers to an event or condition that is disliked by the particular person or persons being examined. As just suggested, there is only partial overlap between objective and subjective strains.

Most of the research on strain theory focuses on objective strains. Researchers ask respondents whether they have experienced events and conditions that are assumed to be disliked. For example, they ask respondents whether they have received failing grades at school. No attempt is made to measure the respondents' subjective evaluation of these events and conditions.

This may cause researchers to underestimate the effect of strains on crime, because objective strains are not always disliked by the individuals being examined. Some people, for example, may not be particularly bothered by the fact that they have received failing grades. It is therefore desirable for criminologists to measure both the individual's exposure to objective strains and the individual's **subjective evaluation** of these strains (e.g., ask individuals whether they have received failing grades *and*, if so, how much they dislike such grades).

Experienced, Vicarious, and Anticipated Strains   GST focuses on the individual's personal experiences with strains; that is, did the individual personally experience disliked events or conditions. For example, was the individual physically assaulted. Personal experiences with strains should bear the strongest relationship to crime. However, it is sometimes important to consider the individual's vicarious and anticipated experiences with strains as well.

Vicarious strains refer to the strains experienced by others around the individual, especially close others like family members and friends. For example, were any of the individual's family members or friends physically assaulted. Vicarious strains can also upset the individual and lead to criminal coping. Agnew (2002), for example, found that individuals were more likely to engage in crime if they reported that their family members and friends had been the victims of serious assaults. This held true even after Agnew took account of other factors, like the individual's own victimization experiences and prior criminal history. Vicarious strains may have increased the likelihood of crime for several reasons. For example, perhaps individuals were seeking revenge against those who had victimized their family and friends. Or perhaps individuals were seeking to prevent the perpetrators from causing further harm. Vicarious strains are most likely to cause crime when they are serious, involve someone that the individual cares about and has assumed responsibility for protecting, involve unjust treatment, and pose a threat to others.

It is also sometimes important to consider anticipated experiences with strains. **Anticipated strains** refer to the individual's expectation that his or her current strains will continue into the future or that new strains will be experienced. For example, individuals may anticipate that they will be the victims of physical assault. Like vicarious strains, anticipated strains may upset individuals and lead to criminal coping. Individuals may engage in crime to prevent anticipated strains from occurring, to seek revenge against those who might inflict such strains, or to alleviate negative emotions. To illustrate, many adolescents, particularly in high-crime communities, anticipate that they will be the victims of violence. They often (illegally) carry weapons as a result and they may even engage in violence against others in an effort to reduce the likelihood that they will be victimized. In this area, Anderson (1999) argues that the young men in very poor, high-crime communities often try to reduce the likelihood they will be victimized by adopting a tough demeanor and responding to even minor shows of disrespect with violence. Anticipated strains are most likely to result in crime when individuals believe that they have a high probability of occurring in the near future, they will be serious in nature, and they will involve unjust treatment by others.

What Are the Characteristics of Those Types of Strains Most Likely to Cause Crime?

Not all strains result in crime. In fact, the experience of some strains may reduce the likelihood of crime. This is the case, for example, with parental discipline that is consistent and fair. Juveniles may not like such discipline, but much data suggest that it reduces the likelihood of crime. Strains are most likely to cause crime when they: (1) are seen as high in magnitude, (2) are seen as unjust, (3) are associated with low social control, and (4) create some pressure or incentive to engage in criminal coping. Strains with these characteristics are more likely to elicit strong negative emotions, reduce the ability to engage in legal coping, reduce the perceived costs of crime, and create a disposition for crime.

The Strain is Seen as High in Magnitude Imagine that you are chatting with a group of acquaintances and someone reacts to a remark you make by stating "you don't know what you're talking about." Now imagine the same situation, but this time someone reacts to your remark by stating "you're an asshole" and then shoving you. Both reactions are likely to upset you, but I think most people would agree that the second reaction is more likely to lead to crime. Part of the reason for this is that the second reaction is more severe than the first. Generally speaking, strains that are seen as more severe or higher in magnitude are more likely to result in crime. The severity of the strain refers to the extent to which the strain is negatively evaluated; that is, the extent to which it is disliked and viewed as having a negative impact on one's life. Among other things, severe strains are more likely to elicit strong negative emotions, which create pressure for corrective action, and they are more difficult to cope with in a legal manner (e.g., it is more difficult to legally cope with a large rather than a small need for money).

A strain is more likely to be seen as severe if: (a) it is high in degree or size (e.g., a large versus a small financial loss); (b) it is frequent, recent, of long duration, and expected to continue in the future; and (c) it threatens the *core* goals, needs, values, activities, and/or identities of the individual (e.g., does the strain threaten a core identity, perhaps one's masculine identity, or a secondary identity, perhaps one's identity as a good chess player?).

The Strain is Seen as Unjust   Imagine you are walking down the street. Someone accidentally trips on a crack in the sidewalk, bumps into you, and knocks you to the ground. Now imagine that you are walking down the street and someone deliberately shoves you aside, knocking you to the ground. Both incidents qualify as strains; being knocked to the ground is disliked by most people. But I think most people would agree that the second incident is more likely to result in crime. Even though both incidents involve the same amount of physical harm, the behavior in the second incident is more likely to be seen as unjust. Unjust strains are more likely to lead to crime for several reasons; most notably the fact that they make individuals more angry.

A strain is more likely to be seen as unjust when it involves the voluntary and intentional violation of a relevant justice norm or rule. Most strains involve a perpetrator who does something to a victim. We are more likely to view the perpetrator's behavior as unjust if the perpetrator freely chose to treat the victim in a way that he or she knew would probably be disliked (the "voluntary and intentional" part). We are less likely to view the behavior as unjust if it is the result of such things as reasonable accident or chance. That is why we are less upset by an accidental bump than by a deliberate shove, even though both may cause the same physical harm. We are also less likely to view strains as unjust if they are the result of our own behavior (e.g., we injure ourselves while behaving in a reckless manner) or natural forces (e.g., our home suffers damage during a storm).

Unjust behavior, however, involves more than a voluntary and intentional effort to harm someone. For example, parents voluntarily and intentionally punish their children on a regular basis, but we usually do not view their behavior as unjust. In order for voluntary and intentional behaviors to be seen as unjust, they must also violate a relevant justice rule. In particular, researchers have discovered that most people employ certain rules to determine whether a particular behavior is just or unjust. For example, the voluntary and intentional infliction of strain is more likely to be seen as unjust when victims believe that the strain they have experienced is undeserved and not in the service of some greater good (e.g., God or country).

The Strain is Associated with Low Social Control   Consider the following two strains. First, someone is unemployed for a long period of time. Second, a well-paid lawyer has to work long hours on a regular basis, often performing difficult and complex tasks. I think most people would agree that the first strain is more likely to result in crime. This example highlights a third factor affecting the likelihood that strains will lead to crime. Strains are more likely to lead to crime when they are associated with low social control.

There are several types of social control, with each referring to a factor or set of factors that restrains the individual from crime. There is **direct control,** which refers to the extent to which others set rules that prohibit crime, monitor the individual's behavior, and consistently sanction the individual for rule violations. There is the individual's **emotional bond or attachment to conventional others,** such as family members and teachers. There is the **individual's investment in conventional institutions,** such as school and work. It is easier to engage in crime when emotional bonds and investments are weak, since there is less to lose through crime. And there is the individual's **beliefs regarding crime.** It is easier to engage in crime when one does not believe that it is wrong to do so.

Certain strains are associated with low levels of social control. For example, this is the case with parental rejection. Children who are rejected by their parents probably have little emotional bond to them and are probably subject to little direct control by them. To give another example, those strains involving unemployment and work in the secondary labor market ("bad jobs") are associated with a low investment in conventional institutions. Strains associated with low social control are more likely to result in crime because they reduce the costs of crime, among other things.

The Strain Creates Some Pressure or Incentive for Criminal Coping   A final factor affecting the likelihood that a strain will lead to crime is the extent to which the strain creates some incentive or pressure to engage in criminal coping. Certain strains are more easily resolved through crime and/or less easily resolved through legal channels than other strains. As a consequence, individuals have more incentive to cope with these strains through crime. For example, that type of strain involving a desperate need for money is more easily resolved through crime than is that type involving the inability to achieve educational success. It is much easier to get money through crime than it is to get educational success. Also, certain strains are associated with exposure to others who model crime, reinforce crime, teach beliefs favorable to crime, or otherwise try to pressure or entice the individual into crime. For example, individuals who experience child abuse are exposed to criminal models—who may foster the belief that crime is an appropriate way to deal with one's problems. To give another example, many interpersonal disputes occur before an audience, with the audience members often urging or pressuring the disputants to engage in violence.

What Specific Strains are Most Likely to Cause Crime?   Drawing on the above discussion, it is predicted that the following specific strains are most likely to cause crime:

- Parental rejection.
- Supervision/discipline that is erratic, excessive, and/or harsh.
- Child abuse and neglect.

- Negative secondary school experiences (e.g., low grades, negative relations with teachers, the experience of school as boring and a waste of time).
- Abusive peer relations (e.g., insults, threats, physical assaults).
- Work in the secondary labor market (i.e., "bad jobs" that pay little, have few benefits, little opportunity for advancement, and unpleasant working conditions).
- Chronic unemployment.
- Marital problems.
- The failure to achieve selected goals, including thrills/excitement, high levels of autonomy, masculine status, and the desire for much money in a short period of time.
- Criminal victimization.
- Residence in economically deprived communities.
- Homelessness.
- Discrimination based on characteristics such as race/ethnicity and gender.

Data support these predictions, although the effect of certain of these strains—such as abusive peer relations and discrimination—has not been well examined.

It should also be noted that crime is especially likely when individuals experience two or more strains close together in time. Experiencing several strains at once is especially likely to generate negative emotions and tax the individual's ability to cope in a legal manner. Unfortunately, it is not uncommon for strains to occur together. One strain, such as a job loss, frequently leads to other strains—such as family conflict. As a consequence, researchers who test GST should employ cumulative measures, which count the number of different strains conducive to crime that are experienced.

WHY DO STRAINS INCREASE THE LIKELIHOOD OF CRIME?

Strains lead individuals to commit crimes for several reasons, the most important of which involves the effect of strains on negative emotions.

Strains Lead to Negative Emotions

Strains increase the likelihood that individuals will experience a range of negative emotions, including anger, frustration, jealousy, depression, and fear. These negative emotions increase the likelihood of crime for several reasons. Most notably, they create pressure for corrective action. Individuals feel bad and want to do something about it. As indicated above, crime is one possible response. Crime may be used to reduce or escape from strains, obtain revenge, and alleviate negative emotions (through illegal drug use). Negative emotions may also reduce the ability to cope in a legal manner, reduce the perceived costs of crime, and create a disposition for crime. For example, angry individuals have more trouble reasoning with others, are less aware of and concerned about the costs of crime, and have a desire for revenge. Most research on GST has focused on the emotion of anger, and studies suggest that anger partly explains the effect of strains on crime—particularly violent crime. Other emotions may be more relevant to other types of crime. For example, frustration may be most relevant to the explanation of property crime, while depression may help explain drug use....

Strains May Reduce Levels of Social Control

Strains may also lead to reductions in the major types of social control. This argument has been made by most of the leading strain theorists. Many strains involve negative treatment by conventional others, like parents, spouses, teachers, and employers. Further, these strains are often chronic or occur on a repeated basis. Such strains include child abuse, harsh discipline by parents, demeaning treatment by teachers, the receipt of low grades, conflict with spouses, unemployment, and work in "bad" jobs. These strains may reduce one's emotional bond to conventional others. Child abuse, for example, is likely to reduce the child's bond to parents. These strains may also reduce one's investment in conventional activities. Chronic unemployment, for example, represents a major reduction in one's investment in conventional activities. Further, these strains may reduce levels of direct control by causing individuals to retreat from conventional others, like parents and teachers. And these effects may reduce the individual's belief that crime is wrong, since the individual's ties to those who teach this belief are weakened. Also, individuals are less likely to accept societal norms condemning crime when they fail to reap the benefits that society has to offer, like a good education, a good job, and a loving family. Individuals who are very poor, for example, are less likely to condemn criminal methods for obtaining money. Data provide some support for these arguments.

Strains May Foster the Social Learning of Crime

Finally, strains may foster the social learning of crime. Most notably, strains increase the likelihood that individuals will join or form criminal groups, like delinquent peer groups and gangs. The members of such groups, in turn, model crime, reinforce crime, and teach beliefs favorable to crime. As just indicated, strains often reduce levels of social control, which increases the likelihood that individuals will come in contact with criminal groups and frees individuals to associate with such groups. Strains also increase the appeal of criminal groups. In particular, the victims of strains often view criminal groups as a solution to their strains. For example, individuals who cannot achieve status through conventional channels, like educational and occupational success, often join criminal gangs because the gang makes them feel important, respected, and/or feared. Further, strains directly increase the likelihood that individuals will come to view crime as a desirable, justifiable, or at least excusable form of behavior. For example, many strains involve unjust treatment by others. Such treatment may foster the belief that crime is justified since it is being used to "right" an injustice. Studies also provide some support for these arguments.

WHY ARE SOME PEOPLE MORE LIKELY THAN OTHERS TO COPE WITH STRAINS THROUGH CRIME?

Not all people cope with strains through crime. Most people, in fact, cope in a legal manner. For example, they negotiate with the people who irritate or harass them, they file complaints against the people who wrong them, or they alleviate their negative emotions by exercising or listening to music. A number of factors influence how individuals cope with the strains and negative emotions they experience. Criminal coping is most likely when:

Individuals Lack the Ability to Cope with Strains in a Legal Manner   Some individuals are less able to cope with strains in a legal manner than others. Their ability to cope in a legal manner is partly a function of their individual traits, like their intelligence, social and problem-solving skills, and personality traits. It is partly a function of the resources they possess, including their financial resources. And it is partly a function of their level of conventional social support. Are there conventional others, such as parents and friends, that they can turn to for aid and comfort?

The Costs of Criminal Coping are Low   Many individuals avoid criminal coping because the costs of crime are high for them. There is a good chance that they will be sanctioned by others if they engage in crime. They also have a lot to lose if they engage in crime; they might get expelled from school, lose their job, or jeopardize their relationship with people they care about. Further, engaging in crime will make them feel guilty, because they believe that crime is wrong. But for other individuals, the costs of criminal coping are low. They are in environments where the likelihood of sanction for crime is small. Perhaps they are poorly supervised by their parents, their friends do not care if they engage in crime, and neighborhood residents seldom report crimes to the police. They do not have jobs to lose or close relationships with others that might be jeopardized by crime. And they do not believe that crime is wrong. Such individuals, then, are more likely to cope with strains through crime.

Individuals are Disposed to Crime   Some individuals are more disposed than others to respond to strains with crime. They may possess personality traits which increase their inclination to crime, such as the traits of negative emotionality and low constraint. Also, some individuals may believe that crime is an appropriate response to certain strains, like disrespectful treatment by others. Further, some individuals may associate with criminal others, who model and reinforce crime. This too increases their disposition to respond to strains with crime.

Studies provide mixed support for these arguments. Some studies find that certain of the above factors increase the likelihood that individuals will respond to strains with crime, while other studies do not. Agnew (2006) discusses some possible reasons for these mixed findings.

CONCLUSION

Why do individuals engage in crime according to GST? They experience strains, become upset as a result, and may cope with their strains and negative emotions through crime. Criminal coping is especially likely if they lack the ability to cope in a legal manner, their costs of crime are low, and they are disposed to crime. Crime may allow them to reduce or escape from their strains, obtain revenge, or alleviate their negative emotions (through illegal drug use).

While GST focuses on the explanation of individual differences in offending, it can also shed light on patterns of offending over the life course, group differences in crime, and community and societal differences in crime. Such patterns and differences can be partly explained in terms of differences in the exposure to strains conducive to crime and in the possession of those factors that influence the likelihood of criminal coping. For example, gender differences in crime are partly due to the fact that males are more often exposed to strains conducive to crime and are more likely to cope with strains through crime.

GST also provides recommendations for reducing crime. Crime can be reduced by reducing individuals' exposure to strains conducive to crime and their likelihood of responding to strains with crime. Strategies for reducing the exposure to strains include (a) eliminating strains conducive to crime, (b) altering strains so as to make them less conducive to crime (e.g., reducing their magnitude or perceived injustice), (c) removing individuals from strains conducive to crime, (d) equipping individuals with the traits and skills to avoid strains conducive to crime, and (e) altering the perceptions and goals of individuals to reduce subjective strains. Strategies for reducing the likelihood that individuals will respond to strains with crime include (a) improving conventional coping skills and resources, (b) increasing social support, (c) increasing social control, and (d) reducing association with delinquent peers and beliefs favorable to crime.

GST presents a rather different explanation of crime than that offered by the other leading crime theories. GST is the only theory to focus explicitly on **negative relationships with others,** relationships in which others take the individual's valued possessions, treat the individual in an aversive manner, or prevent the individual from achieving his or her goals. Further, GST is the only theory to argue that individuals are **pressured into crime** by these negative relationships and the emotions that result from them. At the same time, GST is intimately related to the other leading theories of crime. As argued earlier, strains may contribute to personality traits conducive to crime, reduce social control, and foster the social learning of crime. Further, the factors associated with all of these theories interact with one another in their effect on crime. In particular, the effect of strains on crime is influenced or conditioned by personality traits, level of social control, and those factors that foster the social learning of crime—like association with criminal peers.

REFERENCES

Agnew, Robert. 2002. "Experienced, Vicarious, and Anticipated Strain: An Exploratory Study Focusing on Physical Victimization and Delinquency." *Justice Quarterly* 19: 603–32.

———. 2006. *Pressured into Crime: An Overview of General Strain Theory*. Los Angeles: Roxbury.

Anderson, Elijah. 1999. *Code of the Street*. New York: Norton.

Wheaton, Blair. 1990. "Where Work and Family Meet: Stress across Social Roles." Pp. 153–74 in J. Eckenrode and S. Gore (eds.), *Stress between Work and Family*. New York: Plenum.

8

Differential Association

EDWIN H. SUTHERLAND AND DONALD R. CRESSEY

Like Durkheim, Sutherland and Cressey do not regard crime and deviance as the result of either pathology in society or pathological behavior patterns. Crime, they argue, is learned in much the same way as all ordinary behavior and represents the expression of the same behavioral needs and values as other behavior. Crime is less likely to be learned, in fact, from frightening or suspicious outsiders than from people's own intimate associates. In this way, Sutherland and Cressey cast the learning of crime and deviance as a normal process and note that it is a likely occurrence as people become surrounded with increasing numbers of deviant friends. Sutherland and Cressey place the learning of deviance within people's most intense and personal relations: families and peer groups. By repeatedly watching others modeling crime, by learning from them how to effectively do it, and by becoming convinced by their rationalizations and neutralizations that such behaviors are acceptable, individuals move into criminal or deviant behavior patterns.

The following statements refer to the process by which a particular person comes to engage in criminal behavior.

1. *Criminal behavior is learned.* Negatively, this means that criminal behavior is not inherited, as such; also, the person who is not already trained in crime does not invent criminal behavior, just as a person does not make mechanical inventions unless he has had training in mechanics.

2. *Criminal behavior is learned in interaction with other persons in a process of communication.* This communication is verbal in many respects but includes also "the communication of gestures."

3. *The principal part of the learning of criminal behavior occurs within intimate personal groups.* Negatively, this means that the impersonal agencies of communication, such as movies and newspapers, play a relatively unimportant part in the genesis of criminal behavior.

4. *When criminal behavior is learned, the learned includes (a) techniques of committing the crime, which are sometimes very complicated, sometimes very simple; (b) the specific direction of motive, drives, rationalizations, and attitudes.*

From Edwin H. Sutherland, Donald Cressey, and David Luckenbill, *Principles of Criminology,* (11th edition) (pp. 88–90), 1992. Reprinted by permission of Alta Mira Press.

5. *The specific direction of motives and drives is learned from definitions of the legal codes as favorable or unfavorable.* In some societies an individual is surrounded by persons who invariably define the legal codes as rules to be observed, while in others he is surrounded by persons whose definitions are favorable to the violation of the legal codes. In our American society these definitions are almost always mixed, with the consequence that we have culture conflict in relation to the legal codes.

6. *A person becomes delinquent because of an excess of definitions favorable to violation of law over definitions unfavorable to violation of law.* This is the principle of differential association. It refers to both criminal and anticriminal associations and has to do with counteracting forces. When persons become criminal, they do so because of contacts with criminal patterns and also because of isolation from anticriminal patterns. Any person inevitably assimilates the surrounding culture unless other patterns are in conflict; a southerner does not pronounce *r* because other southerners do not pronounce *r*. Negatively, this proposition of differential association means that associations which are neutral so far as crime is concerned have little or no effect on the genesis of criminal behavior. Much of the experience of a person is neutral in this sense, for example, learning to brush one's teeth. This behavior has no negative or positive effect on criminal behavior except as it may be related to associations which are concerned with the legal codes. This neutral behavior is important especially as an occupier of the time of a child so that he is not in contact with criminal behavior during the time he is so engaged in the neutral behavior.

7. *Differential associations may vary in frequency, duration, priority, and intensity.* This means that associations with criminal behavior and also associations with anticriminal behavior vary in those respects. "Frequency" and "duration" as modalities of associations are obvious and need no explanation. "Priority" is assumed to be important in the sense that lawful behavior developed in early childhood may persist throughout life, and also that delinquent behavior developed in early childhood may persist throughout life. This tendency, however, has not been adequately demonstrated, and priority seems to be important principally through its selective influence. "Intensity" is not precisely defined, but it has to do with such things as the prestige of the source of a criminal or anticriminal pattern and with emotional reactions related to the associations. In a precise description of the criminal behavior of a person, these modalities would be rated in quantitative form and a mathematical ratio reached. A formula in this sense has not been developed, and the development of such a formula would be extremely difficult.

8. *The process of learning criminal behavior by association with criminal and anticriminal patterns involves all of the mechanisms that are involved in any other learning.* Negatively, this means that the learning of criminal behavior is not restricted to the process of imitation. A person who is seduced, for instance, learns criminal behavior by association, but this process would not ordinarily be described as imitation.

9. *While criminal behavior is an expression of general needs and values, it is not explained by those general needs and values, since noncriminal behavior is an expression of the same needs and values.* Thieves generally steal in order to secure money, but likewise honest laborers work in order to secure money. The attempts by many scholars to explain criminal behavior by general drives and values, such as the happiness principle, striving for social status, the money motive, or frustration, have been, and must continue to be, futile, since they explain lawful behavior as completely as they explain criminal behavior. They are similar to respiration, which is necessary for any behavior, but which does not differentiate criminal from noncriminal behavior.

It is not necessary, at this level of explanation, to explain why a person has the associations he has; this certainly involves a complex of many things. In an area where the delinquency rate is high, a boy who is sociable, gregarious, active, and athletic is very likely to come in contact with other boys in the neighborhood, learn delinquent behavior patterns from them, and become a criminal; in the same neighborhood the psychopathic boy who is isolated, introverted, and inert may remain at home, not become acquainted with the other boys in the neighborhood, and not become delinquent. In another situation, the sociable, athletic, aggressive boy may become a member of a scout troop and not become involved in delinquent behavior. The person's associations are determined in a general context of social organization. A child is ordinarily reared in a family; the place of residence of the family is determined largely by family income; and the delinquency rate is in many respects related to the rental value of the houses. Many other aspects of social organization affect the kinds of associations a person has.

The preceding explanation of criminal behavior purports to explain the criminal and noncriminal behavior of individual persons. It is possible to state sociological theories of criminal behavior which explain the criminality of a community, nation, or other group. The problem, when thus stated, is to account for variations in crime rates and involves a comparison of the crime rates of various groups or the crime rates of a particular group at different times. The explanation of a crime rate must be consistent with the explanation of the criminal behavior of the person, since the crime rate is a summary statement of the number of persons in the group who commit crimes and the frequency with which they commit crimes. One of the best explanations of crime rates from this point of view is that a high crime rate is due to social disorganization. The term *social disorganization* is not entirely satisfactory, and it seems preferable to substitute for it the term *differential social organization.* The postulate on which this theory is based, regardless of the name, is that crime is rooted in the social organization and is an expression of that social organization. A group may be organized for criminal behavior or organized against criminal behavior. Most communities are organized for both criminal and anticriminal behavior, and in that sense the crime rate is an expression of the differential group organization. Differential group organization as an explanation of variations in crime rates is consistent with the differential association theory of the processes by which persons become criminals.

9

Control Theory

TRAVIS HIRSCHI

Hirschi focuses on the delinquency of youths, the age where most deviation from the norms generally occurs. In contrast to previous perspectives, Hirschi's model is a more individualistic one, looking at the bond between each person and society. Conforming behavior is reinforced by individuals' attachment to norm-abiding members of society, the commitment and investment they have made in a legitimate life and identity (educational credentials, respectable reputation), their level of involvement in legitimate activities and organizations, and their subscription to the commonly held beliefs and values characterizing normative society. People who violate norms have a flaw in one or more of these bonds to society and can be brought back into the normative ranks by strengthening and reinforcing these weak bonds. Hirschi's perspective is thus more social psychological than structural.

Control theories assume that delinquent acts result when an individual's bond to society is weak or broken. Since these theories embrace two highly complex concepts, the *bond* of the individual to *society,* it is not surprising that they have at one time or another formed the basis of explanations of most forms of aberrant or unusual behavior. It is also not surprising that control theories have described the elements of the bond to society in many ways, and that they have focused on a variety of units as the point of control. . . .

ELEMENTS OF THE BOND

Attachment

In explaining conforming behavior, sociologists justly emphasize sensitivity to the opinion of others.[1] Unfortunately, . . . they tend to suggest that man *is* sensitive to the opinion of others and thus exclude sensitivity from their explanations of deviant behavior. In explaining deviant behavior, psychologists, in contrast, emphasize insensitivity to the opinion of others.[2] Unfortunately, they too tend to ignore variation, and, in addition, they tend to tie sensitivity inextricably to other variables, to make it part of a syndrome or "type," and thus seriously to reduce its

From Travis Hirschi, *Causes of Delinquency,* pp. 16–26, Berkeley: University of California Press, 1969. Reprinted by permission of the author.

value as an explanatory concept. The psychopath is characterized only in part by "deficient attachment to or affection for others, a failure to respond to the ordinary motivations founded in respect or regard for one's fellow";[3] he is also characterized by such things as "excessive aggressiveness," "lack of superego control," and "an infantile level of response."[4] Unfortunately, too, the behavior that psychopathy is used to explain often becomes part of the *definition* of psychopath. As a result, in Barbara Wootton's words: "[The psychopath] is ... *par excellence,* and without shame or qualification, the model of the circular process by which mental abnormality is inferred from anti-social behavior while anti-social behavior is explained by mental abnormality."[5]

The problems of diagnosis, tautology, and name-calling are avoided if the dimensions of psychopathy are treated as causally and therefore problematically interrelated, rather than as logically and therefore necessarily bound to each other. In fact, it can be argued that all of the characteristics attributed to the psychopath follow from, are effects of, his lack of attachment to others. To say that to lack attachment to others is to be free from moral restraints is to use lack of attachment to explain the guiltlessness of the psychopath, the fact that he apparently has no conscience or superego. In this view, lack of attachment to others is not merely a symptom of psychopathy, it *is* psychopathy; lack of conscience is just another way of saying the same thing; and the violation of norms is (or may be) a consequence.

For that matter, given that man is an animal, "impulsivity" and "aggressiveness" can also be seen as natural consequences of freedom from moral restraints. However, since the view of man as endowed with natural propensities and capacities like other animals is peculiarly unpalatable to sociologists, we need not fall back on such a view to explain the amoral man's aggressiveness.[6] The process of becoming alienated from others often involves or is based on active interpersonal conflict. Such conflict could easily supply a reservoir of *socially derived* hostility sufficient to account for the aggressiveness of those whose attachments to others have been weakened.

Durkheim said it many years ago: "We are moral beings to the extent that we are social beings."[7] This may be interpreted to mean that we are moral beings to the extent that we have "internalized the norms" of society. But what does it mean to say that a person has internalized the norms of society? To violate a norm is, therefore, to act contrary to the wishes and expectations of other people. If a person does not care about the wishes and expectations of other people—that is, if he is insensitive to the opinion of others—then he is to that extent not bound by the norms. He is free to deviate.

The essence of internalization of norms, conscience, or superego thus lies in the attachment of the individual to others.[8] This view has several advantages over the concept of internalization. For one, explanations of deviant behavior based on attachment do not beg the question, since the extent to which a person is attached to others can be measured independently of his deviant behavior. Furthermore, change or variation in behavior is explainable in a way that it is not when notions of internalization or superego are used. For example, the divorced man is more likely after divorce to commit a number of deviant acts, such as suicide or forgery. If we explain these acts by reference to the superego (or internal control), we are forced to say that the man "lost his conscience" when he got a divorce; and, of course, if he remarries, we have to conclude that he gets his conscience back.

This dimension of the bond to conventional society is encountered in most social control-oriented research and theory. F. Ivan Nye's "internal control" and "indirect control" refer to the same element, although we avoid the problem of explaining changes over time by locating the "conscience" in the bond to others rather than making it part of the personality.[9] Attachment to others is just one aspect of Albert J. Reiss's "personal controls"; we avoid his problems of tautological empirical *observations* by making the relationship between attachment and delinquency problematic rather than definitional.[10] Finally, Scott Briar and Irving Piliavin's "commitment" or "stake in conformity" subsumes attachment, as their discussion illustrates, although the terms they use are more closely associated with the next element to be discussed.[11]

Commitment

"Of all passions, that which inclineth men least to break the laws, is fear. Nay, excepting some generous natures, it is the only thing, when there is the appearance of profit or pleasure by breaking the laws, that makes men keep them."[12] Few would deny that men on occasion obey the rules simply from fear of the consequences. This rational component in conformity we label commitment. What does it mean to say that a person is committed to conformity? In Howard S. Becker's formulation it means the following:

> First, the individual is in a position in which his decision with regard to some particular line of action has consequences for other interests and activities not necessarily [directly] related to it. Second, he has placed himself in that position by his own prior actions. A third element is present though so obvious as not to be apparent; the committed person must be aware [of these other interests] and must recognize that his decision in this case will have ramifications beyond it.[13]

The idea, then, is that the person invests time, energy, himself, in a certain line of activity—say, getting an education, building up a business, acquiring a reputation for virtue. When or whenever he considers deviant behavior, he must consider the costs of this deviant behavior, the risk he runs of losing the investment he has made in conventional behavior.

If attachment to others is the sociological counterpart of the superego or conscience, commitment is the counterpart of the ego or common sense. To the person committed to conventional lines of action, risking one to ten years in prison for a ten-dollar holdup is stupidity, because to the committed person the costs and risks obviously exceed ten dollars in value. (To the psychoanalyst, such an act exhibits failure to be governed by the "reality-principle.") In the sociological control theory, it can be and is generally assumed that the decision to commit a criminal act may well be rationally determined—that the actor's decision was not irrational given the risks and costs he faces. Of course, as Becker points out, if the actor is capable of in some sense calculating the costs of a line of action, he is also capable of calculational errors: ignorance and error return, in the control theory, as possible explanations of deviant behavior.

The concept of commitment assumes that the organization of society is such that the interest of most persons would be endangered if they were to engage in criminal acts. Most people, simply by the process of living in an organized society, acquire goods, reputations, prospects that they do not want to risk losing. These accumulations are society's insurance that they will abide by the rules. Many hypotheses about the antecedents of delinquent behavior are based on this premise. For example, Arthur L. Stinchcombe's hypothesis that "high school rebellion ... occurs when future status is not clearly related to present performance"[14] suggests that one is committed to conformity not only by what one has but also by what one hoped to obtain. Thus "ambition" and/or "aspiration" play an important role in producing conformity. The person becomes committed to a conventional line of action, and he is therefore committed to conformity.

Most lines of action in a society are of course conventional. The clearest examples are educational and occupational careers. Actions thought to jeopardize one's chances in these areas are presumably avoided. Interestingly enough, even nonconventional commitments may operate to produce conventional conformity. We are told, at least, that boys aspiring to careers in the rackets or professional thievery are judged by their "honesty" and "reliability"—traits traditionally in demand among seekers of office boys.[15]

Involvement

Many persons undoubtedly owe a life of virtue to a lack of opportunity to do otherwise. Time and energy are inherently limited: "Not that I would not, if I could, be both handsome and fat and well dressed, and a great athlete, and make a million a year, be a wit, a bon vivant, and a lady killer, as well as a philosopher, a philanthropist, a statesman, warrior, and African explorer, as well as a 'tone-poet' and saint. But the thing is simply impossible."[16] The things that William James here says he would like to be or do are all, I suppose, within the realm of conventionality, but if he were to include illicit actions he would still have to eliminate some of them as simply impossible.

Involvement or engrossment in conventional activities is thus often part of a control theory. The assumption, widely shared, is that a person may be simply too busy doing conventional things to find time to engage in deviant behavior. The person involved in conventional activities is tied to appointments, deadlines, working hours, plans, and the like, so the opportunity to commit deviant acts rarely arises. To the extent that he is engrossed in conventional activities, he cannot even think about deviant acts, let alone act out his inclinations.[17]

This line of reasoning is responsible for the stress placed on recreational facilities in many programs to reduce delinquency, for much of the concern with the high school dropout, and for the idea that boys should be drafted into the army to keep them out of trouble. So obvious and persuasive is the idea that involvement in conventional activities is a major deterrent to delinquency that it was accepted even by Sutherland: "In the general area of juvenile delinquency it is probable that the most significant difference between juveniles who engage in delinquency and those who do not is that the latter are provided abundant opportunities of a conventional type for satisfying their recreational interests, while the former lack those opportunities or facilities."[18]

The view that "idle hands are the devil's workshop" has received more sophisticated treatment in recent sociological writings on delinquency. David Matza and Gresham M. Sykes, for example, suggest that delinquents have the values of a leisure class, the same values ascribed by Veblen to *the* leisure class: a search for kicks, disdain of work, a desire for the big score, and acceptance of aggressive toughness as proof of masculinity.[19] Matza and Sykes explain delinquency by reference to this system of values, but they note that adolescents at all class levels are "to some extent" members of a leisure class, that they "move in a limbo between earlier parental domination and future integration with the social structure through the bonds of work and marriage."[20] In the end, then, the leisure of the adolescent produces a set of values, which, in turn, leads to delinquency.

Belief

Unlike the cultural deviance theory, the control theory assumes the existence of a common value system within the society or group whose norms are being violated. If the deviant is committed to a value system different from that of conventional society, there is, within the context of the theory, nothing to explain. The question is, "Why does a man violate the rules in which he believes?" It is not, "Why do men differ in their beliefs about what constitutes good and desirable conduct?" The person is assumed to have been socialized (perhaps imperfectly) into the group whose rules he is violating; deviance is not a question of one group imposing its rules on the members of another group. In other words, we not only assume the deviant *has* believed the rules, we assume he believes the rules even as he violates them.

How can a person believe it is wrong to steal at the same time he is stealing? In the strain theory, this is not a difficult problem. (In fact, . . . the strain theory was devised specifically to deal with this question.) The motivation to deviance adduced by the strain theorist is so strong that we can well understand the deviant act even assuming the deviator believes strongly that it is wrong.[21] However, given the control theory's assumptions about motivation, if both the deviant and the nondeviant believe the deviant act is wrong, how do we account for the fact that one commits it and the other does not?

Control theories have taken two approaches to this problem. In one approach, beliefs are treated as mere words that mean little or nothing if the other forms of control are missing. "Semantic dementia," the dissociation between rational faculties and emotional control which is said to be characteristic of the psychopath, illustrates this way of handling the problem.[22] In short, beliefs, at least insofar as they are expressed in words, drop out of the picture; since they do not differentiate between deviants and nondeviants, they are in the same class as "language" or any other characteristic common to all members of the group. Since they represent no real obstacle to the commission of delinquent acts, nothing need be said about how they are handled by those committing such acts. The control theories that do not mention beliefs (or values), and many do not, may be assumed to take this approach to the problem.

The second approach argues that the deviant rationalizes his behavior so that he can at once violate the rule and maintain his belief in it. Donald R. Cressey had advanced this argument with respect to embezzlement,[23] and Sykes and Matza have advanced it with respect to delinquency.[24] In both Cressey's and Sykes and Matza's treatments, these rationalizations (Cressey calls them "verbalizations," Sykes and Matza term them "techniques of neutralization") occur prior to the commission of the deviant act. If the neutralization is successful, the person is free to commit the act(s) in question. Both in Cressey and in Sykes and Matza, the strain that prompts the effort at neutralization also provides the motive force that results in the subsequent deviant act. Their theories are thus, in this sense, strain theories. Neutralization is difficult to handle within the context of a theory that adheres closely to control theory assumptions, because in the control theory there is no special motivational force to account for the neutralization. This difficulty is especially noticeable in Matza's later treatment of this topic, where the motivational component, the "will to delinquency," appears *after* the moral vacuum has been created by the techniques of neutralization.[25] The question thus becomes: Why neutralize?

In attempting to solve a strain-theory problem with control-theory tools, the control theorist is thus led into a trap. He cannot answer the crucial question. The concept of neutralization assumes the existence of moral obstacles to the commission of deviant acts. In order plausibly to account for a deviant act, it is necessary to generate motivation to deviance that is at least equivalent in force to the resistance provided by these moral obstacles. However, if the moral obstacles are removed, neutralization and special motivation are no longer required. We therefore follow the implicit logic of control theory and remove these moral obstacles by hypothesis. Many persons do not have an attitude of respect toward the rules of society; many persons feel no moral obligation to conform regardless of personal advantage. Insofar as the values and beliefs of these persons are consistent with their feelings, and there should be tendency toward consistency, neutralization is unnecessary; it has already occurred.

Does this merely push the question back a step and at the same time produce conflict with the assumption of a common value system? I think not. In the first place, we do not assume, as does Cressey, that neutralization occurs in order to make a specific criminal act possible.[26] We do not assume, as do Sykes and Matza, that neutralization occurs to make many delinquent acts possible. We do not assume, in other words, that the person constructs a system of rationalizations in order to justify commission of acts he *wants* to commit. We assume, in contrast, that the beliefs that free a man to commit deviant acts are *unmotivated* in the sense that he does not construct or adopt them in order to facilitate the attainment of illicit ends. In the second place, we do not assume, as does Matza, that "delinquents concur in the conventional assessment of delinquency."[27] We assume, in contrast, that there is *variation* in the extent to which people believe they should obey the rules of society, and, furthermore, that the less a person believes he should obey the rules, the more likely he is to violate them.[28]

In chronological order, then, a person's beliefs in the moral validity of norms are, for no teleological reason, weakened. The probability that he will commit delinquent acts is therefore increased. When and if he commits a delinquent act, we may justifiably use the weakness of his beliefs in explaining it, but no special motivation is required to explain either the weakness of his beliefs or, perhaps, his delinquent act.

The keystone of this argument is of course the assumption that there is variation in belief in the moral validity of social rules. This assumption is amenable to direct empirical test and can thus survive at least until its first confrontation with data. For the present, we must return to the idea of a common value system with which this section was begun.

The idea of a common (or perhaps better, a single) value system is consistent with the fact, or presumption, of variation in the strength of moral beliefs. We have not suggested that delinquency is based on beliefs counter to conventional morality; we have not suggested that delinquents do not believe delinquent acts are wrong. They may well believe these acts are wrong, but the meaning and efficacy of such beliefs are contingent on other beliefs and, indeed, on the strength of other ties to the conventional order.[29]

NOTES

1. Books have been written on the increasing importance of interpersonal sensitivity in modern life. According to this view, controls from within have become less important than controls from without in *producing* conformity. Whether or not this observation is true as a description of historical trends, it is true that interpersonal sensitivity has become more important in *explaining* conformity. Although logically it should also have become more important in explaining nonconformity, the opposite has been the case, once again showing that Cohen's observation that an explanation of conformity should be an explanation of deviance cannot be translated as "an explanation of conformity has to be an explanation of deviance." For the view that interpersonal sensitivity currently plays a greater role than formerly in producing conformity, see William J. Goode, "Norm Commitment and Conformity to Role-Status Obligations," *American Journal of Sociology* LXVI (1960): 246–258. And, of course, also see David Riesman, Nathan Glazer, and Reuel Denney, *The Lonely Crowd* (Garden City, New York: Doubleday, 1950), especially Part I.

2. The literature on psychopathy is voluminous. See William McCord and Joan McCord, *The Psychopath* (Princeton: D. Van Nostrand, 1964).

3. John M. Martin and Joseph P. Fitzpatrick, *Delinquent Behavior* (New York: Random House, 1964), p. 130.

4. *Ibid.* For additional properties of the psychopath, see McCord and McCord, *The Psychopath*, pp. 1–22.

5. Barbara Wootton, *Social Science and Social Pathology* (New York: Macmillan, 1959), p. 250.

6. "The logical untenability [of the position that there are forces in man 'resistant to socialization'] was ably demonstrated by Parsons over 30 years ago, and it is widely recognized that the position is empirically unsound because it assumes [!] some universal biological drive system distinctly separate from socialization and social context—a basic and intransigent human nature" (Judith Blake and Kingsley Davis, "Norms, Values, and Sanctions," *Handbook of Modern Sociology*, ed. Robert E. L. Paris [Chicago: Rand McNally, 1964], p. 471).

7. Emile Durkheim, *Moral Education*, trans. Everett K. Wilson and Herman Schnurer (New York: The Free Press, 1961), p. 64.

8. Although attachment alone does not exhaust the meaning of internalization, attachments and beliefs combined would appear to leave only a small residue of "internal control" not susceptible in principle to direct measurement.

9. F. Ivan Nye, *Family Relationships and Delinquent Behavior* (New York: Wiley, 1958), pp. 5–7.

10. Albert J. Reiss, Jr., "Delinquency as the Failure of Personal and Social Controls," *American Sociological Review* XVI (1951): 196–207. For example, "Our observations show . . . that delinquent recidivists are less often persons with mature ego ideals or nondelinquent social roles" (p. 204).

11. Scott Briar and Irving Piliavin, "Delinquency, Situational Inducements, and Commitment to Conformity," *Social Problems* XIII (1965): 41–42. The concept "stake in conformity" was introduced by Jackson Toby in his "Social Disorganization and Stake in Conformity: Complementary Factors in the Predatory Behavior of Hoodlums," *Journal of Criminal Law, Criminology and Police Science* XLVIII (1957): 12–17. See also his "Hoodlum or Business Man: An American Dilemma," in *The Jews*, ed. Marshall Sklare (New York: The Free Press, 1958), pp. 542–550. Throughout the text, I occasionally use "stake in conformity" in speaking in general of the strength of the bond to conventional society. So used, the concept is somewhat broader than is true for either Toby or Briar and Piliavin, where the concept is roughly equivalent to what is here called "commitment."

12. Thomas Hobbes, *Leviathan* (Oxford: Basil Blackwell, 1957), p. 195.

13. Howard S. Becker, "Notes on the Concept of Commitment," *American Journal of Sociology* LXVI (1960): 35–36.

14. Arthur L. Stinchcombe, *Rebellion in a High School* (Chicago: Quadrangle, 1964), p. 5.

15. Richard A. Cloward and Lloyd E. Ohlin, *Delinquency and Opportunity* (New York: The Free Press, 1960), p. 147, quoting Edwin H. Sutherland, ed., *The Professional Thief* (Chicago: University of Chicago Press, 1937), pp. 211–213.

16. William James. *Psychology* (Cleveland: World Publishing Co., 1948), p. 186.

17. Few activities appear to be so engrossing that they rule out contemplation of alternative lines of behavior, at least if estimates of the amount of time men spend plotting sexual deviations have any validity.

18. *The Sutherland Papers*, ed. Albert K. Cohen et al. (Bloomington: Indiana University Press, 1956), p. 37.

19. David Matza and Gresham M. Sykes, "Juvenile Delinquency and Subterranean Values," *American Sociological Review* XXVI (1961): 712–719.

20. *Ibid.*, p. 718.

21. The starving man stealing the loaf of bread is the image evoked by most strain theories. In this image, the starving man's belief in the wrongness of his act is clearly not something that must be explained away. It can be assumed to be present without causing embarrassment to the explanation.

22. McCord and McCord, *The Psychopath*, pp. 12–15.

23. Donald R. Cressey, *Other People's Money* (New York: The Free Press, 1953).

24. Gresham M. Sykes and David Matza, "Techniques of Neutralization: A Theory of Delinquency," *American Sociological Review* XXII (1957), 664–670.

25. David Matza, *Delinquency and Drift* (New York: Wiley, 1964), pp. 181–191.

26. In asserting that Cressey's assumption is invalid with respect to delinquency, I do not wish to suggest that it is invalid for the question of embezzlement, where the problem faced by the deviator is fairly specific and he can reasonably be assumed to be an upstanding citizen. (Although even here the fact that the embezzler's non-sharable financial problem often results from some sort of hanky-panky suggests that "verbalizations" may be less necessary than might otherwise be assumed.)

27. *Delinquency and Drift,* p. 43.

28. This assumption is not, I think, contradicted by the evidence presented by Matza against the existence of a delinquent subculture. In comparing the attitudes and actions of delinquents with the picture painted by delinquent subculture theorists, Matza emphasizes—and perhaps exaggerates—the extent to which delinquents are tied to the conventional order. In implicitly comparing delinquents with a supermoral man, I emphasize—and perhaps exaggerate—the extent to which they are not tied to the conventional order.

29. The position taken here is therefore somewhere between the "semantic dementia" and the "neutralization" positions. Assuming variation, the delinquent is, at the extremes, freer than the neutralization argument assumes. Although the possibility of wide discrepancy between what the delinquent professes and what he practices still exists, it is presumably much rarer than is suggested by studies of articulate "psychopaths."

10

Feminist Theory

MEDA CHESNEY-LIND

*Chesney-Lind speaks strongly for the feminist perspective in pointing out that
most theories of crime and deviance have overfocused on a male model of offending.
She points out that girls and women hold a very different structural position in
society, and the experience they are likely to encounter along with the opportunities
(or lack thereof) they face are often markedly at odds with those of boys and men.
Girls grow up in America facing vastly different sets of pressures, encounters, and
forms of social control than their male counterparts.*

    *Gender, she notes, is a master status, which means that most of girls'
experiences in society are filtered through this lens. From early youth, their role in
society is affected by the structure of male domination. When we consider the effect
of patriarchy on women, the special pathways that girls take into crime and
deviance are illuminated. Girls are more vulnerable to physical and sexual abuse
in the home than boys and lack the resources to rebuff or escape them, not only
interpersonally, within the family, but also due to the double standard of sexuality
and sexual control embedded within the juvenile justice system. Runaways
who escape from abusive families are systematically returned right back into them.
Girls who stay and remain subjected to further abuse often marry at a young age to
get out but often find themselves domestically abused by their partners. Girls
who are strong enough to fight for survival escape abusive family contexts by going
to the street, where their opportunities (both legitimate and within the world
of crime) are extremely limited. Once again, then, they find themselves subject
to the domination and abuse of male predators. In these situations, girls' background experiences and lack of other survival techniques lead them into adult
situations of abuse to which they voluntarily and involuntarily subject themselves
in order to stay fed, sheltered, and clothed.*

    *This cycle of victimization to criminalization characterizes the gender-related
pathway of girls into deviance and crime and illustrates the structural disadvantage
faced by young and older women in society.*

There is considerable question as to whether existing theories that were admittedly developed to explain male delinquency can adequately explain female
delinquency. Clearly, these theories were much influenced by the notion that

From *Crime & Delinquency*, 35(1), pp. 10–11, pp. 19–27, copyright © 1989 by Sage
Publications. Reprinted by Permission of Sage Publications Inc.

class and protest masculinity were at the core of delinquency. Will the "add women and stir approach" be sufficient? Are these really theories of delinquent behavior as some have argued?

This article will suggest that they are not. The extensive focus on male delinquency and the inattention to the role played by patriarchal arrangements in the generation of adolescent delinquency and conformity has rendered the major delinquency theories fundamentally inadequate to the task of explaining female behavior. There is, in short, an urgent need to rethink current models in light of girls' situation in patriarchal society.

. . . This discussion will also establish that the proposed overhaul of delinquency theory is not, as some might think, solely an academic exercise. Specifically, it is incorrect to assume that because girls are charged with less serious offenses, they actually have few problems and are treated gently when they are drawn into the juvenile justice system. Indeed, the extensive focus on disadvantaged males in public settings has meant that girls' victimization and the relationship between that experience and girls' crime has been systematically ignored. Also missed has been the central role played by the juvenile justice system in the sexualization of girls' delinquency and the criminalization of girls' survival strategies. Finally, it will be suggested that the official actions of the juvenile justice system should be understood as major forces in girls' oppression as they have historically served to reinforce the obedience of all young women to demands of patriarchal authority no matter how abusive and arbitrary. . . .

TOWARD A FEMINIST THEORY OF DELINQUENCY

To sketch out completely a feminist theory of delinquency is a task beyond the scope of this article. It may be sufficient, at this point, simply to identify a few of the most obvious problems with attempts to adapt male-oriented theory to explain female conformity and deviance. Most significant of these is the fact that all existing theories were developed with no concern about gender stratification.

Note that this is not simply an observation about the power of gender roles (though this power is undeniable). It is increasingly clear that gender stratification in patriarchal society is as powerful a system as is class. A feminist approach to delinquency means construction of explanations of female behavior that are sensitive to its patriarchal context. Feminist analysis of delinquency would also examine ways in which agencies of social control—the police, the courts, and the persons—act in ways to reinforce woman's place in male society. Efforts to construct a feminist model of delinquency must first and foremost be sensitive to the situations of girls. Failure to consider the existing empirical evidence on girls' lives and behavior can quickly lead to stereotypical thinking and theoretical dead ends.

An example of this sort of flawed theory building was the early fascination with the notion that the women's movement was causing an increase in women' crime; a notion that is now more or less discredited. A more recent example of the same sort of thinking can be found in recent work on the "power-control" model of delinquency. Here, the authors speculate that girls commit less delinquency in part because their behavior is more closely controlled by the patriarchal family. The authors' promising beginning quickly gets bogged down in a very limited definition of patriarchal control (focusing on parental supervision and variations in power within the family). Ultimately, the authors' narrow formulation of patriarchal control results in their arguing that mother's work force participation (particularly in high status occupations) leads to increases in daughters' delinquency since these girls find themselves in more "egalitarian families."

This is essentially a not-too-subtle variation on the earlier "liberation" hypothesis. Now, mother's liberation causes daughter's crime. Aside from the methodological problems with the study (e.g., the authors argue that female-headed households are equivalent to upper-class "egalitarian" families where both parents work, and they measure delinquency using a six-item scale that contains no status offense items), there is a more fundamental problem with the hypothesis. There is no evidence to suggest that as women's labor force participation has increased, girls' delinquency has increased. Indeed, during the last decade when both women's labor force participation accelerated and the number of female-headed households soared, aggregate female delinquency measured both by self-report and official statistics either declined or remained stable.

By contrast, a feminist model of delinquency would focus more extensively on the few pieces of information about girls' actual lives and the role played by girls' problems, including those caused by racism and poverty, in their delinquency behavior. Fortunately, a considerable literature is now developing on girls' lives and much of it bears directly on girls' crime.

CRIMINALIZING GIRLS' SURVIVAL

It has long been understood that a major reason for girls' presence in juvenile courts was the fact that their parents insisted on their arrest. In the early years, conflicts with parents were by far the most significant referral source; in Honolulu 44% of the girls who appeared in court in 1929 through 1930 were referred by parents.

Recent national data, while slightly less explicit, also show that girls are more likely to be referred to court by "sources other than law enforcement agencies" (which would include parents). In 1983, nearly a quarter (23%) of all girls but only 16% of boys charged with delinquent offenses were referred to court by non-law enforcement agencies. The pattern among youth referred for status offenses (for which girls are overrepresented) was even more pronounced. Well over half (56%) of the girls charged with these offenses and 45% of the boys were referred by sources other than law enforcement.

The fact that parents are often committed to two standards of adolescent behavior is one explanation for such a disparity—and one that should not be discounted as a major source of tension even in modern families. Despite expectations to the contrary, gender-specific socialization patterns have not changed very much and this is especially true for parents' relationships with their daughters. It appears that even parents who oppose sexism in general feel "uncomfortable tampering with existing traditions" and "do not want to risk their children becoming misfits." Clearly, parental attempts to adhere to and enforce these traditional notions will continue to be a source of conflict between girls and their elders. Another important explanation for girls' problems with their parents, which has received attention only in more recent years, is the problem of physical and sexual abuse. Looking specifically at the problem of childhood sexual abuse, it is increasingly clear that this form of abuse is a particular problem for girls.

Girls are, for example, much more likely to be the victims of child sexual abuse than are boys. Finkelhor and Baron estimate from a review of community studies that roughly 70% of the victims of sexual abuse are female, they are more likely than boys to be assaulted by a family member (often a stepfather), and, as a consequence, their abuse tends to last longer than male sexual abuse. All of these factors are associated with more severe trauma—causing dramatic short- and long-term effects in victims. The effects noted by researchers in this area move from the more well known "fear, anxiety, depression, anger and hostility, and inappropriate sexual behavior" to behaviors of greater familiarity to criminologists, including running away from home, difficulties in school, truancy, and early marriage. Herman's study of incest survivors in therapy found that they were more likely to have run away from home than a matched sample of women whose fathers were "seductive" (33% compared to 5%). Another study of women patients found that 50% of the victims of child sexual abuse, but only 20% of the nonvictim group, had left home before the age of 18.

Not surprisingly, then, studies of girls on the streets or in court populations are showing high rates of both physical and sexual abuse. Silbert and Pines (1981, p. 409) found, for example, that 60% of the street prostitutes they interviewed had been sexually abused as juveniles. Girls at an Arkansas diagnostic unit and school who had been adjudicated for either status or delinquent offenses reported similarly high-levels of physical abuse; 53% indicated they had been sexually abused, 25% recalled scars, 38% recalled bleeding from abuse, and 51% recalled bruises. A sample survey of girls in the juvenile justice system in Wisconsin revealed that 79% had been subjected to physical abuse that resulted in some form of injury, and 32% had been sexually abused by parents or other persons who were closely connected to their families. Moreover, 50% had been sexually assaulated ("raped" or forced to participate in sexual acts). Even higher figures were reported by McCormack and her associates in their study of youth in a runaway shelter in Toronto. They found that 73% of the females and 38% of the males had been sexually abused. Finally, a study of youth charged with running away, truancy, or listed as missing persons in Arizona found that 55% were incest victims.

Many young women, then, are running away from profound sexual victimization at home, and once on the streets they are forced further into crime in order to survive. Interviews with girls who have run away from home show, very clearly, that they do not have a lot of attachment to their delinquent activities. In fact, they are angry about being labeled as delinquent, yet all engaged in illegal acts. The Wisconsin study found that 54% of the girls who ran away found it necessary to steal money, food, and clothing in order to survive. A few exchanged sexual contact for money, food, and/or shelter. In their study of runaway youth, McCormack, Janus and Burgess found that sexually abused female runaways were significantly more likely than their nonabused counterparts to engage in delinquent or criminal activities such as substance abuse, petty theft, and prostitution. No such pattern was found among male runaways.

Research on the backgrounds of adult women in prison underscores the important links between women's childhood victimizations and their later criminal careers. The interviews revealed that virtually all of this sample were the victims of physical and/or sexual abuse as youngsters; over 60% had been sexually abused and about half had been raped as young women. This situation promoted these women to run away from home (three-quarters had been arrested for status offenses) where once on the streets they began engaging in prostitution and other forms of petty property crime. They also begin what becomes a lifetime problem with drugs. As adults, the women continue in these activities since they possess truncated educational backgrounds and virtually no marketable occupational skills.

Confirmation of the consequences of childhood sexual and physical abuse on adult female criminal behavior has also recently come from a large quantitative study of 908 individuals with substantiated and validated histories of these victimizations. Widom (1988) found that abused or neglected females were twice as likely as a matched group of controls to have an adult record (16% compared to 7.5). The difference was also found among men, but it was not as dramatic (42% compared to 33%). Men with abuse backgrounds were also more likely to contribute to the "cycle of violence" with more arrests for violent offenses as adult offenders than the control group. In contrast, when women with abuse backgrounds did become involved with the criminal justice system, their arrests tended to involve property and order offenses (such as disorderly conduct, curfew, and loitering violations).

Given this information, a brief example of how a feminist perspective on the causes of female delinquency might look seems appropriate. First, like young men, girls are frequently the recipients of violence and sexual abuse. But unlike boys, girls' victimization and their response to that victimization is specifically shaped by their status as young women. Perhaps because of the gender and sexual scripts found in patriarchal families, girls are much more likely than boys to be victims of family-related sexual abuse. Men, particularly men with traditional attitudes toward women, are likely to define their daughters or stepdaughters as their sexual property. In a society that idealizes inequality in male/female relationships and venerates youth in women, girls are easily defined as sexually attractive by older men. In addition, girls' vulnerability to both physical and sexual abuse is heightened by norms that require that they stay at home where their victimizers have access to them.

Moreover, their victimizers (usually males) have the ability to invoke official agencies of social control in their efforts to keep young women at home and vulnerable. That is to say, abusers have traditionally been able to utilize the uncritical commitment of the juvenile justice system toward parental authority to force girls to obey them. Girls' complaints about abuse were, until recently, routinely ignored. For this reason, statutes that were originally placed in law to "protect" young people have, in the case of girls' delinquency, criminalized their survival strategies. As they run away from abusive homes, parents have been able to employ agencies to enforce their return. If they persisted in their refusal to stay in that home, however intolerable, they were incarcerated.

Young women, a large number of whom are on the run from homes characterized by sexual abuse and parental neglect, are forced by the very statutes designed to protect them into the lives of escaped convicts. Unable to enroll in school or take a job to support themselves because they fear detection, young female runaways are forced into the streets. Here they engage in panhandling, petty theft, and occasional prostitution in order to survive. Young women in conflict with their parents (often for very legitimate reasons) may actually be forced by present laws into petty criminal activity, prostitution, and drug use.

In addition, the fact that young girls (but not necessarily young boys) are defined as sexually desirable and, in fact, more desirable then their older sisters due to the double standard of aging means that their lives on the streets (and their survival strategies) take on unique shape—once again shaped by patriarchal values. It is no accident that girls on the run from abusive homes, or on the streets because of profound poverty, get involved in criminal activities that exploit their sexual object status. American society has defined as desirable youthful, physically perfect women. This means that girls on the streets, who have little else of value to trade, are encouraged to utilize this "resource." It also means that the criminal sub-culture views them from this perspective.

FEMALE DELINQUENCY, PATRIARCHAL AUTHORITY, AND FAMILY COURTS

The early insights into male delinquency were largely gleaned by intensive field observation of delinquent boys. Very little of this sort of work has been done in the case of girls' delinquency, though it is vital to an understanding of girls' definitions of their own situations, choices, and behavior. Time must be spent listening to girls. Fuller research on the settings, such as families and schools, that girls find themselves in and the impact of variations in those settings should also be undertaken. A more complete understanding of how poverty and racism shape girls' lives is also vital.

Finally, current qualitative research on the reaction of official agencies to girls' delinquency must be conducted. This latter task, admittedly more difficult, is particularly critical to the development of delinquency theory that is as sensitive to gender as it is to race and class.

It is clear that throughout most of the court's history, virtually all female delinquency has been placed within the larger context of girls' sexual behavior. One explanation for this pattern is that familial control over girls' sexual capital has historically been central to the maintenance of patriarchy. The fact that young women have relatively more of this capital has been one reason for the excessive concern that both families and official agencies of social control have expressed about youthful female defiance (otherwise much of the behavior of criminal justice personnel makes virtually no sense). Only if one considers the role of women's control over their sexuality at the point in their lives that their value to patriarchal society is so pronounced, does the historic pattern of jailing of huge numbers of girls guilty of minor misconduct make sense.

This framework also explains the enormous resistance that the movement to curb the juvenile justice system's authority over status offenders encountered. Supporters of this change were not really prepared for the political significance of giving youth the freedom to run. Horror stories told by the opponents of deinstitutionalization about victimized youth, youthful prostitution, and youthful involvement in pornography all neglected the unpleasant reality that most of these behaviors were often in direct response to earlier victimization, frequently by parents, that officials had, for years, routinely ignored. What may be at stake in efforts to roll back deinstitutionalization efforts is not so much "protection" of youth as it is curbing the right of young women to defy patriarchy.

In sum, research in both the dynamics of girls' delinquency and official reactions to that behavior is essential to the development of theories of delinquency that are sensitive to its patriarchal as well as class and racial context.

REFERENCES

Silbert, Mimi. H., and Ayala Pines. 1981. "Sexual Child Abuse as an Antecedent to Prostitution." *International Journal of Child Abuse and Neglect* 5: 407–411.

Widom, Cathy Spatz. 1988. "Sampling Biases and Implications for Child Abuse Research." *American Journal of Orthopsychiatry* 58(2): 260–270.

11

The Constructionist Stance

JOEL BEST

Best, one of the leading practitioners of the social constructionist approach, offers a historical analysis of the way theories of deviance have evolved. He notes the rise and decline in significance of several approaches to deviance theory, showing how social constructionism arose from its early roots in the sociology of deviance and moved into explaining social problems. The constructionist perspective represents a wedding of the views of labeling and conflict theories, as we noted in the General Introduction. It joins the microanalysis of the former by looking at how individuals encounter societal reactions and become labeled as deviants with the broader, more structural, social power contribution of the latter, isolating certain groups as more likely to have social reactions and definitions formed and applied by and against them. It is this social constructionist stance that will frame the organization and selections comprising the remainder of this book, as we examine the process by which groups vie for power in society and try to legislate their views into morality and then focus on how people develop deviant identities and manage their stigma as a result of those definitions and enforcements.

What does it mean to say that deviance is "socially constructed?" Some people assume that social construction is the opposite of real, but this is a mistake. Reality, that is everything we understand about the world, is socially constructed. The term calls attention to the processes by which people make sense of the world: we create—or construct—meaning. When we define some behavior as deviant, we are socially constructing deviance. The constructionist approach recognizes that people can only understand the world in terms of words and categories that they create and share with one another.

THE EMERGENCE OF CONSTRUCTIONISM

The constructionist stance had its roots in two developments. The first was the publication of Peter L. Berge and Thomas Luckmann's (1966) *The Social Construction of Reality.* Berger and Luckmann were writing about the sociology of knowledge—how social life shapes everything that people know. Their book introduced the term "social construction" to a wide sociological audience, and soon other sociologists were writing about the construction of science, news, and other sorts of knowledge, including what we think about deviance.

Second, labeling theory, which had become the leading approach to studying deviance during the 1960s, came under attack from several different directions by the mid-1970s. Conflict theorists charged that labeling theory ignored how elites shaped definitions of deviance and social control policies. Feminists complained that labeling ignored the victimization of women at the hands of both male offenders and male-dominated social control agencies. Activists for gay rights and disability rights insisted that homosexuals and the disabled should be viewed as political minorities, rather than deviants. At the same time, mainstream sociologists began challenging labeling's claims about the ways social control operated and affected deviants' identities.

THE CONSTRUCTIONIST RESPONSE

In response to these attacks, some sociologists sympathetic to the labeling approach moved away from studying deviance. Led by John I. Kitsuse, a sociologist whose work had helped shape labeling theory, these sociologists of deviance turned to studying the sociology of social problems. With Malcolm Spector, Kitsuse published *Constructing Social Problems* (1977)—a book that would inspire many sociologists to begin studying how and why particular social problems emerged as topics of public concern. They argued that sociologists ought to redefine social problems as claims that various conditions constituted social problems; therefore, the constructionist approach involved studying claims and those who made them—the claimsmakers. In this view, sociologists ought to study how and why particular issues such as date rape or binge drinking on college campuses suddenly became the focus of attention and concern. How were these problems constructed?

There were several advantages to studying social problems. First, constructionists had the field virtually to themselves. Although many sociology departments taught social problems courses, there were no rival well-established, coherent theories of social problems. In contrast, labeling had to struggle against functionalism, conflict theory, and other influential approaches to studying deviance.

The constructionist approach was also flexible. Analysts of social problems construction might concentrate on various actors: some examined the power of political and economic elites in shaping definitions of social problems; others focused on the role of activists in bringing attention to problems; and still others concentrated on how media coverage shaped the public's and policymakers' understandings of problems. This flexibility meant that constructionists might criticize some claims as exaggerated, distorted, or unfounded (the sort of critique found in several studies of claims about the menace of Satanism), but they might also celebrate the efforts of claimsmakers to draw attention to neglected problems (for example, researchers tended to treat claims about domestic violence sympathetically).

Again, it is important to appreciate that "socially constructed" is not a synonym for erroneous or mistaken. All knowledge is socially constructed; to say that a social problem is socially constructed is not to imply that it does not exist, but rather that it is through social interaction that the problem is assigned particular meanings.

THE RETURN TO DEVIANCE

Although constructionists studied the emergence and evolution of many different social problems, ranging from global warming to homelessness, much of their work remained focused on deviance. They studied the construction of rape, child abduction, illicit drugs, family violence, and other forms of deviance.

Closely related to the rise of constructionism were studies of medicalization (Conrad and Schneider 1980). Medicalization—defining deviance as a form of illness requiring medical treatment—was one popular, contemporary way of constructing deviance. By the end of the twentieth century, medical language—"disease," "symptom," "therapy," and so on—was used, not only by medical authorities, but even by amateurs (for example, in the many Twelve-Step programs of the recovery movement).

A large share of constructionist studies traced the rise of social problems to national attention; for example, the construction of the federal War on Drugs was studied by several constructionist researchers. However, other sociologists began studying how deviance was constructed in smaller settings, through interpersonal interaction. In particular, they examined social problems work (Holstein and Miller 1993). Even after claimsmakers have managed to draw attention to some social problem and shape the creation of social policies to deal with it, those claims must be translated into action. Police officers, social workers, and other social problems workers must apply broad constructions to particular cases. Thus, after wife abuse is defined as a social problem, it is still necessary for the police officer investigating a domestic disturbance call to define—or construct—these particular events as an instance of wife abuse (Loseke 1992). Studies of this sort of social problems work are a continuation of earlier research on the labeling process.

CONSTRUCTIONISM'S DOMAIN

Social constructionism, then, has become an influential stance for thinking about deviance, particularly for understanding how concerns about particular forms of deviance emerge and evolve, and for studying how social control agents construct particular acts as deviance and individuals as deviants. Constructionism emphasizes the role of interpretation, of people assigning meaning, or making sense of the behaviors they classify as deviant. This can occur at a societal level, as when the mass media draw attention to a new form of deviance and legislators pass laws against it, but it can also occur in face-to-face interaction, when one individual expresses disapproval of anther's rule breaking. Deviance, like all reality, is constantly being constructed.

REFERENCES

Berger, Peter L., and Thomas Luckmann. 1966. *The Social Construction of Reality*. New York: Doubleday.

Conrad, Peter, and Joseph W. Schneider. 1980. *Deviance and Medicalization*. St. Louis: Mosby.

Holstein, James A., and Gale Miller. 1993. "Social Constructionist and Social Problems Work." Pp.151–72 in *Reconsidering Social Constructionism*, edited by James Holstein and Gale Miller. Hawthorne, NY: Aldine de Gruyter.

Loseke, Donileen R. 1992. *The Battered Woman and Shelters*. Albany: State University of New York Press.

Spector, Malcolm, and John I. Kitsuse. 1977. *Constructing Social Problems*. Menlo Park, CA: Cummings.

Studying Deviance

Accurate and reliable knowledge about deviance is critically important to many groups in society. First, policy makers are concerned with deviant groups such as the homeless and transient, the chronically mentally ill, high school drop-outs, criminal offenders, prostitutes, juvenile delinquents, gang members, run-aways, and other members of disadvantaged and disenfranchised populations. These people pose social problems that lawmakers and social welfare agencies want to alleviate. Second, sociologists and other researchers have an interest in deviance based on their goal of understanding human nature, human behavior, and human society. Deviants are a critical group to this enterprise because they reside near the margins of social definition: They help define the boundaries of what is acceptable and unacceptable by given groups.

In this part we will discuss information about deviance coming from three primary data sources. Government officials and employees of social service agencies routinely collect information about their clients as they process them. This information includes arrest data that are compiled by the police and pub-lished by the FBI (the *Uniform Crime Reports*), census data on various shifting pop-ulations (such as the homeless), victim data from helping agencies (such as battered women's shelters), medical data from emergency rooms (such as DAWN, the Drug Abuse Warning Network) or from state public health agencies (such as the coroner or medical examiner offices), and prosecution data on cases that are tried in the courts. These precollected **official statistics** are then compiled by the various government organizations responsible for collecting them and made available to the public. Official statistics are connected to the absolutist perspective on defining deviance because they are considered an objec-tive source of measurement.

Another source of statistical data about deviance is **survey research.** Rather than relying on information the government collects, sociologists gather their own data through large-scale questionnaire surveys. Prominent ones include the National Youth Survey, a self-report questionnaire about delinquent behavior, the annual survey "Monitoring the Future," conducted by the Institute for Social Research at the University of Michigan on the drug use of high school seniors (in which many of you reading this book have probably participated), and some of the Kinsey surveys about sexual behavior.

A third kind of information, richly descriptive and analytical rather than numerical, comes from sociologists who conduct **participant–observation field research** (also called ethnography) on deviance. Much like anthropologists who go out to live among native peoples, sociological field-workers live with members of deviant groups and become intimately familiar with their lives. This type of research yields information more deeply based on the subjects' own perspectives, detailing how they see the world, the allure of deviance for them, the problems they encounter, the ways they resolve them, the significant individuals and groups in their lives, and their role among these others. Unlike other types of research, participant observation is generally a longitudinal method, entailing years of involvement with subjects. Researchers must gain acceptance by group members, develop meaningful relationships with them, and learn about their deepest thoughts and emotions.

There are many differences among these types of data and among the methods used to gather them. Each has its advantages and disadvantages, and each may do a better job than the others of answering certain research questions. Thus, depending on people's particular needs, they may turn to one type of data or use a mixed-method approach. Official statistics have the advantage of being inexpensive to gather and quick to access, since they are already collected and published. They intend to include the entire population of those they address, not some subsample, that is, all criminals, all victims, all emergency room admittees. Records about these occurrences can be accessed back as far as the official statistics have been collected, potentially a rather long time. Yet they have certain validity problems and tend to be inaccurate in patterned and systematic ways.

Official police statistics, the *Uniform Crime Reports,* for example, fail to include a host of crimes for several reasons: Crimes may be unrecognized by victims who do not notice their occurrence or lack the power to define them as deviant; crimes may be unreported by individuals who see no gain by reporting them or fear embarrassment, censure, or retaliation from calling police attention to their victimization; or crimes may be unrecorded by police officers who use their broad discretion to handle problems informally. Official statistics on suicide, determined and collected by coroners and medical examiners, also fluctuate (Whitt 2006). They may be unreliable in rural areas because officials know families, making them loathe to render a verdict on the cause of death that would stigmatize a family or impede their collecting an insurance settlement. In urban areas they may rise and fall due to political pressures, personnel turnovers, fluctuating resources, or policy changes or being falsely indicated as "found" (such as in single-car accidents).

Other types of official statistics vastly underrepresent criminal activity for similar and other reasons and may be problematic because the categories used to conceive of them and the way they are assembled and reported change over time, making comparisons over the years frustrating. In sum, although official statistics yield information about a broad spectrum of people, they may be fairly shallow and unreliable in nature. Besharov and Laumann-Billings' chapter on child abuse statistics discusses the dramatic rise in the number of reported cases of child abuse and some of the sociological factors that have accounted for this wild swing in the official statistics. They discuss factors that artificially both inflate and deflate our official estimates of child abuse and their consequences for the protection of abused children.

Survey research lets social scientists collect data on the topics of their choice, but it is a more expensive and time-consuming enterprise. Through careful sampling procedures, researchers can gather data about a smaller population and generalize from it to a much larger group with a high degree of accuracy (external reliability). Strict controls over standardization of procedures and detachment of data gatherers makes this a relatively objective method. Correlations (although not causal relations) between social factors can be established carefully. But survey research, like official statistics, has internal validity problems: It may not yield an accurate portrait of the sample group it is studying, especially for topics as sensitive as deviance. First, it is unlikely that people, especially deviants, will fill out a questionnaire and readily disclose information about the hidden aspects of their lives. Second, in responding to the questionnaire, subjects may not define their behavior the same way or use the same terms as the researchers who are writing the questions (prostitutes' conceptions of a "date" may be different from those of survey researchers, and runaways may mean different things when they refer to their "home" than researchers intend). Researchers are then likely to misinterpret the nature and extent of behavior from the answers they receive. Third, sometimes the correlational connections of survey research, that is, what trends occur together (such as deviance and divorce, or violence in the media and violence in everyday life), are mistakenly assumed to be causal connections. But survey research cannot tell us *why* or *how* people act; it can only tell us *what* people are doing, even if these trends range across a broad spread of the population. A much heralded study of sexual behavior is featured in Edward Laumann and his associates' description of their survey, a major, highly professionally designed and conducted study, that gives us a glimpse into the problems and creative adaptations that can arise when a comprehensive effort is made to conduct large-scale survey research into Americans' sexual practices.

Participant observation, in contrast, cannot reach as many people but yields deeper and more accurate information about research subjects, backed by researchers' own direct observations, to enhance its internal validity. Participant observers spend long hours in the field becoming close to the people they study and learning how their subjects perceive, interpret, and act upon the complex and often contradictory nature of their social worlds. In contrast to the detached and objective relationships between survey researchers and their subjects, participant observers rely on the subjectivity and strength of the close personal relationships they forge with the people they study to get behind false fronts and to find out what is really going on. Depth of understanding is especially important when studying a topic such as deviance, where so much behavior is hidden due to its stigma and illicit status. Also critically important is the ability of participant observation to study deviance as it occurs *in situ*, in its natural setting, not via the structural constraints of police reporting or the interpretation and recollection of questionnaire research (Polsky 1967). But although often less costly than survey research to conduct, it is very time consuming, as rapport and trust with subjects take a long time to develop. Field research also lacks the generalizability of careful survey research, as subjects tend to be gathered through a referral ("snowball") technique or because they are members of a common "scene." The assurance of randomization and objectivity, then, is not as strong as in other methodologies. We share with readers our own experiences with participant observation in the chapter on field research, where we talk about what it is like to carry out such research with a criminal, and potentially dangerous, group.

Figure 1 offers a comparison of the strengths and weakness of these three methods or sources of data.

The empirical chapters that fill the remainder of this book are primarily based on participant observation studies of deviance for two main reasons. First, as Becker (1973) remarked, participant observation is the method of the interactionist perspective; it offers direct access to the way definitions and laws are socially constructed, to the way people's actions are influenced by their associates, and

FIGURE 1   **Strengths and Weakness of the Three Sources of Data**

	Survey Research	Field Research	Official Statistics
Cost	High	Low	Free
Time	Medium	Long	Short
Approach	Objective	Subjective	Clerical
Generalizability	High	Low	High
Accuracy	Medium	High	Low

to the way people's identities are affected by the deviant labels cast on them. Second, these types of studies offer a deeper view of people's feelings, experiences, motivations, and social psychological states, which give a richer and more vivid portrayal of deviance than charts of numbers.

12

Child Abuse Reporting

DOUGLAS J. BESHAROV WITH LISA A. LAUMANN-BILLINGS

Besharov and Laumann-Billings discuss official statistics in our first chapter on the varieties of ways that deviance is studied. They note the spectacular rise in our official rates of child abuse, with figures increasing by 300 percent over a recent 30-year period. Such a dramatic change cannot be solely attributable to changes in deviant behavior but must also involve a measurement artifact. They root the increase in child abuse statistics in three factors: mandatory reporting laws, the media campaigns surrounding child abuse, and the changed social definition of what constitutes abuse. Besharov and Laumann-Billings discuss two ironically opposing problems associated with child abuse statistics, the presence of both unreported and unsubstantiated cases. On the one hand, they claim, we are still unaware of many cases of child abuse because it tends to be hidden, defined as a private family matter, and regarded as "normal" childrearing practice.

Yet at the same time, the way we, as a society, tumultuously attacked this "discovered" social problem and deputized numerous social groups to document it, resulted in cases that could not be substantiated. Some of these were unfounded because they were investigated and found to be lacking in substance, but others were unprovable because the families could not be located; the child abuse, when investigated, was able to remain hidden; or the huge increase in the number of cases requiring investigation overburdened the dockets of social service agencies and diminished their ability to resolve all allegations. Some desperate situations are being attended to, but others are slipping through the cracks due to overreporting problems. These cases signal continued ambiguity over definitions of child abuse.

Together these problems cast light on the work of social welfare agents to gather official statistics and some of the enormous inaccuracies inherent in these kinds of data. Official statistics have the advantage of being inexpensive and quick to gather (they are precollected by people in the course of doing their jobs), going backwards in time over long periods, and bypassing sampling to contain information about the entire population of interest, but they are notoriously inaccurate, relying on the carelessness of disinterested data gatherers, the personal and political pressures to skew official reports, and the limited resources of public agencies to collect these data.

F or 30 years, advocates, program administrators, and politicians have joined to encourage even more reports of suspected child abuse and neglect. Their efforts have been spectacularly successful, with about three million cases of suspected child abuse having been reported in 1993. Large numbers of endangered children still go unreported, but an equally serious problem has developed: Upon investigation, as many as 65 percent of the reports now being made are determined to be "unsubstantiated," raising serious civil liberties concerns and placing a heavy burden on already overwhelmed investigative staffs.

These two problems—nonreporting and inappropriate reporting—are linked and must be addressed together before further progress can be made in combating child abuse and neglect. To lessen both problems, there must be a shift in priorities—away from simply seeking more reports and toward encouraging better reports.

REPORTING LAWS

Since the early 1960s, all states have passed laws that require designated professionals to report specified types of child maltreatment. Over the years, both the range of designated professionals and the scope of reportable conditions have been steadily expanded.

Initially, mandatory reporting laws applied only to physicians, who were required to report only "serious physical injuries" and "nonaccidental injuries." In the ensuing years, however, increased public and professional attention, sparked in part by the number of abused children revealed by these initial reporting laws, led many states to expand their reporting requirements. Now almost all states have laws that require the reporting of all forms of suspected child maltreatment, including physical abuse, physical neglect, emotional maltreatment, and of course, sexual abuse and exploitation.

Under threat of civil and criminal penalties, these laws require most professionals who serve children to report suspected child abuse and neglect. About twenty states require all citizens to report, but in every state, any citizen is permitted to report.

These reporting laws, associated public awareness campaigns, and professional education programs have been strikingly successful. In 1993, there were about three million reports of children suspected of being abused or neglected. This is a twenty-fold increase since 1963, when about 150,000 cases were reported to the authorities. (As we will see, however, this figure is bloated by reports that later turn out to be unfounded.)

Many people ask whether this vast increase in reporting signals a rise in the incidence of child maltreatment. Recent increases in social problems such as out-of-wedlock births, inner-city poverty, and drug abuse have probably raised the underlying rates of child maltreatment, at least somewhat. Unfortunately, so many maltreated children previously went unreported that earlier reporting statistics do not provide a reliable baseline against which to make comparisons.

One thing is clear, however: The great bulk of reports now received by child protective agencies would not be made but for the passage of mandatory reporting laws and the media campaigns that accompanied them.

This increase in reporting was accompanied by a substantial expansion of prevention and treatment programs. Every community, for example, is now served by specialized child protective agencies that receive and investigate reports. Federal and state expenditures for child protective programs and associated foster care services now exceed $6 billion a year. (Federal expenditures for foster care, child welfare, and related services make up less than 50 percent of total state and federal expenditures for these services; in 1992, they amounted to a total of $2,773.7 million. In addition, states may use a portion of the $2.8 billion federal Social Services Block Grant for such services, though detailed data on these expenditures are not available. Beginning in 1994, additional federal appropriations funded family preservation and support services.)

As a result, many thousands of children have been saved from serious injury and even death. The best estimate is that over the past twenty years, child abuse and neglect deaths have fallen from over 3,000 a year—and perhaps as many as 5,000—to about 1,000 a year. In New York State, for example, within five years of the passage of a comprehensive reporting law, which also created specialized investigative staffs, there was a 50 percent reduction in child fatalities, from about two hundred a year to less than one hundred. (This is not meant to minimize the remaining problem. Even at this level, maltreatment is the sixth largest cause of death for children under fourteen.)

UNREPORTED CASES

Most experts agree that reports have increased over the past 30 years because professionals and laypersons have become more likely to report apparently abusive and neglectful situations. But the question remains: How many more cases still go unreported?

Two studies performed for the National Center on Child Abuse and Neglect by Westat, Inc., provide a partial answer. In 1980 and then again in 1986, Westat conducted national studies of the incidence of child abuse and neglect. (A third Westat incidence study is now underway.) Each study used essentially the same methodology: In a stratified sample of counties, a broadly representative sample of professionals who serve children was asked whether, during the study period, the children they had seen in their professional capacities appeared to have been abused or neglected. (Actually, the professionals were not asked the ultimate question of whether the children appeared to be "abused" or "neglected" Instead, they were asked to identify children with certain specified harms or conditions, which were then decoded into a count of various types of child abuse and neglect.)

Because the information these selected professionals provided could be matched against pending cases in the local child protective agency, Westat was able to estimate rates of nonreporting among the surveyed professionals. It could not, of course, estimate the level of unintentional nonreporting, since there is no way to know of the situations in which professionals did not recognize signs of possible maltreatment. There is also no way to know how many children the professionals recognized as being maltreated but chose not to report to the study. Obviously, since the study methodology involved asking professionals about children they had seen in their professional capacities, it also did not allow Westat to estimate the number of children seen by nonprofessionals, let alone their nonreporting rate.

Westat found that professionals failed to report many of the children they saw who had observable signs of child abuse and neglect. Specifically, it found that in 1986, 56 percent of apparently abused or neglected children, or about 500,000 children, were not reported to the authorities. This figure, however, seems more alarming than it is: Basically, the more serious the case, the more likely the report. For example, the surveyed professionals reported over 85 percent of the fatal or serious physical abuse cases they saw, 72 percent of the sexual abuse cases, and 60 percent of the moderate physical abuse cases. In contrast, they only reported 15 percent of the educational neglect cases they saw, 24 percent of the emotional neglect cases, and 25 percent of the moderate physical neglect cases.

Nevertheless, there is no reason for complacency. Translating these raw percentages into actual cases means that in 1986, about 2,000 children with observable physical injuries severe enough to require hospitalization were not reported and that more than 100,000 children with moderate physical injuries went unreported, as did more than 30,000 apparently sexually abused children. And these are the rates of nonreporting among relatively well-trained professionals. One assumes that nonreporting is higher among less-well-trained professionals and higher still among laypersons.

Obtaining and maintaining a high level of reporting requires a continuation of the public education and professional training begun 30 years ago. But, now, such efforts must also address a problem as serious as nonreporting: inappropriate reporting.

At the same time that many seriously abused children go unreported, an equally serious problem further undercuts efforts to prevent child maltreatment: The nation's child protective agencies are being inundated by inappropriate reports. Although rules, procedures, and even terminology vary—some states use the phrase "unfounded," others "unsubstantiated" or not indicated—an "unfounded" report, in essence, is one that is dismissed after an investigation finds insufficient evidence upon which to proceed.

UNSUBSTANTIATED REPORTS

Nationwide, between 60 and 65 percent of all reports are closed after an initial investigation determines that they are "unfounded" or "unsubstantiated." This is in sharp contrast to 1974, when only about 45 percent of all reports were unfounded.

A few advocates, in a misguided effort to shield child protective programs from criticism, have sought to quarrel with estimates that I and others have made that the national unfounded rate is between 60 and 65 percent. They have grasped at various inconsistencies in the data collected by different organizations to claim either that the problem is not so bad or that it has always been this bad.

To help settle this dispute, the American Public Welfare Association (APWA) conducted a special survey of child welfare agencies in 1989. The APWA researchers found that between fiscal year 1986 and fiscal year 1988, the weighted average for the substantiation rates in 31 states declined 6.7 percent—from 41.8 percent in fiscal year 1986 to 39 percent in fiscal year 1988.

Most recently, the existence of this high unfounded rate was reconfirmed by the annual Fifty State Survey of the National Committee to Prevent Child Abuse (NCPCA), which found that in 1993 only about 34 percent of the reports received by child protective agencies were substantiated.

The experience of New York City indicates what these statistics mean in practice. Between 1989 and 1993, as the number of reports received by the city's child welfare agency increased by over 30 percent (from 40,217 to 52,472), the percentage of substantiated reports fell by about 47 percent (from 45 percent to 24 percent). In fact, the number of substantiated cases—a number of families were reported more than once—actually fell by about 41 percent, from 14,026 to 8,326. Thus, 12,255 additional families were investigated, while 5,700 fewer families received child protective help.

The determination that a report is unfounded can only be made after an unavoidably traumatic investigation that is inherently a breach of parental and family privacy. To determine whether a particular child is in danger, caseworkers must inquire into the most intimate personal and family matters. Often it is necessary to question friends, relatives, and neighbors, as well as school teachers, day-care personnel, doctors, clergy, and others who know the family.

Laws against child abuse are an implicit recognition that family privacy must give way to the need to protect helpless children. But in seeking to protect children, it is all too easy to ignore the legitimate rights of parents. Each year, about 700,000 families are put through investigations of unfounded reports. This is a massive and unjustified violation of parental rights.

Few unfounded reports are made maliciously. Studies of sexual abuse reports, for example, suggest that, at most, from 4 to 10 percent of these reports are knowingly false. Many involve situations in which the person reporting, in a well-intentioned effort to protect a child, overreacts to a vague and often misleading possibility that the child may be maltreated. Others involve situations of poor child care that, though of legitimate concern, simply do not amount to child abuse or neglect. In fact, a substantial proportion of unfounded cases are referred to other agencies for them to provide needed services for the family.

Moreover, an unfounded report does not necessarily mean that the child was not actually abused or neglected. Evidence of child maltreatment is hard to obtain and might not be uncovered when agencies lack the time and resources to complete a thorough investigation or when inaccurate information is given to the investigator. Other cases are labeled unfounded when no services are available to help the family. Some cases must be closed because the child or family cannot be located.

A certain proportion of unfounded reports, therefore, is an inherent—and legitimate—aspect of reporting *suspected* child maltreatment and is necessary to ensure adequate child protection. Hundreds of thousands of strangers report their suspicions; they cannot all be right. But unfounded rates of the current magnitude go beyond anything reasonably needed. Worse, they endanger children who are really abused.

The current flood of unfounded reports is overwhelming the limited resources of child protective agencies. For fear of missing even one abused child, workers perform extensive investigations of vague and apparently unsupported reports. Even when a home visit based on an anonymous report turns up no evidence of maltreatment, they usually interview neighbors, school teachers, and day-care personnel to make sure that the child is not abused. And even repeated anonymous and unfounded reports do not prevent a further investigation. But all this takes time.

As a result, children in real danger are getting lost in the press of inappropriate cases. Forced to allocate a substantial portion of their limited resources to unfounded reports, child protective agencies are less able to respond promptly and effectively when children are in serious danger. Some reports are left uninvestigated for a week and even two weeks after they are received. Investigations often miss key facts, as workers rush to clear cases, and dangerous home situations receive inadequate supervision, as workers must ignore pending cases as they investigate the new reports that arrive daily on their desks. Decision making also suffers. With so many cases of unsubstantiated or unproven risk to children, caseworkers are desensitized to the obvious warning signals of immediate and serious danger.

These nationwide conditions help explain why from 25 to 50 percent of child abuse deaths involve children previously known to the authorities. In 1993, the NCPCA reported that of the 1,149 child maltreatment deaths, 42 percent had already been reported to the authorities. Tens of thousands of other children suffer serious injuries short of death while under child protective agency supervision.

In a 1992 New York City case, for example, five-month-old Jeffrey Harden died from burns caused by scalding water and three broken ribs while under the supervision of New York City's Child Welfare Administration. Jeffrey Harden's family had been known to the administration for more than a year and a half. Over this period, the case had been handled by four separate caseworkers, each conducting only partial investigations before resigning or being reassigned to new cases. It is unclear whether Jeffrey's death was caused by his mother or her boyfriend, but because of insufficient time and overburdened caseloads, all four workers failed to pay attention to a whole host of obvious warning signals: Jeffrey's mother had broken her parole for an earlier conviction of child sexual abuse, she had a past record of beating Jeffrey's older sister, and she had a history of crack addiction and past involvement with violent boyfriends.

Here is how two of the Hardens' caseworkers explained what happened: Their first caseworker could not find Ms. Harden at the address she had listed in her files. She commented, "It was an easy case. We couldn't find the mother so we closed it." Their second caseworker stated that he was unable to spend a sufficient amount of time investigating the case, let alone make the minimum monthly visits because he was tied down with an overabundance of cases and paperwork. He stated, "It's impossible to visit these people within a month. They're all over New York City." Just before Jeffrey's death every worker who had been on the case had left the department. Ironically, by weakening the system's ability to respond, unfounded reports actually discourage appropriate ones. The sad fact is that many responsible individuals are not reporting endangered children because they feel that the system's response will be so weak that reporting will do no good or may even make things worse. . . .

13

Survey of Sexual Behavior
of Americans

EDWARD O. LAUMANN, JOHN H. GAGNON,
ROBERT T. MICHAEL, AND STUART MICHAELS

*A much-heralded study of sexual behavior is featured in this selection by
Laumann and his associates. This major, highly professionally designed and
conducted study gives us a glimpse into the problems and creative adaptations that
arise when a comprehensive effort is made to conduct large-scale survey research
into Americans' sexual practices. This selection outlines the procedures for
conceiving and carrying out the study, from initial conceptualization to sampling,
administration, interviewer training, questionnaire design, and issues of privacy
and confidentiality. Readers can get some feel for both the generic features of survey
research, the specific decisions as to how this project was implemented, and the
strengths and weaknesses associated with this mode of gathering data about
deviance. Survey research is an objective methodology, scientifically controlled
through the standardization of the interview questionnaire and the careful training
of interviewers so that they do not lead respondents into or away from particular
answers. When careful probability sampling is used, such as we see here, survey
research holds the greatest potential for generalizability from the sample population
back to the larger population of interest. Political, voting, and public opinion polls,
for example, use this kind of methodology and (with careful sampling) have a high
degree of accurately predicting the attitudes and behavior of large swaths of people.
Yet surveys have their disadvantages as well, especially in their problems of
internal validity, or accuracy. People regularly misrepresent themselves on surveys
because they can't understanding the meaning of the questions, because they
misremember their past attitudes or behavior, or just plain intentionally. When we
look at the cost of survey research, we see that this approach can be rather
expensive, making a solid, social-scientific study unaffordable without grant
funding. Survey research represents one of the major sources of information about
deviance and is preferred by public policy analysts because of its numbers and use
of the rhetoric of science. Its strength lies in gathering moderately shallow levels of
information about less sensitive subjects from a broad spectrum of people.*

M ost people with whom we talked when we first broached the idea of a national survey of sexual behavior were skeptical that it could be done. Scientists and laypeople alike had similar reactions: "Nobody will agree to participate in such a study." "Nobody will answer questions like these, and, even if they do, they won't tell the truth." "People don't know enough about sexual practices as they relate to disease transmission or even to pleasure or physical and emotional satisfaction to be able to answer questions accurately." It would be dishonest to say that we did not share these and other concerns. But our experiences over the past seven years, rooted in extensive pilot work, focus-group discussions, and the fielding of the survey itself, resolved these doubts, fully vindicating our growing conviction that a national survey could be conducted according to high standards of scientific rigor and replicability. . . .

The society in which we live treats sex and everything related to sex in a most ambiguous and ambivalent fashion. Sex is at once highly fascinating, attractive, and, for many at certain stages in their lives, preoccupying, but it can also be frightening, disturbing, or guilt inducing. For many, sex is considered to be an extremely private matter, to be discussed only with one's closest friends or intimates, if at all. And, certainly for most if not all of us, there are elements of our sexual lives never acknowledged to others, reserved for our own personal fantasies and self-contemplation. It is thus hardly surprising that the proposal to study sex scientifically, or any other way for that matter, elicits confounding and confusing reactions. Mass advertising, for example, unremittingly inundates the public with explicit and implicit sexual messages, eroticizing products and using sex to sell. At the same time, participants in political discourse are incredibly squeamish when handling sexual themes, as exemplified in the curious combination of horror and fascination displayed in the public discourse about Long Dong Silver and pubic hairs on pop cans during the Senate hearings in September 1991 on the appointment of Clarence Thomas to the Supreme Court. We suspect, in fact, that with respect to discourse on sexuality there is a major discontinuity between the sensibilities of politicians and other self-appointed guardians of the moral order and those of the public at large, who, on the whole, display few hang-ups in discussing sexual issues in appropriately structured circumstances. This work is a testament to that proposition.

The fact remains that, until quite recently, scientific research on sexuality has been taboo and therefore to be avoided or at best marginalized. While there is a visible tradition of (in)famous sex research, what is, in fact, most striking is how little prior research exists on sexuality in the general population. Aside from the research on adolescence, premarital sex, and problems attendant to sex such as fertility, most research attention seems to have been directed toward those believed to be abnormal, deviant, criminal, perverted, rare, or unusual, toward sexual pathology, dysfunction, and sexually transmitted disease—the label used typically reflecting the way in which the behavior or condition in question is to be regarded. "Normal sex" was somehow off limits, perhaps because it was considered too ordinary, trivial, and self-evident to deserve attention. To be fair, then, we cannot blame the public and the politicians entirely for the lack of sustained work on sexuality at large—it also reflects the prejudices and understandings of researchers about what are "interesting" scientific questions. There has simply been a dearth of mainstream scientific thinking and speculation about sexual issues. We have repeatedly encountered this relative lack of systematic thinking about sexuality to guide us in interpreting and understanding the many findings reported in this book.

... In order to understand the results of our survey, the National Health and Social Life Survey (NHSLS), one must understand how these results were generated. To construct a questionnaire and field a large-scale survey, many research design decisions must be made. To understand the decisions made, one needs to understand the multiple purposes that underlie this research project. Research design is never just a theoretical exercise. It is a set of practical solutions to a multitude of problems and considerations that are chosen under the constraints of limited resources of money, time, and prior knowledge.

SAMPLE DESIGN

The sample design for the NHSLS is the most straightforward element of our methodology because nothing about probability sampling is specific to or changes in a survey of sexual behavior....

Probability sampling, that is, sampling where every member of a clearly specified population has a known probability of selection—what lay commentators often somewhat inaccurately call random sampling—is the sine qua non of modern survey research (see Kish 1965, the classic text on the subject). There is no other scientifically acceptable way to construct a representative sample and thereby to be able to generalize from the actual sample on which data are collected to the population that that sample is designed to represent. Probability sampling as practiced in survey research is a highly developed practical application of statistical theory to the problem of selecting a sample. Not only does this type of sampling avoid the problems of bias introduced by the researcher or by subject self-selection bias that come from more casual techniques, but it also allows one to quantify the variability in the estimates derived from the sample.

In order to determine how large a sample size for a given study should be, one must first decide how precise the estimates to be derived need to be. To illustrate this reasoning process, let us take one of the simplest and most commonly used statistics in survey research, the proportion. Many of the most important results reported are proportions. For example, what proportion of the population had more than five sex partners in the last year? What proportion engaged in anal intercourse? With condoms? Estimates based on our sample will differ from the true proportion in the population because of sampling error (i.e., the random fluctuations in our estimates that are due to the fact that they are based on samples rather than on complete enumerations or censuses). If one drew repeated samples using the same methodology, each would produce a slightly different estimate. If one looks at the distribution of these *estimates,* it turns out that they will be normally distributed (i.e., will follow the famous bell-shaped curve known as the Gaussian or normal distribution) and centered around the true proportion in the population. The larger the sample size, the tighter the distribution of estimates will be.

This analysis applies to an estimate of a single proportion based on the whole sample. In deciding the sample size needed for a study, one must consider the subpopulations for which one will want to construct estimates. For example, one almost always wants to know not just a single parameter for the whole population but parameters for subpopulations such as men and women, whites, blacks, and Hispanics, and younger people and older people. Furthermore, one is usually interested in the intersections of these various breakdowns of the population, for example, young black women. The size of the interval estimate for a proportion based on a subpopulation depends on the size of that group in the sample (sometimes called the *base* "*N*," i.e., the number in the sample on which the estimate is based). It is actually this kind of number that one needs to consider in determining the sample size for a study.

When we were designing the national survey of sexual behavior in the United States for the NICHD (National Institute of Child Health and Development), we applied just these sorts of considerations to come to the conclusion that we needed a sample size of about 20,000 people. . . .

GAINING COOPERATION: THE RESPONSE RATE

First, let us consider the cooperation or response rate. No survey of any size and complexity is able to get every sampling-designated respondent to complete an interview. Individuals can have many perfectly valid reasons why they cannot participate in the survey: being too ill, too busy, or always absent when an effort to schedule an interview is made or simply being unwilling to grant an interview. While the face-to-face or in-person survey is considerably more expensive than other techniques, such as mail or telephone surveys, it usually gets the highest response rate. Even so, a face-to-face, household-based survey such as the General Social Survey successfully interviews, on the average, only about 75 percent of the target sample (Davis and Smith 1991). The missing 25 percent pose a serious problem for the reliability and validity of a survey: is there some systematic (i.e., nonrandom) process at work that distinguishes respondents from nonrespondents? That is, if the people who refuse to participate or who can never be reached to be interviewed differ systematically in terms of the issues being researched from those who are interviewed, then one will not have a representative sample of the population from which the sample was drawn. If the respondents and nonrespondents do not differ systematically, then the results will not be affected. Unfortunately, one usually has no (or only minimal) information about nonrespondents. It is thus a challenge to devise ways of evaluating the extent of bias in the selection of respondents and nonrespondents. Experience tells us that, in most well-studied fields in which survey research has been applied, such moderately high response rates as 75 percent do not lead to biased results.

And it is difficult and expensive to push response rates much higher than that. Experience suggests that a response rate close to 90 percent may well represent a kind of upper limit.

Because of our subject matter and the widespread skepticism that survey methods would be effective, we set a completion rate of 75 percent as the survey organization's goal. In fact, we did much better than this; our final completion rate was close to 80 percent. We have extensively investigated whether there are detectable participation biases in the final sample.... To summarize these investigations, we have compared our sample and our results with other surveys of various sorts and have been unable to detect systematic biases of any substantive significance that would lead us to qualify our findings at least with respect to bias due to sampling.

One might well ask what the secret was of our remarkably high response rate, by far the highest of any national sexual behavior survey conducted so far. There is no secret. Working closely with the NORC (National Opinion Research Center) senior survey and field management team, we proceeded in the same way as one would in any other national area probability survey. We did not scrimp on interviewer training or on securing a highly mobilized field staff that was determined to get respondent participation in a professional and respectful manner. It was an expensive operation: the average cost of a completed interview was approximately $450.

We began with an area probability sample, which is a sample of households, that is, of addresses, not names. Rather than approach a household by knocking on the door without advance warning, we followed NORC's standard practice of sending an advance letter, hand addressed by the interviewer, about a week before the interviewer expected to visit the address. In this case, the letter was signed by the principal investigator, Robert Michael, who was identified as the dean of the Irving B. Harris Graduate School of Public Policy Studies of the University of Chicago. The letter briefly explained the purpose of the survey as helping "doctors, teachers, and counselors better understand and prevent the spread of diseases like AIDS and better understand the nature and extent of harmful and of healthy sexual behavior in our country." The intent was to convince the potential respondent that this was a legitimate scientific study addressing personal and potentially sensitive topics for a socially useful purpose. AIDS was the original impetus for the research, and it certainly seemed to provide a timely justification for the study. But any general purpose approach has drawbacks. One problem that the interviewers frequently encountered was potential respondents who did not think that AIDS affected them and therefore that information about their sex lives would be of little use.

Mode of Administration: Face-to-Face,
Telephone, or Self-Administered

Perhaps the most fundamental design decision, one that distinguishes this study from many others, concerned how the interview itself was to be conducted. In survey research, this is usually called the *mode* of interviewing or of questionnaire administration. We chose face-to-face interviewing, the most costly mode, as the primary vehicle for data collection in the NHSLS. What follows is the reasoning behind this decision.

A number of recent sex surveys have been conducted over the telephone, . . . The principal advantage of the telephone survey is its much lower cost. Its major disadvantages are the length and complexity of a questionnaire that can be realistically administered over the telephone and problems of sampling and sample control. . . . The NHSLS, cut to its absolute minimum length, averaged about ninety minutes. Extensive field experience suggests an upper limit of about forty-five minutes for phone interviews of a cross-sectional survey of the population at large. Another disadvantage of phone surveys is that it is more difficult to find people at home by phone and, even once contact has been made, to get them to participate. . . . One further consideration in evaluating the phone as a mode of interviewing is its unknown effect on the quality of responses. Are people more likely to answer questions honestly and candidly or to dissemble on the telephone as opposed to face-to-face? Nobody knows for sure.

The other major mode of interviewing is through self-administered forms distributed either face to face or through the mail.[1] When the survey is conducted by mail, the questions must be self-explanatory, and much prodding is typically required to obtain an acceptable response rate. . . . This procedure has been shown to produce somewhat higher rates of reporting socially undesirable behaviors, such as engaging in criminal acts and substance abuse. We adopted the mixed-mode strategy to a limited extent by using four short, self-administered forms, totaling nine pages altogether, as part of our interview. When filled out, these forms were placed in a "privacy envelope" by the respondent so that the interviewer never saw the answers that were given to these questions. . . .

The fundamental disadvantage of self-administered forms is that the questions must be much simpler in form and language than those that an interviewer can ask. Complex skip patterns must be avoided. Even the simplest skip patterns are usually incorrectly filled out by some respondents on self-administered forms. One has much less control over whether (and therefore much less confidence that) respondents have read and understood the questions on a self-administered form. The NHSLS questionnaire (discussed below) was based on the idea that questions about sexual behavior must be framed as much as possible in the specific contexts of particular patterns and occasions. We found that it is impossible to do this using self-administered questions that are easily and fully comprehensible to people of modest educational attainments.

To summarize, we decided to use face-to-face interviewing as our primary mode of administration of the NHSLS for two principal reasons: it was most likely to yield a substantially higher response rate for a more inclusive cross section of the population at large, and it would permit more complex and detailed questions to be asked. While by far the most expensive approach, such a strategy provides a solid benchmark against which other modes of interviewing can and should be judged. The main unresolved question is whether another mode has an edge over face-to-face interviewing when highly sensitive questions likely to be upsetting or threatening to the respondent are being asked. As a partial control and test of this question, we have asked a number of sensitive questions in both formats so that an individual's responses can be systematically compared. . . . Suffice it to say at this point that there is a stunning consistency in the responses secured by the different modes of administration.

Recruiting and Training Interviewers

Gaining respondents' cooperation requires mastery of a broad spectrum of techniques that successful interviewers develop with experience, guidance from the research team, and careful field supervision. This project required extensive training before entering the field. While interviewers are generally trained to be neutral toward topics covered in the interview, this was especially important when discussing sex, a topic that seems particularly likely to elicit emotionally freighted sensitivities both in the respondents and in the interviewers. Interviewers needed to be fully persuaded about the legitimacy and importance of the research. Toward this end, almost a full day of training was devoted to presentations and discussions with the principal investigators in addition to the extensive advance study materials to read and comprehend. Sample answers to frequently asked questions by skeptical respondents and brainstorming about strategies to convert reluctant respondents were part of the training exercises. A set of endorsement letters from prominent local and national notables and refusal conversion letters were also provided to interviewers. A hotline to the research office at the University of Chicago was set up to allow potential respondents to call in with their concerns. Concerns ranged from those about the legitimacy of the survey, most fearing that it was a commercial ploy to sell them something, to fears that the interviewers were interested in robbing them. Ironically, the fact that the interviewer initially did not know the name of the respondent (all he or she knew was the address) often led to behavior by the interviewer that appeared suspicious to the respondent. For example, asking neighbors for the name of the family in the selected household and/or questions about when the potential respondent was likely to be home induced worries that had to be assuaged. Another major concern was confidentiality—respondents wanted to know how they had come to be selected and how their answers were going to be kept anonymous.

THE QUESTIONNAIRE

The questionnaire itself is probably the most important element of the study design. It determines the content and quality of the information gathered for analysis. Unlike issues related to sample design, the construction of a questionnaire is driven less by technical precepts and more by the concepts and ideas motivating the research. It demands even more art than applied sampling design requires.

Before turning to the specific forms that this took in the NHSLS, we should first discuss several general problems that any survey questionnaire must address. The essence of survey research is to ask a large sample of people from a defined population the *same set of questions*. To do this in a relatively short period of time, many interviewers are needed. In our case, about 220 interviewers from all over the country collected the NHSLS data. The field period, beginning on 14 February 1992 and ending in September, was a time in which over 7,800 households were contacted (many of which turned out to be ineligible for the study) and 3,432 interviews were completed. Central to this effort was gathering comparable information on the same attributes from each and every one of these respondents. The attributes measured by the questionnaire become the variables used in the data analysis. They range from demographic characteristics (e.g., gender, age, and race/ethnicity) to sexual experience measures (e.g., numbers of sex partners in given time periods, frequency of particular practices, and timing of various sexual events) to measures of mental states (e.g., attitudes toward premarital sex, the appeal of particular techniques like oral sex, and levels of satisfaction with particular sexual relationships).

Very early in the design of a national sexual behavior survey, in line with our goal of not reducing this research to a simple behavioral risk inventory, we faced the issue of where to draw the boundaries in defining the behavioral domain that would be encompassed by the concept of sex. This was particularly crucial in defining sexual activity that would lead to the enumeration of a set of sex partners. There are a number of activities that commonly serve as markers for sex and the status of sex partner, especially intercourse and orgasm. While we certainly wanted to include these events and their extent in given relationships and events, we also felt that using them to define and ask about sexual activity might exclude transactions or partners that should be included. Since the common meaning and uses of the term *intercourse* involve the idea of the intromission of a penis, intercourse in that sense as a defining act would at the very least exclude a sexual relationship between two women. There are also many events that we would call sexual that may not involve orgasm on the part of either or both partners.

Another major issue is what sort of language is appropriate in asking questions about sex. It seemed obvious that one should avoid highly technical language because it is unlikely to be understood by many people. One tempting alternative is to use colloquial language and even slang since that is the only language that some people ever use in discussing sexual matters. There is even some evidence that one can improve reporting somewhat by allowing respondents to select their own preferred terminology (Blair et al. 1977; Bradburn, Sudman et al. 1979; Bradburn and Sudman 1983). Slang and other forms of colloquial speech, however, are likely to be problematic in several ways. First, the use of slang can produce a tone in the interview that is counterproductive because it downplays the distinctiveness of the interviewing situation itself. An essential goal in survey interviewing, especially on sensitive topics like sex, is to create a neutral, nonjudgmental, and confiding atmosphere and to maintain a certain professional distance between the interviewer and the respondent. A key advantage that the interviewer has in initiating a topic for discussion is being a stranger or an outsider who is highly unlikely to come in contact with the respondent again. It is not intended that a longer-term bond between the interviewer and the respondent be formed, whether as an advice giver or a counselor or as a potential sex partner.[2]

The second major shortcoming of slang is that it is highly variable across class and education levels, ages, regions, and other social groupings. It changes meanings rapidly and is often imprecise. Our solution was to seek the simplest possible language—standard English—that was neither colloquial nor highly technical. For example, we chose to use the term *oral sex* rather than the slang *blow job* and *eating pussy* or the precise technical but unfamiliar terms *fellatio* and *cunnilingus*. Whenever possible, we provided definitions when terms were first introduced in a questionnaire—that is, we tried to train our respondents to speak about sex in our terms. Many terms that seemed clear to us may not, of course, be universally understood; for example, terms like *vaginal* or *heterosexual* are not understood very well by substantial portions of the population. Coming up with simple and direct speech was quite a challenge because most of the people working on the questionnaire were highly educated, with strong inclinations toward the circum-locutions and indirections of middle-class discourse on sexual themes. Detailed reactions from field interviewers and managers and extensive pilot testing with a broad cross section of recruited subjects helped minimize these language problems.

ON PRIVACY, CONFIDENTIALITY, AND SECURITY

Issues of respondent confidentiality are at the very heart of survey research. The willingness of respondents to report their views and experiences fully and hon-estly depends on the rationale offered for why the study is important and on the assurance that the information provided will be treated as confidential. We offered respondents a strong rationale for the study, our interviewers made great efforts to conduct the interview in a manner that protected respondents' privacy, and we went to great lengths to honor the assurances that the informa-tion would be treated confidentially. The subject matter of the NHSLS makes the issues of confidentiality especially salient and problematic because there are so many easily imagined ways in which information voluntarily disclosed in an interview might be useful to interested parties in civil and criminal cases involv-ing wrongful harm, divorce proceedings, criminal behavior, or similar matters.

NOTES

1. We ruled out the idea of a mail survey because its response rate is likely to be very much lower than any other mode of interviewing (see Bradburn, Sudman et al. 1979).

2. Interviewers are not there to give information or to correct misinformation. But such information is often requested in the course of an interview. Interviewers are given training in how to avoid answering such questions (other than clarification of the meaning of particular questions). They are not themselves experts on the topics raised and often do not know the correct answers to questions. For this reason, and also in case emotionally freighted issues for the respondent were raised during the interview process, we provided inter-viewers with a list of toll-free phone numbers for a variety of professional sex- and health-related referral services (e.g., the National AIDS Hotline, an STD hot-line, the National Child Abuse Hotline, a domestic violence hotline, and the phone number of a national rape and sexual assault organization able to provide local referrals).

REFERENCES

Blair, Ellen, Seymour Sudman, Norman M. Bradburn, and Carol Stacking. 1977. "How to Ask Questions About Drinking and Sex: Response Effects in Measuring Consumer Behavior." *Journal of Marketing Research* 14: 316–321.

Bradburn, Norman M., and Seymour Sudman. 1983. *Asking Questions: A Practical Guide to Questionnaire Design*. San Francisco: Jossey-Bass.

Bradburn, Norman M., Seymour Sudman, Ed Blair, and Carol Stacking. 1979. *Improving Interview Method and Design*. San Francisco: Jossey-Bass.

Davis, James Allan, and Tom W. Smith. 1991. *General Social Surveys, 1972–1991: Cumulative Codebook*. Chicago: National Opinion Research Center.

Kish, Leslie. 1965. *Survey Sampling*. New York: Wiley.

14

Researching Dealers
and Smugglers

PATRICIA A. ADLER

Adler offers us a glimpse of what it is like to carry out participant-observation research with a deviant group in this description of her study of upper-level drug traffickers. This natural history carefully explains the process used in field research, the relationships formed with setting members, and the feelings researchers experience. This stage-by-stage analysis of the activities, pitfalls, mishaps, intimacies, and relationships Adler encountered shows the connection and overlap between the development of research ties and those found in natural, everyday life. Field research, we learn, cannot be carried out by just anyone in every setting but is dependent on researchers' ability to build a bridge of understanding, rapport, and trust between themselves and their subjects. Adler's experiences vividly show us the dangers posed to field-workers in criminal and deviant settings and the intimacy of the connections forged there. These kinds of ties, which form the basis of the data gathering, make participant observation a subjective, rather than an objective, type of research. The strength of the data rests on the real-world bonds forged in the field as well as on researchers' ability to not only hear what their subjects have to say but also to see them in action, and to cross-check their self-presentations against hard facts, the accounts of others, and common sense.

Field research, while costing considerably less than survey research, takes much longer to conduct, requiring years to find deviants, develop trust and relationships, and obtain deeply meaningful information. This approach not only offers us a better idea of the sequential development of causal forces in the field but also gives us the best insight into what is really going on in a deviant scene inhabited by a hidden population. Thus, while field research may not yield the ability to generalize with as much scientific precision as survey research, it is the preferred approach for those who want to deeply and accurately understand the perceptions, interpretations, analyses, life worlds, and unfolding careers of secretive deviants.

From Patricia A. Adler, *Wheeling and Dealing* (New York: Columbia University Press, 1985). Reprinted by permission of the publisher.

I strongly believe that investigative field research (Douglas 1976), with emphasis on direct personal observation, interaction, and experience, is the only way to acquire accurate knowledge about deviant behavior. Investigative techniques are especially necessary for studying groups such as drug dealers and smugglers because the highly illegal nature of their occupation makes them secretive, deceitful, mistrustful, and paranoid. To insulate themselves from the straight world, they construct multiple false fronts, offer lies and misinformation, and withdraw into their group. In fact, detailed, scientific information about upper-level drug dealers and smugglers is lacking precisely because of the difficulty sociological researchers have had in penetrating into their midst. As a result, the only way I could possibly get close enough to these individuals to discover what they were doing and to understand their world from their perspectives (Blumer 1969) was to take a membership role in the setting. While my different values and goals precluded my becoming converted to complete membership in the subculture, and my fears prevented my ever becoming "actively" involved in their trafficking activities, I was able to assume a "peripheral" membership role (Adler and Adler 1987). I became a member of the dealers' and smugglers' social world and participated in their daily activities on that basis. In this chapter, I discuss how I gained access to this group, established research relations with members, and how personally involved I became in their activities.

GETTING IN

When I moved to Southwest County [California] in the summer of 1974, I had no idea that I would soon be swept up in a subculture of vast drug trafficking and unending partying, mixed with occasional cloak-and-dagger subterfuge. I had moved to California with my husband, Peter, to attend graduate school in sociology. We rented a condominium town house near the beach and started taking classes in the fall. We had always felt that socializing exclusively with academicians left us nowhere to escape from our work, so we tried to meet people in the nearby community. One of the first friends we made was our closest neighbor, a fellow in his late twenties with a tall, hulking frame and gentle expression. Dave, as he introduced himself, was always dressed rather casually, if not sloppily, in T-shirts and jeans. He spent most of his time hanging out or walking on the beach with a variety of friends who visited his house, and taking care of his two young boys, who lived alternately with him and his estranged wife. He also went out of town a lot. We started spending much of our free time over at his house, talking, playing board games late into the night, and smoking marijuana together. We were glad to find someone from whom we could buy marijuana in this new place, since we did not know too many people. He also began treating us to a fairly regular supply of cocaine, which was a thrill because this was a drug we could rarely afford on our student budgets. We noticed right away, however, that there was something unusual about his use and knowledge of drugs: while he always had a plentiful supply and was fairly expert about marijuana and cocaine, when we tried to buy a small bag of marijuana from him he had little idea of the going price. This incongruity piqued our curiosity and raised suspicion. We wondered if he might be dealing in larger quantities. Keeping our suspicions to ourselves, we began observing Dave's activities a little more closely. Most of his friends were in their late twenties and early thirties and, judging by their lifestyles and automobiles, rather wealthy. They came and left his house at all hours, occasionally extending their parties through the night and the next day into the following night. Yet throughout this time we never saw Dave or any of his friends engage in any activity that resembled a legitimate job. In most places this might have evoked community suspicion, but few of the people we encountered in Southwest County seemed to hold traditionally structured jobs. Dave, in fact, had no visible means of financial support. When we asked him what he did for a living, he said something vague about being a real estate speculator, and we let it go at that. We never voiced our suspicions directly since he chose not to broach the subject with us.

We did discuss the subject with our mentor, Jack Douglas, however. He was excited by the prospect that we might be living among a group of big dealers, and urged us to follow our instincts and develop leads into the group. He knew that the local area was rife with drug trafficking, since he had begun a life history case study of two drug dealers with another graduate student several years previously. That earlier study was aborted when the graduate student quit school, but Jack still had many hours of taped interviews he had conducted with them, as well as an interview that he had done with an undergraduate student who had known the two dealers independently, to serve as a cross-check on their accounts. He therefore encouraged us to become friendlier with Dave and his friends. We decided that if anything did develop out of our observations of Dave, it might make a nice paper for a field methods class or independent study.

Our interests and background made us well suited to study drug dealing. First, we had already done research in the field of drugs. As undergraduates at Washington University we had participated in a nationally funded project on urban heroin use (see Cummins et al. 1972). Our role in the study involved using fieldwork techniques to investigate the extent of heroin use and distribution in St. Louis. In talking with heroin users, dealers, and rehabilitation personnel, we acquired a base of knowledge about the drug world and the subculture of drug trafficking. Second, we had a generally open view toward soft drug use, considering moderate consumption of marijuana and cocaine to be generally nondeviant. This outlook was partially etched by our 1960s-formed attitudes, as we had first been introduced to drug use in an environment of communal friendship, sharing, and counterculture ideology. It also partially reflected the widespread acceptance accorded to marijuana and cocaine use in the surrounding local culture. Third, our age (mid-twenties at the start of the study) and general appearance gave us compatibility with most of the people we were observing.

We thus watched Dave and continued to develop our friendship with him. We also watched his friends and got to know a few of his more regular visitors. We continued to build friendly relations by doing, quite naturally, what Becker (1963), Polsky (1969), and Douglas (1972) had advocated for the early stages of field research: we gave them a chance to know us and form judgments about our trust-worthiness by jointly pursuing those interests and activities which we had in common.

Then one day something happened which forced a breakthrough in the research. Dave had two guys visiting him from out of town and, after snorting quite a bit of cocaine, they turned their conversation to a trip they had just made from Mexico, where they piloted a load of marijuana back across the bor-der in a small plane. Dave made a few efforts to shift the conversation to another subject, telling them to "button their lips," but they apparently thought that he was joking. They thought that anybody as close to Dave as we seemed to be undoubtedly knew the nature of his business. They made further allusions to his involvement in the operation and discussed the outcome of the sale. We could feel the wave of tension and awkwardness from Dave when this conversa-tion began, as he looked toward us to see if we understood the implications of what was being said, but then he just shrugged it off as done. Later, after the two guys left, he discussed with us what happened. He admitted to us that he was a member of a smuggling crew and a major marijuana dealer on the side. He said that he knew he could trust us, but that it was his practice to say as little as possible to outsiders about his activities. This inadvertent slip, and Dave's sub-sequent opening up, were highly significant in forging our entry into Southwest County's drug world. From then on he was open in discussing the nature of his dealing and smuggling activities with us.

He was, it turned out, a member of a smuggling crew that was importing a ton of marijuana weekly and 40 kilos of cocaine every few months. During that first winter and spring, we observed Dave at work and also got to know the other members of his crew, including Ben, the smuggler himself. Ben was also very tall and broad shouldered, but his long black hair, now flecked with gray, bespoke his earlier membership in the hippie subculture. A large physical stature, we observed, was common to most of the male participants involved in this drug community. The women also had a unifying physical trait: they were extremely attractive and stylishly dressed. This included Dave's ex-wife, Jean, with whom he reconciled during the spring. We therefore became friendly with Jean and through her met a number of women ("dope chicks") who hung around the dealers and smugglers. As we continued to gain the friendship of Dave and Jean's associates, we were progressively admitted into their inner circle and apprised of each person's dealing or smuggling role.

Once we realized the scope of Ben's and his associates' activities, we saw the enormous research potential in studying them. This scene was different from any analysis of drug trafficking that we had read in the sociological literature because of the amounts they were dealing and the fact that they were importing it them-selves. We decided that, if it was at all possible, we would capitalize on this sit-uation, to "opportunistically" (Riemer 1977) take advantage of our prior expertise and of the knowledge, entrée, and rapport we had already developed with several key people in this setting. We therefore discussed the idea of doing a study of the general subculture with Dave and several of his closest friends (now becoming our friends). We assured them of the anonymity, confidentiality, and innocuousness of our work. They were happy to reciprocate our friendship by being of help to our professional careers. In fact, they basked in the subsequent attention we gave their lives.

We began by turning first Dave, then others, into key informants and collecting their life histories in detail. We conducted a series of taped, in-depth interviews with an unstructured, open-ended format. We questioned them about such topics as their backgrounds, their recruitment into the occupation, the stages of their dealing careers, their relations with others, their motivations, their lifestyle, and their general impressions about the community as a whole.

We continued to do taped interviews with key informants for the next six years until 1980, when we moved away from the area. After that, we occasionally did follow-up interviews when we returned for vacation visits. These later interviews focused on recording the continuing unfolding of events and included detailed probing into specific conceptual areas, such as dealing networks, types of dealers, secrecy, trust, paranoia, reputation, the law, occupational mobility, and occupational stratification. The number of taped interviews we did with each key informant varied, ranging between 10 and 30 hours of discussion.

Our relationship with Dave and the others thus took on an added dimension—the research relationship. As Douglas (1976), Henslin (1972), and Wax (1952) have noted, research relationships involve some form of mutual exchange. In our case, we offered everything that friendship could entail. We did routine favors for them in the course of our everyday lives, offered them insights and advice about their lives from the perspective of our more respectable position, wrote letters on their behalf to the authorities when they got in trouble, testified as character witnesses at their non-drug-related trials, and loaned them money when they were down and out. When Dave was arrested and brought to trial for check-kiting, we helped Jean organize his defense and raise the money to pay his fines. We spelled her in taking care of the children so that she could work on his behalf. When he was eventually sent to the state prison we maintained close ties with her and discussed our mutual efforts to buoy Dave up and secure his release. We also visited him in jail. During Dave's incarceration, however, Jean was courted by an old boyfriend and gave up her reconciliation with Dave. This proved to be another significant turning point in our research because, desperate for money, Jean looked up Dave's old dealing connections and went into the business herself. She did not stay with these marijuana dealers and smugglers for long, but soon moved into the cocaine business. Over the next several years her experiences in the world of cocaine dealing brought us into contact with a different group of people. While these people knew Dave and his associates (this was very common in the Southwest County dealing and smuggling community), they did not deal with them directly. We were thus able to gain access to a much wider and more diverse range of subjects than we would have had she not branched out on her own.

Dave's eventual release from prison three months later brought our involvement in the research to an even deeper level. He was broke and had nowhere to go. When he showed up on our doorstep, we took him in. We offered to let him stay with us until he was back on his feet again and could afford a place of his own.

He lived with us for seven months, intimately sharing his daily experiences with us. During this time we witnessed, firsthand, his transformation from a scared ex-con who would never break the law again to a hard-working legitimate employee who only dealt to get money for his children's Christmas presents, to a full-time dealer with no pretensions at legitimate work. Both his process of changing attitudes and the community's gradual reacceptance of him proved very revealing.

We socialized with Dave, Jean, and other members of Southwest County's dealing and smuggling community on a near-daily basis, especially during the first four years of the research (before we had a child). We worked in their legitimate businesses, vacationed together, attended their weddings, and cared for their children. Throughout their relationship with us, several participants became co-opted to the researcher's perspective[1] and actively sought out instances of behavior which filled holes in the conceptualizations we were developing. Dave, for one, became so intrigued by our conceptual dilemmas that he undertook a "natural experiment" entirely on his own, offering an unlimited supply of drugs to a lower-level dealer to see if he could work up to higher levels of dealing, and what factors would enhance or impinge upon his upward mobility.

In addition to helping us directly through their own experiences, our key informants aided us in widening our circle of contacts. For instance, they let us know when someone in whom we might be interested was planning on dropping by, vouching for our trustworthiness and reliability as friends who could be included in business conversations. Several times we were even awakened in the night by phone calls informing us that someone had dropped by for a visit, should we want to "casually" drop over too. We rubbed the sleep from our eyes, dressed, and walked or drove over, feeling like sleuths out of a television series. We thus were able to snowball, through the active efforts of our key informants,[2] into an expanded study population. This was supplemented by our own efforts to cast a research net and befriend other dealers, moving from contact to contact slowly and carefully through the domino effect.

THE COVERT ROLE

The highly illegal nature of dealing in illicit drugs and dealers' and smugglers' general level of suspicion made the adoption of an overt research role highly sensitive and problematic. In discussing this issue with our key informants, they all agreed that we should be extremely discreet (for both our sakes and theirs). We carefully approached new individuals before we admitted that we were studying them. With many of these people, then, we took a covert posture in the research setting. As nonparticipants in the business activities which bound members together into the group, it was difficult to become fully accepted as peers. We therefore tried to establish some sort of peripheral, social membership in the general crowd, where we could be accepted as "wise" (Goffman 1963) individuals and granted a courtesy membership. This seemed an attainable goal, since we

had begun our involvement by forming such relationships with our key informants. By being introduced to others in this wise rather than overt role, we were
able to interact with people who would otherwise have shied away from us.
Adopting a courtesy membership caused us to bear a courtesy stigma,[3] however,
and we suffered since we, at times, had to disguise the nature of our research from
both lay outsiders and academicians.

In our overt posture we showed interest in dealers' and smugglers' activities,
encouraged them to talk about themselves (within limits, so as to avoid acting
like narcs), and ran home to write field notes. This role offered us the advantage
of gaining access to unapproachable people while avoiding researcher effects, but
it prevented us from asking some necessary, probing questions and from tape
recording conversations.[4] We therefore sought, at all times, to build toward a
conversion to the overt role. We did this by working to develop their trust.

DEVELOPING TRUST

Like achieving entrée, the process of developing trust with members of unorganized deviant groups can be slow and difficult. In the absence of a formal structure separating members from outsiders, each individual must form his or her
own judgment about whether new persons can be admitted to their confidence.
No gatekeeper existed to smooth our path to being trusted, although our key
informants acted in this role whenever they could by providing introductions
and references. In addition, the unorganized nature of this group meant that
we met people at different times and were constantly at different levels in our
developing relationships with them. We were thus trusted more by some people
than by others, in part because of their greater familiarity with us. But as Douglas
(1976) has noted, just because someone knew us or even liked us did not automatically guarantee that they would trust us.

We actively tried to cultivate the trust of our respondents by tying them to us
with favors. Small things, like offering the use of our phone, were followed with
bigger favors, like offering the use of our car, and finally really meaningful favors,
like offering the use of our home. Here we often trod a thin line, trying to ensure
our personal safety while putting ourselves in enough of a risk position, along with
our research subjects, so that they would trust us. While we were able to build a
"web of trust" (Douglas 1976) with some members, we found that trust, in large
part, was not a simple status to attain in the drug world. Johnson (1975) has pointed
out that trust is not a one-time phenomenon, but an ongoing developmental process. From my experiences in this research I would add that it cannot be simply
assumed to be a one-way process either, for it can be diminished, withdrawn, rein-
stated to varying degrees, and re-questioned at any point. Carey (1972) and Douglas
(1972) have remarked on this waxing and waning process, but it was especially
pronounced for us because our subjects used large amounts of cocaine over an
extended period of time. This tended to make them alternately warm and cold to
us. We thus lived through a series of ups and downs with the people we were trying
to cultivate as research informants.

THE OVERT ROLE

After this initial covert phase, we began to feel that some new people trusted us. We tried to intuitively feel when the time was right to approach them and go overt. We used two means of approaching people to inform them that we were involved in a study of dealing and smuggling: direct and indirect. In some cases our key informants approached their friends or connections and, after vouching for our absolute trustworthiness, convinced these associates to talk to us. In other instances, we approached people directly, asking for their help with our project. We worked our way through a progression with these secondary contacts, first discussing the dealing scene overtly and later moving to taped life history interviews. Some people reacted well to us, but others responded skittishly, making appointments to do taped interviews only to break them as the day drew near, and going through fluctuating stages of being honest with us or putting up fronts about their dealing activities. This varied, for some, with their degree of active involvement in the business. During the times when they had quit dealing, they would tell us about their present and past activities, but when they became actively involved again, they would hide it from us.

This progression of covert to overt roles generated a number of tactical difficulties. The first was the problem of *coming on too fast* and blowing it. Early in the research we had a dealer's old lady (we thought) all set up for the direct approach. We knew many dealers in common and had discussed many things tangential to dealing with her without actually mentioning the subject. When we asked her to do a taped interview of her bohemian lifestyle, she agreed without hesitation. When the interview began, though, and she found out why we were interested in her, she balked, gave us a lot of incoherent jumble, and ended the session as quickly as possible. Even though she lived only three houses away we never saw her again. We tried to move more slowly after that.

A second problem involved simultaneously *juggling our overt and covert roles* with different people. This created the danger of getting our cover blown with people who did not know about our research (Henslin 1972). It was very confusing to separate the people who knew about our study from those who did not, especially in the minds of our informants. They would make occasional veiled references in front of people, especially when loosened by intoxicants, that made us extremely uncomfortable. We also frequently worried that our snooping would someday be mistaken for police tactics. Fortunately, this never happened.

CROSS-CHECKING

The hidden and conflictual nature of the drug dealing world made me feel the need for extreme certainty about the reliability of my data. I therefore based all my conclusions on independent sources and accounts that we carefully verified. First, we tested information against our own common sense and general knowledge of the scene. We adopted a hard-nosed attitude of suspicion, assuming people were up to more than they would originally admit. We kept our attention especially riveted on "reformed" dealers and smugglers who were living better than they could outwardly afford, and were thereby able to penetrate their public fronts.

Second, we checked out information against a variety of reliable sources. Our own observations of the scene formed a primary reliable source, since we were involved with many of the principals on a daily basis and knew exactly what they were doing. Having Dave live with us was a particular advantage because we could contrast his statements to us with what we could clearly see was happening. Even after he moved out, we knew him so well that we could generally tell when he was lying to us or, more commonly, fooling himself with optimistic dreams. We also observed other dealers' and smugglers' evasions and misperceptions about themselves and their activities. These usually occurred when they broke their own rules by selling to people they did not know, or when they commingled other people's money with their own. We also cross-checked our data against independent, alternative accounts. We were lucky, for this purpose, that Jean got reinvolved in the drug world. By interviewing her, we gained additional insight into Dave's past, his early dealing and smuggling activities, and his ongoing involvement from another person's perspective. Jean (and her connections) also talked to us about Dave's associates, thereby helping us to validate or disprove their statements. We even used this pincer effect to verify information about people we had never directly interviewed. This occurred, for instance, with the tapes that Jack Douglas gave us from his earlier study. After doing our first round of taped interviews with Dave, we discovered that he knew the dealers Jack had interviewed. We were excited by the prospect of finding out what had happened to these people and if their earlier stories checked out. We therefore sent Dave to do some investigative work. Through some mutual friends he got back in touch with them and found out what they had been doing for the past several years.

Finally, wherever possible, we checked out accounts against hard facts: newspaper and magazine reports; arrest records; material possessions; and visible evidence. Throughout the research, we used all these cross-checking measures to evaluate the veracity of new information and to prod our respondents to be more accurate (by abandoning both their lies and their self-deceptions).[5]

After about four years of near-daily participant observation, we began to diminish our involvement in the research. This occurred gradually, as first pregnancy and then a child hindered our ability to follow the scene as intensely and spontaneously as we had before. In addition, after having a child, we were less willing to incur as many risks as we had before; we no longer felt free to make decisions based solely on our own welfare. We thus pulled back from what many have referred to as the "difficult hours and dangerous situations" inevitably present in field research on deviants (see Becker 1963; Carey 1972; Douglas 1972). We did, however, actively maintain close ties with research informants (those with whom we had gone overt), seeing them regularly and periodically doing follow-up interviews.

PROBLEMS AND ISSUES

Reflecting on the research process, I have isolated a number of issues which I believe merit additional discussion. These are rooted in experiences which have the potential for greater generic applicability.

The first is the *effect of drugs on the data-gathering process*. Carey (1972) has elaborated on some of the problems he encountered when trying to interview respondents who used amphetamines, while Wax (1952, 1957) has mentioned the difficulty of trying to record field notes while drinking sake. I found that marijuana and cocaine had nearly opposite effects from each other. The latter helped the interview process, while the former hindered it. Our attempts to interview respondents who were stoned on marijuana were unproductive for a number of reasons. The primary obstacle was the effects of the drug. Often, people became confused, sleepy, or involved in eating to varying degrees. This distracted them from our purpose. At times, people even simulated overreactions to marijuana to hide behind the drug's supposed disorienting influence and thereby avoid divulging information. Cocaine, in contrast, proved to be a research aid. The drug's warming and sociable influence opened people up, diminished their inhibitions, and generally increased their enthusiasm for both the interview experience and us.

A second problem I encountered involved *assuming risks while doing research*. As I noted earlier, dangerous situations are often generic to research on deviant behavior. We were most afraid of the people we studied. As Carey (1972), Henslin (1972), and Whyte (1955) have stated, members of deviant groups can become hostile toward a researcher if they think that they are being treated wrongfully. This could have happened at any time from a simple occurrence, such as a misunderstanding, or from something more serious, such as our covert posture being exposed. Because of the inordinate amount of drugs they consumed, drug dealers and smugglers were particularly volatile, capable of becoming malicious toward each other or us with little warning. They were also likely to behave erratically owing to the great risks they faced from the police and other dealers. These factors made them moody, and they vacillated between trusting us and being suspicious of us.

At various times we also had to protect our research tapes. We encountered several threats to our collection of taped interviews from people who had granted us these interviews. This made us anxious, since we had taken great pains to acquire these tapes and felt strongly about maintaining confidences entrusted to us by our informants. When threatened, we became extremely frightened and shifted the tapes between different hiding places. We even ventured forth one rainy night with our tapes packed in a suitcase to meet a person who was uninvolved in the research at a secret rendezvous so that he could guard the tapes for us.

We were fearful, lastly, of the police. We often worried about local police or drug agents discovering the nature of our study and confiscating or subpoenaing our tapes and field notes. Sociologists have no privileged relationship with their subjects that would enable us legally to withhold evidence from the authorities should they subpoena it.[6] For this reason we studiously avoided any publicity about the research, even holding back on publishing articles in scholarly journals until we were nearly ready to move out of the setting. The closest we came to being publicly exposed as drug researchers came when a former sociology graduate student (turned dealer, we had heard from inside sources) was arrested at the scene of a cocaine deal. His lawyer wanted us to testify about the dangers of doing drug-related research, since he was using his research status as his defense. Fortunately, the crisis was averted when his lawyer succeeded in suppressing evidence and had the case dismissed before the trial was to have begun. Had we been exposed, however, our respondents would have acquired guilt by association through their friendship with us.

Our fear of the police went beyond our concern for protecting our research subjects, however. We risked the danger of arrest ourselves through our own violations of the law. Many sociologists (Becker 1963; Carey 1972; Polsky 1969; Whyte 1955) have remarked that field researchers studying deviance must inevitably break the law in order to acquire valid participant observation data. This occurs in its most innocuous form from having "guilty knowledge": information about crimes that are committed. Being aware of major dealing and smuggling operations made us an accessory to their commission, since we failed to notify the police. We broke the law, secondly, through our "guilty observations," by being present at the scene of a crime and witnessing its occurrence (see also Carey 1972). We knew it was possible to get caught in a bust involving others, yet buying and selling was so pervasive that to leave every time it occurred would have been unnatural and highly suspicious. Sometimes drug transactions even occurred in our home, especially when Dave was living there, but we finally had to put a stop to that because we could not handle the anxiety. Lastly, we broke the law through our "guilty actions," by taking part in illegal behavior ourselves. Although we never dealt drugs (we were too scared to be seriously tempted), we consumed drugs and possessed them in small quantities. Quite frankly, it would have been impossible for a nonuser to have gained access to this group to gather the data presented here. This was the minimum involvement necessary to obtain even the courtesy membership we achieved. Some kind of illegal action was also found to be a necessary or helpful component of the research by Becker (1963), Carey (1972), Johnson (1975), Polsky (1969), and Whyte (1955).

Another methodological issue arose from the *cultural clash between our research subjects and ourselves*. While other sociologists have alluded to these kinds of differences (Humphreys 1970; Whyte 1955), few have discussed how the research relationships affected them. Relationships with research subjects are unique because they involve a bond of intimacy between persons who might not ordinarily associate together, or who might otherwise be no more than casual friends. When fieldworkers undertake a major project, they commit themselves to maintaining a long-term relationship with the people they study. However, as researchers try to get depth involvement, they are apt to come across fundamental differences in character, values, and attitudes between their subjects and themselves. In our case, we were most strongly confronted by differences in present versus future orientations, a desire for risk versus security, and feelings of spontaneity versus self-discipline. These differences often caused us great frustration. We repeatedly saw dealers act irrationally, setting themselves up for failure. We wrestled with our desire to point out their patterns of foolhardy behavior and offer advice, feeling competing pulls between our detached, observer role which advised us not to influence the natural setting, and our involved, participant role which called for us to offer friendly help whenever possible.[7]

Each time these differences struck us anew, we gained deeper insights into our core, existential selves. We suspended our own taken-for-granted feelings and were able to reflect on our culturally formed attitudes, character, and life choices from the perspective of the other. When comparing how we might act in situations faced by our respondents, we realized where our deepest priorities lay. These revelations had the effect of changing our self-conceptions: whereas we, at one time, had thought of ourselves as what Rosenbaum (1981) has called "the hippest of non-addicts" (in this case nondealers), we were suddenly faced with being the straightest members of the crowd. Not only did we not deal, but we had a stable, long-lasting marriage and family life, and needed the security of a reliable monthly paycheck. Self-insights thus emerged as one of the unexpected outcomes of field research with members of a different cultural group.

The final issue I will discuss involved the various *ethical problems* which arose during this research. Many fieldworkers have encountered ethical dilemmas or pangs of guilt during the course of their research experiences (Carey 1972; Douglas 1976; Humphreys 1970; Johnson 1975; Klockars 1977, 1979; Rochford 1985). The researchers' role in the field makes this necessary because they can never fully align themselves with their subjects while maintaining their identity and personal commitment to the scientific community. Ethical dilemmas, then, are directly related to the amount of deception researchers use in gathering the data, and the degree to which they have accepted such acts as necessary and therefore neutralized them.

Throughout the research, we suffered from the burden of intimacies and confidences. Guarding secrets which had been told to us during taped interviews was not always easy or pleasant. Dealers occasionally revealed things about themselves or others that we had to pretend not to know when interacting with their close associates. This sometimes meant that we had to lie or build elaborate stories to cover for some people. Their fronts therefore became our fronts, and we had to weave our own web of deception to guard their performances. This became especially disturbing during the writing of the research report, as I was torn by conflicts between using details to enrich the data and glossing over description to guard confidences.[8]

Using the covert research role generated feelings of guilt, despite the fact that our key informants deemed it necessary, and thereby condoned it. Their own covert experiences were far more deeply entrenched than ours, being a part of their daily existence with non-drug world members. Despite the universal presence of covert behavior throughout the setting, we still felt a sense of betrayal every time we ran home to write research notes on observations we had made under the guise of innocent participants.

We also felt guilty about our efforts to manipulate people. While these were neither massive nor grave manipulations, they involved courting people to procure information about them. Our aggressively friendly postures were based on hidden ulterior motives: we did favors for people with the clear expectation that they could only pay us back with research assistance. Manipulation bothered us in two ways: immediately after it was done, and over the long run. At first, we felt awkward, phony, almost ashamed of ourselves, although we believed our rationalization that the end justified the means. Over the long run, though, our feelings were different. When friendship became intermingled with research goals, we feared that people would later look back on our actions and feel we were exploiting their friendship merely for the sake of our research project.

The last problem we encountered involved our feelings of whoring for data. At times, we felt that we were being exploited by others, that we were putting more into the relationship than they, that they were taking us for granted or using us. We felt that some people used a double standard in their relationship with us: they were allowed to lie to us, borrow money and not repay it, and take advantage of us, but we were at all times expected to behave honorably. This was undoubtedly an outgrowth of our initial research strategy where we did favors for people and expected little in return. But at times this led to our feeling bad. It made us feel like we were selling ourselves, our sincerity, and usually our true friendship, and not getting treated right in return.

CONCLUSIONS

The aggressive research strategy I employed was vital to this study. I could not just walk up to strangers and start hanging out with them as Liebow (1967) did, or be sponsored to a member of this group by a social service or reform organization as Whyte (1955) was, and expect to be accepted, let alone welcomed. Perhaps such a strategy might have worked with a group that had nothing to hide, but I doubt it. Our modern, pluralistic society is so filled with diverse subcultures whose interests compete or conflict with each other that each subculture has a set of knowledge which is reserved exclusively for insiders. In order to serve and prosper, they do not ordinarily show this side to just anyone. To obtain the kind of depth insight and information I needed, I had to become like the members in certain ways. They dealt only with people they knew and trusted, so I had to become known and trusted before I could reveal my true self and my research interests. Confronted with secrecy, danger, hidden alliances, misrepresentations, and unpredictable changes of intent, I had to use a delicate combination of overt and covert roles. Throughout, my deliberate cultivation of the norm of reciprocal exchange enabled me to trade my friendship for their knowledge, rather than waiting for the highly unlikely event that information would be delivered into my lap. I thus actively built a web of research contacts, used them to obtain highly sensitive data, and carefully checked them out to ensure validity.

Throughout this endeavor I profited greatly from the efforts of my husband, Peter, who served as an equal partner in this team field research project. It would have been impossible for me to examine this social world as an unattached female and not fall prey to sex role stereotyping which excluded women from business dealings. As a couple, our different genders allowed us to relate in different ways to both men and women (see Warren and Rasmussen 1977). We also protected each other when we entered the homes of dangerous characters, buoyed each other's initiative and courage, and kept the conversation going when one of us faltered. Conceptually, we helped each other keep a detached and analytical eye on the setting, provided multiperspectival insights, and corroborated, clarified, or (most revealingly) contradicted each other's observations and conclusions.

Finally, I feel strongly that to ensure accuracy, research on deviant groups must be conducted in the settings where it naturally occurs. As Polsky (1969: 115–16) has forcefully asserted:

> This means—there is no getting away from it—the study of career criminals *au natural,* in the field, the study of such criminals as they normally go about their work and play, the study of "uncaught" criminals and the study of others who in the past have been caught but are not caught at the time you study them.... Obviously we can no longer afford the convenient fiction that in studying criminals in their natural habitat, we would discover nothing really important that could not be discovered from criminals behind bars.

By studying criminals in their natural habitat I was able to see them in the full variability and complexity of their surrounding subculture, rather than within the artificial environment of a prison. I was thus able to learn about otherwise inaccessible dimensions of their lives, observing and analyzing first-hand the nature of their social organization, social stratification, lifestyle, and motivation.

NOTES

1. Gold (1958) discouraged this methodological strategy, cautioning against overly close friendship or intimacy with informants, lest they lose their ability to act as informants by becoming too much observers. Whyte (1955), in contrast, recommended the use of informants as research aides, not for helping in conceptualizing the data but for their assistance in locating data which supports, contradicts, or fills in the researcher's analysis of the setting.

2. See also Biernacki and Waldorf 1981; Douglas 1976; Henslin 1972; Hoffman 1980; McCall 1980; and West 1980 for discussions of "snowballing" through key informants.

3. See Kirby and Corzine 1981; Birenbaum 1970; and Henslin 1972 for more detailed discussion of the nature, problems, and strategies for dealing with courtesy stigmas.

4. We never considered secret tapings because, aside from the ethical problems involved, it always struck us as too dangerous.

5. See Douglas (1976) for a more detailed account of these procedures.

6. A recent court decision, where a federal judge ruled that a sociologist did not have to turn over his field notes to a grand jury investigating a suspicious fire at a restaurant where he worked, indicates that this situation may be changing (Fried 1984).

7. See Henslin 1972 and Douglas 1972, 1976 for further discussions of this dilemma and various solutions to it.

8. In some cases I resolved this by altering my descriptions of people and their actions as well as their names so that other members of the dealing and smuggling community would not recognize them. In doing this, however, I had to keep a primary concern for maintaining the sociological integrity of my data so that the generic conclusions I drew from them would be accurate. In places, then, where my attempts to conceal people's identities from people who know them have been inadequate, I hope that I caused them no embarrassment. See also Polsky 1969; Rainwater and Pittman 1967; and Humphreys 1970 for discussions of this problem.

REFERENCES

Adler, Patricia A., and Peter Adler. 1987. *Membership Roles in Field Research*. Beverly Hills, CA: Sage.

Becker, Howard. 1963. *Outsiders*. New York: Free Press.

Biernacki, Patrick, and Dan Waldorf. 1981. "Snowball sampling." *Sociological Methods and Research* 10: 141–63.

Birenbaum, Arnold. 1970. "On managing a courtesy stigma." *Journal of Health and Social Behavior* 11: 196–206.

Blumer, Herbert. 1969. *Symbolic Interactionism*. Englewood Cliffs, NJ: Prentice Hall.

Carey, James T. 1972. "Problems of access and risk in observing drug scenes." In Jack D. Douglas, ed., *Research on Deviance*, pp. 71–92. New York: Random House.

Cummins, Marvin, et al. 1972. *Report of the Student Task Force on Heroin Use in Metropolitan Saint Louis*. Saint Louis: Washington University Social Science Institute.

Douglas, Jack D. 1972. "Observing deviance." In Jack D. Douglas, ed., *Research on Deviance*, pp. 3–34. New York: Random House.

———. 1976. *Investigative Social Research*. Beverly Hills, CA: Sage.

Fried, Joseph P. 1984. "Judge protects waiter's notes on fire inquiry." *New York Times*, April 8: 47.

Goffman, Erving. 1963. *Stigma*. Englewood Cliffs, NJ: Prentice Hall.

Gold, Raymond. 1958. "Roles in sociological field observations." *Social Forces* 36: 217–23.

Henslin, James M. 1972. "Studying deviance in four settings: research experiences with cabbies, suicides, drug users and abortionees." In Jack D. Douglas, ed., *Research on Deviance*, pp. 35–70. New York: Random House.

Hoffman, Joan E. 1980. "Problems of access in the study of social elites and boards of directors." In William B. Shaffir, Robert A. Stebbins, and Allan Turowetz, eds., *Fieldwork Experience*, pp. 45–56. New York: St. Martin's.

Humphreys, Laud. 1970. *Tearoom Trade*. Chicago: Aldine.

Johnson, John M. 1975. *Doing Field Research*. New York: Free Press.

Kirby, Richard, and Jay Corzine. 1981. "The contagion of stigma." *Qualitative Sociology* 4: 3–20.

Klockars, Carl B. 1977. "Field ethics for the life history." In Robert Weppner, ed., *Street Ethnography*, pp. 201–26. Beverly Hills, CA: Sage.

———. 1979. "Dirty hands and deviant subjects." In Carl B. Klockars and Finnbarr W. O'Connor, eds., *Deviance and Decency*, pp. 261–82. Beverly Hills, CA: Sage.

Liebow, Elliott. 1967. *Tally's Corner*. Boston: Little, Brown.

McCall, Michal. 1980. "Who and where are the artists?" In William B. Shaffir, Robert A. Stebbins, and Allan Turowetz, eds., *Fieldwork Experience*, pp. 145–58. New York: St. Martin's.

Polsky, Ned. 1969. *Hustlers, Beats, and Others*. New York: Doubleday.

Rainwater, Lee R., and David J. Pittman. 1967. "Ethical problems in studying a politically sensitive and deviant community." *Social Problems* 14: 357–66.

Riemer, Jeffrey W. 1977. "Varieties of opportunistic research." *Urban Life* 5: 467–77.

Rochford, E. Burke, Jr. 1985. *Hare Krishna in America*. New Brunswick, NJ: Rutgers University Press.

Rosenbaum, Marsha. 1981. *Women on Heroin*. New Brunswick, NJ: Rutgers University Press.

Warren, Carol A. B., and Paul K. Rasmussen. 1977. "Sex and gender in field research." *Urban Life* 6: 349–69.

Wax, Rosalie. 1952. "Reciprocity as a field technique." *Human Organization* 11: 34–37.

———. 1957. "Twelve years later: An analysis of a field experience." *American Journal of Sociology* 63: 133–42.

West, W. Gordon. 1980. "Access to adolescent deviants and deviance." In William B. Shaffir, Robert A. Stebbins, and Allan Turowetz, eds., *Fieldwork Experience*, pp. 31–44. New York: St. Martin's.

Whyte, William F. 1955. *Street Corner Society*. Chicago: University of Chicago Press.

PART IV

Constructing Deviance

As we noted in the General Introduction, the Social Constructionist perspective suggests that deviance should be regarded as lodged in a process of definition, rather than in some objective feature of an object, person, or act. It therefore guides us to look at the process by which a society constructs definitions of deviance and applies them to specific groups of people associated with these objects or acts. The dynamics of these deviance-defining processes may sometimes eclipse the factual grounding on which they rest in their significance for the rise of collective moral sentiments.

MORAL ENTREPRENEURS: CAMPAIGNING

The process of constructing and applying definitions of deviance can be understood as a moral enterprise. That is, it involves the constructions of moral meanings and the association of them with specific acts or conditions. The way people "make" deviance is similar to the way they manufacture anything else, but because deviance is an abstract concept rather than a tangible product, this process involves individuals drawing on the power and resources of organizations, institutions, agencies, symbols, ideas, communication, and audiences. Becker (1963) has suggested that we call the people involved in these activities **moral entrepreneurs.** The deviance-making enterprise has two facets: rule creating (without which there would be no deviant behavior) and rule enforcing (applying these rules to specific groups of people). We have two kinds of moral entrepreneurs: rule creators and rule enforcers. Rule creators would include such people as politicians, crusading public figures, teachers, parents, school administrators, and CEOs of business organizations. When we think of rule enforcers, we immediately imagine the police/courts/judges, but these can also be dormitory RAs, members of neighborhood associations, the Inter-Fraternity Council, and (here as well) parents.

Rule creating can be done by individuals acting either alone or in groups. Prominent individuals who have been influential in campaigning for definitions of deviance include former First Lady Nancy Reagan for her "Just Say No" and D.A.R.E. antidrug campaigns; actor Charlton Heston for his presidency of the National Rifle Association (NRA); John Walsh for founding the Missing and Exploited Children's Network and the television show "America's Most Wanted"; and filmmaker Michael Moore for his documentaries about General Motors (*Roger and Me*), the gun lobby (*Bowling for Columbine*), the oil industry/ Bush administration (*Fahrenheit 9/11*), and the health care industry (*Sicko*). More commonly, however, individuals band together to use their collective energy and resources to change social definitions and to create norms and rules. Groups of moral entrepreneurs represent interest groups that can be galvanized and activated into pressure groups, such as Mothers Against Drunk Driving (MADD), the Group Against Smoking Pollution (GASP), the National Organization for the Reform of Marijuana Laws (NORML), and Focus on the Family (a right-wing, Christian pro-family group). Rule creators ensure that our society is supplied with a constant stock of deviance by defining the behavior of others as immoral. They do this because they perceive people as threats and feel fearful, distrustful, and suspicious of their behavior. In so doing, they seek to transform private troubles into public issues and their private morality into the normative order.

Moral entrepreneurs manufacture public morality through a multistage process. Their first goal is to generate broad **awareness** of a problem. They do this through a process of what Spector and Kitsuse (1977) called "claims making." Claims makers draw our attention to given issues by asserting "danger messages." Not only do they use these messages to create a sense that certain conditions are problematic and pose a present or future potential danger to society, but they also usually have specific solutions that they recommend. Issues about which we have recently seen danger messages raised include second-hand smoke, drunken driving, hate crimes, college binge drinking, illegal immigrants (and their link to terrorists), outsourcing, guns in schools, junk food, politics in the classroom, and obesity. Because no rules exist to deal with the threatening condition, claims makers construct the impression that these are necessary.

In so doing, they draw on the testimonials of various "experts" in the field, such as scholars, doctors, eyewitnesses, ex-participants (professional -ex's), and others with specific knowledge of the situation. Issues are framed in these testimonials and disseminated to society via the media as "typical" to promote specific examples, orientations, causes, and solutions. We see the Surgeon General warning about the dangers of second-hand smoke, college administrators talking about student drinking and deaths, psychiatrists writing about the dangers of self-injury and eating disorders, sociologists discussing the spread of date and acquaintance rape, national conservative watchdogs such as David Horowitz monitoring liberal bias in the classroom, and ex-FLDS (Fundamentalist Latter Day Saints) members warning about the forced marriage (statutory rape) of young girls, the expulsion of teenage boys, and the isolation and subservience of whole communities to the point of brainwashing.

Several rhetorical techniques are used to package and present these facts in the most compelling way. Statistics may show the rise in incidence of a given behavior, or its correlation with other social problems. For example, suicide statistics may be contrasted with homicide statistics to draw attention to the importance of suicide, which occurs at a rate nearly 50 percent higher than homicide. Growth in the rates of obesity and its relation to diabetes, heart disease, stroke, and worker absenteeism may be documented. We may hear about the rising tide of illegal immigrants and their financial drain on our public schools and medical services. Free trade agreements may be tied to increasing rates of unemployment. Dramatic case examples can paint a picture of horror in the public's mind, inspiring fear and loathing. Particular cases are usually selected because they have no moral ambiguity and feature the dimension of the problem that is being highlighted.

Thus, for missing children we want to see stranger abductors rather than snatchings by noncustodial parents, while for AIDS we want to see innocent children who are sick rather than gay men. New syndromes can be advanced, packaging different issues together into a behavioral pattern portrayed as dangerous, such as "Internet addiction disorder" (caution: people are abandoning their homes and families to spend all their time and money lost in chat rooms); "centerfold syndrome" (caution: men who read pornography objectify, commodify, and victimize women, seek unattainable trophy figurines, and are unable to engage in meaningful relationships); "road rage disorder" (caution: traffic induces people to cut us off and give us the finger); "compulsive hoarding" (caution: practitioners have excessive clutter, difficulty categorizing, organizing, making decisions about, and throwing away possessions, and fears about needing items that could be thrown away); "chronic procrastination" (caution: you may be an "arousal procrastinator," where you put things off for the last-minute rush, or an "avoider" and put off dealing with things out of fear of failure or success); and ADHD (caution: you could have this medical problem if you daydream or are slow to complete tasks, fidget, have poor concentration, or are distractible, hyperactive, impulsive, and/or reckless). Finally, rhetoric requires that each side seek the (usually competing) "moral high ground" in their assertions and attacks on each other, disavowing special interests and pursuing only the purest public good. For example, people who are opposed to gay marriage are upholding the bible, decency, and the sanctity of marriage, while people who support it are upholding individual civil liberties and fighting unjust discrimination.

Second, rule creators must bring about a **moral conversion,** convincing others of their views. With the problem outlined, they have to convert neutral parties and previous opponents into supporting partisans. Their successful conversion of others further legitimates their own beliefs. To effect a moral conversion, rule creators must compete for space in the public arena, often a limited resource. Hilgartner and Bosk (1988) have suggested that only so many issues can claim widespread attention, and they do so at the expense of others. As Durkheim noted for deviance, only a limited number of public concerns can be supported in society at any given time. Moral entrepreneurs must draw on elements of drama, novelty, politics, and deep mythic themes of the culture to gain the visibility they need. To gain visibility, moral entrepreneurs must attract the media attention necessary to spread word of their campaign widely. They may orchestrate this through hunger strikes (students protesting the manufacturing of university-labeled clothing in Third World "sweat shops"); demonstrations (antiwar, proimmigration); civil disobedience (such as environmentalists tying themselves to trees to prevent de-forestation or blocking highways to prevent nuclear waste from being stored in their state); marches (the Gay Pride and Civil Rights movements); and strikes and picketing (against deunionization).

They must also enlist the support of sponsors, opinion leaders who need not have expert knowledge on any particular subject but are liked and respected, to provide them with public endorsements. Moral entrepreneurial campaigners often turn to athletes, actors, musicians, religious leaders, and media personalities for such endorsements. Former Democratic presidential candidate Al Gore is notable for his campaign highlighting the growth of global warming (*An Inconvenient Truth*), actor Tom Cruise has actively promoted the dangers of antidepressants, and actor Michael J. Fox campaigned for Democratic candidates to endorse stem cell research.

Finally, they look to different groups in society to form alliances or coalitions to support their campaigns. Alliances are made up of long-term allies, such the Christian Coalition, Focus on the Family, conservative Republican politicians, and big business. Coalitions, on the other hand, represent groups that do not normally lobby together but are bonded by their mutual interests in a single issue, such as we see in the strange union when family groups, conservative Republicans, religious leaders, and, ironically, radical feminists come together to campaign against pornography.

At times, the efforts of moral entrepreneurs are so successful that they create a "moral panic." A threat to society is depicted and concerned individuals promoting the problem, reacting legislators, and sensationalist news media whip the public into a "feeding frenzy." Moral panics—most recently witnessed in cases of school shootings, priesthood pedophilia, drinking and rioting on college campuses, Internet predators, obesity, and domestic terrorists—tend to develop a life of their own, often moving in exaggerated propulsion beyond their original impetus. To be successful, they usually have to be triggered by a specific event, occur during a ripe historical period, draw attention to a specific individual or group as a target, have meaty content that gets revealed, and become heightened by the spread of the panic through the mass media, grassroots communication, Internet warnings, and/or public presentations. But all moral panics die out eventually, since they represent inflated fears and cannot be sustained indefinitely. They may founder of their own accord or be replaced by new panics, but they usually leave in their wake some residual effect. For example, flying in America will never be the same as it was prior to 9/11.

Once the public viewpoint has been swayed and a majority (or a vocal and powerful enough minority) of people have adopted a social definition, it may remain at the level of a norm or become elevated to the status of law through a legislative effort. For example, although obesity has been defined as disgusting and unhealthy it is not illegal, while antismoking campaigns have been successful in banning cigarette use in most public places.

After norms or rules have been enacted, rule enforcers ensure that they are applied. In our society, this process often tends to be selective. Various individuals or groups have greater or lesser power to resist the enforcement of rules against them due to their socioeconomic, racial, religious, gender, political, or other status. Whole battles may begin anew over individuals' or groups' strength to apply or resist the enforcement of norms and laws, with this arena becoming once again a moral entrepreneurial combat zone. President George W. Bush's efforts, after the invasion of Iraq went sour, to rebuff the investigatory efforts of the Democratic Party into charges of manipulated intelligence over his claims about WMDs (weapons of mass destruction) illustrates core issues of social power.

DIFFERENTIAL SOCIAL POWER: LABELING

Specific behavioral acts are not the only things that can be constructed as deviant; this definition can also be applied to a social status, demographic characteristic, or lifestyle. When entire groups of people become relegated to a deviant status through their condition (especially if it is ascribed through birth rather than voluntarily achieved), we see the force of inequality and differential social power in operation. This dynamic has been discussed earlier in reference to both conflict theory and social constructionism, as we noted that those who control the resources in society (politics, social status, gender, wealth, religious beliefs, mobilization of the masses) have the ability to dominate, both materially and ideologically, over the subordinate groups. Thus, certain kinds of laws and enforcement are a product of political action by moral entrepreneurial interest groups that are connected to society's power base. Dominant groups use their strength and position to label and to subjugate the weak.

A range of different factors give certain groups greater **social power** in society to construct definitions of deviance and to apply those labels onto others. Money is one of the clearest elements, with its potential influence being felt at least two ways. Big businesses can use it to make campaign contributions and sway political candidates, to fund research favorable to their products (studies paid for by the soft drink industry were recently found eight time more likely to find no harmful health influences of drinking soda than those otherwise funded), to lobby against unfavorable legislation, and to fight restrictive lawsuits. Money also defines individuals' social class, and while rich people do not have as much cash available to do what businesses can afford, it is much harder to define the practices of the middle and upper classes as deviant than those of the lower, working, and underclasses. Second, race and ethnicity influence social power, so that the behaviors of the dominant white population are less likely to be defined and enforced than those of Hispanics and African Americans. Gender is a third element of social power, with men dominating over women politically, economically, historically, religiously, occupationally, culturally, and, hence, interpersonally.

Fourth, people's age affects their relative power in society, with young people (up to the age of 30) and older people (65 and older) holding less respect, influence, attention, and command than their middle-aged counterparts. Fifth, greater numbers and organization can empower groups, as positions backed by larger populations often hold sway over smaller ones. Yet at the same time, well-organized groups, even if they are in the minority, may dominate over bigger, unorganized masses. We acknowledge education as a sixth element of social power since, as the chapters in this section argue, well-educated professionals have the ability to speak as experts, to organize moral entrepreneurial campaigns, to advocate for their positions, and to argue from a legitimate base of knowledge. Finally, social status (apart from social class) generates power through the prestige, tradition, and respectability associated with various positions in society. For example, religious people have greater social status in contemporary society than atheists, heterosexual people command greater legitimacy than homosexuals, and married individuals hold greater sway, as a group, than single ones. There are, of course, many more elements of social power that could be articulated as having an influence over labeling individuals, groups, and their characteristics as deviant, but these, to us, represent the key ones.

In a society characterized by striving for social influence, status, and power, one way to attain this is to pass and enforce rules that define others' behavior as deviant. Thus, the deviant labeling of attitudes, behaviors, and conditions such as minority ethnic or racial status, female gender, lower social class, youthful age, homosexual orientation, and a criminal record (as some of the following readings show), if taken in this light, can be seen to reflect the application of differential social power in our society. Individuals in these groups may find themselves discriminated against or blocked from the mainstream of society by virtue of this basic feature of their existence, unrelated to any particular situation or act. This application of the deviant label emphatically illustrates the role of power in the deviance-defining enterprise, as those positioned closer to the center of society, holding the greater social, economic, political, and moral resources, can turn the force of the deviant stigma onto others less fortunately placed. In so doing, they use the definition of deviance to reinforce their own favored position. This politicization of deviance and the power associated with its use serve to remind us that deviance is not a category inhabited only by those on the marginal outskirts of society: the exotics, erotics, and neurotics. Instead, any group can be pushed into this category by the exercise of another group's greater power.

DIFFERENTIAL SOCIAL POWER: RESISTING LABELING

On the other side of the coin, powerful groups may be successful in working to resist the application of definitions of deviance to them. Like better-looking people, they are granted a "halo effect" that leads others to think highly of them. Higher-statused groups in society are less likely to be perceived as deviant whether they actively work to fight the label or not.

In some cases they undertake proactive collective identity protection as a group. Many organizations such as pharmaceutical companies and the manufacturers of cigarettes or alcoholic beverages, as well as groups organized around specific issues such as the National Rifle Association (NRA), work to build and sustain a positive social image. They may hire claims makers, lobbyists, contribute to politicians' campaigns, or fund research showing that they are upstanding individuals and organizations. As mentioned above, one comparative study recently showed that research funded by the soft drink industry was eight times as likely to find no negative correlation between consuming soft drinks and poor health as those otherwise financed.

But even when they do not become specifically involved in protecting their images, members of more powerful and respected groups in society are less likely to become tainted by deviant labels. Part of the reason for this is that people hold preconceived biases in their favor and assume that they are responsible and prosocial, whether they are or are not. People also make perceptual biases toward them based on their appearances, occupations, behavior, and/or associations, forming instantaneous judgments about them that are positive. Members of such protected groups are often unaware of their privileged status in society and do not realize the discrimination routinely encountered by underprivileged populations.

This differential social power may be applied either directly or comparatively, as when society judges the behavior of one group against another, or when individuals or a group are judged on their own. Together, these groups of people continue to receive treatment from society as either deviant or nondeviant that reinforces social inequality and the status quo.

15

The Social Construction of Drug Scares

CRAIG REINARMAN

In this overview of America's social policies, Reinarman tackles moral and legal attitudes toward illicit drugs. He briefly offers a history of drug scares, the major players engineering them, and the social contexts that have enhanced their development and growth. He then outlines seven factors common to drug scares. These enable him to dissect the essential processes in the rule creation and enforcement phases of drug scares, despite the contradictory cultural values of temperance and hedonistic consumption. From this selection, we can see how drugs have been scapegoated to account for a wide array of social problems and used to keep some groups down by defining their actions as deviant. It is clear that despite our society's views on the negative features associated with all illicit drugs, our moral entrepreneurial and enforcement efforts have been concentrated more stringently against the drugs used by members of the powerless underclass and minority racial groups.

Drug "wars," anti-drug crusades, and other periods of marked public concern about drugs are never merely reactions to the various troubles people can have with drugs. These drug scares are recurring cultural and political phenomena *in their own right* and must, therefore, be understood sociologically on their own terms. It is important to understand why people ingest drugs and why some of them develop problems that have something to do with having ingested them. But the premise of this chapter is that it is equally important to understand patterns of acute societal concern about drug use and drug problems. This seems especially so for U.S. society, which has had *recurring* anti-drug crusades and a *history* of repressive anti-drug laws.

Many well-intentioned drug policy reform efforts in the U.S. have come face to face with staid and stubborn sentiments against consciousness-altering substances. The repeated failures of such reform efforts cannot be explained solely in terms of ill-informed or manipulative leaders. Something deeper is involved,

Reprinted by permission of Craig Reinarman.

something woven into the very fabric of American culture, something which explains why claims that some drug is the cause of much of what is wrong with the world are *believed* so often by so many. The origins and nature of the *appeal* of anti-drug claims must be confronted if we are ever to understand how "drug problems" are constructed in the U.S. such that more enlightened and effective drug policies have been so difficult to achieve.

In this chapter I take a step in this direction. First, I summarize briefly some of the major periods of anti-drug sentiment in the U.S. Second, I draw from them the basic ingredients of which drug scares and drug laws are made. Third, I offer a beginning interpretation of these scares and laws based on those broad features of American culture that make *self-control* continuously problematic.

DRUG SCARES AND DRUG LAWS

What I have called drug scares (Reinarman and Levine, 1989a) have been a recurring feature of U.S. society for 200 years. They are relatively autonomous from whatever drug-related problems exist or are said to exist.[1] I call them "scares" because, like Red Scares, they are a form of moral panic ideologically constructed so as to construe one or another chemical bogeyman, à la "communists," as the core cause of a wide array of preexisting public problems.

The first and most significant drug scare was over drink. Temperance movement leaders constructed this scare beginning in the late 18th and early 19th century. It reached its formal end with the passage of Prohibition in 1919.[2] As Gusfield showed in his classic book *Symbolic Crusade* (1963), there was far more to the battle against booze than long-standing drinking problems. Temperance crusaders tended to be native born, middle-class, non-urban Protestants who felt threatened by the working-class, Catholic immigrants who were filling up America's cities during industrialization.[3] The latter were what Gusfield termed "unrepentant deviants" in that they continued their long-standing drinking practices despite middle-class W.A.S.P. norms against them. The battle over booze was the terrain on which was fought a cornucopia of cultural conflicts, particularly over whose morality would be the dominant morality in America.

In the course of this century-long struggle, the often wild claims of Temperance leaders appealed to millions of middle-class people seeking explanations for the pressing social and economic problems of industrializing America. Many corporate supporters of Prohibition threw their financial and ideological weight behind the Anti-Saloon League and other Temperance and Prohibitionist groups because they felt that traditional working-class drinking practices interfered with the new rhythms of the factory, and thus with productivity and profits (Rumbarger, 1989). To the Temperance crusaders' fear of the bar room as a breeding ground of all sorts of tragic immorality, Prohibitionists added the idea of the saloon as an alien, subversive place where unionists organized and where leftists and anarchists found recruits (Levine, 1984).

This convergence of claims and interests rendered alcohol a scapegoat for most of the nation's poverty, crime, moral degeneracy, "broken" families, illegitimacy, unemployment, and personal and business failure—problems whose sources lay in broader economic and political forces. This scare climaxed in the first two decades of this century, a tumultuous period rife with class, racial, cultural, and political conflict brought on by the wrenching changes of industrialization, immigration, and urbanization (Levine, 1984; Levine and Reinarman, 1991).

America's first real drug law was San Francisco's anti–opium den ordinance of 1875. The context of the campaign for this law shared many features with the context of the Temperance movement. Opiates had long been widely and legally available without a prescription in hundreds of medicines (Brecher, 1972; Musto, 1973; Courtwright, 1982; cf. Baumohl, 1992), so neither opiate use nor addiction was really the issue. This campaign focused almost exclusively on what was called the "Mongolian vice" of opium *smoking* by Chinese immigrants (and white "fellow travelers") in dens (Baumohl, 1992). Chinese immigrants came to California as "coolie" labor to build the railroad and dig the gold mines. A small minority of them brought along the practice of smoking opium—a practice originally brought to China by British and American traders in the 19th century. When the railroad was completed and the gold dried up, a decade-long depression ensued. In a tight labor market, Chinese immigrants were a target. The white Workingman's Party fomented racial hatred of the low-wage "coolies" with whom they now had to compete for work. The first law against opium smoking was only one of many laws enacted to harass and control Chinese workers (Morgan, 1978).

By calling attention to this broader political-economic context I do not wish to slight the specifics of the local political-economic context. In addition to the Workingman's Party, downtown businessmen formed merchant associations and urban families formed improvement associations, both of which fought for more than two decades to reduce the impact of San Francisco's vice districts on the order and health of the central business district and on family neighborhoods (Baumohl, 1992).

In this sense, the anti–opium den ordinance was not the clear and direct result of a sudden drug scare alone. The law was passed against a specific form of drug use engaged in by a disreputable group that had come to be seen as threatening in lean economic times. But it passed easily because this new threat was understood against the broader historical backdrop of long-standing local concerns about various vices as threats to public health, public morals, and public order. Moreover, the focus of attention were dens where it was suspected that whites came into intimate contact with "filthy, idolatrous" Chinese (see Baumohl, 1992). Some local law enforcement leaders, for example, complained that Chinese men were using this vice to seduce white women into sexual slavery (Morgan, 1978). Whatever the hazards of opium smoking, its initial criminalization in San Francisco had to do with both a general context of recession, class conflict, and racism, and with specific local interests in the control of vice and the prevention of miscegenation.

A nationwide scare focusing on opiates and cocaine began in the early 20th century. These drugs had been widely used for years, but were first criminalized when the addict population began to shift from predominantly white, middle-class, middle-aged women to young, working-class males, African-Americans in particular. This scare led to the Harrison Narcotics Act of 1914, the first federal anti-drug law (see Duster, 1970).

Many different moral entrepreneurs guided its passage over a six-year campaign: State Department diplomats seeking a drug treaty as a means of expanding trade with China, trade which they felt was crucial for pulling the economy out of recession; the medical and pharmaceutical professions whose interests were threatened by self-medication with unregulated proprietary tonics, many of which contained cocaine or opiates; reformers seeking to control what they saw as the deviance of immigrants and Southern Blacks who were migrating off the farms; and a pliant press which routinely linked drug use with prostitutes, criminals, transient workers (e.g., the Wobblies), and African-Americans (Musto, 1973). In order to gain the support of Southern Congressmen for a new federal law that might infringe on "states' rights," State Department officials and other crusaders repeatedly spread unsubstantiated suspicions, repeated in the press, that, e.g., cocaine induced African-American men to rape white women (Musto, 1973: 6–10, 67). In short, there was more to this drug scare, too, than mere drug problems.

In the Great Depression, Harry Anslinger of the Federal Narcotics Bureau pushed Congress for a federal law against marijuana. He claimed it was a "killer weed" and he spread stories to the press suggesting that it induced violence—especially among Mexican-Americans. Although there was no evidence that marijuana was widely used, much less that it had any untoward effects, his crusade resulted in its criminalization in 1937—and not incidentally a turnaround in his Bureau's fiscal fortunes (Dickson, 1968). In this case, a new drug law was put in place by a militant moral-bureaucratic entrepreneur who played on racial fears and manipulated a press willing to repeat even his most absurd claims in a context of class conflict during the Depression (Becker, 1963). While there was not a marked scare at the time, Anslinger's claims were never contested in Congress because they played upon racial fears and widely held Victorian values against taking drugs solely for pleasure.

In the drug scare of the 1960s, political and moral leaders somehow reconceptualized this same "killer weed" as the "drop out drug" that was leading America's youth to rebellion and ruin (Himmelstein, 1983). Bio-medical scientists also published uncontrolled, retrospective studies of very small numbers of cases suggesting that, in addition to poisoning the minds and morals of youth, LSD produced broken chromosomes and thus genetic damage (Cohen et al., 1967). These studies were soon shown to be seriously misleading if not meaningless (Tijo et al., 1969), but not before the press, politicians, the medical profession, and the National Institute of Mental Health used them to promote a scare (Weil, 1972: 44–46).

I suggest that the reason even supposedly hard-headed scientists were drawn into such propaganda was that dominant groups felt the country was at war—and not merely with Vietnam. In this scare, there was not so much a "dangerous class" or threatening racial group as multi-faceted political and cultural conflict, particularly between generations, which gave rise to the perception that middle-class youth who rejected conventional values were a dangerous threat.[4] This scare resulted in the Comprehensive Drug Abuse Control Act of 1970, which criminalized more forms of drug use and subjected users to harsher penalties.

Most recently we have seen the crack scare, which began in earnest *not* when the prevalence of cocaine use quadrupled in the late 1970s, nor even when thousands of users began to smoke it in the more potent and dangerous form of freebase. Indeed, when this scare was launched, crack was unknown outside of a few neighborhoods in a handful of major cities (Reinarman and Levine, 1989a) and the prevalence of illicit drug use had been dropping for several years (National Institute on Drug Use, 1990). Rather, this most recent scare began in 1986 when freebase cocaine was renamed crack (or "rock") and sold in precooked, inexpensive units on ghetto streetcorners (Reinarman and Levine, 1989b). Once politicians and the media linked this new form of cocaine use to the inner-city, minority poor, a new drug scare was underway and the solution became more prison cells rather than more treatment slots.

The same sorts of wild claims and Draconian policy proposals of Temperance and Prohibition leaders resurfaced in the crack scare. Politicians have so outdone each other in getting "tough on drugs" that each year since crack came on the scene in 1986 they have passed more repressive laws providing billions more for law enforcement, longer sentences, and more drug offenses punishable by death. One result is that the U.S. now has more people in prison than any industrialized nation in the world—about half of them for drug offenses, the majority of whom are racial minorities.

In each of these periods more repressive drug laws were passed on the grounds that they would reduce drug use and drug problems. I have found no evidence that any scare actually accomplished those ends, but they did greatly expand the quantity and quality of social control, particularly over subordinate groups perceived as dangerous or threatening. Reading across these historical episodes one can abstract a recipe for drug scares and repressive drug laws that contains the following *seven ingredients:*

1. **A Kernel of Truth** Humans have ingested fermented beverages at least since human civilization moved from hunting and gathering to primitive agriculture thousands of years ago. The pharmacopoeia has expanded exponentially since then. So, in virtually all cultures and historical epochs, there has been sufficient ingestion of consciousness-altering chemicals to provide some basis for some people to claim that it is a problem.

2. **Media Magnification** In each of the episodes I have summarized and many others, the mass media has engaged in what I call the *routinization of caricature*—rhetorically recrafting worst cases into typical cases and the episodic into the epidemic. The media dramatize drug problems, as they do other problems, in the course of their routine news-generating and

sales-promoting procedures (see Brecher, 1972: 321–34; Reinarman and Duskin, 1992; and Molotch and Lester, 1974).

3. **Politico-Moral Entrepreneurs** I have added the prefix "politico" to Becker's (1963) seminal concept of moral entrepreneur in order to emphasize the fact that the most prominent and powerful moral entrepreneurs in drug scares are often political elites. Otherwise, I employ the term just as he intended: to denote the *enterprise,* the work, of those who create (or enforce) a rule against what they see as a social evil.[5]

    In the history of drug problems in the U.S., these entrepreneurs call attention to drug using behavior and define it as a threat about which "something must be done." They also serve as the media's primary source of sound bites on the dangers of this or that drug. In all the scares I have noted, these entrepreneurs had interests of their own (often financial) which had little to do with drugs. Political elites typically find drugs a functional demon in that (like "outside agitators") drugs allow them to deflect attention from other, more systemic sources of public problems for which they would otherwise have to take some responsibility. Unlike almost every other political issue, however, to be "tough on drugs" in American political culture allows a leader to take a firm stand without risking votes or campaign contributions.

4. **Professional Interest Groups** In each drug scare and during the passage of each drug law, various professional interests contended over what Gusfield (1981: 10–15) calls the "ownership" of drug problems—"the ability to create and influence the public definition of a problem" (1981: 10), and thus to define what should be done about it. These groups have included industrialists, churches, the American Medical Association, the American Pharmaceutical Association, various law enforcement agencies, scientists, and most recently the treatment industry and groups of those former addicts converted to disease ideology.[6] These groups claim for themselves, by virtue of their specialized forms of knowledge, the legitimacy and authority to name what is wrong and to prescribe the solution, usually garnering resources as a result.

5. **Historical Context of Conflict** This trinity of the media, moral entrepreneurs, and professional interests typically interact in such a way as to inflate the extant "kernel of truth" about drug use. But this interaction does not by itself give rise to drug scares or drug laws without underlying conflicts which make drugs into functional villains. Although Temperance crusaders persuaded millions to pledge abstinence, they campaigned for years without achieving alcohol control laws. However, in the tumultuous period leading up to Prohibition, there were revolutions in Russia and Mexico, World War I, massive immigration and impoverishment, and socialist, anarchist, and labor movements, to say nothing of increases in routine problems such as crime. I submit that all this conflict made for a level of

cultural anxiety that provided fertile ideological soil for Prohibition. In each of the other scares, similar conflicts—economic, political, cultural, class, racial, or a combination—provided a context in which claims makers could viably construe certain classes of drug users as a threat.

6. **Linking a Form of Drug Use to a "Dangerous Class"** Drug scares are never about drugs *per se,* because drugs are inanimate objects without social consequence until they are ingested by humans. Rather, drug scares are about the use of a drug by particular groups of people who are, typically, *already* perceived by powerful groups as some kind of threat (see Duster, 1970; Himmelstein, 1978). It was not so much alcohol problems *per se* that most animated the drive for Prohibition but the behavior and morality of what dominant groups saw as the "dangerous class" of urban, immigrant, Catholic, working-class drinkers (Gusfield, 1963; Rumbarger, 1989). It was *Chinese* opium smoking dens, not the more widespread use of other opiates, that prompted California's first drug law in the 1870s. It was only when smokable cocaine found its way to the African-American and Latino underclass that it made headlines and prompted calls for a drug war. In each case, politico-moral entrepreneurs were able to construct a "drug problem" by linking a substance to a group of users perceived by the powerful as disreputable, dangerous, or otherwise threatening.

7. **Scapegoating a Drug for a Wide Array of Public Problems** The final ingredient is scapegoating, i.e., blaming a drug or its alleged effects on a group of its users for a variety of preexisting social ills that are typically only indirectly associated with it. Scapegoating may be the most crucial element because it gives great explanatory power and thus broader resonance to claims about the horrors of drugs (particularly in the conflictual historical contexts in which drug scares tend to occur).

Scapegoating was abundant in each of the cases noted previously. To listen to Temperance crusaders, for example, one might have believed that without alcohol use, America would be a land of infinite economic progress with no poverty, crime, mental illness, or even sex outside marriage. To listen to leaders of organized medicine and the government in the 1960s, one might have surmised that without marijuana and LSD there would have been neither conflict between youth and their parents nor opposition to the Vietnam War. And to believe politicians and the media in the past six years is to believe that without the scourge of crack the inner cities and the so-called underclass would, if not disappear, at least be far less scarred by poverty, violence, and crime. There is no historical evidence supporting any of this.

In short, drugs are richly functional scapegoats. They provide elites with fig leaves to place over unsightly social ills that are endemic to the social system over which they preside. And they provide the public with a restricted aperture of attribution in which only a chemical bogeyman or the lone deviants who ingest it are seen as the cause of a cornucopia of complex problems.

TOWARD A CULTURALLY SPECIFIC
THEORY OF DRUG SCARES

Various forms of drug use have been and are widespread in almost all societies comparable to ours. A few of them have experienced limited drug scares, usually around alcohol decades ago. However, drug scares have been *far* less common in other societies, and never as virulent as they have been in the U.S. (Brecher, 1972; Levine, 1992; MacAndrew and Edgerton, 1969). There has never been a time or place in human history without drunkenness, for example, but in *most* times and places drunkenness has not been nearly as problematic as it has been in the U.S. since the late 18th century. Moreover, in comparable industrial democracies, drug laws are generally less repressive. Why then do claims about the horrors of this or that consciousness–altering chemical have such unusual power in American culture?

Drug scares and other periods of acute public concern about drug use are not just discrete, unrelated episodes. There is a historical pattern in the U.S. that cannot be understood in terms of the moral values and perceptions of individual anti-drug crusaders alone. I have suggested that these crusaders have benefited in various ways from their crusades. For example, making claims about how a drug is damaging society can help elites increase the social control of groups perceived as threatening (Duster, 1970), establish one class's moral code as dominant (Gusfield, 1963), bolster a bureaucracy's sagging fiscal fortunes (Dickson, 1968), or mobilize voter support (Reinarman and Levine, 1989a, b). However, the recurring character of pharmaco-phobia in U.S. history suggests that there is something about our *culture* which makes citizens more vulnerable to anti-drug crusaders' attempts to demonize drugs. Thus, an answer to the question of America's unusual vulnerability to drug scares must address why the scapegoating of consciousness–altering substances regularly *resonates* with or appeals to substantial portions of the population.

There are three basic parts to my answer. The first is that claims about the evils of drugs are especially viable in American culture in part because they provide a welcome *vocabulary of attribution* (cf. Mills, 1940). Armed with "DRUGS" as a generic scapegoat, citizens gain the cognitive satisfaction of having a folk devil on which to blame a range of bizarre behaviors or other conditions they find troubling but difficult to explain in other terms. This much may be true of a number of other societies, but I hypothesize that this is particularly so in the U.S. because in our political culture individualistic explanations for problems are so much more common than social explanations.

Second, claims about the evils of drugs provide an especially serviceable vocabulary of attribution in the U.S. in part because our society developed from a *temperance culture* (Levine, 1992). American society was forged in the fires of ascetic Protestantism and industrial capitalism, both of which demand *self-control*. U.S. society has long been characterized as the land of the individual "self-made man." In such a land, self-control has had extraordinary importance. For the middle-class Protestants who settled, defined, and still dominate the U.S., self-control was both central to religious world views and a characterological necessity for economic survival and success in the capitalist market (Weber, 1930 [1985]). With Levine (1992), I hypothesize that in a culture in which self-control is inordinately important, drug-induced altered states of consciousness are especially likely to be experienced as "loss of control," and thus to be inordinately feared.[7]

Drunkenness and other forms of drug use have, of course, been present everywhere in the industrialized world. But temperance cultures tend to arise only when industrial capitalism unfolds upon a cultural terrain deeply imbued with the Protestant ethic.[8] This means that only the U.S., England, Canada, and parts of Scandinavia have Temperance cultures, the U.S. being the most extreme case.

It may be objected that the influence of such a Temperance culture was strongest in the 19th and early 20th century and that its grip on the American *Zeitgeist* has been loosened by the forces of modernity and now, many say, post-modernity. The third part of my answer, however, is that on the foundation of a Temperance culture, advanced capitalism has built a *postmodern, mass consumption culture* that exacerbates the problem of self-control in new ways.

Early in the 20th century, Henry Ford pioneered the idea that by raising wages he could simultaneously quell worker protests and increase market demand for mass-produced goods. This mass consumption strategy became central to modern American society and one of the reasons for our economic success (Marcuse, 1964; Aronowitz, 1973; Ewen, 1976; Bell, 1978). Our economy is now so fundamentally predicated upon mass consumption that theorists as diverse as Daniel Bell and Herbert Marcuse have observed that we live in a mass consumption culture. Bell (1978), for example, notes that while the Protestant work ethic and deferred gratification may still hold sway in the workplace, Madison Avenue, the media, and malls have inculcated a new indulgence ethic in the leisure sphere in which pleasure-seeking and immediate gratification reign.

Thus, our economy and society have come to depend upon the constant cultivation of new "needs," the production of new desires. Not only the hardware of social life such as food, clothing, and shelter but also the software of the self—excitement, entertainment, even eroticism—have become mass consumption commodities. This means that our society offers an increasing number of incentives for indulgence—more ways to lose self-control—and a decreasing number of countervailing reasons for retaining it.

In short, drug scares continue to occur in American society in part because people must constantly manage the contradiction between a Temperance culture that insists on self-control and a mass consumption culture which renders self-control continuously problematic. In addition to helping explain the recurrence of drug scares, I think this contradiction helps account for why in the last dozen years millions of Americans have joined 12-Step groups, more than 100 of which have nothing whatsoever to do with ingesting a drug (Reinarman, 1995). "Addiction," or the generalized loss of self-control, has become the meta-metaphor for a staggering array of human troubles. And, of course, we also seem to have a staggering array of politicians and other moral entrepreneurs who take advantage of such cultural contradictions to blame new chemical bogeymen for our society's ills.

NOTES

1. In this regard, for example, Robin Room wisely observes "that we are living at a historic moment when the rate of (alcohol) dependence as a cognitive and existential experience is rising, although the rate of alcohol consumption and of heavy drinking is falling." He draws from this a more general hypothesis about "long waves" of drinking and societal reactions to them: "[I]n periods of increased questioning of drinking and heavy drinking, the trends in the two forms of dependence, psychological and physical, will tend to run in opposite directions. Conversely, in periods of a "wettening" of sentiments, with the curve of alcohol consumption beginning to rise, we may expect the rate of physical dependence . . . to rise while the rate of dependence as a cognitive experience falls" (1991: 154).

2. I say "formal end" because Temperance ideology is not merely alive and well in the War on Drugs but is being applied to all manner of human troubles in the burgeoning 12-Step Movement (Reinarman, 1995).

3. From Jim Baumohl I have learned that while the Temperance movement attracted most of its supporters from these groups, it also found supporters among many others (e.g., labor, the Irish, Catholics, former drunkards, women), each of which had its own reading of and folded its own agenda into the movement.

4. This historical sketch of drug scares is obviously not exhaustive. Readers interested in other scares should see, e.g., Brecher's encyclopedic work *Licit and Illicit Drugs* (1972), especially the chapter on glue sniffing, which illustrates how the media actually created a new drug problem by writing hysterical stories about it. There was also a PCP scare in the 1970s in which law enforcement officials claimed that the growing use of this horse tranquilizer was a severe threat because it made users so violent and gave them such super-human

strength that stun guns were necessary. This, too, turned out to be unfounded and the "angel dust" scare was short-lived (see Feldman et al., 1979). The best analysis of how new drugs themselves can lead to panic reactions among users is Becker (1967).

5. Becker wisely warns against the "one-sided view" that sees such crusaders as merely imposing their morality on others. Moral entrepreneurs, he notes, do operate "with an absolute ethic," are "fervent and righteous," and will use "any means" necessary to "do away with" what they see as "totally evil." However, they also "typically believe that their mission is a holy one," that if people do what they want it "will be good for them." Thus, as in the case of abolitionists, the crusades of moral entrepreneurs often "have strong humanitarian overtones" (1963: 147–8). This is no less true for those whose moral enterprise promotes drug scares. My analysis, however, concerns the character and consequences of their efforts, not their motives.

6. As Gusfield notes, such ownership sometimes shifts over time, e.g., with alcohol problems, from religion to criminal law to medical science. With other drug problems, the shift in ownership has been away from medical science toward criminal law. The most insightful treatment of the medicalization of alcohol/drug problems is Peele (1989).

7. See Baumohl's (1990) important and erudite analysis of how the human will was valorized in the therapeutic temperance thought of 19th-century inebriate homes.

8. The third central feature of Temperance cultures identified by Levine (1992), which I will not dwell on, is predominance of spirits drinking, i.e., more concentrated alcohol than wine or beer and thus greater likelihood of drunkenness.

REFERENCES

Aronowitz, Stanley. 1973. *False Promises: The Shaping of American Working Class Consciousness.* New York: McGraw-Hill.

Baumohl, Jim. 1990. "Inebriate Institutions in North America, 1840–1920." *British Journal of Addiction* 85: 1187–1204.

Baumohl, Jim. 1992. "The 'Dope Fiend's Paradise' Revisited: Notes from Research in Progress on Drug Law Enforcement in San Francisco, 1875–1915." *Drinking and Drug Practices Surveyor* 24: 3–12.

Becker, Howard S. 1963. *Outsiders: Studies in the Sociology of Deviance.* Glencoe, IL: Free Press.

———. 1967. "History, Culture, and Subjective Experience: An Exploration of the Social Bases of Drug-Induced Experiences." *Journal of Health and Social Behavior* 8: 162–176.

Bell, Daniel. 1978. *The Cultural Contradictions of Capitalism.* New York: Basic Books.

Brecher, Edward M. 1972. *Licit and Illicit Drugs.* Boston: Little Brown.

Cohen, M. M., K. Hirshorn, and W. A. Frosch. 1967. "In Vivo and in Vitro Chromosomal Damage Induced by LSD-25." *New England Journal of Medicine* 227: 1043.

Courtwright, David. 1982. *Dark Paradise: Opiate Addiction in America Before 1940.* Cambridge, MA: Harvard University Press.

Dickson, Donald. 1968. "Bureaucracy and Morality." *Social Problems* 16: 143–156.

Duster, Troy. 1970. *The Legislation of Morality: Law, Drugs, and Moral Judgement.* New York: Free Press.

Ewen, Stuart. 1976. *Captains of Consciousness: Advertising and the Social Roots of Consumer Culture.* New York: McGraw-Hill.

Feldman, Harvey W., Michael H. Agar, and George M. Beschner. 1979. *Angel Dust.* Lexington, MA: Lexington Books.

Gusfield, Joseph R. 1963. *Symbolic Crusade: Status Politics and the American Temperance Movement.* Urbana: University of Illinois Press.

———. 1981. *The Culture of Public Problems: Drinking-Driving and the Symbolic Order.* Chicago: University of Chicago Press.

Himmelstein, Jerome. 1978. "Drug Politics Theory." *Journal of Drug Issues* 8.

———. 1983. *The Strange Career of Marihuana.* Westport, CT: Greenwood Press.

Levine, Harry Gene. 1984. "The Alcohol Problem in America: From Temperance to Alcoholism." *British Journal of Addiction* 84: 109–119.

———. 1992. "Temperance Cultures: Concern About Alcohol Problems in Nordic and English-Speaking Cultures." In G. Edwards et al., eds., *The Nature of Alcohol and Drug Related Problems.* New York: Oxford University Press.

Levine, Harry Gene, and Craig Reinarman. 1991. "From Prohibition to Regulation: Lessons from Alcohol Policy for Drug Policy." *Milbank Quarterly* 69: 461–494.

MacAndrew, Craig, and Robert Edgerton. 1969. *Drunken Comportment.* Chicago: Aldine.

Marcuse, Herbert. 1964. *One-Dimensional Man: Studies in the Ideology of Advanced Industrial Society.* Boston: Beacon Press.

Mills, C. Wright. 1940. "Situated Actions and Vocabularies of Motive." *American Sociological Review* 5: 904–913.

Molotch, Harvey, and Marilyn Lester. 1974. "News as Purposive Behavior: On the Strategic Uses of Routine Events, Accidents, and Scandals." *American Sociological Review* 39: 101–112.

Morgan, Patricia. 1978. "The Legislation of Drug Law: Economic Crisis and Social Control." *Journal of Drug Issues* 8: 53–62.

Musto, David. 1973. *The American Disease: Origins of Narcotic Control.* New Haven, CT: Yale University Press.

National Institute on Drug Abuse. 1990. *National Household Survey on Drug Abuse: Main Findings 1990.* Washington, DC: U.S. Department of Health and Human Services.

Peele, Stanton. 1989. *The Diseasing of America: Addiction Treatment Out of Control.* Lexington, MA: Lexington Books.

Reinarman, Craig. 1995. "The 12-Step Movement and Advanced Capitalist Culture: Notes on the Politics of Self-Control in Postmodernity." In B. Epstein, R. Flacks, and M. Darnovsky, eds., *Contemporary Social Movements and Cultural Politics.* New York: Oxford University Press.

Reinarman, Craig, and Ceres Duskin. 1992. "Dominant Ideology and Drugs in the Media." *International Journal on Drug Policy* 3: 6–15.

Reinarman, Craig, and Harry Gene Levine. 1989a. "Crack in Context: Politics and Media in the Making of a Drug Scare." *Contemporary Drug Problems* 16: 535–577.

———. 1989b. "The Crack Attack: Politics and Media in America's Latest Drug Scare." In Joel Best, ed., *Images of Issues: Typifying Contemporary Social Problems,* pp.115–137. New York: Aldine de Gruyter.

Room, Robin G. W. 1991. "Cultural Changes in Drinking and Trends in Alcohol Problems Indicators: Recent U.S. Experience." In Walter B. Clark and Michael E. Hilton, eds., *Alcohol in America: Drinking Practices and Problems,* pp. 149–162. Albany: State University of New York Press.

Rumbarger, John J. 1989. *Profits, Power, and Prohibition: Alcohol Reform and the Industrializing of America, 1800–1930.* Albany: State University of New York Press.

Tijo, J. H., W. N. Pahnke, and A. A. Kurland. 1969. "LSD and Chromosomes: A Controlled Experiment." *Journal of the American Medical Association* 210: 849.

Weber, Max. 1985 (1930). *The Protestant Ethic and the Spirit of Capitalism.* London: Unwin.

Weil, Andrew. 1972. *The Natural Mind.* Boston: Houghton Mifflin.

16

Blowing Smoke: Status Politics and the Smoking Ban

JUSTIN L. TUGGLE AND MALCOLM D. HOLMES

Tuggle and Holmes's chapter on the struggle over cigarette smoking in America expands Reinarman's consideration to the licit drug realm, examining the struggle and counterstruggle over tobacco between the moral entrepreneurs and the status quo defenders. They note the medical, ethical, and socioeconomic arguments raised to sway public opinion and demonize public consumption of cigarettes. They mention the spate of claims put forward, pitting antismoking groups that have argued that secondhand smoke is toxic and that smokers should not be allowed to inflict their pollution onto others against opponents who have argued that the government should not legislate morality. These issues illustrate the concern that frequently arises in deviance, where society must balance the right of individual freedoms (the desire to smoke) against the needs of the common good (public health). This reading also shows the relation between claims making and social power, tracing the status of the social groups on each side of this campaign. Its fundamental message lies in how moral entrepreneurs use their status to attach deviant labels to the behavior of others, and, in so doing, to keep those others in a subordinate status position. They thus show deviance, like Quinney does, as a tool by which higher status groups retain and enforce their interests over subordinate groups.

O ver the past half century, perceptions of tobacco and its users have changed dramatically. In the 1940s and 1950s, cigarette smoking was socially accepted and commonly presumed to lack deleterious effects (see, e.g., Ram 1941). Survey data from the early 1950s showed that a minority believed cigarette smoking caused lung cancer (Viscusi 1992). By the late 1970s, however, estimates from survey data revealed that more than 90% of the population thought that this link existed (Roper Organization 1978). This and other harms associated with tobacco consumption have provided the impetus for an antismoking crusade that aims to normatively redefine smoking as deviant behavior (Markle and Troyer 1979).

There seems to be little question that tobacco is a damaging psychoactive substance characterized by highly adverse chronic health effects (Steinfeld 1991). In this regard, the social control movement probably makes considerable sense in terms of public policy. At the same time, much as ethnicity and religion played a significant role in the prohibition of alcohol (Gusfield 1963), social status may well play a part in this latest crusade.

Historically, attempts to control psychoactive substances have linked their use to categories of relatively powerless people. Marijuana use was associated with Mexican Americans (Bonnie and Whitebread 1970), cocaine with African Americans (Ashley 1975), opiates with Asians (Ben-Yehuda 1990), and alcohol with immigrant Catholics (Gusfield 1963). During the heyday of cigarette smoking, it was thought that

> Tobacco's the one blessing that nature has left for all humans to enjoy. It can be consumed by both the "haves" and "have nots" as a common leveler, one that brings all humans together from all walks of life regardless of class, race, or creed. (Ram 1941, p. 125)

But in contrast to this earlier view, recent evidence has shown that occupational status (Ferrence 1989; Marcus et al. 1989; Covey et al. 1992), education (Ferrence 1989; Viscusi 1992) and family income (Viscusi 1992) are related negatively to current smoking. Further, the relationships of occupation and education to cigarette smoking have become stronger in later age cohorts (Ferrence 1989). Thus we ask, *is the association of tobacco with lower-status persons a factor in the crusade against smoking in public facilities?* Here we examine that question in a case study of a smoking ban implemented in Shasta County, California.

STATUS POLITICS AND THE CREATION OF DEVIANCE

Deviance is socially constructed. Complex pluralistic societies have multiple, competing symbolic-moral universes that clash and negotiate (Ben-Yehuda 1990). Deviance is relative, and social morality is continually restructured. Moral, power, and stigma contests are ongoing, with competing symbolic-moral universes striving to legitimize particular lifestyles while making others deviant (Schur 1980; Ben-Yehuda 1990).

The ability to define and construct reality is closely connected to the power structure of society (Gusfield 1963). Inevitably, then, the distribution of deviance is associated with the system of stratification. The higher one's social position, the greater one's moral value (Ben-Yehuda 1990). Differences in lifestyles and moral beliefs are corollaries of social stratification (Gusfield 1963; Zurcher and Kirkpatrick 1976; Luker 1984). Accordingly, even though grounded in the system of stratification, status conflicts need not be instrumental; they may also be symbolic.

Social stigma may, for instance, attach to behavior thought indicative of a weak will (Goffman 1963). Such moral anomalies occasion status degradation ceremonies, public denunciations expressing indignation not at a behavior per se, but rather against the individual motivational type that produced it (Garfinkel 1956). The denouncers act as public figures, drawing upon communally shared experience and speaking in the name of ultimate values. In this respect, status degradation involves a reciprocal element: Status conflicts and the resultant condemnation of a behavior characteristic of a particular status category symbolically enhances the status of the abstinent through the degradation of the participatory (Garfinkel 1956; Gusfield 1963).

Deviance creation involves political competition in which moral entrepreneurs originate moral crusades aimed at generating reform (Becker 1963; Schur 1980; Ben-Yehuda 1990). The alleged deficiencies of a specific social group are revealed and reviled by those crusading to define their behavior as deviant. As might be expected, successful moral crusades are generally dominated by those in the upper social strata of society (Becker 1963). Research on the anti-abortion (Luker 1984) and antipornography (Zurcher and Kirkpatrick 1976) crusades has shown that activists in these movements are of lower socioeconomic status than their opponents, helping explain the limited success of efforts to redefine abortion and pornography as deviance.

Moral entrepreneurs' goals may be either assimilative or coercive reform (Gusfield 1963). In the former instance, sympathy to the deviants' plight engenders integrative efforts aimed at lifting the repentant to the superior moral plane allegedly held by those of higher social status. The latter strategy emerges when deviants are viewed as intractably denying the moral and status superiority of the reformers' symbolic-moral universe. Thus, whereas assimilative reform may employ educative strategies, coercive reform turns to law and force for affirmation.

Regardless of aim, the moral entrepreneur cannot succeed alone. Success in establishing a moral crusade is dependent on acquiring broader public support. To that end, the moral entrepreneur must mobilize power, create a perceived threat potential for the moral issue in question, generate public awareness of the issue, propose a clear and acceptable solution to the problem, and overcome resistance to the crusade (Becker 1963; Ben-Yehuda 1990).

THE STATUS POLITICS OF CIGARETTE SMOKING

The political dynamics underlying the definition of deviant behaviors may be seen clearly in efforts to end smoking in public facilities. Cigarettes were an insignificant product of the tobacco industry until the end of the 19th century, after which they evolved into its staple (U.S. Department of Health and Human Services 1992). Around the turn of the century, 14 states banned cigarette smoking and all but one other regulated sales to and possession by minors (Nuehring and Markle 1974). Yet by its heyday in the 1940s and 1950s, cigarette smoking was almost universally accepted, even considered socially desirable (Nuehring and Markle 1974; Steinfeld 1991). Per capita cigarette consumption in the United States peaked at approximately 4,300 cigarettes per year in the early 1960s, after which it declined to about 2,800 per year by the early 1990s (U.S. Department of Health and Human Services 1992). The beginning of the marked decline in cigarette consumption corresponded to the publication of the report to the surgeon general on the health risks of smoking (U.S. Department of Health, Education and Welfare 1964). Two decades later, the hazards of passive smoking were being publicized (e.g., U.S. Department of Health and Human Services 1986).

Increasingly, the recognition of the apparent relationship of smoking to health risks has socially demarcated the lifestyles of the smoker and nonsmoker, from widespread acceptance of the habit to polarized symbolic-moral universes. Attitudes about smoking are informed partly by medical issues, but perhaps even more critical are normative considerations (Nuehring and Markle 1974); more people have come to see smoking as socially reprehensible and deviant, and smokers as social misfits (Markle and Troyer 1979). Psychological assessments have attributed an array of negative evaluative characteristics to smokers (Markle and Troyer 1979). Their habit is increasingly thought unclean and intrusive.

Abstinence and bodily purity are the cornerstones of the nonsmoker's purported moral superiority (Feinhandler 1986). At the center of their symbolic-moral universe, then, is the idea that people have the right to breathe clean air in public spaces (Goodin 1989). Smokers, on the other hand, stake their claim to legitimacy in a precept of Anglo-Saxon political culture—the right to do whatever one wants unless it harms others (Berger 1986). Those sympathetic to smoking deny that environmental tobacco smoke poses a significant health hazard to the nonsmoker (Aviado 1986). Yet such arguments have held little sway in the face of counterclaims from authoritative governmental agencies and high status moral entrepreneurs.

The development of the antismoking movement has targeted a lifestyle particularly characteristic of the working classes (Berger 1986). Not only has there been an overall decline in cigarette smoking, but, as mentioned above, the negative relationships of occupation and education to cigarette smoking have become more pronounced in later age cohorts (Ferrence 1989). Moreover, moral entrepreneurs crusading against smoking are representatives of a relatively powerful "knowledge class," comprising people employed in areas such as education and the therapeutic and counseling agencies (Berger 1986).

Early remedial efforts focused on publicizing the perils of cigarette smokers, reflecting a strategy of assimilative reform (Neuhring and Markle 1974; Markle and Troyer 1979). Even many smokers expressed opposition to cigarettes and a generally repentant attitude. Early educative efforts were thus successful in decreasing cigarette consumption, despite resistance from the tobacco industry. Then, recognition of the adverse effects of smoking on nonusers helped precipitate a turn to coercive reform measures during the mid 1970s (Markle and Troyer 1979). Rather than a repentant friend in need of help, a new definition of the smoker as enemy emerged. Legal abolition of smoking in public facilities became one focus of social control efforts, and smoking bans in public spaces have been widely adopted in recent years (Markle and Troyer 1979; Goodin 1989).

The success of the antismoking crusade has been grounded in moral entrepreneurs' proficiency at mobilizing power, a mobilization made possible by highly visible governmental campaigns, the widely publicized health risks of smoking, and the proposal of workable and generally acceptable policies to ameliorate the problem. The success of this moral crusade has been further facilitated by the association of deviant characteristics with those in lower social strata, whose stigmatization reinforces existing relations of power and prestige. Despite the formidable resources and staunch opposition of the tobacco industry, the tide of public opinion and policy continues to move toward an antismoking stance.

RESEARCH PROBLEM

The study presented below is an exploratory examination of the link between social status and support for a smoking ban in public facilities. Based on theorizing about status politics, as well as evidence about patterns of cigarette use, it was predicted that supporters of the smoking ban would be of higher status than those who opposed it. Further, it was anticipated that supporters of the ban would be more likely to make negative normative claims denouncing the allegedly deviant qualities of smoking, symbolically enhancing their own status while lowering that of their opponents.

The site of this research was Shasta County, California. The population of Shasta County is 147,036, of whom 66,462 reside in its only city, Redding (U.S. Bureau of the Census 1990). This county became the setting for the implementation of a hotly contested ban on smoking in public buildings.

In 1988, California voters passed Proposition 99, increasing cigarette taxes by 25 cents per pack. The purpose of the tax was to fund smoking prevention and treatment programs. Toward that end, Shasta County created the Shasta County Tobacco Education Program. The director of the program formed a coalition with officials of the Shasta County chapters of the American Cancer Society and American Lung Association to propose a smoking ban in all public buildings. The three groups formed an organization to promote that cause, Smoke-Free Air For Everyone (SAFE). Unlike other bans then in effect in California, the proposed ban included restaurants and bars, because its proponents considered these to be places in which people encountered significant amounts of secondhand smoke. They procured sufficient signatures on a petition to place the measure on the county's general ballot in November 1992.

The referendum passed with a 56% majority in an election that saw an 82% turnout. Subsequently, the Shasta County Hospitality and Business Alliance, an antiban coalition, obtained sufficient signatures to force a special election to annul the smoking ban. The special election was held in April 1993. Although the turnout was much lower (48%), again a sizable majority (58.4%) supported the ban. The ordinance went into effect on July 1, 1993.

ANALYTIC STRATEGY

... [D]ata were analyzed in our effort to ascertain the moral and status conflicts underlying the Shasta County smoking ban ... [based on] interviews with five leading moral entrepreneurs and five prominent status quo defenders.[1] These individuals were selected through a snowball sample, with the original respondents identified through interviews with business owners or political advertisements in the local mass media. The selected respondents repeatedly surfaced as the leading figures in their respective coalitions. Semistructured interviews were conducted to determine the reasons underlying their involvement. These data were critical to understanding how the proposed ban was framed by small groups of influential proponents and opponents; it was expected that their concerns would be reflected in the larger public debate about the ban.

FINDINGS

Moral Entrepreneur/Status Quo Defender Interviews

The moral entrepreneurs and status quo defenders interviewed represented clearly different interests. The former group included three high-level administrators in the county's chapters of the American Cancer Society and American Lung Association. A fourth was an administrator for the Shasta County Tobacco Education Project. The last member of this group was a pulmonary physician affiliated with a local hospital. The latter group included four bar and/or restaurant owners and an attorney who had been hired to represent their interests. Thus the status quo defenders were small business owners who might see their economic interests affected adversely by the ban. Importantly, they were representatives of a less prestigious social stratum than the moral entrepreneurs.

The primary concern of the moral entrepreneurs was health. As one stated,

> I supported the initiative to get the smoking ban on the ballot because of all the health implications that secondhand smoke can create. Smoking and secondhand smoke are the most preventable causes of death in this nation.

Another offered that

> On average, secondhand smoke kills 53,000 Americans each year. And think about those that it kills in other countries! It contains 43 cancer-causing chemical agents that have been verified by the Environmental Protection Agency. It is now listed as a Type A carcinogen, which is the same category as asbestos.

Every one of the moral entrepreneurs expressed concern about health issues during the interviews. This was not the only point they raised, however. Three of the five made negative normative evaluations of smoking, thereby implicitly degrading the status of smokers. They commented that "smoking is no longer an acceptable action," that "smoke stinks," or that "it is just a dirty and annoying habit." Thus, whereas health was their primary concern, such comments revealed the moral entrepreneurs' negative view of smoking irrespective of any medical issues. Smokers were seen as engaging in unclean and objectionable behavior—stigmatized qualities defining their deviant social status.

The stance of the status quo defenders was also grounded in two arguments. All of them expressed concern about individual rights. As one put it,

> I opposed that smoking ban because I personally smoke and feel that it is an infringement of my rights to tell me where I can and cannot smoke. Smoking is a legal activity, and therefore it is unconstitutional to take that right away from me.

Another argued that

> Many people have died for us to have these rights in foreign wars and those also fought on American soil. Hundreds of thousands of people thought that these rights were worth dying for, and now some small group of people believe that they can just vote away these rights.

Such symbolism implies that smoking is virtually a patriotic calling, a venerable habit for which people have been willing to forfeit their lives in time of war. In the status quo defenders' view, smoking is a constitutionally protected right.

At the same time, each of the status quo defenders was concerned about more practical matters, namely business profits. As one stated, "my income was going to be greatly affected." Another argued,

> If these people owned some of the businesses that they are including in this ban, they would not like it either. By taking away the customers that smoke, they are taking away the mainstay of people from a lot of businesses.

The competing viewpoints of the moral entrepreneurs and status quo defenders revealed the moral issues—health versus individual rights—at the heart of political conflict over the smoking ban. Yet it appears that status issues also fueled the conflict. On the one hand, the moral entrepreneurs denigrated smoking, emphasizing the socially unacceptable qualities of the behavior and symbolically degrading smokers' status. On the other hand, status quo defenders were concerned that their livelihood would be affected by the ban. Interestingly, the occupational status of the two groups differed, with the moral entrepreneurs representing the new knowledge class, the status quo defenders a lower stratum of small business owners. Those in the latter group may not have been accorded the prestige and trust granted those in the former (Berger 1986). Moreover, the status quo defenders' concern about business was likely seen as self-aggrandizing.

SUMMARY AND DISCUSSION

This research has examined the moral and status politics underlying the implementation of a smoking ban in Shasta County, California. Moral entrepreneurs crusading for the ban argued that secondhand smoke damages health, implicitly grounding their argument in the principle that people have a right to a smoke-free environment. Status quo defenders countered that smokers have a constitutional right to indulge wherever and whenever they see fit. Public discourse echoed these themes, as seen in the letters to the editor of the local newspaper. Thus debate about the smoking ban focused especially on health versus smokers' rights; yet evidence of social status differences between the competing symbolic-moral universes also surfaced. Competing symbolic-moral universes are defined not only by different ethical viewpoints on a behavior, but also by differences in social power—disparities inevitably linked to the system of stratification (Ben-Yehuda 1990). Those prevailing in moral and stigma contests typically represent the higher socioeconomic echelons of society.

The moral entrepreneurs who engineered the smoking ban campaign were representatives of the prestigious knowledge class, including among their members officials from the local chapters of respected organizations at the forefront of the national antismoking crusade. In contrast, the small business owners who were at the core of the opposing coalition, of status quo defenders, represented the traditional middle class. Clearly, there was an instrumental quality to the restaurant and bar owners' stance, because they saw the ban as potentially damaging to their business interests. But they were unable to shape the public debate, as demonstrated by the letters to the editor.

In many respects, the status conflicts involved in the passage of the Shasta County smoking ban were symbolic. The moral entrepreneurs focused attention on the normatively undesirable qualities of cigarette smoking, and their negative normative evaluations of smoking were reflected in public debate about the ban. Those who wrote in support of the ban more frequently offered negative normative evaluations than antiban writers; their comments degraded smoking and, implicitly, smokers. Since the advent of the antismoking crusade in the United States, smoking has come to be seen as socially reprehensible, smokers as social misfits characterized by negative psychological characteristics (Markle and Troyer 1979).

Ultimately, a lifestyle associated with the less educated, less affluent, lower occupational strata was stigmatized as a public health hazard and targeted for coercive reform. Its deviant status was codified in the ordinance banning smoking in public facilities, including restaurants and bars. The ban symbolized the deviant status of cigarette smokers, the prohibition visibly demonstrating the community's condemnation of their behavior. Further, the smoking ban symbolically amplified the purported virtues of the abstinent lifestyle. A political victory such as the passage of a law is a prestige-enhancing symbolic triumph that is perhaps even more rewarding than its end result (Gusfield 1963). The symbolic nature of the ban serendipitously surfaced in another way during one author's unstructured observations in 42 restaurants and 21 bars in the area: Whereas smoking was not observed in a single restaurant, it occurred without sanction in all but one of the bars. Although not deterring smoking in one of its traditional bastions, the ban called attention to its deviant quality and, instrumentally, effectively halted it in areas more commonly frequented by the abstemious.

Although more systematic research is needed, the findings of this exploratory case study offer a better understanding of the dynamics underlying opposition to smoking and further support to theorizing about the role of status politics in the creation of deviant types. Denunciation of smoking in Shasta County involved not only legitimate allegations about public health, but negative normative evaluations of those engaged in the behavior. In the latter regard, the ban constituted a status degradation ceremony, symbolically differentiating the pure and abstinent from the unclean and intrusive. Not coincidentally, the stigmatized were more likely found among society's lower socioeconomic strata, their denouncers among its higher echelons.

Certainly the class and ethnic antipathies underlying attacks on cocaine and opiate users earlier in the century were more manifest than those revealed in the crusade against cigarette smoking. But neither are there manifest status conflicts in the present crusades against abortion (Luker 1984) and pornography (Zurcher and Kirkpatrick 1976); yet the underlying differences of status between opponents in those movements are reflected in their markedly different symbolic-moral universes, as was the case in the present study.

This is not to suggest that smoking should be an approved behavior. The medical evidence seems compelling: Cigarette smoking is harmful to the individual smoker and to those exposed to secondhand smoke. However, the objective harms of the psychoactive substance in question are irrelevant to the validity of our analysis, just as they were to Gusfield's (1963) analysis of the temperance movement's crusade against alcohol use. Moreover, it is not our intention to imply that the proban supporters consciously intended to degrade those of lower social status. No doubt they were motivated primarily by a sincere belief that smoking constitutes a public health hazard. In the end, however, moral indignation and social control flowed down the social hierarchy. Thus we must ask: Would cigarette smoking be defined as deviant if there were a positive correlation between smoking and socioeconomic status?

NOTE

1. Although the term moral entrepreneur is well established in the literature on deviance, there seems to be little attention to or consistency in a corresponding term for the interest group(s) opposing them. Those that have been employed, such as "forces for the status quo" (Markle and Troyer 1979), tend to be awkward, "Status quo defenders" is used here for lack of a simpler or more common term.

REFERENCES

Ashley, Richard. 1975. *Cocaine: Its History, Uses, and Effects.* New York: St. Martin's Press.

Aviado, Domingo M. 1986. "Health Issues Relating to 'Passive' Smoking." Pp.137–165 in *Smoking and Society: Toward a More Balanced Assessment,* edited by Robert D. Tollison. Lexington, MA: Lexington Books.

Becker, Howard S. 1963. *Outsiders: Studies in the Sociology of Deviance.* New York: Free Press.

Ben-Yehuda, Nachman. 1990. *The Politics and Morality of Deviance: Moral Panics, Drug Abuse, Deviant Science, and Reversed Stigmatization.* Albany, NY: State University of New York Press.

Berger, Peter L. 1986. "A Sociological View of the Antismoking Phenomenon." Pp. 225–240 in *Smoking and Society: Toward a More Balanced Assessment,* edited by Robert D. Tollison. Lexington, MA: Lexington Books.

Bonnie, Richard J., and Charles H. Whitebread II. 1970. "The Forbidden Fruit and the Tree of Knowledge: An Inquiry into the Legal History of American Marihuana Prohibition." *Virginia Law Review* 56: 971–1203.

Covey, Lirio S., Edith A. Zang, and Ernst L. Wynder. 1992. "Cigarette Smoking and Occupational Status: 1977 to 1990." *American Journal of Public Health* 82: 1230–1234.

Feinhandler, Sherwin J. 1986. *The Social Role of Smoking.* Pp. 167–187 in *Smoking and Society: Toward a More Balanced Assessment,* edited by Robert D. Tollison. Lexington, MA: Lexington Books.

Ferrence, Roberta G. 1989. *Deadly Fashion: The Rise and Fall of Cigarette Smoking in North America.* New York: Garland.

Garfinkel, Harold. 1956. "Conditions of Successful Degradation Ceremonies." *American Journal of Sociology* 61: 402–424.

Goffman, Erving. 1963. *Stigma: Notes on the Management of Spoiled Identity.* Englewood Cliffs, NJ: Prentice Hall.

Goodin, Robert E. 1989. *No Smoking: The Ethical Issues.* Chicago: University of Chicago Press.

Gusfield, Joseph R. 1963. *Symbolic Crusade: Status Politics and the American Temperance Movement.* Urbana, IL: University of Illinois Press.

Luker, Kristin. 1984. *Abortion and the Politics of Motherhood.* Berkeley, CA: University of California.

Marcus, Alfred C., Donald R. Shopland, Lori A. Crane, and William R. Lynn. 1989. "Prevalence of Cigarette Smoking in United States: Estimates from the 1985 Current Population Survey." *Journal of the National Cancer Institute* 81: 409–414.

Markle, Gerald E., and Ronald J. Troyer. 1979. "Smoke Gets in Your Eyes: Cigarette Smoking as Deviant Behavior." *Social Problems* 26: 611–625.

Neuhring, Elane, and Gerald E. Markle. 1974. "Nicotine and Norms: The Re-Emergence of a Deviant Behavior." *Social Problems* 21: 513–526.

Ram, Sidney P. 1941. *How to Get More Fun Out of Smoking.* Chicago: Cuneo.

Roper Organization. 1978, May. *A Study of Public Attitudes Toward Cigarette Smoking and the Tobacco Industry in 1978, Volume 1.* New York: Roper.

Schur, Edwin M. 1980. *The Politics of Deviance: Stigma Contests and the Uses of Power.* New York: Random House.

Steinfeld, Jesse. 1991. "Combating Smoking in the United States: Progress Through Science and Social Action." *Journal of the National Cancer Institute* 83: 1126–1127.

U.S. Bureau of the Census. 1990. *General Population Characteristics.* Washington, DC: U.S. Government Printing Office.

U.S. Department of Health, Education and Welfare. 1964. *Smoking and Health: Report of the Advisory Committee to the Surgeon General of the Public Health Service.* Washington, DC: U.S. Government Printing Office.

U.S. Department of Health and Human Services. 1986. *The Health Consequences of Involuntary Smoking. A Report of the Surgeon General.* Washington, DC: U.S. Government Printing Office.

U.S. Department of Health and Human Services. 1992. *Smoking and Health in the Americas. A 1992 Report of the Surgeon General, in Collaboration with the Pan American Health Organization.* Washington, DC: U.S. Government Printing Office.

Viscusi, W. Kip. 1992. *Smoking: Making the Risky Decision.* New York: Oxford University Press.

Zurcher, Louis A. Jr., and R. George Kirkpatrick. 1976. *Citizens for Decency: Antipornography Crusades as Status Defense.* Austin, TX: University of Texas Press.

17

The Cyberporn and Child Sexual Predator Moral Panic

ROBERTO HUGH POTTER AND LYNDY A. POTTER

Potter and Potter discuss another aspect of constructing deviance in their discussion of the cyberporn moral panic. Moral panics arise during times of unsettling social change and represent irrational and overblown reactions to new phenomena in society. Often built on some kernel of truth as Reinarman notes, they escalate out of proportion and draw members of the public into their extreme scare. We have all seen the concern that has arisen over the range of information and social contacts available to Internet users, even children. Parents have become especially worried about the unfettered access their children have to immoral and unsafe influences through their computers. These kinds of concerns grow exponentially with the ongoing development of Usenet groups, MySpace sites, YouTube videos, child sexual predators, and unsavory influences.

In this chapter Potter and Potter examine some of the claims fostered in the media and public thinking about Internet pornography and child predators and contrast these with actual facts. Not surprisingly, they discover that the cyberporn moral panic attributes greater availability of pornography to casual Web browsers than is actually true. They then examine, item by item, the research on sexual solicitation of children over the Internet such as unwanted exposure to sexual material, cyberstalking, and people who traveled to meet Internet "friends" face-to-face, showing these also to be less than reported. They conclude that the urban mythology surrounding cyberporn moral panic is rooted in the pressure felt by middle-class parents to meet high parenting standards at a time when many are working full time and cannot constantly police their children's academic and recreational behavior.

Cyberporn is the term given to "pornographic" bulletin boards, digitized images, and "interactive" sites available through locations on the World-Wide Web (WWW). More specifically, the term refers to such information located on "nodes" of the broader information superhighway. Some of the information and images available there are nothing more than what is available in most news

Reprinted by permission of Transaction Publishers. "The Moral Panic Over Internet Cyberporn and the Sexual Exploitation of Children," Roberto Hugh Potter and Lyndy A. Potter, *Sexuality & Culture*, 5(3), 31–48. Copyright © 2001 by Transaction Publishers.

agencies throughout Australia or at news counters in North American areas. Others clearly violate the guidelines for on-line services in Australia, and apparently at least some U.S. states. Unlike the concepts of "pornography," "obscenity," or "(in)decency," the components of the Internet are definable. Beyond that, however, they become a bit more confusing. Here we will briefly outline some of the major components related to the issue of cyberporn for the interested reader.

The Australian version of *Time* (Elmer-Dewitt, 1995)... published a cover story on Cyberporn that featured prominently stories of "unsolicited" thumbnail pictures of pornographic acts being emailed to pre-teen children and the contact of young boys by pedophiles seeking prey. Stories of library computers being used by school boys to find and down-load "pornography" were then being featured in the Australian media as almost staple "filler" for slow news days. Generally these referred to the same incident at a metropolitan school, with requisite interviews with local school librarians who would agree that, yes, the same thing *could* happen here. To a person, however, they always noted that it had not yet happened as far as they knew. In fact, there were very few documented cases of school-aged children accessing "pornographic" sites either at school, public libraries, or in their homes.

It was at this point that we became interested in the role that children, "pornography," and the Internet were playing in the shaping of public opinion ahead of governmental policy on the regulation of the Internet in Australia. This was toward the end of the Australian Federal Parliamentary inquiry into the regulation of On-Line Services. Upon returning to the United States, the debate over the Communications Decency Act (CDA) had just resulted in the passing of that act. Shortly after, the law suits began that ultimately led to the Supreme Court of the United States nullifying the CDA. What became interesting in both instances was the "generation gap" in knowledge about the new technology between children and parents and the role that sexual content played in establishing the "danger" of the new technology. That there was precious little systematic empirical evidence to back the isolated "horror stories" led us to examine the role that "mythology" about the sexual content of the Internet was playing in warning parents about the dangers of allowing their children unsupervised and unregulated access to this new technology. Therein lies the crux of this article.

The *Time* (Elmer-Dewitt, 1995) article began by noting a very important, though not necessarily recognized, qualification about the infamous Rimm (1995) study, namely, that most of the data about pornography were collected from Usenet, not the Internet. If that begins to confuse you, you are among most casual users of the information superhighway. These are important distinctions because each has a set of concerns about content that often become fused into an overall concern with something called the "Internet," though it may be something related, but distinct.

There are many "dangerous" ideas, products, and people that might be encountered by an individual "surfing" the web. Among these are instructions in violence, information from hate groups and cults, gambling, invasion of privacy, infringement of intellectual property rights, gender imbalance of gender equality, cultural imperialism, and the focus of this paper, forms of sexual information and representation. In short, every conceivable form of dangerous information, activity, and persons found in society might also be found in cyberspace....

CYBERPORN AS A MEDIA-INDUCED MORAL PANIC

Fear of cyberporn or pornography on the Internet is a form of "moral panic." The term "moral panic" is taken from the work of Stanley Cohen (1972). Cohen described a condition, situation or group of persons who become defined as a threat to social values and interests as providing the grounds for a moral panic. Their activities, real or imagined, are then presented in the mass media in a stereotypic and stylized manner. "Moral entrepreneurs," such as religious leaders, politicians, "other right-thinking people" and "experts" are then called upon to provide appropriate answers to the problem. Some form of solution to the problem is developed and the problem then disappears from sight for some period of time.

Cohen pointed out that the source of the threat may be quite novel, such as the Internet. Or the problem may be something that has been there for quite a while but now, for whatever reason, becomes publicly a problem. Some moral panics pass just as quickly as they appear on the scene. Others may have serious impacts with long-lasting effects. These effects may be found in the formulation of social and legal policy which have far-reaching implications. The attention paid to the issues of cyberporn over the past several years by the North American and Australasian media suggests that it is a full-blown moral panic. Almost all television news programs and current affairs magazines on these two continents have carried some item warning of the possibility that school students or children will find and download "pornographic" materials from the Internet. Or, in a more sinister vein, that pornography will impose itself on child users of the Internet, or that children will be lured away by pedophiles.

In the United States, Stallings (1990) and Best (1990) have outlined the ways in which media organizations construct patterns and degrees of risk by focusing on similarities among a wide range of individual incidents to construct a risk pattern, while simultaneously ignoring evidence which would render a pattern problematic. Both of these writers demonstrate the activities of news organizations in an attempt to define the situation for their consumers. In this way media organizations become "secondary claims-makers," choosing which "primary" claims they will "legitimize" and which reforms they will push. Best (1990) again develops the idea of a cultural "tool kit" of "templates," which media organizations can fit onto a putative social problem in order to render it understandable to the general public. We will return to the application of this template and why it might occur later....

Anecdotes and media reports abound of sexual predators using the Internet to contact and lure children and teens for sexual exploitation, yet there remains very little empirical research on the subject. An early systematic excursion into the sexual content of the Web was provided in the lead-up to the "Investigation into the Content of On-line Services." The Australian Office of Film and Literature Classification (OFLC) conducted a 27-hour targeted search for WWW sites containing sexually explicit materials for the ABA. The OFLC is responsible for classifying all videos, films, and video games shown or hired in Australia. These classifications include an "X" rating for videos and Categories 1 (e.g., Playboy) and 2 (i.e., "hard core," sexually explicit) for print literature, as well as the reasons why material might be "refused classification" (banned), which specify the level of sexual content allowable. The results of this research included (ABA, 1995: 39–40):

No child pornography located after seven hours of specific searching;
An extensive list of bomb-making recipes was found after one hour of searching;

Categories 1 and 2 material was advertised as being available or was suggested to be available on the Web. However, after searching for seven hours, none was able to be immediately downloaded;

Approximately 90 percent of the category 1 and category 2 level listings on the Web had on-screen warnings of content and an age recommendation, while 10 percent had no specific warnings but contained material of an unrestricted (non-sexual) nature when accessed;

Approximately 60 percent of sites required the searcher to apply to an address with details and/or credit information before the material could be obtained; ...

In the United States, careful empirical research on children and the dangers of the Internet has been rare. Responding to national interest, the National Center for Missing and Exploited Children (NCMEH) contracted with the Crimes Against Children Research Center at the University of New Hampshire to conduct a national survey of adolescent Internet users. Finkelhor, Mitchell, and Wolak (2000) reported the results of the first U.S. national study of on-line victimization of minors (n = 1501; ages 10–17) using the Internet on a regular basis. The range of victimizations included sexual solicitations and approaches, unsolicited receipt of sexually explicit images or messages, and other forms of threat or harassment (Finkelhor et al., 2000: x). They found that 19 percent of the respondents received a sexual solicitation or approach via the Internet in the previous year. These included situations where someone tried to get the teens to talk about sex when the teen did not want to, or asked unwanted questions about the teen's intimate life, solicitations to do sexual things the teen did not want to do, invitations to run away (for vulnerable youths), and close friendships formed with adults on-line when these had included "sexual overtures."

Two-thirds (65%) of the sexual overture incidents occurred in "chat rooms," with just under one-quarter (24%) coming through "instant message" facilities. In 10 percent of the cases, the perpetrator asked for a physical meeting, and a handful attempted other forms of communication. "In most instances, the youth ended the solicitations, using a variety of strategies like logging off, leaving the site, or blocking the person" (Finkelhor et al., 2000: 4). Just under half (49%) of the youths stated that they had told no one about the incident, and only 10 percent of these incidents were reported to any authority figure or organization. Three-quarters (75%) of the youth said they had no or minor reactions ("not very upset or afraid") as a result of the incident. Among those solicited five or more times, 17 percent demonstrated five or more symptoms of depression at the time of the interview.

A key ingredient in the template of Internet predator stories is the adult soliciting a child for sexual purposes. In the Finkelhor et al. (2000: 6–7) study, 3 percent of the youths reported forming "close online friendships" with adults, mostly in the young adult age range (18–25). Only two of the friendships formed "may have had sexual aspects." The authors point out that, while minors may be forming friendships online, the proportion that become sexual appears to be very small. Seven youths (0.4% of the sample) reported being encouraged to run away from home by someone they met via the Internet. Adults comprised one-quarter of the sexual solicitations reported by the teen respondents, with most of those reported to be in the young adult age group. Only a handful of the sexual solicitors were over the age of 25. Juveniles accounted for 48 percent of the identified perpetrators (27% of unknown age), with around two-thirds being male. Overall, it would appear that the levels of sexual solicitation and the profile of the solicitors are similar to those encountered in physical space by teens.

One-quarter (25%) of the youths reported receiving unwanted exposure to sexual material while online. This exposure consisted primarily of images of naked people (94%), while just over one-third (38%) received images of people having sex (it was possible to have more than one experience). Exposure occurred in two primary manners. First, 281 (19%) of the youths had been exposed to unwanted materials while surfing the net. Just under half (47%) of these were discovered while engaged in a search for other materials, and 17 percent each as either a misspelled address or as a result of clicking on an address within a site. Just over one-third (36%) of those who stumbled upon the images found the experience distressing, while about one-quarter (24%) of those in the latter category were distressed. Among those who received unsolicited materials via their private email address (n = 112; 7%), over half (58%), found the experience distressing (Finkelhor et al., 2000: 13–19).

"Cyberstalking" as harassment or using the Internet to spread rumors about someone (Ogilvie, 2000) was experienced by 95 of the respondents (6%). One-third of these (n = 37) reported the incident to be very distressing (Finkelhor et al., 2000: 21–26). Among the perpetrators of the cyberstalking, 63 percent were reported to be other minors, and 14 percent aged 18 or over. While most of the harassed teens had little or no reaction to the event, nearly one-third (31%) were very/extremely upset, and one-fifth (19%) were very/extremely afraid by the incident. Finkelhor et al. suggest that this area of harassment, which we have included under cyberstalking, is an area that deserves further attention with regard to children.

Additionally, the NCMEH identified 785 "traveler cases" in 1999; "cases in which a child or adult traveled to physically meet with someone he or she had first encountered on the Internet" (Allen, in Finkelhor et al., 2000: vi). These reports consisted of 302 from the Federal Bureau of Investigation (FBI), 272 from local police agencies, 186 reported directly to the NCMEH, and an additional 25 from news organizations (Allen, in Finkelhor et al., 2000: vii). Allen did allow that some of these were duplicate cases. Unfortunately, he did not report how many of these physical contact situations eventuated in actual sexual contact, consensual or non-consensual, between the "travelers." His information also failed to identify the age relationship between the travelers, which is a deficit given the results of the study to which his comments are an introduction.

These findings are consistent with what has been reported in non-sensationalized journalistic accounts of cyberporn and the efforts of one of the current authors to access such sites. With this seeming lack of supporting evidence for claims of pernicious pornography leaping out at uninterested Internet users, one is left to wonder why there has been such an emphasis on this topic in the media and the resultant regulatory efforts such as the Communications Decency Act.

THE BEGINNINGS OF THE MORAL PANICS

"Pornography" has been linked to every technological innovation affecting visual communication from cave paintings to cyberspace. Steven Marcus (1966) recounts the legend that the second set of books off the Guttenberg press was a collection of pornography. As printing became more economical, not only pornography became of concern. Duncan Chappell (1993) has analyzed the Australian concern over video game content as a continuation of the concern with "penny dreadful" novels ("pulp fiction") which became economically feasible in the early part of the 20th century thanks to advances in printing and paper processing. The content of many of these novels was considered obscene and distasteful to middle-class society in North America and the United Kingdom. It would seem that every technological innovation leads to a concern with sex (especially "pornography") and children.

Durkin and Bryant (1995) have employed Ogburn's concept of "culture lag" to discuss the impact of technological change on sexual activity and attitudes toward sexual behavior. They argue that each technological innovation produces new ways of experiencing sexuality. Among the examples they cite are the opportunities for sexual liaisons presented by the mobility of the automobile, the proliferation of sexually explicit videos following the mass marketing of video cassette players, and the development of "phone sex" following developments in telephone services. They go on to speculate about the future of "cybersex" as virtual reality technology improves with computer design changes. Throughout their discussion they note the anxiety each innovation produces in segments of the population regarding the danger posed by the new developments. However, they do not address how these perceived dangers are expressed by the public generally and moral entrepreneurs specifically. For a move from the more concrete to the symbolic, we turn to the work of Gary Alan Fine.

CYBERPORN AS MORAL MYTHOLOGY FOR MIDDLE-CLASS PARENTS IN THE COMPUTER AGE

Fine (1979, 1981), has pointed out how many urban myths focus on the impersonality of modern life and rapid technological change. In his analysis, such myths often help us to face the contradictions of the modern world, thus reducing the strain we feel from conflicting sets of values and demands. In two essays, one dealing with "Kentucky Fried Rat" and one on "Cokelore" (about soft drinks), Fine argues that we are dealing with a move from a more personal and home-centered existence to the world of mass production and impersonal service. Best and Horiuchi (1985) have employed these arguments to examine the rise of primarily suburban folk tales regarding the dangers of Halloween "trick or treating." Similar myths are not far from the surface in either the United States or Australia. Australians often take the top off of a pie (meat or vegetable) to see whether or not there are materials there that shouldn't be. There have been mass-produced meat pie, canned tomato, pineapple and other such scares in Australia within the past decade.

Fine argues that urban myths help us to expect such anomalies in the world of impersonally produced food and other products formally associated with home life. Another substantial component of his argument has to do with the notion that many of these products are being produced by technology that is beyond the control of most of us. This is where the argument is especially germane to the cyberporn moral panic. The recognition that parents are not as acquainted with information technology as their children are is always highlighted. Almost all of the media items regarding cyberporn end with admonitions to parents about monitoring their children's activities with computers more closely. In short, these moral panics are forming the base of a new urban mythology about the dangers of the increasing generation gap regarding information technology and computers between today's parents and their children. This is essentially the same message parents received about their knowledge of rock and roll and sex in the mid-1950s (and again with PMRC), then drugs (the mid-1960s). Same message, different topics. Both of these developments were tied to proliferating and changing technology in the recording and broadcast sectors.

The clear message is that parents need to monitor their children's activities and react to them when they see their children heading in an objectionable direction. The equally clear signal is that children are growing up faster than they did in "our" day and we need to stay one step ahead of them in order to control them. In a sense, the message remains the same, only the technology changes.

New technology brings uncertainty and danger if parents do not police its employment in the home.

The urban mythology surrounding cyberporn is also essentially focused on a middle-class moral panic. Although North America and Australia may have some of the highest proportional participation rates in Internet usage, it is still primarily a middle-class phenomenon. This is especially true when we speak of home-based computer usage.... [where] such households [are] likely to turn a blind eye to increased telephone bills and credit card charges for downloaded information from a bulletin board in some other part of the world. It is much more likely that middle-class parents would not question such charges showing up on their bills. And that is the point of the urban mythology—such parents should be taking a closer notice of their children's activity.

Like much historically recent mythology, the urban mythology that is emerging from the moral panic surrounding cyberporn is aimed at reinforcing "good parenting" in the present time. Cyberporn represents the newest danger in a long line of dangers to the innocence of childhood. It is also the latest in a long line of challenges to parental authority and notions of what it is to be a "good" parent. In a time when many parents are finding it difficult to spend any time with their children, let alone "quality time," cyberporn is a moral prick to the conscience of those who would be good parents.

REFERENCES

ABA (Australian Broadcasting Association). 1995. "Code of Practice for On-line Services." Canberra: Australian Communications and Media Authority.

Best, Joel. 1990. *Threatened Children*. Chicago: University of Chicago Press.

Best, Joel, and Gerald T. Horiuchi. 1985. "The Razor in the Apple: The Social Construction of Urban Legends." *Social Problems* 32: 488–99.

Chappell, Duncan. 1993. "The Cost of Crime." *Criminology Australia* 4(2): 20–24.

Cohen, Stanley. 1972. *Folk Devils and Moral Panics*. London: MacGibbon and Kee.

Durkin, Keith F., and Clifton D. Bryant. 1995. "'Log on to Sex': Some Notes on the Carnal Computer and Erotic Cyberspace as an Emerging Research Frontier." *Deviant Behavior* 16: 179–200.

Elmer-Dewitt, Phillip. 1995. "On a Screen Near You: Cyberporn." *Time*, July 3: 38–45.

Fine, Gary Alan. 1979. "Cokelore and Coke Law: Urban Belief Tales and the Problem of Multiple Origins." *Journal of American Folklore* 92: 477–82.

———. 1981. "The Kentucky Fried Rat: Legends and Modern Society." *Journal of the Folklore Institute* 192: 222–43.

Finkelhor, David., Kimberly J. Mitchell, and Janis Wolak. 2000. *Online Victimization: A Report on the Nation's Youth*, Washington, DC: National Center for Missing and Exploited Children.

Marcus, Steven. 1966. *The Other Victorians*. New York: Basic.

Ogilvie, Emma. 2000. "Cyberstalking." *Trends and Issues in Crime and Criminal Justice*. No. 166. Canberra: The Australian Institute of Criminology.

Rimm, Marty, 1995. "Marketing Pronography on the Information Superhighway." *Georgetown Law Journal* 83: 184–193.

Stallings, Robert A. 1990. "Media Discourse and the Social Construction of Risk." *Social Problems* 37: 80–95.

18

The Police and the Black Male

ELIJAH ANDERSON

In this excerpt from his book Streetwise, *Anderson gives us a glimpse into the perspective of ghetto inhabitants as they are handled by police: the agents of social control. Race (African American), gender (male), and age (youth) are the disempowering features of this population. For young African American men living in America's inner city, life may be made even more difficult by police who assume that, if there is trouble in the neighborhood, it must be caused by these youngsters. Black men, walking alone at night or cruising a neighborhood in a car, may be stopped, harassed, questioned, or beaten, even if they have done nothing wrong. Merely due to their ascribed status as African American, their activities are scrutinized in different ways than others in society. They may be subject to the "cycle of oppression," as they are color-coded (racially profiled), stopped and questioned, arrested without substantive cause, and assigned to a public defender who most often leads them into a plea bargain, resulting in their having a record that shows up and verifies their deviant status the next time they are arbitrarily stopped. Drawing on conflict theory, we see here how deviant status is used by social control agents to subordinate less powerful groups.*

The police, in the Village-Northton [neighborhood] as elsewhere, represent society's formal, legitimate means of social control.[1] Their role includes protecting law-abiding citizens from those who are not law-abiding, by preventing crime and by apprehending likely criminals. Precisely how the police fulfill the public's expectations is strongly related to how they view the neighborhood and the people who live there. On the streets, color-coding often works to confuse race, age, class, gender, incivility, and criminality, and it expresses itself most concretely in the person of the anonymous black male. In doing their job, the police often become willing parties to this general color-coding of the public environment, and related distinctions, particularly those of skin color and gender, come to convey definite meanings. Although such coding may make the work of the police more manageable, it may also fit well with their own presuppositions regarding race and class relations, thus shaping officers' perceptions of crime "in the city." Moreover, the anonymous black male is usually an ambiguous figure who arouses the utmost caution and is generally considered dangerous until he proves he is not. . . .

There are some who charge— . . . perhaps with good reason—that the police are primarily agents of the middle class who are working to make the area more hospitable to middle-class people at the expense of the lower classes. It is obvious that the police assume whites in the community are at least middle class and are trustworthy on the streets. Hence the police may be seen primarily as protecting "law-abiding" middle-class whites against anonymous "criminal" black males.

To be white is to be seen by the police—at least superficially—as an ally, eligible for consideration and for much more deferential treatment than that accorded blacks in general. This attitude may be grounded in the backgrounds of the police themselves.[2] Many have grown up in Eastern City's "ethnic" neighborhoods. They may serve what they perceive as their own class and neighborhood interests, which often translates as keeping blacks "in their place"—away from neighborhoods that are socially defined as "white." In trying to do their job, the police appear to engage in an informal policy of monitoring young black men as a means of controlling crime, and often they seem to go beyond the bounds of duty. The following field note shows what pressures and racism young black men in the Village may endure at the hands of the police:

At 8:30 on a Thursday evening in June I saw a police car stopped on a side street near the Village. Beside the car stood a policeman with a young black man. I pulled up behind the police car and waited to see what would happen. When the policeman released the young man, I got out of my car and asked the youth for an interview.

"So what did he say to you when they stopped you? What was the problem?" I asked. "I was just coming around the corner, and he stopped me, asked me what was my name, and all that. And what I had in my bag. And where I was coming from. Where I lived, you know, all the basic stuff, I guess. Then he searched me down and, you know, asked me who were the supposedly tough guys around here? That's about it. I couldn't tell him who they are. How do I know? Other gang members could, but I'm not from a gang, you know. But he tried to put me in a gang bag, though. "How old are you?" I asked. "I'm seventeen, I'll be eighteen next month." "Did he give any reason for stopping you?" "No, he didn't. He just wanted my address, where I lived, where I was coming from, that kind of thing. I don't have no police record or nothin'. I guess he stopped me on principle, 'cause I'm black." "How does that make you feel?" I asked. "Well, it doesn't bother me too much, you know, as long as I know that I hadn't done nothin', but I guess it just happens around here. They just stop young black guys and ask 'em questions, you know. What can you do?"

On the streets late at night, the average young black man is suspicious of others he encounters, and he is particularly wary of the police. If he is dressed in the uniform of the "gangster," such as a black leather jacket, sneakers, and a "gangster cap," if he is carrying a radio or a suspicious bag (which may be confiscated), or if he is moving too fast or too slow, the police may stop him. As part of the routine, they search him and make him sit in the police car while they run a check to see whether there is a "detainer" on him. If there is nothing, he is allowed to go on his way. After this ordeal the youth is often left afraid, sometimes shaking, and uncertain about the area he had previously taken for granted. He is upset in part because he is painfully aware of how close he has come to being in "big trouble." He knows of other youths who have gotten into a "world of trouble" simply by being on the streets at the wrong time or when the police were pursuing a criminal. In these circumstances, particularly at night, it is relatively easy for one black man to be mistaken for another. Over the years, while walking through the neighborhood I have on occasion been stopped and questioned by police chasing a mugger, but after explaining myself I was released.

Many youths, however, have reason to fear such mistaken identity or harassment, since they might be jailed, if only for a short time, and would have to post bail money and pay legal fees to extricate themselves from the mess (Anderson 1986). When law-abiding blacks are ensnared by the criminal justice system, the scenario may proceed as follows. A young man is arbitrarily stopped by the police and questioned. If he cannot effectively negotiate with the officer(s), he may be accused of a crime and arrested. To resolve this situation he needs financial resources, which for him are in short supply. If he does not have money for an attorney, which often happens, he is left to a public defender who may be more interested in going along with the court system than in fighting for a poor black person. Without legal support, he may well wind up "doing time" even if he is innocent of the charges brought against him. The next time he is stopped for questioning he will have a record, which will make detention all the more likely.

Because the young black man is aware of many cases when an "innocent" black person was wrongly accused and detained, he develops an "attitude" toward the police. The street word for police is "the man," signifying a certain machismo, power, and authority. He becomes concerned when he notices "the man" in the community or when the police focus on him because he is outside his own neighborhood. The youth knows, or soon finds out, that he exists in a legally precarious state. Hence he is motivated to avoid the police, and his public life becomes severely circumscribed.

To obtain fair treatment when confronted by the police, the young man may wage a campaign for social regard so intense that at times it borders on obsequiousness. As one streetwise black youth said: "If you show a cop that you nice and not a smartass, they be nice to you. They talk to you like the man you are. You gonna get ignorant like a little kid, they gonna get ignorant with you." Young black males often are particularly deferential toward the police even when they are completely within their rights and have done nothing wrong. Most often this is not out of blind acceptance or respect for the "law," but because they know the police can cause them hardship. When confronted or arrested, they adopt a particular style of behavior to get on the policeman's good side. Some simply "go limp" or politely ask, "What seems to be the trouble, officer?" This pose requires a deference that is in sharp contrast with the youth's more usual image, but many seem to take it in stride or not even to realize it. Because they are concerned primarily with staying out of trouble, and because they perceive the police as arbitrary in their use of power, many defer in an equally arbitrary way. Because of these pressures, however, black youths tend to be especially mindful of the police and, when they are around, to watch their own behavior in public. Many have come to expect harassment and are inured to it; they simply tolerate it as part of living in the Village-Northton.

After a certain age, say twenty-four, a black man may no longer be stopped so often, but he continues to be the object of policy scrutiny. As one twenty-seven-year-old black college graduate speculated:

> I think they see me with my little bag with papers in it. They see me with penny loafers on. I have a tie on, some days. They don't stop me so much now. See, it depends on the circumstances. If something goes down, and they hear that the guy had on a big black coat, I may be the one. But when I was younger, they could just stop me, carte blanche, any old time. Name taken, searched, and this went on endlessly. From the time I was about twelve until I was sixteen or seventeen, endlessly, endlessly. And I come from a lower-middle-class black neighborhood, OK, that borders a white neighborhood. One neighborhood is all black, and one is all white. OK, just because we were so close to that neighborhood, we were stopped endlessly. And it happened even more when we went up into a suburban community. When we would ride up and out to the suburbs, we were stopped every time we did it.
>
> If it happened today, now that I'm older, I would really be upset. In the old days when I was younger, I didn't know any better. You just expected it, you knew it was gonna happen. Cops would come up, "What you doing, where you coming from?" Say things to you. They might even call you nigger.

Such scrutiny and harassment by local police makes black youths see them as a problem to get beyond, to deal with, and their attempts affect their overall behavior. To avoid encounters with "the man," some streetwise young men camouflage themselves, giving up the urban uniform and emblems that identify them as "legitimate" objects of police attention. They may adopt a more conventional presentation of self, wearing chinos, sweat suits, and generally more conservative dress. Some youths have been known to "ditch" a favorite jacket if they see others wearing one like it, because wearing it increases their chances of being mistaken for someone else who may have committed a crime.

But such strategies do not always work over the long run and must be constantly modified. For instance, because so many young ghetto blacks have begun to wear Fila and Adidas sweat suits as status symbols, such dress has become incorporated into the public image generally associated with young black males. These athletic suits, particularly the more expensive and colorful ones, along with high-priced sneakers, have become the leisure dress of successful drug dealers, and other youths will often mimic their wardrobe to "go for bad" in the quest for local esteem. Hence what was once a "square" mark of distinction approximating the conventions of the wider culture has been adopted by a neighborhood group devalued by that same culture. As we saw earlier, the young black male enjoys a certain power over fashion: whatever the collective peer group embraces can become "hip" in a manner the wider society may not desire (see Goffman 1963). These same styles then attract the attention of the agents of social control.

THE IDENTIFICATION CARD

Law-abiding black people, particularly those of the middle class, set out to approximate middle-class whites in styles of self-presentation in public, including dress and bearing. Such middle-class emblems, often viewed as "square," are not usually embraced by young working-class blacks. Instead, their connections with and claims on the institutions of the wider society seem to be symbolized by the identification card. The common identification card associates its holder with a firm, a corporation, a school, a union, or some other institution of substance and influence. Such a card, particularly from a prominent establishment, puts the police and others on notice that the youth is "somebody," thus creating an important distinction between a black man who can claim a connection with the wider society and one who is summarily judged as "deviant." Although blacks who are established in the middle class might take such cards for granted, many lower-class blacks, who continue to find it necessary to campaign for civil rights denied them because of skin color, believe that carrying an identification card brings them better treatment than is meted out to their less fortunate brothers and sisters. For them this link to the wider society, though often tenuous, is psychically and socially important. The young college graduate continues:

> I know [how] I used to feel when I was enrolled in college last year, when I had an ID card. I used to hear stories about the blacks getting stopped over by the dental school, people having trouble sometimes. I would see that all the time. Young black male being stopped by the police. Young black male in handcuffs. But I knew that because I had that ID card that I would not be mistaken for just somebody snatching a pocketbook, or just somebody being where maybe I wasn't expected to be. See, even though I was intimidated by the campus police—I mean, the first time I walked into the security office to get my ID they all gave me the double-take to see if I was somebody they were looking for. See, after I got the card, I was like, well, they can think that now, but I have this [ID card]. Like, see, late at night when I be walking around, and the cops be checking me out, giving me the looks, you know. I mean, I know guys, students, who were getting stopped all the time, sometimes by the same officer, even though they had the ID. And even they would say, "Hey, I got the ID, so why was I stopped?"

The cardholder may believe he can no longer be treated summarily by the police, that he is no longer likely to be taken as a "no count," to be prejudicially confused with that class of blacks "who are always causing trouble on the trolley." Furthermore, there is a firm belief that if the police stop a person who has a card, they cannot "do away with him without somebody coming to his defense." This concern should not be underestimated. Young black men trade stories about mistreatment at the hands of the police; a common one involves policemen who transport youths into rival gang territories and release them, telling them to get home the best way they can. From the youth's perspective, the card signifies a certain status in circumstances where little recognition was formerly available.

"DOWNTOWN" POLICE AND LOCAL POLICE

In attempting to manage the police—and by implication to manage themselves—some black youths have developed a working connection of the police in certain public areas of the Village-Northton. Those who spend a good amount of their time on these corners, and thus observing the police, have come to distinguish between the "downtown" police and the "regular" local police.

The local police are the ones who spend time in the area; normally they drive around in patrol cars, often one officer to a car. These officers usually make a kind of working peace with the young men on the streets; for example, they know the names of some of them and may even befriend a young boy. Thus they offer an image of the police department different from that displayed by the "downtown" police. The downtown police are distant, impersonal, and often actively looking for "trouble." They are known to swoop down arbitrarily on gatherings of black youths standing on a street corner; they might punch them around, call them names, and administer other kinds of abuse, apparently for sport. A young Northton man gave the following narrative about his experiences with the police.

> And I happen to live in a violent part. There's a real difference between the violence level in the Village and the violence level in Northton. In the nighttime it's more dangerous over there.
>
> It's so bad now, they got downtown cops over there now. They doin' a good job bringin' the highway patrol over there. Regular cops don't like that. You can tell that. They even try to emphasize to us the certain category. Highway patrol come up, he leave, they say somethin' about it. "We can do our job over here." We call [downtown police] Nazis. They about six feet eight, seven feet. We walkin', they jump out. "You run, and we'll blow your nigger brains out." I hate bein' called a nigger. I want to say somethin' but get myself in trouble.
>
> When a cop do somethin', nothing happen to 'em. They come from downtown. From what I heard some of 'em don't even wear their real badge numbers. So you have to put up with that. Just keep your mouth shut when they stop you, that's all. Forget about questions, get against the wall, just obey 'em. "Put all that out right there"—might

get rough with you now. They snatch you by the shirt, throw you against the wall, pat you hard, and grab you by the arms, and say, "Get outta here." They call you nigger this and little black this, and things like that. I take that. Some of the fellas get mad. It's a whole different world.

Yeah, they lookin' for trouble. They gotta look for trouble when you got five, eight police cars together and they laughin' and talkin', start teasin' people. One night we were at a bar, we read in the paper that the downtown cops comin' to straighten things out. Same night, three police cars, downtown cops with their boots on, they pull the sticks out, beatin' around the corner, chase into bars. My friend Todd, one of 'em grabbed him and knocked the shit out of him. He punched 'im, a little short white guy. They start a riot. Cops started that shit. Everybody start seein' how wrong the cops was—they start throwin' bricks and bottles, cussin' 'em out. They lock my boy up; they had to let him go. He was just standin' on the corner, they snatch him like that.

One time one of 'em took a gun and began hittin' people. My boy had a little hickie from that. He didn't know who the cop was, because there was no such thing as a badge number. They have phony badge numbers. You can tell they're tougher, the way they dress, plus they're bigger. They have boots, trooper pants, blond hair, blue eyes, even black [eyes]. And they seven feet tall, and six foot six inches and six foot eight inches. Big! They are the rough cops. You don't get smart with them or they beat the shit out of you *in front of everybody,* they don't care.

We call 'em Nazis. Even the blacks among them. They ride along with 'em. They stand there and watch a white cop beat your brains out. What takes me out is the next day you don't see 'em. Never see 'em again, go down there, come back, and they ride right back downtown, come back, do their little dirty work, go back downtown, and put their real badges on. You see 'em with a forty-five or fifty-five number: "Ain't no such number here, I'm sorry, son." Plus, they got unmarked cars. No sense takin' 'em to court. But when that happened at that bar, another black cop from the sixteenth [local] district, ridin' a real car, came back and said, "Why don't y'all go on over to the sixteenth district and file a complaint? Them musclin' cops was wrong. Beatin' people." So about ten people went over there; sixteenth district knew nothin' about it. They come in unmarked cars, they must have been downtown cops. Some of 'em do it. Some of 'em are off duty, on their way home. District commander told us they do that. They have a patrol over there, but them cops from downtown have control of them cops. Have bigger ranks and bigger guns. They carry .357s and regular cops carry little .38s. Downtown cops are all around. They carry magnums.

Two cars the other night. We sittin' on the steps playing cards. Somebody called the cops. We turn around and see four regular police cars and two highway police cars. We drinkin' beer and playin' cards.

Police get out and say you're gamblin'. We say we got nothin' but cards here, we got no money. They said all right, got back in their cars, and drove away. Downtown cops dressed up like troopers. That's intimidation. Damn!

You call a cop, they don't come. My boy got shot, we had to take him to the hospital ourselves. A cop said, "You know who did it?" We said no. He said, "Well, I hope he dies if y'all don't say nothin'." What he say that for? My boy said, "I hope your mother die," he told the cop right to his face. And I was grabbin' another cop, and he made a complaint about that. There were a lot of witnesses. Even the nurse behind the counter said the cop had no business saying nothin' like that. He said it loud, "I hope he dies." Nothin' like that should be comin' from a cop.

Such behavior by formal agents of social control may reduce the crime rate, but it raises questions about social justice and civil rights. Many of the old-time liberal white residents of the Village view the police with some ambivalence. They want their streets and homes defended, but many are convinced that the police manhandle "kids" and mete out an arbitrary form of "justice." These feelings make many of them reluctant to call the police when they are needed, and they may even be less than completely cooperative after a crime has been committed. They know that far too often the police simply "go out and pick up some poor black kid." Yet they do cooperate, if ambivalently, with these agents of social control.

In an effort to gain some balance in the emerging picture of the police in the Village-Northton, I interviewed local officers. The following edited conversation with Officer George Dickens (white) helps place in context the fears and concerns of local residents, including black males:

I'm sympathetic with the people who live in this neighborhood [the Village-Northton], who I feel are victims of drugs. There are a tremendous number of decent, hardworking people who are just trying to live their life in peace and quiet, not cause any problems for their neighbors, not cause any problems for themselves. They just go about their own business and don't bother anyone. The drug situation as it exists in Northton today causes them untold problems. And some of the young kids are involved in one way or another with this drug culture. As a result, they're gonna come into conflict even with the police they respect and have some rapport with.

We just went out last week on Thursday and locked up ten young men on Cherry Street, because over a period of about a week, we had undercover police officers making drug buys from those young men. This was very well documented and detailed. They were videotaped selling the drugs. And as a result, right now, if you walk down Cherry Street, it's pretty much a ghost town; there's nobody out. [Before, Cherry Street was notorious for drug traffic.] Not only were people buying drugs there, but it was a very active street. There's been some shock value as a result of all those arrests at one time.

Now, there's two reactions to that. The [television] reporters went out and interviewed some people who said, "Aw, the police overreacted, they locked up innocent people. It was terrible, it was harassment." One of the neighbors from Cherry Street called me on Thursday, and she was outraged. Because she said, "Officer, it's not fair. We've been working with the district for well over a year trying to solve some of the problems on Cherry Street." But most of the neighbors were thrilled that the police came and locked all those kids up. So you're getting two conflicting reactions here. One from the people that live there that just wanta be left alone, alright? Who are really being harassed by the drug trade and everything that's involved in it. And then you have a reaction from the people that are in one way or another either indirectly connected or directly connected, where they say, "You know, if a young man is selling drugs, to him that's a job." And if he gets arrested, he's out of a job. The family's lost their income. So they're not gonna pretty much want anybody to come in there to make arrests. So you've got contradicting elements of the community there. My philosophy is that we're going to try to make Northton livable. If that means we have to arrest some of the residents of Northton, that's what we have to do.

You talk to Tyrone Pitts, you know the group that they formed was formed because of a reaction to complaints against one of the officers of how the teenagers were being harassed. And it turned out that basically what he [the officer] was doing was harassing drug dealers. When Northton against Drugs actually formed and seemed to jell, they developed a close working relationship with the police here. For that reason, they felt the officer was doing his job.

I've been here eighteen months. I've seen this neighborhood go from . . . let me say, this is the only place I've ever worked where I've seen a rapport between the police department and the general community like the one we have right now. I've never seen it any place else before coming here. And I'm not gonna claim credit because this happened while I happened to be here. I think a lot of different factors were involved. I think the community was ready to work with the police because of the terrible situation in reference to crack. My favorite expression when talking about crack is "crack changed everything." Crack changed the rules of how the police and the community have to interact with each other. Crack changed the rules about how the criminal justice system is gonna work, whether it works well or poorly. Crack is causing the prisons to be overcrowded. Crack is gonna cause the people that do drug rehabilitation to be overworked. It's gonna cause a wide variety of things. And I think the reason the rapport between the police and the community in Northton developed at the time it did is very simply that drugs to a certain extent made many areas in this city unlivable.

In effect the officer is saying that the residents, regardless of former attitudes, are now inclined to be more sympathetic with the police and to work with them. And at the same time, the police are more inclined to work with the residents. Thus, not only are the police and the black residents of Northton working together, but different groups in the Village and Northton are working with each other against drugs. In effect, law-abiding citizens are coming together, regardless of race, ethnicity, and class. He continues:

> Both of us [police and the community] are willing to say, "Look, let's try to help each other." The nice thing about what was started here is that it's spreading to the rest of the city. If we don't work together, this problem is gonna devour us. It's gonna eat us alive. It's a state of emergency, more or less.

In the past there was significant negative feeling among young black men about the "downtown" cops coming into the community and harassing them. In large part these feelings continue to run strong, though many young men appear to "know the score" and to be resigned to their situation, accommodating and attempting to live with it. But as the general community feels under attack, some residents are willing to forgo certain legal and civil rights and undergo personal inconvenience in hopes of obtaining a sense of law and order. The officer continues:

> Today we don't have too many complaints about police harassment in the community. Historically there were these complaints, and in almost any minority neighborhood in Eastern City where I ever worked there was more or less a feeling of that [harassment]. It wasn't just Northton; it was a feeling that the police were the enemy. I can honestly say that for the first time in my career I don't feel that people look at me like I'm the enemy. And it feels nice; it feels real good not to be the enemy, ha-ha. I think we [the police] realize that a lot of problems here [in the Village-Northton] are related to drugs. I think the neighborhood realizes that too. And it's a matter of "Who are we gonna be angry with? Are we gonna be angry with the police because we feel like they're this army of occupation, or are we gonna argue with these people who are selling drugs to our kids and shooting up our neighborhoods and generally causing havoc in the area? Who deserves the anger more?" And I think, to a large extent, people of the Village-Northton decided it was the drug dealers and not the police.
> I would say there are probably isolated incidents where the police would stop a male in an area where there is a lot of drugs, and this guy may be perfectly innocent, not guilty of doing anything at all. And yet he's stopped by the police because he's specifically in that area, on that street corner where we know drugs are going hog wild. So there may be isolated incidents of that. At the same time, I'd say I know for a fact that our complaints against police in this division, the

whole division, were down about 45 percent. If there are complaints,
if there are instances of abuse by the police, I would expect that
our complaints would be going up. But they're not; they're dropping.

Such is the dilemma many Villagers face when they must report a crime or
deal in some direct way with the police. Stories about police prejudice against
blacks are often traded at Village get-togethers. Cynicism about the effectiveness
of the police mixed with community suspicion of their behavior toward blacks
keeps middle-class Villagers from embracing the notion that they must rely
heavily on the formal means of social control to maintain even the minimum
freedom of movement they enjoy on the streets.

Many residents of the Village, especially those who see themselves as the "old
guard" or "old-timers," who were around during the good old days when
antiwar and antiracist protest was a major concern, sigh and turn their heads
when they see the criminal justice system operating in the ways described
here. They express hope that "things will work out," that tensions will ease,
that crime will decrease and police behavior will improve. Yet as incivility and
crime become increasing problems in the neighborhood, whites become less
tolerant of anonymous blacks and more inclined to embrace the police as their
heroes.

Such criminal and social justice issues, crystallized on the streets, strain
relations between the newcomers and many of the old guard, but in the present
context of drug-related crime and violence in the Village-Northton, many of the
old-timers are adopting a "law and order" approach to crime and public safety,
laying blame more directly on those they see as responsible for such crimes,
though they retain some ambivalence. Newcomers can share such feelings
with an increasing number of old-time "liberal" residents. As one middle-aged
white woman who has lived in the Village for fifteen years said:

When I call the police, they respond. I've got no complaints. They
are fine for me. I know they sometimes mistreat black males. But
let's face it, most of the crime is committed by them, and so they can
simply tolerate more scrutiny. But that's them.

Gentrifiers and the local old-timers who join them, and some traditional
residents continue to fear, care more for their own safety and well-being than
for the rights of young blacks accused of wrong-doing. Yet reliance on the
police, even by an increasing number of former liberals, may be traced to a
general feeling of oppression at the hands of street criminals, whom many believe
are most often black. As these feelings intensify and as more yuppies and students
inhabit the area and press the local government for services, especially police
protection, the police may be required to "ride herd" more stringently on the
youthful black population. Thus young black males are often singled out as
the "bad" element in an otherwise healthy diversity, and the tensions between
the lower-class black ghetto and the middle- and upper-class white community
increase rather than diminish.

NOTES

1. See Rubinstein (1973); Wilson (1968); Fogelson (1977); Reiss (1971); Bittner (1967); Banton (1964).

2. For an illuminating typology of police work that draws a distinction between "fraternal" and "professional" codes of behavior, see Wilson (1968).

REFERENCES

Anderson, Elijah. 1986. "Of old heads and young boys: Notes on the urban black experience." Unpublished paper commissioned by the National Research Council, Committee on the Status of Black Americans.

Banton, Michael. 1964. *The policeman and the community*. New York: Basic Books.

Bittner, Egon. 1967. The police on Skid Row. *American Sociological Review* 32 (October): 699–715.

Fogelson, Robert. 1977. *Big city police*. Cambridge: Harvard University Press.

Goffman, Erving. 1963. *Behavior in public places*. New York: Free Press.

Reiss, Albert J. 1971. *The police and the public*. New Haven: Yale University Press.

Rubinstein, Jonathan. 1973. *City police*. New York: Farrar, Straus and Giroux.

Wilson, James Q. 1968. "The police and the delinquent in two cities." In *Controlling delinquents*, ed. Stanton Wheeler. New York: John Wiley.

19

Homophobia and Women's Sport

ELAINE M. BLINDE AND DIANE E. TAUB

Blinde and Taub explore the role of attributed sexual orientation in disempowering women who violate gender norms: varsity female collegiate athletes. By challenging the gender order and opposing male domination, these women intrude into a traditional male sanctum and threaten the male domain of physicality and strength. By casting the lesbian label on women athletes, society stigmatizes them as masculine and as sexual perverts. While the homosexual label is routinely used to degrade male athletes who fail to live up to the hypermasculine ideal, the lesbian label is used to divide and silence female athletes. They may adopt the perspective of their oppressors and demean their teammates as lesbians, thus destroying team solidarity, and/or shun the label, but be forced to acknowledge its demeaning power as they attempt to escape it. The forceful effect of the lesbian label applied to women athletes shows the dominance not only of heterosexuals over homosexuals, but also of men over women.

Central to the preservation of a patriarchal and heterosexist society is a well-established gender order with clearly defined norms and sanctions governing the behavior of men and women. This normative gender system is relayed to and installed in members of society through a pervasive socialization network that is evident in both everyday social interaction and social institutions (Schur 1984). Conformity to established gender norms contributes to the reproduction of male dominance and heterosexual privilege (Lenskyj 1991; Stockard and Johnson 1980).

Despite gender role socialization, not all individuals engage in behavior consistent with gender expectations. Recognizing the potential threat of such aberrations, various mechanisms exist that encourage compliance with the normative gender order. Significant in such processes are the stigmatization and devaluation of those whose behavior deviates from the norm (Schur 1984).

Women's violation of traditional gender role norms represents a particularly serious threat to the patriarchal and heterosexist society because this deviant behavior resists women's subordinate status (Schur 1984). When women engage

"Homophobia and Women's Sport: The Disempowerment of Athletes," by Elaine M. Blinde and Diane E. Taub, *Sociological Focus,* the Journal of the North Central Sociological Association, Vol. 25, No. 2, May 1992. Reprinted by permission of the North Central Sociological Association.

in behavior that challenges the established gender order, and thus opposes male domination, attempts are often made by those most threatened to devalue these women and ultimately control their actions. One means of discrediting women who violate gender norms and thereby questioning their "womanhood" is to label them lesbian (Griffin 1987).

The accusation of lesbianism is a powerful controlling mechanism given the homophobia that exists within American society. Homophobia, representing a fear of or negative reaction to homosexuality (Pharr 1988), results in stigmatization directed at those assumed to violate sexuality norms. Lesbianism, in particular, is viewed as threatening to the established patriarchal order and heterosexual family structure since lesbians reject their "natural" gender role, as well as resist economic, emotional, and sexual dependence on men (Gartrell 1984; Lenskyj 1991).

As a means for both discouraging homosexuality and maintaining a patriarchal and heterosexist gender order (Pharr 1988), homophobia controls behavior through contempt for purported norm violators (Koedt, Levine, and Rapone 1973). One method of control is the frequent application of the lesbian label to women who move into traditional male-dominated fields such as politics, business, or the military (Lenskyj 1991). This "lesbian baiting" (Pharr 1988:19) suggests that women's advancement into these arenas is inappropriate. Such messages are particularly potent since they are lodged in a society that condemns, devalues, oppresses, and victimizes individuals labeled as homosexuals (Lenskyj 1990).

Another male arena in which women have made significant strides, and thus risk damaging accusation and innuendo, is that of sport (Blinde and Taub 1992; Lenskyj 1990). Sport is a particularly susceptible arena for lesbian labeling due to the historical linkage of masculinity with athleticism (Birrell 1988). When women enter the domain of sport they are viewed as violating the docile female gender role and therefore extending culturally constructed boundaries of femininity (Cobban 1982; Lenskyj 1986; Watson 1987). The attribution of masculine qualities to women who participate in sport leads to a questioning of their sexuality and subsequently makes athletes targets of homophobic accusations (Lenskyj 1986). . . .

Therefore, the present study explores the stereotyping of women athletes as lesbians and the accompanying homophobia fostering this label. General themes and processes which inform us of how these individuals handle the lesbian issue are identified. These dynamics are grounded in the contextual experiences of women athletes and relayed through their voices.

Athletic directors at seven large Division I universities were contacted by telephone and asked to participate in a study examining various aspects of the sport experience of female college athletes. These administrators were requested to provide a list of the names and addresses of all varsity women athletes for the purpose of contacting them for telephone interviews. . . . Interested athletes were encouraged to return an informed consent form indicating their willingness to participate in a tape-recorded telephone interview. Based on this initial contact, a total of 16 athletes agreed to be in the study.

In order to increase the sample size to the desired 20 to 30 respondents, the names of 30 additional athletes were randomly and proportionately selected from the three lists. Eight of these athletes agreed to be interviewed, resulting in a final sample size of 24. Athletes in the sample were currently participating in a variety of women's intercollegiate varsity sports—basketball (n = 5), track and field (n = 4), volleyball (n = 3), swimming (n = 3), soft-ball (n = 3), tennis (n = 2), diving (n = 2), and gymnastics (n = 2). With an average age of 20.2 years and overwhelmingly Caucasian (92%), the sample contained 2 freshmen, 9 sophomores, 5 juniors, and 8 seniors. A majority of the athletes (n = 22) were recipients of an athletic scholarship. . . .

Semistructured telephone interviews were conducted by two trained female interviewers. All interviews were tape-recorded and lasted from 50 to 90 minutes. Questions were open-ended in nature so that athletes would not feel constrained in discussing those issues most relevant to their experiences. Follow-up questions were utilized to probe how societal perceptions of women athletes impact their behavior and experiences.

RESULTS

Examination of the responses of athletes revealed two prevailing themes related to the presence of the lesbian stereotype in women's sport—(a) a silence surrounding the issue of lesbianism in women's sport, and (b) athletes' internalization of societal stereotypes concerning lesbians and women athletes. It is suggested that these two processes disempower women athletes and thus are counterproductive to the self-actualizing capability of sport participation (Theberge 1987).

Silence Surrounding Lesbianism
in Women's Sport

One of the most pervasive themes throughout the interviews related to the general silence associated with the lesbian stereotype in women's sport. Although a topic of which athletes are cognizant, reluctance to discuss and address lesbianism in women's sport was evident. Based on the responses of athletes, this silence was manifested in several ways: (a) athletes' difficulty in discussing lesbian topic, (b) viewing lesbianism as a personal and irrelevant issue, (c) disguising athletic identity to avoid lesbian label, (d) team difficulty in addressing lesbian issue, and (e) administrative difficulty in addressing lesbian issue.

Athletes' Difficulty in Discussing
Lesbian Topic

Initial indication of silencing was illustrated by the difficulty and uneasiness many athletes experienced in discussing the lesbian stereotype. Some respondents were initially reluctant to mention the topic of lesbianism; discussion of the issue was frequently preceded by awkward or long pauses suggesting feelings of uneasiness or discomfort. Athletes were most likely to introduce this topic when questions were asked about societal perceptions of women's sport and female athletes, as well as inquiries about the existence of stereotypes associated with women athletes. Moreover, the lesbian issue was sometimes discussed without specifically using the term lesbian. For example, some athletes evaded the issue by making indirect references to lesbianism (e.g., using the word "it" rather than a more descriptive term). . . .

Respondents' approach to the topic of lesbianism indicates the degree to which women athletes have been socialized into a cycle of silence. Such silence highlights the suppressing effects of homophobia. Moreover, athletes' reluctance to discuss topics openly related to lesbianism may be to avoid what Goffman (1963) has termed "courtesy stigma," a stigma conferred despite the absence of usual qualifying behavior.

Viewing Lesbianism as a Personal and Irrelevant Issue

A second indicator of the silence surrounding the lesbian stereotype was reflected in athletes' general comments about lesbianism. Many respondents indicated that sexual orientation was a very personal issue and thus represented a private and extraneous aspect of an individual's life. These athletes felt it was inappropriate for others to be concerned about the sexual orientation of women athletes.

Although such a manifestation of silence might reflect the path of least resistance by relieving athletes of the need to discuss or disclose their sexual orientation (Lenskyj 1991), it does not eliminate the stigma and stress experienced by women athletes. Also, making lesbianism a private issue does not confront or challenge the underlying homophobia that allows the label to carry such significance. The strategy of making sexual orientation a personal issue depoliticizes lesbianism and ignores broader societal issues.

Disguising Athletic Identity to Avoid Lesbian Label

A third form of silence surrounding the lesbian stereotype was the tendency for athletes to hide their athletic identities. Nearly all respondents indicated that despite feeling pride in being an athlete, there were situations where they preferred that others not know their athletic identity. Although not all athletes indicated that this concealment was to prevent being labeled a lesbian, it was obvious that there was a perceived stigma associated with athletics that many women wanted to avoid (e.g., masculine women, women trying to be men, jock image). In most cases, respondents indicated that disguising their athletic identity was either directly or indirectly related to the lesbian stereotype. . . .

Athletes also stated that they (or other athletes they knew) accentuated certain behaviors in order to reduce the possibility of being labeled a lesbian. Being seen with men, having a boyfriend, or even being sexually promiscuous with men were commonly identified strategies to reaffirm an athlete's heterosexuality. As one athlete commented. "If you are a female athlete and do not have a boyfriend, you are labeled [lesbian]."

As reflected in the responses of athletes, the role of sport participant was often intentially de-emphasized in order to reduce the risk of being labeled lesbian. Modification of athletes' behavior, even to the point of denying critical aspects of self, was deemed necessary for protection from the negativism attached to the lesbian label. This disguising of athletic identity exemplifies what Kitzinger (1987:92) termed "role inversion." In such a situation individuals attempt to demonstrate that their group stereotype is inaccurate by accentuating traits that are in opposition to those commonly associated with the group (in the case of women athletes, stressing femininity and heterosexuality).

Team Difficulty in Addressing Lesbian Issue

Not only did the silence surrounding lesbianism impact certain aspects of the lives of individual athletes, but it also affected interpersonal relationships among team members. This silence was often counterproductive to the development of positive group dynamics (e.g., team cohesion, open lines of communication).

As was often true at the individual level, women's sport teams were unable collectively to discuss, confront, or challenge the labeling of women athletes as lesbians. One factor complicating the ability of women athletes to confront the lesbian stereotype was the divisive nature of the label itself (Gentile 1982); the lesbian issue sometimes split teams into factions or served as the basis for clique formation.

Heterosexual and lesbian athletes often had limited interaction with each other outside the sport arena. Moreover, athletes established distance between themselves and those athletes most likely to be labeled lesbian (i.e., those possessing "masculine" physical or personality characteristics). . . .

From the interviews, there was little evidence that lesbian and nonlesbian athletes collectively pooled their efforts to confront or challenge the lesbian stereotype so prevalent in women's sport. The silence surrounding lesbianism creates divisions among women athletes; this dissension has the effect of preventing female bonding and camaraderie (Lenskyj 1986). Rather than recognizing their shared interests, women athletes focus on their differences and thus deny the formation of "alliance" (Pheterson 1986:149). This difficulty in attaining team cohesion is unfortunate since women's sport is an activity where women as a group can strive for common goals (Lenskyj 1990). The lesbian stereotype not only limits female solidarity, but also minimizes women's ability to challenge collectively the patriarchal and heterosexist system in which they reside (Bennett et al. 1987).

Administrative Difficulty
in Addressing Lesbian Issue

Another manifestation of silence relayed in the responses of athletes was the apparent unwillingness of coaches and athletic directors to confront openly the lesbian stereotype. As was found with individual athletes and teams, those in leadership positions in women's sport refused to address or challenge this stereotype. Reluctance to confront the lesbian issue at the administrative level undoubtedly influenced the manner in which athletes handled the stereotype. . . .

Because the women's intercollegiate sport system is homophobic and predominately male-controlled (i.e., over half of coaches and four-fifths of administrators are men) (Acosta and Carpenter 1992), it is assumed that survival in women's sport requires collusion in a collective strategy of silence about and denial of lesbianism (Griffin 1987). Coaches and administrators fear that openly addressing the lesbian issue may result in women's sport losing the recent gains made in such areas as fan support, budgets, sponsorship, and credibility (Griffin 1987). Therefore, leaders yield to this fear as they strive to achieve acceptability for women's sport. Such accommodation to the patriarchal, heterosexist sport structure not only contributes to isolation as coaches and administrators are afraid to discuss lesbianism, but also limits their identification with feminist and women's issues (Duquin 1981; Hargreaves 1990; Pharr 1988; Zipter 1988).

CONCLUSIONS

Based on athletes' responses, it was evident that the silence surrounding the lesbian issue in women's sport was deeply ingrained at all levels of the women's intercollegiate sport structure. Such widespread silencing reflects the negativism and fear associated with lesbianism that are so prevalent in a homophobic society. This strategy of silence or avoidance, however, is counterproductive to efforts to dispel or minimize the impact of the lesbian stereotype. Not only does silence disallow a direct confrontation with those who label athletes lesbian, but it also perpetuates the power of the label by leaving unchallenged rumors and insinuations. Moreover, the fear, ignorance, and negative images that are frequently associated with women athletes are reinforced by this silence (Zipter 1988).

Numerous aspects of women's experience in sport are ignored due to the silence surrounding the subject of lesbianism. For example, refusing to address this issue has limited understanding of the dimensionality and complexity of women's sport participation. Moreover, since the stigma associated with the lesbian label inhibits athletes from discussing this topic with each other, these women frequently do not realize that they possess shared experiences that would provide the foundation for female bonding. Without an "alliance" among athletes, little progress is made in improving their plight (Pheterson 1986). Finally, as a result of this preoccupation with silence, women athletes often engage in self-denial as they hide their athletic identity.

Athletes' Internalization
of Societal Stereotypes

A second major theme reflected in the responses of athletes was a general internalization of stereotypic representations of lesbians and women athletes. As argued by Kitzinger (1987) and Pheterson (1986), members of oppressed and socially marginalized groups often find themselves accepting the stereotypes and prejudices held by the dominant society. Representing "internalized oppression" (Pheterson 1986:148), the responses of athletes revealed an identification with the aggressor, self-concealment, and dependence on others for self-definition (Kitzinger 1987; Pheterson 1986). Acceptance of these societal representations by a disadvantaged group (in this case women athletes) grants legitimacy to the position of those who oppress and contributes to the continued subordination of the oppressed (Wolf 1986). Based on our interviews, athletes' internalization of stereotypes and prejudices were reflected by three categories of responses: (1) acceptance of lesbian stereotypes, (2) acceptance of women's sport team stereotypes, and (3) acceptance of negative images of lesbianism.

Acceptance of Lesbian Stereotypes

In response to various open-ended questions, it was apparent that athletes were able to identify a variety of factors that they felt led others to label women athletes as lesbians (e.g., physical appearance, dress, personality characteristics, nature of sport activity). Given that the attribution of homosexuality is most likely to be associated with traits and behaviors judged to be more appropriate for members of the opposite sex (Dunbar, Brown, and Amoroso 1973; Dunkle and Francis 1990), it was not surprising that athletes' rationale for the lesbian label included such attributes as muscularity, short hair, masculine clothing, etc.

When athletes were asked about the validity of the lesbian label in women's sport, affirmative replies were frequently based on conjecture. For example, to provide support for why they felt there was a basis for labeling women athletes as lesbians, respondents made such comments as "there are masculine girls on some teams," "it is really obvious," or "you can just tell that some athletes are lesbians."

These explanations tend to reflect an acceptance of societal definitions of lesbianism—beliefs that are largely male-centered and supportive of a patriarchal, heterosexist system (e.g., "girls who look like guys"). Indeed, previous research has shown that people associate physical appearance with homosexuality (Levitt and Klassen 1974; McArthur 1982; Unger, Hilderbrand, and Madar 1982). For example, attractiveness is equated with heterosexuality and a larger, muscular body build is identified with lesbianism.

Moreover, the remarks of athletes demonstrate that the very group that is oppressed (in this case women athletes) accepts societal stereotypes about lesbians and has incorporated these images into their managing of the situation. As suggested by Gartrell (1984) and certainly evident in this sample of women athletes, cultural myths about lesbianism perpetuated in a homophobic society are often firmly ingrained in the thinking of affected individuals.

Acceptance of Women's Sport
Team Stereotypes

Relative to providing a rationale for why the lesbian label was more likely to be associated with athletes in certain sports, respondents again demonstrated an understanding and internalization of societal stereotypes. The sports most commonly identified with the lesbian label were Softball, field hockey, and basketball. In attempting to explain why these team sports were singled out, athletes mentioned such factors as the nature of bodily contact or amount of aggression in the sport, as well as the body build, muscularity, or athleticism needed to play the sport.

Respondents often relied on the "masculine" and "feminine" stereotypes to differentiate sports in which participating women were more or less likely to be subjected to the lesbian label. Although participants in team sports were more likely than individual sports (e.g., gymnastics, swimming, tennis, golf) to be associated with the lesbian label, it was interesting to note that volleyball was often exempt from the connotations of lesbianism.

The higher incidence of lesbian labeling found in team sports (as opposed to individual sports) may also be related to the potential that team sports provide for interpersonal interactions. As previously mentioned, emphasizing teamwork and togetherness, team sports allow women rare opportunities to bond collectively in pursuit of a group goal (Lenskyj 1990). Recognition of this power of female bonding is often reflected by male opposition to women-only activities (Lenskyj 1990).

Acceptance of Negative
Images of Lesbianism

During the course of the interviews, a large majority of athletes made comments about lesbians which reflected an internalization of the negativism associated with lesbianism. Respondents also demonstrated a similar acceptance when they relayed conversations they had had with both teammates and outsiders.

One form of negativism was reflected by statements that specifically "put down" lesbians. Athletes' negative comments about lesbians were included in conversations with outsiders so others would not associate the lesbian label with them. Representing a form of projection (Gross 1978), some athletes attempted to disassociate from traits that they saw in themselves (e.g., strength, muscularity, aggressiveness). . . .

It is ironic that athletes rarely directed their anger or condemnations at the homophobic society that restricts the actions of women athletes, including the nonlesbian athlete. Rather, by focusing on athletes as lesbians, a blame-the-victim approach diverts attention from the cause of the oppression

(Pharr 1988). As is often true of oppressed groups, a blame-the-victim philosophy results in an acceptance of the belief system of the oppressor (in this case a patriarchal, heterosexist society) (Pharr 1988). Like other marginalized groups, women athletes accept the normative definitions of their deviance (Kitzinger 1987); in effect, such responses represent a form of collusion with the oppressive forces (Pheterson 1986). Interestingly, no mention was made by respondents about attempts to engage the assistance or support of units on campus sympathetic to gay and lesbians issues (e.g., feminist groups, gay and lesbian organization, affirmative action offices).

SUMMARY

From the interview responses, it was evidence that athletes had internalized societal stereotypes related to lesbians and women athletes, as well as the negativism directed toward lesbianism. This acceptance was so ingrained in these athletes that they were generally unaware of the political ramifications of both lesbianism and the accompanying lesbian stereotype as applied to women athletes. Despite their gender norm violation as athletes, these women often had a superficial understanding of gender issues. Such a lack of awareness may be due in part to the absence of a feminist consciousness in athletes (Boutilier and SanGiovanni 1983; Kaplan 1979) and their open disavowal of being a "feminist," "activist," or "preacher of women's liberation." Accepting societal definitions of their deviance, as well as the inability to see their personal experiences as political in nature, attests to this limited consciousness (Boutilier and SanGiovanni 1983). Athletes' responses are indicative of the degree to which they exhibit internal homophobia so common in American society.

Only a few athletes possessed deeper insight into factors that may underlie the labeling of women athletes as lesbians. For example, one responent felt women athletes were a "threat to men since they can stand on their own feet." Or, in another situation, an athlete viewed lesbian labeling as a means to devalue women athletes or successful women in general. Still another respondent suggested the label stemmed from jealousy and thus was used as a means to "get back" at women athletes. These rare remarks by respondents transcend the blame-the-victim view held by the majority of athletes. Such commitments indicate a deeper understanding of how homophobia and patriarchal ideology limit or control women's activities and their bodies.

Disempowerment

Given the silence surrounding the lesbian issue and the degree to which athletes have internalized societal images of lesbians and women athletes, the presence of the lesbian stereotype has negative ramifications for women athletes. Although sport participation possesses the potential for creativity and physical excellence (Theberge 1987), women modify their behavior so they will not be viewed as "stepping out of line." Women athletes become disempowered (Pharr 1988) through processes that detract from or reduce the self-actualizing potential of the sport experience.

Attaching the label of lesbian to women who engage in sport diminishes the sporting accomplishments of athletes. Women athletes are seen as something less than "real women" because they do not exemplify traditional female qualities (e.g., dependency, weakness, passivity); thus their accomplishments are not viewed as threatening to men (Birrell 1988). Interestingly, the athlete interviewed believed that the specific group most likely to engage in lesbian labeling was male athletes.

Discrediting women with the label of lesbian works further to control the number of females in sport, particularly in a homophobic society where prejudice against lesbians is intense (Birrell 1988; Zipter 1988). Keeping women out of sport, in turn, prevents females from discovering the power and joy of their own physicality (Birrell 1988) and experiencing the potential of their body. Moreover, discouraging women from participating in sport disempowers them by removing an arena where women can bond together (Birrell 1988; Cobban 1982)....

Another form of disempowerment occurs for those athletes who are lesbians. Intense homophobia often forces lesbians to deny their very essence, thus making the lesbian athlete invisible. Concealment, although protecting the lesbian athletes' identity, imposes psychological strain and can undermine positive self-conceptions (Schur 1984). Misrepresenting their sexuality, lesbian athletes are not in a position to confront the homophobia so prevalent in women's sport. Consequently, this ideology not only remains intact, but also is strengthened (Ettore 1980).

REFERENCES

Acosta, R. Vivian and Linda Jean Carpenter. 1992. "Women in Intercollegiate Sport: A Longitudinal Study—Fifteen Year Update 1977–1992." Unpublished manuscript, Brooklyn College, Department of Physical Education, Brooklyn.

Bennett, Roberts S., K. Gail Whitaker, Nina Jo Woolley Smith, and Anne Sablove. 1987. "Changing the Rules of the Game: Reflections Toward a Feminist Analysis of Sport." Women's Studies International Forum 10: 369–386.

Birrell, Susan. 1988. "Discourses on The Gender/Sport Relationship: From Women in Sport to Gender Relations." Pp. 459–502 in Exercise and Sport Science Reviews, vol. 16, edited by K. B. Pandolf. New York: MacMillan.

Blinde, Elaine M. and Diane E. Taub. 1992. "Women Athletes as Falsely Accused Deviants: Managing the Lesbian Stigma." The Sociological Quarterly 33: 521–533.

Boutilier, Mary A. and Lucinda, SanGiovanni. 1983. The Sporting Woman. Champaign, IL: Human Kinetics.

Cobban, Linn Ni. 1982. "Lesbians in Physical Education And Sport. Pp. 179–186 in Lesbian Studies: Present and Future, edited by M. Cruikshank. New York: Feminist Press.

Dunbar, John, Marvin Brown and Donald M. Amoroso. 1973. "Some Correlates of Attitudes Toward Homosexuality." Journal of Social Psychology 89: 271–279.

Dunkle, John H. and Patricia L. Francis. 1990. "The Role of Facial Masculinity/Feminity in The Attribution of Homosexuality." Sex Roles 23: 157–167.

Duquin, Mary E. 1981. "Feminism and Patriarchy in Physical Education." Paper presented at the annual meetings of the North American Society for the Sociology of Sport, Fort Worth, TX.

Ettore, E. M. 1980. Lesbians, Women and Society. London: Routledge and Kegan Paul.

Gartrell, Nanette. 1984. "Combatting Homophobia in the Psychotherapy of Lesbians." Women and Therapy 3: 13–29.

Gentile, S. 1982. "Out of The Kitchen." *City Sports Monthly* 8: 27.

Goffman, Erving. 1963. *Stigma: Notes on the Management of Spoiled Identity.* Englewood Cliffs, NJ: Prentice Hall.

Griffin, Patricia S. 1987. "Homophobia, Lesbians, and Women's Sports: An Exploratory Analysis." Paper presented at the annual meetings of the American Psychological Association, New York.

Gross, Martin L. 1978. *The Psychological Society.* New York: Simon & Schuster.

Hargreaves, Jennifer A. 1990. "Gender on the Sports Agenda." *International Review for Sociology of Sport* 25: 287–308.

Kaplan, Janice. 1979. *Women and Sports.* New York: Viking.

Kitzinger, Celia. 1987. *The Social Construction of Lesbianism.* London: Sage.

Koedt, Anne, Ellen Levine and Anita Rapone. 1973. *Radical Feminism.* New York: Quadrangle.

Lenskyj, Helen. 1986. *Out of Bounds: Women, Sport and Sexuality.* Toronto: Women's Press.

———. 1990. "Power and Play: Gender and Sexuality Issues in Sport and Physical Activity." *International Review for Sociology of Sport* 25: 235–245.

———. 1991. "Combatting Homophobia in Sport and Physical Education." *Sociology of Sport Journal* 8: 61–69.

Levitt, Eugene E. and Albert D. Klassen, Jr. 1974. "Public Attitudes toward Homosexuality: Part of the 1970 National Survey by the Institute for Sex Research." *Journal of Homosexuality* 1: 29–43.

McArthur, Leslie Z. 1982. "Judging A Book by Its Cover: A Cognitive Analysis of the Relationship between Physical Appearance and Stereotyping."

Pp.149–211 in *Cognitive Social Psychology,* edited by Albert H. Hastorf and Alice M. Isen. New York: Elsevier/North-Holland.

Pharr, Suzanne. 1988. *Homophobia: A Weapon of Sexism.* Inverness, CA: Clurdon.

Pheterson, Gail. 1986. "Alliances between Women: Overcoming Internalized Oppression and Internalized Domination." *Signs: Journal of Women in Culture and Society* 12: 146–160.

Schur, Edwin M. 1984. *Labeling Women Deviant: Gender, Stigma, and Social Control.* New York: McGraw-Hill.

Stockard, Jean and Miriam M. Johnson. 1980. *Sex Roles: Sex Inequality and Sex Role Development.* Englewood Cliffs, NJ: Prentice Hall.

Theberge, Nancy. 1987. "Sport And Women's Empowerment." *Women's Studies International Forum* 10: 387–393.

Unger, Rhoda K., Marcia Hilderbrand and Theresa, Madar. 1982. "Physical Attractiveness and Assumptions about Social Deviance: Some Sex-by-Sex Comparisons." *Personality and Social Psychology Bulletin* 8: 293–301.

Watson, Tracey. 1987. "Women Athletes and Athletic Women: The Dilemmas and Contradictions of Managing Incongruent Identities." *Sociological Inquiry* 57: 431–446.

Wolf, Charlotte. 1986. "Legitimation of Oppression: Response and Reflexivity." *Symbolic Interaction* 9: 217–234.

Zipter, Yvonne. 1988. *Diamonds Are a Dyke's Best Friend: Reflections, Reminiscences, and Reports from the Field on the Lesbian National Pastime.* Ithaca, NY: Firebrand Books.

20

The Mark of a Criminal Record

DEVAH PAGER

To measure concretely the interactive effects of race and a criminal record, Pager devised an experimental design where she constructed a fabricated pair of job applicants who were matched on all features except their criminal history. She then sent out these matched pairs, White pairs and Black pairs, to apply for real jobs in Milwaukee and noted how far the candidates got in the interview process. Along the way she recorded the employer's likelihood of dismissing the applicants right away, checking their references, calling them back for further interviews, and offering them the job. In a real demonstration of the effects of race and criminal record on employment opportunities, Pager found that while Whites were offered more jobs than Blacks, and while applicants with no criminal history were offered more jobs than those who had served time, even Whites with criminal pasts were more likely to be hired than Blacks who had led law-abiding lives. She found that employers were more likely to hold stereotypes suspecting Blacks, especially young Black men, of being prone to crime and of being unreliable employees.

While stratification researchers typically focus on schools, labor markets, and the family as primary institutions affecting inequality, a new institution has emerged as central to the sorting and stratifying of young and disadvantaged men: the criminal justice system. With over 2 million individuals currently incarcerated, and over half a million prisoners released each year, the large and growing numbers of men being processed through the criminal justice system raises important questions about the consequences of this massive institutional intervention.

This article focuses on the consequences of incarceration for the employment outcomes of black and white men. While previous survey research has demonstrated a strong *association* between incarceration and employment, there remains little understanding of the mechanisms by which these outcomes are produced. In the present study, I adopt an experimental audit approach to formally test the degree to which a criminal record affects subsequent employment opportunities. By using matched pairs of individuals to apply for real entry-level jobs, it becomes possible to directly measure the extent to which a criminal record—in the absence of other disqualifying characteristics—serves as a barrier to employment among equally qualified applicants. Further, by varying the race of the tester pairs, we can assess the ways in which the effects of race and criminal record interact to produce new forms of labor market inequalities.

TRENDS IN INCARCERATION

Over the past three decades, the number of prison inmates in the United States has increased by more than 600%, leaving it the country with the highest incarceration rate in the world (Bureau of Justice Statistics 2002; Barclay, Tavares, and Siddique 2001). During this time, incarceration has changed from a punishment reserved primarily for the most heinous offenders to one extended to a much greater range of crimes and a much larger segment of the population. Recent trends in crime policy have led to the imposition of harsher sentences for a wider range of offenses, thus casting an ever-widening net of penal intervention.[1]

While the recent "tough on crime" policies may be effective in getting criminals off the streets, little provision has been made for when they get back out. Of the nearly 2 million individuals currently incarcerated, roughly 95% will be released, with more than half a million being released each year (Slevin 2000). According to one estimate, there are currently over 12 million exfelons in the United States, representing roughly 8% of the working-age population (Uggen, Thompson, and Manza 2000). Of those recently released, nearly two-thirds will be charged with new crimes and over 40% will return to prison within three years (Bureau of Justice Statistics 2000). Certainly some of these outcomes are the result of desolate opportunities or deeply ingrained dispositions, grown out of broken families, poor neighborhoods, and little social control (Sampson and Laub 1993; Wilson 1997). But net of these contributing factors, there is evidence that experience with the criminal justice system in itself has adverse consequences for subsequent opportunities. In particular, incarceration is associated with limited future employment opportunities and earnings potential (Freeman 1987; Western 2002), which themselves are among the strongest predictors of recidivism (Shover 1996; Sampson and Laub 1993; Uggen 2000).

The expansion of the prison population has been particularly consequential for blacks. The incarceration rate for young black men in the year 2000 was nearly 10%, compared to just over 1% for white men in the same age group (Bureau of Justice Statistics 2001). Young black men today have a 28% likelihood of incarceration during their lifetime (Bureau of Justice Statistics 1997), a figure that rises above 50% among young black high school dropouts (Pettit and Western 2001). These vast numbers of inmates translate into a large and increasing population of black ex-offenders returning to communities and searching for work. The barriers these men face in reaching economic self-sufficiency are compounded by the stigma of minority status and criminal record. The consequences of such trends for widening racial disparities are potentially profound (see Western and Pettit 1999; Freeman and Holzer 1986).

The objective of this study is to assess whether the effect of a criminal record differs for black and white applicants. Most research investigating the differential impact of incarceration on blacks has focused on the differential *rates* of incarceration and how those rates translate into widening racial disparities. In addition to disparities in the rate of incarceration, however, it is also important to consider possible racial differences in the *effects* of incarceration. Almost none of the existing literature to date has explored this issue, and the theoretical arguments remain divided as to what we might expect.

On one hand, there is reason to believe that the signal of a criminal record should be less consequential for blacks. Research on racial stereotypes tells us that Americans hold strong and persistent negative stereotypes about blacks, with one of the most readily invoked contemporary stereotypes relating to perceptions of violent and criminal dispositions (Smith 1991; Sniderman and Piazza 1993; Devine and Elliott 1995). If it is the case that employers view all blacks as potential criminals, they are likely to differentiate less among those with official criminal records and those without. Actual confirmation of criminal involvement then will provide only redundant information, while evidence against it will be discounted. In this case, the outcomes for all blacks should be worse, with less differentiation between those with criminal records and those without.

On the other hand, the effect of a criminal record may be worse for blacks if employers, already wary of black applicants, are more hesitant when it comes to taking risks on blacks with proven criminal tendencies. The literature on racial stereotypes also tells us that stereotypes are most likely to be activated and reinforced when a target matches on more than one dimension of the stereotype (Quillian and Pager 2002; Darley and Gross 1983; Fiske and Neuberg 1990). While employers may have learned to keep their racial attributions in check through years of heightened sensitivity around employment discrimination, when combined with knowledge of a criminal history, negative attributions are likely to intensify.

A third possibility, of course, is that a criminal record affects black and white applicants equally. The results of this audit study will help to adjudicate between these competing predictions.

STUDY DESIGN

The basic design of this study involves the use of four male auditors (also called testers), two blacks and two whites. The testers were 23-year-old college students from Milwaukee who were matched on the basis of physical appearance and general style of self-presentation. Objective characteristics that were not already identical between pairs—such as educational attainment and work experience— were made similar for the purpose of the applications. Within each team, one auditor was randomly assigned a "criminal record" for the first week; the pair then rotated which member presented himself as the ex-offender for each successive week of employment searches, such that each tester served in the criminal record condition for an equal number of cases. By varying which member of the pair presented himself as having a criminal record, unobserved differences within the pairs of applicants were effectively controlled. No significant differences were found for the outcomes of individual testers or by month of testing.

Job openings for entry-level positions (defined as jobs requiring no previous experience and no education greater than high school) were identified from the Sunday classified advertisement section of the *Milwaukee Journal Sentinel*.[2] In addition, a supplemental sample was drawn from *Jobnet,* a state-sponsored web site for employment listings, which was developed in connection with the W-2 Welfare-to-Work initiatives.[3]

The audit pairs were randomly assigned 15 job openings each week. The white pair and the black pair were assigned separate sets of jobs, with the same-race testers applying to the same jobs. One member of the pair applied first, with the second applying one day later (randomly varying whether the ex-offender was first or second). A total of 350 employers were audited during the course of this study: 150 by the white pair and 200 by the black pair. Additional tests were performed by the black pair because black testers received fewer callbacks on average, and there were thus fewer data points with which to draw comparisons. A larger sample size enabled me to calculate more precise estimates of the effects under investigation.

Immediately following the completion of each job application, testers filled out a six-page response form that coded relevant information from the test. Important variables included type of occupation, metropolitan status, wage, size of establishment, and race and sex of employer.[4] Additionally, testers wrote narratives describing the overall interaction and any comments made by employers (or included on applications) specifically related to race or criminal records.

TESTER PROFILES

In developing the tester profiles, emphasis was placed on adopting characteristics that were both numerically representative and substantively important. In the present study, the criminal record consisted of a felony drug conviction (possession with intent to distribute, cocaine) and 18 months of (served) prison time. A drug crime (as opposed to a violent or property crime) was chosen because of its prevalence, its policy salience, and its connection to racial disparities in incarceration.[5] It is important to acknowledge that the effects reported here may differ depending on the type of offense.

THE EFFECT OF A CRIMINAL RECORD FOR WHITES

I begin with an analysis of the effect of a criminal record among whites. White noncriminals can serve as our baseline in the following comparisons, representing the presumptively nonstigmatized group relative to blacks and those with criminal records. Given that all testers presented roughly identical credentials, the

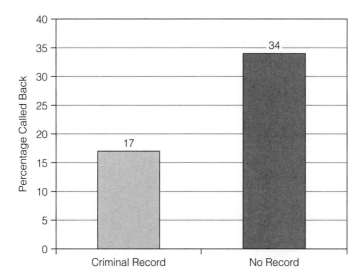

FIGURE 20.1 The effect of a criminal record on employment opportunities for whites. The effect of a criminal record is statistically significant ($P < .01$).

differences experienced among groups of testers can be attributed fully to the effects of race or criminal status.

Figure 20.1 shows the percentage of applications submitted by white testers that elicited callbacks from employers, by criminal status. As illustrated below, there is a large and significant effect of a criminal record, with 34% of whites without criminal records receiving callbacks, relative to only 17% of whites with criminal records. A criminal record thereby reduces the likelihood of a callback by 50%.

There were some fairly obvious examples documented by testers that illustrate the strong reaction among employers to the signal of a criminal record. In one case, a white tester in the criminal record condition went to a trucking service to apply for a job as a dispatcher. The tester was given a long application, including a complex math test, which took nearly 45 minutes to fill out. During the course of this process, there were several details about the application and the job that needed clarification, some of which involved checking with the supervisor about how to proceed. No concerns were raised about his candidacy at this stage. When the tester turned the application in, the secretary brought it into a back office for the supervisor to look over, so that an interview could perhaps be conducted. When the secretary came back out, presumably after the supervisor had a chance to look over the application more thoroughly, he was told the position had already been filled. While, of course, isolated incidents like this are not conclusive, this was not an infrequent occurrence. Often testers reported seeing employers' levels of responsiveness change dramatically once they had glanced down at the criminal record question.

Clearly, the results here demonstrate that criminal records close doors in employment situations. Many employers seem to use the information as a screening mechanism, without attempting to probe deeper into the possible context or complexities of the situation. As we can see here, in 50% of cases, employers were unwilling to consider equally qualified applicants on the basis of their criminal record.

Of course, this trend is not true among all employers, in all situations. There were, in fact, some employers who seemed to prefer workers who had been recently released from prison. One owner told a white tester in the criminal record condition that he "like[d] hiring people who ha[d] just come out of prison because they tend to be more motivated, and are more likely to be hard workers [not wanting to return to prison]." Another employer for a cleaning company attempted to dissuade the white noncriminal tester from applying because the job involved "a great deal of dirty work." The tester with the criminal record, on the other hand, was offered the job on the spot. A criminal record is thus not an obstacle in all cases, but on average, as we see above, it reduces employment opportunities substantially.

THE EFFECT OF RACE

A second major focus of this study concerns the effect of race. African–Americans continue to suffer from lower rates of employment relative to whites, but there is tremendous disagreement over the source of these disparities. The idea that race itself—apart from other correlated characteristics—continues to play a major role in shaping employment opportunities has come under question in recent years (e.g., D'Souza 1995; Steele 1991). The audit methodology is uniquely suited to address this question. While the present study design does not provide the kind of cross-race matched-pair tests that earlier audit studies of racial discrimination have used, the between-group comparisons (white pair vs. black pair) can nevertheless offer an unbiased estimate of the effect of race on employment opportunities.

Figure 20.2 presents the percentage of callbacks received for both categories of black testers relative to those for whites. The effect of race in these findings is strikingly large. Among blacks without criminal records, only 14% received callbacks, relative to 34% of white noncriminals ($P < .01$). In fact, even whites *with* criminal records received more favorable treatment (17%) than blacks *without* criminal records (14%). The rank ordering of groups in this graph is painfully revealing of employer preferences: race continues to play a dominant role in shaping employment opportunities, equal to or greater than the impact of a criminal record.

The magnitude of the race effect found here corresponds closely to those found in previous audit studies directly measuring racial discrimination. Bendick et al. (1994), for example, found that blacks were 24 percentage points less likely to receive a job offer relative to their white counterparts, a finding very close to the 20 percentage point difference (between white and black nonoffenders) found here. Thus in the eight years since the last major employment audit of race was conducted, very little has changed in the reaction of employers to

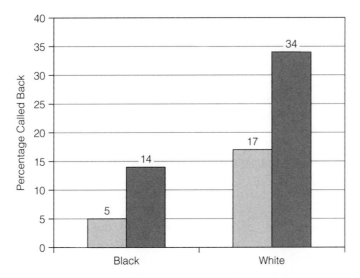

FIGURE 20.2 The effect of a criminal record for black and white job applicants. The main effects of race and criminal record are statistically significant ($P < .01$). The interaction between the two is not significant in the full sample. Light bars represent criminal record; dark bars represent no criminal record.

minority applicants. Despite the many rhetorical arguments used to suggest that direct racial discrimination is no longer a major barrier to opportunity (e.g., D'Souza 1995; Steele 1991), as we can see here, employers, at least in Milwaukee, continue to use race as a major factor in hiring decisions.

RACIAL DIFFERENCES IN THE EFFECTS OF A CRIMINAL RECORD

The final question this study sought to answer was the degree to which the effect of a criminal record differs depending on the race of the applicant. Based on the results presented in Figure 20.2, the effect of a criminal record appears more pronounced for blacks than it is for whites. While this interaction term is not statistically significant, the magnitude of the difference is nontrivial. While the ratio of callbacks for nonoffenders relative to ex-offenders for whites is 2:1, this same ratio for blacks is nearly 3:1. The effect of a criminal record is thus 40% larger for blacks than for whites.

This evidence is suggestive of the way in which associations between race and crime affect interpersonal evaluations. Employers, already reluctant to hire blacks, appear even more wary of blacks with proven criminal involvement. Despite the face that these testers were bright articulate college students with effective styles of self-presentation, the cursory review of entry-level applicants leaves little room for these qualities to be noticed. Instead, the employment barriers of minority status and criminal record are compounded, intensifying the stigma toward this group.

The salience of employers' sensitivity toward criminal involvement among blacks was highlighted in several interactions documented by testers. On three separate occasions, for example, black testers were asked in person (before submitting their applications) whether they had a prior criminal history. None of the white testers were asked about their criminal histories up front.

DISCUSSION

There is serious disagreement among academics, policy makers, and practitioners over the extent to which contact with the criminal justice system—in itself—leads to harmful consequences for employment. The present study takes a strong stand in this debate by offering direct evidence of the causal relationship between a criminal record and employment outcomes. While survey research has produced noisy and indirect estimates of this effect, the current research design offers a direct measure of a criminal record as a mechanism producing employment disparities. Using matched pairs and an experimentally assigned criminal record, this estimate is unaffected by the problems of selection, which plague observational data. While certainly there are additional ways in which incarceration may affect employment outcomes, this finding provides conclusive evidence that mere contact with the criminal justice system, in the absence of any transformative or selective effects, severely limits subsequent employment opportunities. And while the audit study investigates employment barriers to ex-offenders from a microperspective, the implications are far-reaching. The finding that ex-offenders are only one-half to one-third as likely as nonoffenders to be considered by employers suggests that a criminal record indeed presents a major barrier to employment. With over 2 million people currently behind bars and over 12 million people with prior felony convictions, the consequences for labor market inequalities are potentially profound.

Second, the persistent effect of race on employment opportunities is painfully clear in these results. Blacks are less than half as likely to receive consideration by employers, relative to their white counterparts, and black nonoffenders fall behind even whites with prior felony convictions. The powerful effects of race thus continue to direct employment decisions in ways that contribute to persisting racial inequality. In light of these findings, current public opinion seems largely misinformed. According to a recent survey of residents in Los Angeles, Boston, Detroit, and Atlanta, researchers found that just over a quarter of whites believe there to be "a lot" of discrimination against blacks, compared to nearly two-thirds of black respondents (Kluegel and Bobo 2001). Over the past decade, affirmative action has come under attack across the country based on the argument that direct racial discrimination is no longer a major barrier to opportunity. According to this study, however, employers, at least in Milwaukee, continue to use race as a major factor in their hiring decisions. When we combine the effects of race and criminal record, the problem grows more intense. Not only are blacks much more likely to be incarcerated than whites; based on the findings presented here, they may also be more strongly affected by the impact of a criminal record. Previous estimates of the aggregate consequences of incarceration may therefore underestimate the impact on racial disparities.

Finally, in terms of policy implications, this research has troubling conclusions. In our frenzy of locking people up, our "crime control" policies may in fact exacerbate the very conditions that lead to crime in the first place. Research consistently shows that finding quality steady employment is one of the strongest predictors of desistance from crime (Shover 1996; Sampson and Laub 1993; Uggen 2000). The fact that a criminal record severely limits employment opportunities—particularly among blacks—suggests that these individuals are left with few viable alternatives.[6]

As more and more young men enter the labor force from prison, it becomes increasingly important to consider the impact of incarceration on the job prospects of those coming out. No longer a peripheral institution, the criminal justice system has become a dominant presence in the lives of young disadvantaged men, playing a key role in the sorting and stratifying of labor market opportunities. This article represents an initial attempt to specify one of the important mechanisms by which incarceration leads to poor employment outcomes. Future research is needed to expand this emphasis to other mechanisms (e.g., the transformative effects of prison on human and social capital), as well as to include other social domains affected by incarceration (e.g., housing, family formation, political participation, etc.); in this way, we can move toward a more complete understanding of the collateral consequences of incarceration for social inequality.

At this point in history, it is impossible to tell whether the massive presence of incarceration in today's stratification system represents a unique anomaly of the late 20th century, or part of a larger movement toward a system of stratification based on the official certification of individual character and competence. Whether this process of negative credentialing will continue to form the basis of emerging social cleavages remains to be seen.

NOTES

1. For example, the recent adoption of mandatory sentencing laws, most often used for drug offenses, removes discretion from the sentencing judge to consider the range of factors pertaining to the individual and the offense that would normally be taken into account. As a result, the chances of receiving a state prison term after being arrested for a drug offense rose by 547% between 1980 and 1992 (Bureau of Justice Statistics 1995).

2. Occupations with legal restrictions on ex–offenders were excluded from the sample. These include jobs in the health care industry, work with children and the elderly, jobs requiring the handling of firearms (i.e., security guards), and jobs in the public sector. An estimate of the collateral consequences of incarceration

would also need to take account of the wide range of employment formally off-limits to individuals with prior felony convictions.

3. Employment services like *jobnet* have become a much more common method of finding employment in recent years, particularly for difficult-to-employ populations such as welfare recipients and ex-offenders. Likewise, a recent survey by Holzer and Stoll (2001) found that nearly half of Milwaukee employers (46%) use *Jobnet* to advertise vacancies in their companies.

4. See Pager (2002) for a discussion of the variation across each of these dimensions.

5. Over the past two decades, drug crimes were the fastest growing class of offenses. In 1980, roughly one

out of every 16 state inmates was incarcerated for a drug crime; by 1999, this figure had jumped to one out of every five (Bureau of Justice Statistics 2000). In federal prisons, nearly three out of every five inmates are incarcerated for a drug crime (Bureau of justice Statistics 2001). A significant portion of this increase can be attributed to changing policies concerning drug enforcement. By 2000, every state in the country had adopted some form of truth-in-sentencing laws, which impose mandatory sentencing minimums for a range of offenses. These laws have been applied most frequently to drug crimes, leading to more than a fivefold rise in the number of drug arrests that result in incarceration and a doubling of the average length of sentences for drug convictions (Mauer 1999; Blumstein and Beck 1999). While the steep rise in drug enforcement has been felt across the population, this "war on drugs" has had a disproportionate impact on African-Americans. Between 1990 and 1997, the number of black inmates serving time for drug offenses increased by 60%, compared to a 46% increase in the number of whites (Bureau of justice Statistics 1995). In 1999, 26% of all black state inmates were incarcerated for drug offenses, relative to less than half that proportion of whites (Bureau of justice Statistics 2001).

6. There are two primary policy recommendations implied by these results. First and foremost, the widespread use of incarceration, particularly for nonviolent drug crimes, has serious, long-term consequences for the employment problems of young men. The substitution of alternatives to incarceration, therefore, such as drug treatment programs or community supervision, may serve to better promote the well-being of individual offenders as well as to improve public safety more generally through the potential reduction of recidivism. Second, additional thought should be given to the widespread availability of criminal background information. As criminal record databases become increasingly easy to access, this information may be more often used as the basis for rejecting otherwise qualified applicants. If instead criminal history information were suppressed—except in cases that were clearly relevant to a particular kind of job assignment—ex-offenders with appropriate credentials might be better able to secure legitimate employment. While there is some indication that the absence of official criminal background information may lead to a greater incidence of statistical discrimination against blacks (see Bushway 1997; Holzer et al. 2001), the net benefits of this policy change may in fact outweigh the potential drawbacks.

REFERENCES

Barclay, Gordon, Cynthia Tavares, and Arsalaan, Siddique. 2001. "International Comparisons of Criminal Justice Statistics, 1999." London: U.K. Home Office for Statistical Research.

Bendick, Marc, Jr., Charles Jackson, and Victor, Reinoso. 1994. "Measuring Employment Discrimination through Controlled Experiments." *Review of Black Political Economy* 23:25–48.

Blumstein, Alfred, and Allen J. Beck. 1999. "Population Growth in U.S. Prisons, 1980–1996." Pp. 17–62 in *Prisons: Crime and Justice: A Review of Research*, vol. 26. Edited by Michael Tonry and J. Petersilia. Chicago: University of Chicago Press.

Bureau of Justice Statistics. 1995. *Prisoners in 1994*, by Allen J. Beck and Darrell K. Gilliard. Special report. Washington, D.C.: Government Printing Office.

———. 1997. *Lifetime Likelihood of Going to State or Federal Prison*, by Thomas P. Bonczar and Allen J. Beck. Special report, March. Washington, DC.

———. 2000. Bulletin. *Key Facts at a Glance: Number of Persons in Custody of State Correctional Authorities by Most Serious Offense 1980–99*. Washington, D.C.: Government Printing Office.

———. 2001. *Prisoners in 2000*, by Allen J. Beck and Paige M. Harrison. August. Bulletin. Washington, D.C.: NCJ 188207.

———. 2002. *Sourcebook of Criminal Justice Statistics*. Last accessed March 1, 2003. Available http:// www.albany.edu/sourcebook/

Bushway, Shawn D. 1997. "Labor Market Effects of Permitting Employer Access to Criminal History Records." Working paper. University of Maryland, Department of Criminology.

Darley, J. M., and P. H., Gross. 1983. "A Hypothesis-Confirming Bias in Labeling Effects." *Journal of Personality and Social Psychology* 44:20–33.

Devine, P. G., and A. J., Elliot. 1995. "Are Racial Stereotypes Really Fading? The Princeton Trilogy Revisited." *Personality and Social Psychology Bulletin* 21 (11): 1139–50.

D'Souza, Dinesh. 1995. *The End of Racism: Principles for a Multiracial Society*. New York: Free Press.

Fiske, Susan, and Steven Neuberg. 1990. "A Continuum of Impression Formation, from Category-Based to Individuating Processes." Pp. 1–63 in *Advances in Experimental Social Psychology*, vol. 23. Edited by Mark Zanna. New York: Academic Press.

Freeman, Richard B. 1987. "The Relation of Criminal Activity to Black Youth Employment." *Review of Black Political Economy* 16 (1–2): 99–107.

Freeman, Richard B., and Harry J. Holzer, eds. 1986. *The Black Youth Employment Crisis*. Chicago: University of Chicago Press for National Bureau of Economic Research.

Holzer, Harry, Steven Raphael, and Michael, Stoll. 2001. "Perceived Criminality, Criminal Background Checks and the Racial Hiring Practices of Employers." Discussion Paper no. 1254–02. University of Wisconsin—Madison, Institute for Research on Poverty.

Holzer, Harry, and Michael, Stoll. 2001. *Employers and Welfare Recipients: The Effects of Welfare Reform in the Workplace*. San Francisco: Public Policy Institute of California.

Kluegel, James, and Lawrence Bobo. 2001. "Perceived Group Discrimination and Policy Attitudes: The

Sources and Consequences of the Race and Gender Gaps." Pp. 163–216 in *Urban Inequality: Evidence from Four Cities*, edited by Alice O'Connor, Chris Tilly, and Lawrence D. Bobo. New York: Russell Sage Foundation.

Mauer, Marc. 1999. *Race to Incarcerate*. New York: New Press.

Pager, Devah. 2002. "The Mark of a Criminal Record." Doctoral dissertation. Department of Sociology, University of Wisconsin—Madison.

Pettit, Becky, and Bruce, Western. 2001. "Inequality in Lifetime Risks of Imprisonment." Paper presented at the annual meetings of the American Sociological Association. Anaheim, August.

Quillian, Lincoln, and Devah, Pager. 2002. "Black Neighbors, Higher Crime? The Role of Racial Stereotypes in Evaluations of Neighborhood Crime." *American Journal of Sociology* 107 (3): 717–67.

Sampson, Robert J., and John H. Laub. 1993. *Crime in the Making: Pathways and Turning Points through Life*. Cambridge, Mass.: Harvard University Press.

Shover, Neil. 1996. *Great Pretenders: Pursuits and Careers of Presistent Thieves*. Boulder, Colo.: Westview.

Slevin, Peter. 2000. "Life after Prison: Lack of Services Has High Price." *Washington Post*, April 24.

Smith, Tom W. 1991. *What Americans Say about Jews*. New York: American Jewish Committee.

Sniderman, Paul M., and Thomas, Piazza. 1993. *The Scar of Race*. Cambridge, Mass.: Harvard University Press.

Steele, Shelby. 1991. *The Content of Our Character: A New Vision of Race in America*. New York: Harper Perennial.

Uggen, Christopher. 2000. "Work as a Turning Point in the Life Course of Criminals: A Duration Model of Age, Employment, and Recidivism." *American Sociological Review* 65 (4): 529–46.

Uggen, Christopher, Melissa Thompson, and Jeff, Manza. 2000. "Crime, Class, and Reintegration: The Socioeconomic, Familial, and Civic Lives of Offenders." Paper presented at the American Society of Criminology meetings, San Francisco, November 18.

Western, Bruce. 2002. "The Impact of Incarceration on Wage Mobility and Inequality." *American Sociological Review* 67 (4): 526–46.

Western, Bruce, and Becky, Pettit. 1999. "Black–White Earnings Inquality, Employment Rates, and Incarceration." Working Paper no. 150. New York: Russell Sage Foundation.

Wilson, William Julius. 1997. *When Work Disappears: The World of the New Urban Poor*. New York: Vintage Books.

21

The Saints and the Roughnecks

WILLIAM J. CHAMBLISS

Chambliss's description of the Saints and the Roughnecks shows how the power of social class can operate to facilitate groups' resistance of deviant labels. In this classic selection from the sociological literature, Chambliss describes how the Saints engage in as many or more delinquent acts than the Roughnecks, yet are perceived as "good boys," merely engaging in typical adolescent hijinks. The greater social power contained in their higher class background enables the definition of their behavior as socially normative, allowing the police, teachers, community members, and parents to look the other way. On the other hand, the Roughnecks, who come from the "wrong side of the tracks," are perceived to be troublemakers, rabble-rousers, and delinquents. We see conflict and labeling theories in effect here since social class is the determinant of society's reactions. Behavior done by teenagers from upstanding, middle-class families is tolerated, while similar behavior engaged in by lower-class youth is reinforced as deviant. Once again, labels are applied based on status, not on patterns of behavior. In a repeat of the self-fulfilling prophesy, the Roughnecks experience society's expectations of them as deviant boys headed toward a lifetime of criminality and live up to that label while the Saints fulfill their social reinforcement to graduate out of delinquency and move into the white-collar world.

Eight promising young men—children of good, stable, white, upper-middle-class families, active in school affairs, good pre-college students—were some of the most delinquent boys at Hanibal High School. While community residents and parents knew that these boys occasionally sowed a few wild oats, they were totally unaware that sowing wild oats completely occupied the daily routine of these young men. The Saints were constantly occupied with truancy, drinking, wild driving, petty theft, and vandalism. Yet not one was officially arrested for any misdeed during the two years I observed them.

Reprinted by permission of Transaction Publishers. "The Saints and the Roughnecks," by William J. Chambliss, from *Society,* V. 11, No. 1, 1973. Copyright © 1973 Transaction Publishers.

This record was particularly surprising in light of my observations during the same two years of another gang of Hanibal High School students, six lower-class white boys known as the Roughnecks. The Roughnecks were constantly in trouble with police and community even though their rate of delinquency was about equal with that of the Saints. What was the cause of this disparity? the result? The following consideration of the activities, social class, and community perceptions of both gangs may provide some answers.

THE SAINTS FROM MONDAY TO FRIDAY

The Saints' principal daily concern was with getting out of school as early as possible. The boys managed to get out of school with minimum danger that they would be accused of playing hookey through an elaborate procedure for obtaining "legitimate" release from class. The most common procedure was for one boy to obtain the release of another by fabricating a meeting of some committee, program, or recognized club. Charles might raise his hand in his 9:00 chemistry class and ask to be excused—a euphemism for going to the bathroom. Charles would go to Ed's math class and inform the teacher that Ed was needed for a 9:30 rehearsal of the drama club play. The math teacher would recognize Ed and Charles as "good students" involved in numerous school activities and would permit Ed to leave at 9:30. Charles would return to his class, and Ed would go to Tom's English class to obtain his release. Tom would engineer Charles's escape. The strategy would continue until as many of the Saints as possible were freed. After a stealthy trip to the car (which had been parked in a strategic spot), the boys were off for a day of fun.

Over the two years I observed the Saints, this pattern was repeated nearly every day. There were variations on the theme, but in one form or another, the boys used this procedure for getting out of class and then off the school grounds. Rarely did all eight of the Saints manage to leave school at the same time. The average number avoiding school on the days I observed them was five.

Having escaped from the concrete corridors the boys usually went either to a pool hall on the other (lower-class) side of town or to a cafe in the suburbs. Both places were out of the way of people the boys were likely to know (family or school officials), and both provided a source of entertainment. The pool hall entertainment was the generally rough atmosphere, the occasional hustler, the sometimes drunk proprietor, and, of course, the game of pool. The cafe's entertainment was provided by the owner. The boys would "accidentally" knock a glass on the floor or spill cola on the counter—not all the time, but enough to be sporting. They would also bend spoons, put salt in sugar bowls, and generally tease whoever was working in the cafe. The owner had opened the cafe recently and was dependent on the boys' business which was, in fact, substantial since between the horsing around and the teasing they bought food and drinks.

THE SAINTS ON WEEKENDS

On weekends the automobile was even more critical than during the week, for on weekends the Saints went to Big Town—a large city with a population of over a million 25 miles from Hanibal. Every Friday and Saturday night most of the Saints would meet between 8:00 and 8:30 and would go into Big Town. Big Town activities included drinking heavily in taverns or nightclubs, driving drunkenly through the streets, and committing acts of vandalism and playing pranks.

By midnight on Fridays and Saturdays the Saints were usually thoroughly high, and one or two of them were often so drunk they had to be carried to the cars. Then the boys drove around town, calling obscenities to women and girls; occasionally trying (unsuccessfully so far as I could tell) to pick girls up; and driving recklessly through red lights and at high speeds with their lights out. Occasionally they played "chicken." One boy would climb out the back window of the car and across the roof to the driver's side of the car while the car was moving at high speed (between 40 and 50 miles an hour); then the driver would move over and the boy who had just crawled across the car roof would take the driver's seat.

Searching for "fair game" for a prank was the boys' principal activity after they left the tavern. The boys would drive alongside a foot patrolman and ask directions to some street. If the policeman leaned on the car in the course of answering the question, the driver would speed away, causing him to lose his balance. The Saints were careful to play this prank only in an area where they were not going to spend much time and where they could quickly disappear around a corner to avoid having their license plate number taken.

Construction sites and road repair areas were the special province of the Saints' mischief. A soon-to-be-repaired hole in the road inevitably invited the Saints to remove lanterns and wooden barricades and put them in the car, leaving the hole unprotected. The boys would find a safe vantage point and wait for an unsuspecting motorist to drive into the hole. Often, though not always, the boys would go up to the motorist and commiserate with him about the dreadful way the city protected its citizenry.

Leaving the scene of the open hole and the motorist, the boys would then go searching for an appropriate place to erect the stolen barricade. An "appropriate place" was often a spot on a highway near a curve in the road where the barricade would not be seen by an oncoming motorist. The boys would wait to watch an unsuspecting motorist attempt to stop and (usually) crash into the wooden barricade. With saintly bearing the boys might offer help and understanding.

A stolen lantern might well find its way onto the back of a police car or hang from a street lamp. Once a lantern served as a prop for a reenactment of the "midnight ride of Paul Revere" until the "play," which was taking place at 2:00 A.M. in the center of a main street of Big Town, was interrupted by a police car several blocks away. The boys ran, leaving the lanterns on the street, and managed to avoid being apprehended.

Abandoned houses, especially if they were located in out-of-the-way places, were fair game for destruction and spontaneous vandalism. The boys would break windows, remove furniture to the yard and tear it apart, urinate on the walls, and scrawl obscenities inside.

Through all the pranks, drinking, and reckless driving the boys managed miraculously to avoid being stopped by police. Only twice in two years was I aware that they had been stopped by a Big City policeman. Once was for speeding (which they did every time they drove whether they were drunk or sober), and the driver managed to convince the policeman that it was simply an error. The second time they were stopped they had just left a nightclub and were walking through an alley. Aaron stopped to urinate and the boys began making obscene remarks. A foot patrolman came into the alley, lectured the boys, and sent them home. Before the boys got to the car one began talking in a loud voice again. The policeman, who had followed them down the alley, arrested this boy for disturbing the peace and took him to the police station where the other Saints gathered. After paying a $5.00 fine, and with the assurance that there would be no permanent record of the arrest, the boy was released.

The boys had a spirit of frivolity and fun about their escapades. They did not view what they were engaged in as "delinquency," though it surely was by any reasonable definition of that word. They simply viewed themselves as having a little fun and who, they would ask, was really hurt by it? The answer had to be no one, although this fact remains one of the most difficult things to explain about the gang's behavior. Unlikely though it seems, in two years of drinking, driving, carousing, and vandalism no one was seriously injured as a result of the Saints' activities.

THE SAINTS IN SCHOOL

The Saints were highly successful in school. The average grade for the group was "B," with two of the boys having close to a straight "A" average. Almost all of the boys were popular and many of them held offices in the school. One of the boys was vice-president of the student body one year. Six of the boys played on athletic teams.

At the end of their senior year, the student body selected ten seniors for special recognition as the "school wheels"; four of the ten were Saints. Teachers and school officials saw no problem with any of these boys and anticipated that they would all "make something of themselves."

How the boys managed to maintain this impression is surprising in view of their actual behavior while in school. Their technique for covering truancy was so successful that teachers did not even realize that the boys were absent from school much of the time. Occasionally, of course, the system would backfire and then the boy was on his own. A boy who was caught would be most contrite, would plead guilty and ask for mercy. He inevitably got the mercy he sought.

Cheating on examinations was rampant, even to the point of orally communicating answers to exams as well as looking at one another's papers. Since none of the group studied, and since they were primarily dependent on one another for help, it is surprising that grades were so high. Teachers contributed to the deception in their admitted inclination to give these boys (and presumably others like them) the benefit of the doubt. When asked how the boys did in school, and when pressed on specific examinations, teachers might admit that they were disappointed in John's performance, but would quickly add that they "knew that he was capable of doing better," so John was given a higher grade than he had actually earned. How often this happened is impossible to know. During the time that I observed the group, I never saw any of the boys take homework home. Teachers may have been "understanding" very regularly.

One exception to the gang's generally good performance was Jerry, who had a "C" average in his junior year, experienced disaster the next year, and failed to graduate. Jerry had always been a little more nonchalant than the others about the liberties he took in school. Rather than wait for someone to come get him from class, he would offer his own excuse and leave. Although he probably did not miss any more classes than most of the others in the group, he did not take the requisite pains to cover his absences. Jerry was the only Saint whom I ever heard talk back to a teacher. Although teachers often called him a "cut up" or a "smart kid," they never referred to him as a troublemaker or as a kid headed for trouble. It seems likely, then, that Jerry's failure his senior year and his mediocre performance his junior year were consequences of his not playing the game the proper way (possibly because he was disturbed by his parents' divorce). His teachers regarded him as "immature" and not quite ready to get out of high school.

THE POLICE AND THE SAINTS

The local police saw the Saints as good boys who were among the leaders of the youth in the community. Rarely, the boys might be stopped in town for speeding or for running a stop sign. When this happened the boys were always polite, contrite, and pled for mercy. As in school, they received the mercy they asked for. None ever received a ticket or was taken into the precinct by the local police.

The situation in Big City, where the boys engaged in most of their delinquency, was only slightly different. The police there did not know the boys at all, although occasionally the boys were stopped by a patrolman. Once they were caught taking a lantern from a construction site. Another time they were stopped for running a stop sign, and on several occasions they were stopped for speeding. Their behavior was as before: contrite, polite, and penitent. The urban police, like the local police, accepted their demeanor as sincere. More important, the urban police were convinced that these were good boys just out for a lark.

THE ROUGHNECKS

Hanibal townspeople never perceived the Saints' high level of delinquency. The Saints were good boys who just went in for an occasional prank. After all, they were well dressed, well mannered, and had nice cars. The Roughnecks were a different story. Although the two gangs of boys were the same age, and both groups engaged in an equal amount of wild-oat sowing, everyone agreed that the not-so-well-dressed, not-so-well-mannered, not-so-rich boys were heading for trouble. Townspeople would say, "You can see the gang members at the drugstore, night after night, leaning against the storefront (sometimes drunk) or slouching around inside buying cokes, reading magazines, and probably stealing old Mr. Wall blind. When they are outside and girls walk by, even respectable girls, these boys make suggestive remarks. Sometimes their remarks are downright lewd."

From the community's viewpoint, the real indication that these kids were in for trouble was that they were constantly involved with the police. Some of them had been picked up for stealing, mostly small stuff, of course, "but still it's stealing small stuff that leads to big time crimes." "Too bad," people said. "Too bad that these boys couldn't behave like the other kids in town: stay out of trouble, be polite to adults, and look to their future."

The community's impression of the degree to which this group of six boys (ranging in age from 16 to 19) engaged in delinquency was somewhat distorted. In some ways the gang was more delinquent than the community thought; in other ways they were less.

The fighting activities of the group were fairly readily and accurately perceived by almost everyone. At least once a month, the boys would get into some sort of fight, although most fights were scraps between members of the group or involved only one member of the group and some peripheral hanger-on. Only three times in the period of observation did the group fight together: once against a gang from across town, once against two blacks, and once against a group of boys from another school. For the first two fights the group went out "looking for trouble"—and they found it both times. The third fight followed a football game and began spontaneously with an argument on the football field between one of the Roughnecks and a member of the opposition's football team.

Jack had a particular propensity for fighting and was involved in most of the brawls. He was a prime mover of the escalation of arguments into fights.

More serious than fighting, had the community been aware of it, was theft. Although almost everyone was aware that the boys occasionally stole things, they did not realize the extent of the activity. Petty stealing was a frequent event for the Roughnecks. Sometimes they stole as a group and coordinated their efforts; other times they stole in pairs. Rarely did they steal alone.

The thefts ranged from very small things like paperback books, comics, and ballpoint pens to expensive items like watches. The nature of the thefts varied from time to time. The gang would go through a period of systematically shoplifting items from automobiles or school lockers. Types of thievery varied with the whim of the gang. Some forms of thievery were more profitable than others, but all thefts were for profit, not just thrills.

Roughnecks siphoned gasoline from cars as often as they had access to an automobile, which was not very often. Unlike the Saints, who owned their own cars, the Roughnecks would have to borrow their parents' cars, an event which occurred only eight or nine times a year. The boys claimed to have stolen cars for joy rides from time to time.

Ron committed the most serious of the group's offenses. With an unidentified associate the boy attempted to burglarize a gasoline station. Although this station had been robbed twice previously in the same month, Ron denied any involvement in either of the other thefts. When Ron and his accomplice approached the station, the owner was hiding in the bushes beside the station. He fired both barrels of a double-barreled shotgun at the boys. Ron was severely injured; the other boy ran away and was never caught. Though he remained in critical condition for several months, Ron finally recovered and served six months of the following year in reform school. Upon release from reform school, Ron was put back a grade in school, and began running around with a different gang of boys. The Roughnecks considered the new gang less delinquent than themselves, and during the following year Ron had no more trouble with the police.

The Roughnecks, then, engaged mainly in three types of delinquency: theft, drinking, and fighting. Although community members perceived that this gang of kids was delinquent, they mistakenly believed that their illegal activities were primarily drinking, fighting, and being a nuisance to passersby. Drinking was limited among the gang members, although it did occur, and theft was much more prevalent than anyone realized.

Drinking would doubtless have been more prevalent had the boys had ready access to liquor. Since they rarely had automobiles at their disposal, they could not travel very far, and the bars in town would not serve them. Most of the boys had little money, and this, too, inhibited their purchase of alcohol. Their major source of liquor was a local drunk who would buy them a fifth if they would give him enough extra to buy himself a pint of whiskey or a bottle of wine.

The community's perception of drinking as prevalent stemmed from the fact that it was the most obvious delinquency the boys engaged in. When one of the boys had been drinking, even a casual observer seeing him on the corner would suspect that he was high.

There was a high level of mutual distrust and dislike between the Roughnecks and the police. The boys felt very strongly that the police were unfair and corrupt. Some evidence existed that the boys were correct in their perception.

The main source of the boys' dislike for the police undoubtedly stemmed from the fact that the police would sporadically harass the group. From the standpoint of the boys, these acts of occasional enforcement of the law were whimsical and uncalled-for. It made no sense to them, for example, that the police would come to the corner occasionally and threaten them with arrest for loitering when the night before the boys had been out siphoning gasoline from cars and the police had been nowhere in sight. To the boys, the police were stupid on the one hand, for not being where they should have been and catching the boys in a serious offense, and unfair on the other hand, for trumping up "loitering" charges against them.

From the viewpoint of the police, the situation was quite different. They knew, with all the confidence necessary to be a policeman, that these boys were engaged in criminal activities. They knew this partly from occasionally catching them, mostly from circumstantial evidence ("the boys were around when those tires were slashed"), and partly because the police shared the view of the community in general that this was a bad bunch of boys. The best the police could hope to do was to be sensitive to the fact that these boys were engaged in illegal acts and arrest them whenever there was some evidence that they had been involved. Whether or not the boys had in fact committed a particular act in a particular way was not especially important. The police had a broader view: their job was to stamp out these kids' crimes; the tactics were not as important as the end result.

Over the period that the group was under observation, each member was arrested at least once. Several of the boys were arrested a number of times and spent at least one night in jail. While most were never taken to court, two of the boys were sentenced to six months' incarceration in boys' schools.

THE ROUGHNECKS IN SCHOOL

The Roughnecks' behavior in school was not particularly disruptive. During school hours they did not all hang around together, but tended instead to spend most of their time with one or two other members of the gang who were their special buddies. Although every member of the gang attempted to avoid school as much as possible, they were not particularly successful and most of them attended school with surprising regularity. They considered school a burden—something to be gotten through with a minimum of conflict. If they were "bugged" by a particular teacher, it could lead to trouble. One of the boys, Al, once threatened to beat up a teacher and, according to the other boys, the teacher hid under a desk to escape him.

Teachers saw the boys the way the general community did, as heading for trouble, as being uninterested in making something of themselves. Some were also seen as being incapable of meeting the academic standards of the school. Most of the teachers expressed concern for this group of boys and were willing to pass them despite poor performance, in the belief that failing them would only aggravate the problem.

The group of boys had a grade point average just slightly above "C." No one in the group failed a grade, and no one had better than a "C" average. They were very consistent in their achievement or, at least, the teachers were consistent in their perception of the boys' achievement.

Two of the boys were good football players. Herb was acknowledged to be the best player in the school and Jack was almost as good. Both boys were criticized for their failure to abide by training rules, for refusing to come to practice as often as they should, and for not playing their best during practice. What they lacked in sportsmanship they made up for in skill, apparently, and played every game no matter how poorly they had performed in practice or how many practice sessions they had missed.

TWO QUESTIONS

Why did the community, the school, and the police react to the Saints as though they were good, upstanding, nondelinquent youths with bright futures but to the Roughnecks as though they were tough, young criminals who were headed for trouble? Why did the Roughnecks and the Saints in fact have quite different careers after high school—careers which, by and large, lived up to the expectations of the community?

The most obvious explanation for the differences in the community's and law enforcement agencies' reactions to the two gangs is that one group of boys was "more delinquent" than the other. Which group *was* more delinquent? The answer to this question will determine in part how we explain the differential responses to these groups by the members of the community and, particularly, by law enforcement and school officials.

In sheer number of illegal acts, the Saints were the more delinquent. They were truant from school for at least part of the day almost every day of the week. In addition, their drinking and vandalism occurred with surprising regularity. The Roughnecks, in contrast, engaged sporadically in delinquent episodes. While these episodes were frequent, they certainly did not occur on a daily or even a weekly basis.

The difference in frequency of offenses was probably caused by the Roughnecks' inability to obtain liquor and to manipulate legitimate excuses from school. Since the Roughnecks had less money than the Saints, and teachers carefully supervised their school activities, the Roughnecks' hearts may have been as black as the Saints', but their misdeeds were not nearly as frequent.

There are really no clear-cut criteria by which to measure qualitative differences in antisocial behavior. The most important dimension of the difference is generally referred to as the "seriousness" of the offenses.

If seriousness encompasses the relative economic costs of delinquent acts, then some assessment can be made. The Roughnecks probably stole an average of about $5.00 worth of goods a week. Some weeks the figure was considerably higher, but these times must be balanced against long periods when almost nothing was stolen.

The Saints were more continuously engaged in delinquency but their acts were not for the most part costly to property. Only their vandalism and occasional theft of gasoline would so qualify. Perhaps once or twice a month they would siphon a tankful of gas. The other costly items were street signs, construction lanterns, and the like. All of these acts combined probably did not quite average $5.00 a week, partly because much of the stolen equipment was abandoned and presumably could be recovered. The difference in cost of stolen property between the two groups was trivial, but the Roughnecks probably had a slightly more expensive set of activities than did the Saints.

Another meaning of seriousness is the potential threat of physical harm to members of the community and to the boys themselves. The Roughnecks were more prone to physical violence; they not only welcomed an opportunity to fight; they went seeking it. In addition, they fought among themselves frequently. Although the fighting never included deadly weapons, it was still a menace, however minor, to the physical safety of those involved.

The Saints never fought. They avoided physical conflict both inside and outside the group. At the same time, though, the Saints frequently endangered their own and other people's lives. They did so almost every time they drove a car, especially if they had been drinking. Sober, their driving was risky; under the influence of alcohol it was horrendous. In addition, the Saints endangered the lives of others with their pranks. Street excavations left unmarked were a very serious hazard.

Evaluating the relative seriousness of the two gangs' activities is difficult. The community reacted as though the behavior of the Roughnecks was a problem, and they reacted as though the behavior of the Saints was not. But the members of the community were ignorant of the array of delinquent acts that characterized the Saints' behavior. Although concerned citizens were unaware of much of the Roughnecks' behavior as well, they were much better informed about the Roughnecks' involvement in delinquency than they were about the Saints'.

Visibility

Differential treatment of the two gangs resulted in part because one gang was infinitely more visible than the other. This differential visibility was a direct function of the economic standing of the families. The Saints had access to automobiles and were able to remove themselves from the sight of the community. In as routine a decision as to where to go to have a milkshake after school, the Saints stayed away from the mainstream of community life. Lacking transportation, the Roughnecks could not make it to the edge of town. The center of town was the only practical place for them to meet since their homes were scattered throughout the town and any noncentral meeting place put an undue hardship on some members. Through necessity the Roughnecks congregated in a crowded area where everyone in the community passed frequently, including teachers and law enforcement officers. They could easily see the Roughnecks hanging around the drugstore.

The Roughnecks, of course, made themselves even more visible by making remarks to passersby and by occasionally getting into fights on the corner. Meanwhile, just as regularly, the Saints were either at the cafe on one edge of town or in the pool hall at the other edge of town. Without any particular realization that they were making themselves inconspicuous, the Saints were able to hide their time-wasting. Not only were they removed from the mainstream of traffic, but they were almost always inside a building.

On their escapades the Saints were also relatively invisible, since they left Hanibal and travelled to Big City. Here, too, they were mobile, roaming the city, rarely going to the same area twice.

Demeanor

To the notion of visibility must be added the difference in the responses of group members to outside intervention with their activities. If one of the Saints was confronted with an accusing policeman, even if he felt he was truly innocent of a wrongdoing, his demeanor was apologetic and penitent. A Roughneck's attitude was almost the polar opposite. When confronted with a threatening adult authority, even one who tried to be pleasant, the Roughneck's hostility and disdain were clearly observable. Sometimes he might attempt to put up a veneer of respect, but it was thin and was not accepted as sincere by the authority.

School was no different from the community at large. The Saints could manipulate the system by feigning compliance with the school norms. The availability of cars at school meant that once free from the immediate sight of the teacher, the boys could disappear rapidly. And this escape was well enough planned that no administrator or teacher was nearby when the boys left. A Roughneck who wished to escape for a few hours was in a bind. If it were possible to get free from class, downtown was still a mile away, and even if he arrived there, he was still very visible. Truancy for the Roughnecks meant almost certain detection, while the Saints enjoyed almost complete immunity from sanctions.

Bias

Community members were not aware of the transgressions of the Saints. Even if the Saints had been less discreet, their favorite delinquencies would have been perceived as less serious than those of the Roughnecks.

In the eyes of the police and school officials, a boy who drinks in an alley and stands intoxicated on the street corner is committing a more serious offense than is a boy who drinks to inebriation in a nightclub or a tavern and drives around afterwards in a car. Similarly, a boy who steals a wallet from a store will be viewed as having committed a more serious offense than a boy who steals a lantern from a construction site.

Perceptual bias also operates with respect to the demeanor of the boys in the two groups when they are confronted by adults. It is not simply that adults dislike the posture affected by boys of the Roughneck ilk; more important is the conviction that the posture adopted by the Roughnecks is an indication of their devotion and commitment to deviance as a way of life. The posture becomes a cue, just as the type of the offense is a cue, to the degree to which the known transgressions are indicators of the youths' potential for other problems.

Visibility, demeanor, and bias are surface variables which explain the day-to-day operations of the police. Why do these surface variables operate as they do? Why did the police choose to disregard the Saints' delinquencies while breathing down the backs of the Roughnecks?

The answer lies in the class structure of American society and the control of legal institutions by those at the top of the class structure. Obviously, no representative of the upper class drew up the operational chart for the police which led them to look in the ghettoes and on street corners—which led them to see the demeanor of lower-class youth as troublesome and that of upper-middle-class youth as tolerable. Rather, the procedure simply developed from experience—experience with irate and influential upper-middle-class parents insisting that their son's vandalism was simply a prank and his drunkenness only a momentary "sowing of wild oats"—experience with cooperative or indifferent, powerless, lower-class parents who acquiesced to the law's definition of their son's behavior.

ADULT CAREERS OF THE SAINTS
AND THE ROUGHNECKS

The community's confidence in the potential of the Saints and the Roughnecks apparently was justified. If anything the community members underestimated the degree to which these youngsters would turn out "good" or "bad."

Seven of the eight members of the Saints went on to college immediately after high school. Five of the boys graduated from college in four years. The sixth one finished college after two years in the army, and the seventh spent four years in the air force before returning to college and receiving a B.A. degree. Of these seven college graduates, three went on for advanced degrees. One finished law school and is now active in state politics, one finished medical school and is practicing near Hanibal, and one boy is now working for a Ph.D. The other four college graduates entered submanagerial, managerial, or executive training positions with larger firms.

The only Saint who did not complete college was Jerry. Jerry had failed to graduate from high school with the other Saints. During his second senior year, after the other Saints had gone on to college, Jerry began to hang around with what several teachers described as a "rough crowd"—the gang that was heir apparent to the Roughnecks. At the end of his second senior year, when he did graduate from high school, Jerry took a job as a used car salesman, got married, and quickly had a child. Although he made several abortive attempts to go to college by attending night school, when I last saw him (ten years after high school) Jerry was unemployed and had been living on unemployment for almost a year. His wife worked as a waitress.

Some of the Roughnecks have lived up to community expectations. A number of them were headed for trouble. A few were not.

Jack and Herb were the athletes among the Roughnecks and their athletic prowess paid off handsomely. Both boys received unsolicited athletic scholarships to college. After Herb received his scholarship (near the end of his senior year), he apparently did an about-face. His demeanor became very similar to that of the Saints. Although he remained a member in good standing of the Roughnecks, he stopped participating in most activities and did not hang on the corner as often.

Jack did not change. If anything, he became more prone to fighting. He even made excuses for accepting the scholarship. He told the other gang members that the school had guaranteed him a "C" average if he would come to play football—an idea that seems far-fetched, even in this day of highly competitive recruiting.

During the summer after graduation from high school, Jack attempted suicide by jumping from a tall building. The jump would certainly have killed most people trying it, but Jack survived. He entered college in the fall and played four years of football. He and Herb graduated in four years, and both are teaching and coaching in high schools. They are married and have stable families. If anything, Jack appears to have a more prestigious position in the community than does Herb, though both are well respected and secure in their positions.

Two of the boys never finished high school. Tommy left at the end of his junior year and went to another state. That summer he was arrested and placed on probation on a manslaughter charge. Three years later he was arrested for murder; he pleaded guilty to second degree murder and is serving a 30-year sentence in the state penitentiary.

Al, the other boy who did not finish high school, also left the state in his senior year. He is serving a life sentence in a state penitentiary for first degree murder.

Wes is a small-time gambler. He finished high school and "bummed around." After several years he made contact with a bookmaker who employed him as a runner. Later he acquired his own area and has been working it ever since. His position among the bookmakers is almost identical to the position he had in the gang; he is always around but no one is really aware of him. He makes no trouble and he does not get into any. Steady, reliable, capable of keeping his mouth closed, he plays the game by the rules, even though the game is an illegal one.

That leaves only Ron. Some of his former friends reported that they had heard he was "driving a truck up north," but no one could provide any concrete information.

REINFORCEMENT

The community responded to the Roughnecks as boys in trouble, and the boys agreed with that perception. Their pattern of deviancy was reinforced, and breaking away from it became increasingly unlikely. Once the boys acquired an image of themselves as deviants, they selected new friends who affirmed that self-image. As that self-conception became more firmly entrenched, they also became willing to try new and more extreme deviances. With their growing alienation came freer expression of disrespect and hostility for representatives of the legitimate society. This disrespect increased the community's negativism,

perpetuating the entire process of commitment to deviance. Lack of a commitment to deviance works the same way. In either case, the process will perpetuate
itself unless some event (like a scholarship to college or a sudden failure) external
to the established relationship intervenes. For two of the Roughnecks (Herb and
Jack), receiving college athletic scholarships created new relations and culminated
in a break with the established pattern of deviance. In the case of one of the Saints
(Jerry), his parents' divorce and his failing to graduate from high school changed
some of his other relations. Being held back in school for a year and losing his
place among the Saints had sufficient impact on Jerry to alter his self-image
and virtually to assure that he would not go on to college as his peers did.
Although the experiments of life can rarely be reversed, it seems likely in view
of the behavior of the other boys who did not enjoy this special treatment by
the school that Jerry, too, would have "become something" had he graduated
as anticipated. For Herb and Jack outside intervention worked to their advantage;
for Jerry it was his undoing.

Selective perception and labeling—finding, processing, and punishing some
kinds of criminality and not others—means that visible, poor, nonmobile, out-
spoken, undiplomatic "tough" kids will be noticed, whether their actions are
seriously delinquent or not. Other kids, who have established a reputation for
being bright (even though underachieving), disciplined and involved in respectable activities, who are mobile and monied, will be invisible when they deviate
from sanctioned activities. They'll sow their wild oats—perhaps even wider and
thicker than their lower-class cohorts—but they won't be noticed. When it's time
to leave adolescence most will follow the expected path, settling into the ways of
the middle class, remembering fondly the delinquent but unnoticed fling of their
youth. The Roughnecks and others like them may turn around, too. It is more
likely that their noticeable deviance will have been so reinforced by police and
community that their lives will be effectively channeled into careers consistent
with their adolescent background.

22

Doctors' Autonomy and Power

JOHN LIEDERBACH

Doctors are another group with the social status and power to rise above the deviant acts they commit and maintain a prestigious reputation. Malfeasance and misconduct are rampant in the medical professions, as we constantly read in the newspaper, yet we are less likely to suspect doctors of wrongdoing, Liederbach suggests, due to their positive reputation and role as altruistic healers. They are, nonetheless, as subject to the temptations for fraud and abuse as any other individuals. Liederbach describes such common practices as fee splitting, self-referrals, prescription violations, unnecessary treatments, and sexual misconduct that have led insurers and health maintenance organizations (HMOs) to tighten up their scrutiny of the medical industry. Regulation by the government is less conscientious, it appears, and the presence of overpayment, fraud, and abuse of the Medicaid and Medicare systems are more widespread. Liederbach notes that occupational crimes such as these—where people engage in deviance for their own benefit, not that of their organizations—will occur wherever opportunities can be found. This chapter is especially relevant to today's concern about universal health care, the precarious financial footing of Medicaid and Medicare, and the loss of governmental resources (for example, in Hurricane Katrina) due to lack of oversight, carelessness, and fraud.

O ver time, health care has grown into a trillion dollar a year enterprise. The delivery of patient services involves not only physicians, but also large-scale insurance companies, government-financed benefit programs, and Health Maintenance Organizations (HMOs). Estimates of the cost of health-care fraud range from fifty to eighty billion dollars annually (Witkin, Friedman, and Doran 1992). In the Medicare program alone, some seventeen billion dollars a year is lost (Shogren 1995). The financial cost of medical crime has led one observer to characterize the situation as "white-collar wilding" (Witkin, Friedman, and Doran 1992).

The consequences of medical crime are not merely financial. Unnecessary medical procedures, negligent care, prescription violations, and the sexual abuse of patients exact an enormous physical toll as well. Each year some 400,000 patients become victims of negligent mistakes or misdiagnoses. One Harvard researcher estimates that 180,000 patients die every year, due at least in part to negligent care ("Patients, Doctors, Lawyers" 1990). Up to two million patients are needlessly subjected to physical risks through unnecessary operations each year; the resulting price tag approaches four billion dollars (Jesilow, Pontell, and Geis 1993).

THE "PROTECTIVE CLOAK": STATUS, ALTRUISM, AND AUTONOMY

Physicians are recognized as a special group in society—a privileged caste able to decipher puzzling ailments and able to fix broken-down bodies. The privilege is hard won through years of education and exhaustive training. The physician's honored rank, however, sponsors opportunities for doctors to commit crimes within their profession. Attributes synonymous with medical practice, such as high social status, trustworthiness, and professional autonomy, have provided doctors with what some have termed a "protective cloak" that has shielded doctors from scrutiny and legal accountability (Jesilow, Pontell, and Geis 1993; Parsons 1951).

One element of the "cloak," high social status, has helped to afford doctors the protections necessary to commit medical crimes. Doctors' traditional high status derives from two related elements, namely, lucrative salaries and occupational prestige. Physicians remain one of the most highly compensated occupational groups, with median annual incomes exceeding $120,000 (Ruffenach 1988). Aided by the prestige that typically accompanies high salaries in American culture, physicians have been able to retain an elite social position.

Historically, there has also been a general reluctance in American society to use the criminal law against high-status offenders, and criminologists have long recognized the important role that status plays in shaping the criminal opportunities afforded to professional groups. Professionals possess the financial and political wherewithal to influence the manner in which criminal statutes are written and enforced, and they are more apt to "escape arrest and conviction . . . than those who lack such power" (Sutherland 1949). While scholars debate whether this reluctance stems from public apathy and/or ignorance concerning the costs connected to elite crimes (Cullen, Maakestad, and Cavender 1987; Evans, Cullen, and Dubeck 1993; Wilson 1975), the typically lenient sanctions currently imposed on doctors who pillage government benefit programs, provide negligent care, or otherwise physically abuse their patients points to a historical reluctance to treat as criminal even the most egregious forms of physician malfeasance (Jesilow, Pontell, and Geis 1993; Rosoff, Pontell, and Tillman 1998; Tillman and Pontell 1992; Wolfe et al. 1998).

A second protective element is the altruistic and trustworthy image projected by doctors. This image is cemented in the physician's code of ethics. The oath serves to define doctors as selfless professionals who perform an invaluable service without regard to personal financial gain (Jesilow, Pontell, and Geis 1993). The image structures criminal opportunities in several ways: the image creates an assumption of good will on the part of doctors that makes charges of intentional wrongdoing difficult to justify. Prosecutors may find it too challenging to prove intentionally fraudulent or harmful behavior in cases against highly respected and trusted doctors. Also, the physician's altruistic image has traditionally engendered a certain level of trust from patients (McKinlay and Stoeckle 1988; Stoeckle 1989). Trusting patients who are victims of fraudulent medical schemes or negligent care may fail to hold doctors accountable for their crimes. One observer has defined the impact of these factors more generally as a "pattern of deference" to doctors—a prevailing unwillingness to question their presumed trustworthiness (Bucy 1989).

Third, doctors have been relatively immune from legal scrutiny because of the medical professions' historical preference for self-regulation. State medical review boards, whose members are predominantly physicians themselves, are supposed to provide a "first line of defense" against doctors who violate legal or ethical codes (Wolfe et al. 1998). These boards can revoke medical licenses or otherwise discipline doctors who fail to meet professional or legal standards. Doctors argue that self-regulation and autonomy characterize any profession—that is, doctors alone possess the specialized expertise and unique qualifications to judge the actions of other physicians. The medical community regards the imposition of civil and/or criminal penalties as both unwarranted and unnecessary, especially in cases that involve errors in clinical judgement (Abramovsky 1995). The profession's reliance on self-regulation, however, may facilitate criminal opportunities by shielding its members from more effective punishments. State medical boards, for example, have continually failed to identify doctors who are chronically incompetent, and often punish them with "slaps on the wrist" (Wolfe et al. 1998). The case of one New York doctor illustrates the dangers of relying solely on professional controls:

> During the mid-1980s [Dr. Benjamin] was investigated by the Department of Health in connection with numerous medical irregularities. In 1986, after a medical review board convicted him on 38 counts of gross negligence and incompetence, the New York State Health Commissioner asked the Board to revoke his license. . . . The doctor's punishment was reduced, to a three-month suspension and three years' probation. In June 1993 . . . the Department revoked Dr. Benjamin's license for five botched abortions performed in one year. However, Dr. Benjamin was allowed to continue performing abortions pending appeal. Less than a month later (patient) Gaudalupe Negron met her death from another of Dr. Benjamin's botched procedures. (Abramovsky 1995)

The lax enforcement typically provided by state medical boards has created an inviting opportunity structure for doctors to commit fraud and abuse within their profession. The problems related to professional control are exacerbated by the well-documented "code of silence" that exists among medical professionals (Rosoff, Pontell, and Tillman 1998). Doctors are often hesitant to report fraud and abuse for fear of professional recriminations (Karlin 1995; Levy 1995). Still, some in the medical community recognize the extent of the problem: "The profession has done a lousy job of policing its own," acknowledges Arthur Caplan, chairman of the Center for Bioethics at the University of Pennsylvania (Grey 1995).

SELECTED MEDICAL OFFENSES

Medical "Kickbacks": Fee Splitting and Self-Referrals

"Kickbacks" are generally defined as payments from one party to another in exchange for referred business or other income-producing deals. Their acceptance by doctors is considered unethical, and in most cases illegal, because they create a conflict of interest between the physicians' commitment to quality patient care and their own financial self-interest. Doctors who are primarily concerned with financial gain compromise their loyalty to patients, as well as their independent professional judgement. Two well-recognized types of medical kickbacks include fee splitting and self-referrals.

Fee splitting occurs when one physician (usually a general practitioner) receives payment from a surgeon or other specialists in exchange for patient referrals. Fee splitting artificially inflates medical costs, provides incentives for unnecessary tests and specialized treatment, and can also endanger the quality of patient care (Stevens 1971). As Sutherland (1949) explained, the fee-splitting doctor "tends to send his patients to the surgeon who will split the largest fee rather than to the surgeon who will do the best work." Early observers regarded fee splitting as an "almost universal" practice, and estimated that 50 to 90% of physicians split fees (MacEachern 1948; Williams 1948). Despite the advent of more secure payment sources provided by the spread of health insurance coverage in the post-World War II years, congressional investigations in the 1970s continued to recognize fee splitting as a problem (Rodwin 1992; U.S. Congress 1976). While it remains difficult to determine whether the prevalence of fee splitting has increased or decreased over time, it clearly has persisted (Rodwin 1992).

Alternatives to fee splitting have developed more recently, including self-referrals. Self-referrals involve sending patients to specialized medical facilities in which the physician has a financial interest. Between 50,000 and 75,000 doctors have a financial stake in ancillary medical services (quoted in Rosoff, Pontell, and Tillman 1998). Recent research identifies the problems associated with self-referrals, including higher utilization costs and unnecessary services (Hillman et al. 1990; Mitchell and Scott 1992; Rodwin 1992). Self-referring doctors refer patients for laboratory testing at a 45% higher rate than noninvesting physicians (U.S. Department of Health and Human Services 1989). Physicians' utilization of clinical laboratories, diagnostic imaging centers, and rehabilitation facilities was found to be significantly higher when physicians owned these facilities (Hillman et al. 1990).

The medical profession's traditional response to financial conflicts of interest has been less than overwhelming. While most states had declared fee splitting illegal by the mid-1950s, the medical profession did not explicitly address financial conflicts in ethical codes until the 1980s (Rodwin 1992). One newspaper characterized the profession's response with embarrassing clarity: "Fee splitting has been like a venereal disease ... it exist(s), but nice people do not talk about it" (quoted in Rodwin 1992). The legal and professional response to self-referrals has been more ambivalent. Despite recent legislative attempts to prohibit self-referrals, the practice remains legal. Likewise, professional standards have not been effective in curtailing abuses:

> Unlike other professionals who are subject to extensive conflict of interest regulation ... physicians have addressed these issues largely on their own, and have been subject to minimal regulation by state and federal laws or even professional codes ... The American Medical Association addresses these issues primarily by relying on professional norms, individual discretion and subjective standards ... (the profession) lacks an effective means to hold physicians accountable. (Rodwin 1992, 734)

Prescription Violations

Only doctors possess the education and specialized expertise required to safely prescribe dangerous and often addictive drugs, including narcotics, amphetamines, tranquilizers, and other controlled substances. The privilege is entrusted with the legal responsibility to limit access to these drugs on the basis of medical need. While the vast majority of physicians uphold these responsibilities, an alarming number of doctors violate this trust. Between 1988 and 1996, 1,521 doctors were disciplined for misprescribing or overprescribing drugs (Wolfe et al. 1998). Numerous doctors were caught selling blank prescriptions to known addicts. One physician dispensed expired drugs from old, unlabeled spice jars. Another doctor prescribed dangerous weight loss pills to a patient for four years without even examining her—eventually resulting in the patient having a stroke (Wolfe et al. 1998). Some prescription violations are coupled with fraudulent billing schemes designed to maximize profits from illegal prescriptions. Jesilow, Pontell, and Geis (1993) relate one of the most appalling cases:

> In Los Angeles, one investigator reported a Medicaid doctor who saw so many patients daily that red, blue, and yellow lines had been painted on his office floor to expedite traffic. Each color represented a different kind of pill. (22)

Similar to the case of medical kickbacks, the medical profession has also largely failed to adequately discipline doctors who violate prescription laws. At least 69% of the doctors cited between 1988 and 1996 were not even temporarily suspended from practicing medicine (Wolfe et al. 1998).

Unnecessary Treatments

Doctors who intentionally subject patients to medically unnecessary treatments violate the law in two ways. First, unnecessary treatments are fraudulent because they result in compensation that is deceptively gained. More important, unnecessary procedures that are invasive, such as surgery, may be considered a form of assault because they needlessly expose patients to physical risks (Lanza-Kaduce 1980). The highly publicized case of one California ophthalmologist dearly illustrates how unnecessary treatments can result in serious physical harm to patients. The doctor performed unneeded cataract surgery on patients solely to collect a $584 operation fee. In the process, the doctor admitted needlessly "blinding a lot of people" (Pontell, Geis, and Jesilow 1984).

Determining the prevalence of "unnecessary" treatments is often difficult given the inherent uncertainties involved in diagnosing and treating patients. Green (1997) has outlined several methods used by researchers to estimate the extent of unnecessary surgeries: (1) geographical variations in surgical rates, (2) studies of second surgical opinions, (3) variations in surgical rates between payment plans, and (4) expert opinions based on predetermined criteria. Although these studies present mixed findings, wide geographical variations in surgical rates can be used to suggest that unnecessary surgeries occur with some frequency. For example, hysterectomies are performed at an 80% higher rate in Southern states, and 1,000% rate variations in pacemaker operations have been identified in Massachusetts (quoted in Green 1997). Similarly, one government-sponsored study found a 120% higher surgical rate for patients enrolled in fee-for-service plans versus patients enrolled in HMOs (U.S. Department of Health, Education, and Welfare 1971).

Sexual Misconduct

Sexual misconduct by doctors can take a variety of forms (Jacobs 1994). Doctors may engage in sexual misconduct in exchange for professional services. Alternatively, doctors may allow relationships with patients to escalate beyond what is ethically acceptable. Finally, doctors may sexually assault patients while they are under the control of anesthesia or otherwise incapable of consenting to a sexual act (Green 1997). These offenses are especially abhorrent, because they represent an abuse of power by the doctor in situations where the patient is particularly vulnerable. From 1987 to 1996, 393 doctors were disciplined by state medical boards for sexual misconduct with patients. At least 34% of those doctors were not forced to even temporarily stop practicing (Wolfe et al. 1998).

MEDICAID FRAUD AND ABUSE

Medicaid began in 1965 as one of the "Great Society" programs initiated during the Lyndon B. Johnson administration. The program extended health coverage to needy Americans who could not otherwise afford it. Perhaps overshadowed by the nobility of this goal, the program's costs were not considered a primary concern (Jesilow, Pontell, and Geis 1993). Since its inception, fraud and abuse has been endemic to the program. Jesilow, Pontell, and Geis (1993), in their exposé on Medicaid crime, identify several reasons why the introduction of Medicaid has surreptitiously expanded the scope of the medical crime problem.

The medical profession opposed the initial Medicaid legislation. Doctors perceived Medicaid as a threat to their professional autonomy, because the program dictated the price of their professional medical services (Jesilow, Pontell, and Geis 1993). The Medicaid program introduced an unwelcome influence—the government—into decisions that were traditionally left to the independent professional discretion of doctors. This intrusion into the professional autonomy of doctors created widespread dissatisfaction within the medical community, and served to facilitate fraud and abuse within the program in at least two important ways (Jesilow, Pontell, and Geis 1993). First, the medical community's initial opposition led to certain flaws in the program's design that created easy opportunities to violate the law. For example, the original Medicaid legislation did not include provisions for punishing doctors who violated program rules (Jesilow, Pontell, and Geis 1993). The omission was not an accident, but an attempt to placate doctors, without whose cooperation the program could not be launched. As a result, the program lacked the effective sanctions necessary to police an additional design flaw, the program's fee-for-service payment structure. The fee-for-service plan reimbursed doctors a fixed amount for each procedure, but the doctor could earn additional income by double-billing for the same patient, billing for more expensive procedures than those performed, or even charging the program for services that were never done (Jesilow, Pontell, and Geis 1993). A significant number of Medicaid providers could not resist the combination of easy opportunities and lenient sanctions.

Second, the government's intrusion on the professional autonomy of doctors gave rise to an aggressively defiant attitude among some practitioners against the rules that governed Medicaid work. As Jesilow, Pontell, and Geis (1993) explain, this militant defiance "dramatically reduced one of the most powerful deterrents to crime, especially for middle and upper class perpetrators: the sense of guilt and the force of conscience associated with depredations against known human victims." Because they believed Medicaid to be an illegitimate intrusion on their professional autonomy, some Medicaid violators "redefined" their criminal behavior in a positive light. As a result, many of these doctors did not view their actions as wrong (Jesilow, Pontell, and Geis 1993).

This combination of opportunities, motivations, and lax sanctions has produced a dizzying array of violations related to the Medicaid program. Perhaps the most striking example of Medicaid fraud, however, is the discovery of Medicaid "mills" by fraud investigators (Jesilow, Pontell, and Geis 1993).

> Located in dilapidated areas, often in storefronts, and catering almost
> exclusively to patients on Medicaid rolls, the mills resemble clinics in
> that doctors with different specialties are gathered under one roof.
> But the mill's providers often rent space in the building and bill Medicaid
> individually.... Criminal activities flourished in the Medicaid mills.
> Some employed "hawkers" to round up customers. Several catered to

drug traffic. Various government agents, all claiming to be suffering from nothing more than the common cold, had been seen by eighty-five doctors in Medicaid mills. They underwent eighteen electrocardio-grams, eight tuberculosis tests, four allergy tests, as well as a hearing, glaucoma, and electroencephalogram tests. (50)

The inception of the Medicaid program posed one of the first significant challenges to the professional autonomy of physicians. Although this challenge did serve to extend health benefits to many of the nation's most indigent citizens, the program also altered the traditional opportunity structure for medical crime, and created new and unique avenues for physicians to commit offenses relating to their medical practice....

One indication of increasing physician vulnerability is the recent spate of criminal prosecutions against doctors who have killed or maimed patients through negligent or reckless medical care. Because of the protections tradition-ally afforded to doctors, errors in clinical judgment—even if they resulted in the death of patient—have traditionally been sanctioned exclusively through civil actions or peer-oriented sanctions. The filing of *violent* criminal charges, such as manslaughter, assault, reckless homicide, or murder, against doctors who vic-timize patients in the course of their medical practice had been an exceedingly rare occurrence (Abramovsky 1995; Crane 1994; Green 1997). However, at least seven doctors have been criminally prosecuted for their violent victimization of patients over the last ten years (Liederbach, Cullen, Sundt, and Geis 1998). The case against Dr. Joseph Verbrugge is indicative not only because of the gravity of his mistakes, but also because he is believed to be the first doctor to stand trial in Colorado accused of a violent crime related to a medical procedure:

> Verbrugge was charged with reckless manslaughter in connection with the death of eight-year-old Richard Leonard during ear surgery. The normally routine procedure went awry when the boy's heart rate jumped significantly after Verbrugge administered the anesthetic. During the surgery, the patient's breathing became irregular and his temperature soared to 107 degrees. Prosecutors contended that Verbrugge failed to react to those danger signs because he had fallen asleep during the operation. The patient died after three hours in surgery. His reaction to the anesthesia had increased the level of carbon monoxide in his blood to four times the normal level. (Liederbach, Cullen, Sundt, and Geis 1998)

It remains to be seen how prevalent criminal prosecutions against doctors for medical "mistakes" will become. However, the recent cluster of these cases may suggest the beginning of a trend toward increasing physician vulnerability. These cases may be a signpost indicating that the traditional protections afforded to doc-tors have eroded. In particular, as long-term doctor-patient relationships wane and as HMOs increasingly influence the delivery of medical services, patient trust in doctors—especially when things go badly in a medical procedure—will decline, and the ability to see reckless doctors as *criminals* may increase (Liederbach, Cullen, Sundt, and Geis 1998).

REFERENCES

Abramovsky, A. 1995. "Depraved Indifference and the Incompetent Doctor." *New York Law Journal* November 8, pp. 3–10.

Bucy, P. H. 1989. "Fraud by Fright: White Collar Crime by Health Care Providers." *North Carolina Law Review* 67: 855–937.

Crane, M. 1994. "Could Clinical Mistakes Land You in Jail? The Case of Gerald Einaugler." *Medical Economics* 71: 46–52.

Cullen, F. T., W. J. Maakestad, and G. Cavender. 1987. *Corporate Crime under Attack: The Ford Pinto Case and Beyond.* Cincinnati: Anderson.

Evans, D. T., F. T. Cullen, and P. J. Dubeck. 1993. "Public Perceptions of White Collar Crime." Pp. 85–114 in *Understanding Corporate Criminality*, edited by M. B. Blankenship. New York: Garland.

Green, G. S. 1997. *Occupational Crime*, 2nd ed. Chicago: Nelson Hall.

Grey, B. 1995. "Medical Scandal." *Baltimore Sun*, August 21.

Hillman, B. J., V. A. Joseph, M. R. Mabry, J. H. Sunshine, S. D. Kennedy, and M. Noether. 1990. "Frequency and Costs of Diagnostic Imaging in Office Practice—A Comparison of Self-Referring and Radiologist Referring Physicians." *New England Journal of Medicine* 323: 1604–8.

Jacobs, S. 1994. "Social Control of Sexual Assault By Physicians and Lawyers within the Professional Relationship: Criminal and Disciplinary Actions." *American Journal of Criminal Justice* 19(1): 43–60.

Jesilow, P., H. N. Pontell, and G. Geis. 1993. *Prescription for Profit: How Doctors Defraud Medicaid.* Berkeley: University of California Press.

Karlin, R. 1995. "Selective Silence Is under Scrutiny." *The Time Union* (Albany, NY), August 28, p. Bl.

Lanza-Kaduce, L. 1980. "Deviance among Professionals: The Case of Unnecessary Surgery." *Deviant Behavior* 1: 333–59.

Levy, D. 1995. "Physicians Can Run, Hide from Deadly Errors." *USA Today*, September 11, p. 1D.

Liederbach, J., F. T. Cullen, J. Sundt, and G. Geis. 1998. "The Criminalization of Physician Violence: Social Control in Transformation?" Paper presented at the annual meeting of the American Society of Criminology, Washington, DC.

MacEachern, M. 1948. "College Continues Militant Stance against Fee-Splitting and Rebates." *Bulletin of the American College of Surgeons* 33: 65–67.

McKinlay, J. B., and J. D. Stoeckle. 1988. "Corporatization and the Social Transformation of Medicine." *International Journal of Health Services* 18: 191–200.

Mitchell, J. M., and E. Scott. 1992. "New Evidence on the Prevalence and Scope of Physician Joint Ventures." *Journal of the American Medical Association* 268: 80–84.

Parsons, T. 1951. *The Social System*, Glencoe, IL: Free Press.

"Patients, Doctors, Lawyers: Medical Injury, Malpractice Litigation, and Patient Compensation." 1990. *Harvard Medical Practices Study.* Boston: President and Fellows of Harvard University.

Pontell, H. N., G. Geis, and P. D. Jesilow. 1984. "Practitioner Fraud and Abuse in Government Medical Benefit Programs." Washington, DC: U.S. Department of Justice.

Rodwin, M. A. 1992. "The Organized American Medical Profession's Response to Financial Conflicts of Interest: 1890–1992." *Milbank Quarterly* 70(4): 703–41.

Rosoff, S. M., H. N. Pontell, and R. Tillman. 1998. *Profit without Honor: White Collar Crime and the Looting of America.* Upper Saddle River, NJ: Prentice Hall.

Ruffenach, G. 1988. "No Need to Worry, Doctors Do Just Fine." *Wall Street Journal*, October 10.

Shogren, G. 1995. "Rampant Fraud Complicates Medicare Cures." *Los Angeles Times*, October 8, p. 1.

Stevens, R. 1971. *American Medicine and the Public Interest.* New Haven: Yale University Press.

Stoeckle, J. D. 1989. "Reflections on Modern Doctoring." *Milbank Quarterly* 66: 76–89.

Sutherland, E. H. 1949. *White-Collar Crime*, New Haven: Yale University Press.

Tillman, R., and H. N. Pontell. 1992. "Is Justice Color Blind?: Punishing Medicaid Provider Fraud." *Criminology* 30(4): 547–74.

U.S. Congess. Senate Subcommittee on Long-Term Care, Special Committee on Aging. 1976. *Fraud and Abuse among Practitioners Participating in the Medicaid Program.* Washington, DC.

U.S. Department of Health, Education, and Welfare. 1971. *The Federal Employees Health Benefit Program—Enrollment and Utilization of Health Services 1961–1968.* Washington, DC: U.S. Government Printing Office.

U.S. Department of Health and Human Services. 1989. *Financial Arrangements between Physicians and Health Care Businesses.* Washington, DC: U.S. Government Printing Office.

Williams, G. 1948. "The Truth About Fee-Splitting." *Modern Hospital* 70: 43–48 (reprinted in *Reader's Digest*, July 1948).

Wilson, J. Q. 1975. *Thinking about Crime.* New York: Basic Books.

Witkin, G., D. Friedman, and G. Doran. 1992. "Health Care Fraud." *US News and World Report*, February 24, pp. 34–43.

Wolfe, S., K. M. Franklin, P. McCarthy, P. Bame, and B. M. Adler. 1998. *16,638 Questionable Doctors Disciplined by State and Federal Governments.* Washington, DC: Public Citizens Health Research Group.

PART V

Deviant Identity

We have just looked at how some categories of people and behavior become defined as deviant. Yet social constructionism suggests that a deviant classification floating around abstractly in society is not meaningful unless it gets attached to people. Groups in society not only work to create definitions of deviance, but also to create situations where deviance occurs and is labeled. In this section we will examine the way deviance is evoked and shaped. Becoming deviant does not only entail having a definition of deviance and an environment in which it can occur; it also requires that people accept the identity and make it their own. Identities refer to the way people think of themselves. The study of deviant identities has focused on how people develop and manage nonnormative self-conceptions. In Part V we will examine this process: how the concept of deviance becomes applied to individuals and how it affects their self-conception.

IDENTITY DEVELOPMENT

We mentioned earlier that although many people engage in deviance, the label is applied to only a small percentage of them. Such labeling is tied to their formerly "secret deviance" (Becker 1963) becoming exposed, or to an abstract status coming to bear on their personal experience. Thus, Jews may not feel stigmatized unless they experience anti-Semitism, and embezzlers may not think of themselves as thieves until they are caught. When this happens, they enter the pathway to the deviant identity, a pathway that follows a certain trajectory. The process of acquiring a deviant identity unfolds processually as a "deviant" (Becker 1963) or "moral" (Goffman 1961) career, with people passing through stages that move them out of their innocent identities toward one labeled as "different" by society.

In our own work (Adler and Adler 2006), we proposed a model of the seven stages of the **deviant identity career.**

Stage one begins once people are *caught and publicly identified* as deviant; their lives change in several ways. Others start to think of them differently. For example, suppose there has been a rash of thefts in a college dormitory, and Jessica, a first-year student, is finally caught and identified as the culprit. She may or may not be reported to authorities and charged with theft, but regardless, she will experience an informal labeling process. Once she is caught, the news about her is likely to spread. In stage two, people will probably change their attitudes toward her, as they find themselves talking about her behind her back. They may look at her behavior and engage in *"retrospective interpretation"* (Schur 1971) as they think about her differently, reflecting back onto the past to see if her current and earlier behavior can be recast differently in light of their new information. Where did she say she was when the last theft occurred? Where did she say she got the money to buy that new sweater?

In stage three, as this news spreads, either informally or through official agencies of social control, Jessica may develop what Goffman (1963) has called a *"spoiled identity,"* one with a tarnished reputation. Erikson (Chapter 1) noted that news about deviance is of high interest in a community, commanding intense focus from a wide audience. Deviant labeling is hard to reverse, he suggested, and once people's identities are spoiled they are hard to socially rehabilitate. He discussed "commitment ceremonies," such as trials or psychiatric hearings, where individuals are officially labeled as deviant. Few corresponding ceremonies exist, he remarked, to mark the cleansing of people's identities and welcome them back into the normative fold. Individuals may thus find it hard to recover from the lasting effect of such identity labeling, and despite their best efforts, they often find that society expects them to commit further deviance. Merton (1938) referred to this as the "self-fulfilling prophesy," where people tend to enact the labels placed upon them, despite possible intentions otherwise.

Jessica's dormmates and former friends may then, in stage four, engage in what Lemert (1951) has called *"the dynamics of exclusion,"* deriding and ostracizing her from their social group. When she enters the room she may notice that a sudden hush falls over the conversation. People may not feel comfortable leaving her alone in their rooms. They may exclude her from their meal plans and study groups. She may become progressively shut out from nondeviant activities and circles such as honors societies, professional associations, relationships, or jobs. At the same time, in stage five, others may welcome or *include* her in their deviant circles or activities. She may find that she has developed a reputation that, though repelling to some groups, is attractive to others. They may welcome her as "cool" and invite her into their circles. Thus, individuals may find that as they move down the pathway of their deviant careers, they shift friendship circles, being pushed away from the company of some, while being simultaneously welcomed into the company of others.

Sixth, others usually begin to *treat differently* those defined as deviant, indicating through their actions that their feelings and attitudes toward the newly deviant have shifted, often in a negative sense. They may not accord Jessica the same level of credibility they had previously, and they may tighten the margin of social allowance they allot her. Seventh, and finally, people react to this treatment using what Charles Horton Cooley referred to as their "looking glass selves." In the culminating stage of the identity career, they *internalize the deviant label* and come to think of themselves differently. This is likely to affect their future behavior. Although not all people who get caught in deviance progress completely through this full set of stages, Becker (1963) described this process as the effects of labeling.

Once people are labeled as deviant and accept that label into their self-conceptions, a variety of outcomes may ensue. We all juggle a range of identities and social selves through which we relate to people including those of sibling, child, friend, student, neighbor, and customer. Identities also derive from some of our demographic or occupational features, such as race, gender, age, religion, or social class. Hughes (1945) has suggested that some statuses are very dominant, overpowering others and coloring the way people are viewed. Having a known deviant identity may become one of these "**master statuses,**" rising to the top of the hierarchy, infusing people's self-concept and others' reactions and taking precedence over all others. Many social statuses fade in and out of relevance as people move through various situations, but a master status accompanies them into all their contexts, forming the key identity through which others see them. Deviant attributes such as a minority race, heroin addiction, and homosexuality are prime examples of such master statuses. Others, then, may think of an actor, for example, as a Hispanic person, a heroin addict, or as gay, with the individual's occupation or hobbies being seen as secondary to his or her deviant attribute. Hughes noted that master statuses are linked in society to **auxiliary traits,** the common social preconceptions that people associate with these. Self-injurers, for example, may widely be assumed to be adolescent white women from either middle-class backgrounds or disadvantaged youth whose lives are unhappy. They may be thought of as lacking impulse control, seeking attention, abuse survivors, mentally unstable, or as people who seek to cry for help or inject control into their lives. Heroin addicts may be suspected of being prostitutes or thieves and homosexuals may be suspected of being sexually promiscuous or AIDS-infected.

This type of identification spreads the image of deviance to cover the person as a whole and not just one part of him or her.

The relationship between master statuses and their auxiliary traits in society is reciprocal. When people learn that others have a certain deviant master status, they may impute the associated auxiliary traits onto them. Inversely, when people begin to recognize a few traits that they can put together to form the pattern of auxiliary traits associated with a particular deviant master status, they are likely to attribute that master status to others. For example, if parents notice that their children are staying out late with their friends, wearing "alternative" clothing styles, growing dred locks in their hair, dropping out of after-school activities, and hanging out with the "wrong" crowd of friends, they may suspect them of using drugs or committing crimes.

Lemert (1967) asserted another processual depiction of the deviant identity career with his concepts of **primary** and **secondary deviance.** Primary deviance refers to a stage when people commit deviant acts, but their deviance goes unrecognized. As a result, others do not cast the deviant label onto them, and they neither assume it nor perform a deviant role. Their self-conceptions are free of this image. Some people remain at the primary deviance stage throughout the time they are committing deviance, never advancing further. Yet a percentage of them do progress to secondary deviance. The seven stages of the identity career, described above, move people from primary to secondary deviance; their infractions become discovered, others identify them as deviant, and the labeling process ensues, with all of its identity consequences. Others come to regard them as deviant, and they do as well. As they move into secondary deviance, individuals initially deny the label but eventually come to accept it reluctantly as it becomes increasingly pressed upon them. They recognize their own deviance as they are forced to interact through this stigma with others. Sometimes this internalization comes as a justification or social defense to the problems associated with their deviant label, as individuals use it to take the offensive. At any rate, it becomes an identity that significantly affects their role performance. Some people may compartmentalize their deviant identity, but others exhibit "role engulfment" (Schur 1971), becoming totally caught up in this master status.

Most individuals who progress to secondary deviance advance no further, but a subset of them moves on to what Kitsuse (1980) called **tertiary deviance.** In contrast to primary deviants who engage in deviance denial, and secondary deviants who accept their deviant identities, Kitsuse sees tertiary deviants as those who engage in deviance embracement. These are people who decide that their deviance is not a bad thing. They may adopt a relativist perspective and decide that their deviant label is socially constructed by society, not intrinsic to their behavior, such as individuals with learning differences who consider themselves more creative than "typical" people. Or they may hold to an absolutist perspective and embrace their deviant category as intrinsically real, such as gays who "discover" their underlying homosexuality and accept it as natural. They therefore strongly identify with their deviance and fight, usually with the organized help of like others, to combat the deviant label that is applied to them. They may engage in "identity politics" and speak publicly, protest, rally, pursue civil disobedience, educate, raise funds, lobby, and practice various other forms of political advocacy to change society's view of their deviance. Examples of this include people who fight to destigmatize labels such as obesity, prostitution, and race/ethnicity.

All these identity career concepts encompass a progression through several stages. They begin with the commission of the deviance and lead to individuals' apprehension and public identification. They move through the changing expectations of others toward them, marked by shifting social acceptance or rejection by their friends and acquaintances. The breadth, seriousness, and longevity of the deviant identity label are significantly more profound when individuals undergo official labeling processes than when they are merely informally labeled. With their internalization of the deviant label, adoption of the self-identity, and public interaction through it, they ultimately move into groups of differential deviant associates and commit further acts of deviance.

ACCOUNTS

When people say or do things that appear odd to others, they risk being labeled as deviant. We all engage in instances of deviant behavior, but at the same time we desire to maintain a positive self-image in both our own eyes and the eyes of others. In order to avoid the negative consequences of being labeled as deviant and to preserve their untarnished identities, individuals may engage in a variety of interactional strategies designed to normalize their behavior. Mills (1940) suggested that people use "vocabularies of motive" in conversation, where they present legitimate reasons to others around them that explain the meaning of their actions. This motive talk restores a sense of normalcy to interactions that are disrupted by questionable events.

Sykes and Matza (1957: 666) suggested that people commonly make "justifications for deviance that are seen as valid by the delinquent but not by the legal system or society at large." Individuals using these were attempting to resolve the contradictions between what people say and what they do. They offered five "**techniques of neutralization**" through which people rationalize their behavior, either prospectively or retrospectively. Through *denials of responsibility*, individuals suggest that their deviance was due to acts beyond their control ("I couldn't help myself," "it was not my fault"). In *denying injury*, they mitigate their offense by alluding to the lack of consequences, arguing that no one was hurt ("no harm, no foul"). When they make a *denial of the victim*, they legitimate their behavior by suggesting that either no specific victim can be identified ("it's a huge corporation; nobody will notice it"), or that the persons hurt do not deserve victim status ("gays deserve to be beaten up"). Some people *appeal to higher loyalties* by casting their behavior as serving a greater good (loyalty to a friend, to higher principles, to God). Finally, in *condemning the condemners*, people turn the table on the accusers, throwing attention away from themselves by focusing on things their accusers have done wrong ("oh, you think you're so easy to live with?" "police are nothing but pigs").

Scott and Lyman (1968) further refined our conception of accounts by suggesting that all accounts can be seen as either **excuses** or **justifications.** In offering excuses, individuals admit the wrongfulness of their actions but distance themselves from the blame. These excuses are often fairly standard phrases or ideas designed to soften the deviance and relieve individuals of their accountability. These may include *appeals to accidents* ("my computer malfunctioned and lost my file"), *appeals to defeasibility* or misinformation ("I thought my roommate turned my paper in"), *appeals to biological drives* ("men will be men") and *scapegoating* ("she borrowed my notes and I couldn't get them back in time to study for the test").

With justifications, individuals accept responsibility for their actions but seek to have specific instances excused. In so doing, they try to legitimate the acts or their consequences. In drawing on justifications, individuals may invoke *sad tales* ("I am a prostitute so that I can afford to put food on the table to feed my kids" or "I turn tricks because I was sexually abused as a child") or the need for *self-fulfillment* ("taking hallucinogenic drugs expands my consciousness and makes me a more caring person").

Hewitt and Stokes (1975) added to our understanding of accounts by presenting a set of verbal explanations specifically designed to precede the deviant acts that people saw as imminent in their future. They suggested that Lyman and Scott's accounts were primarily retrospective in nature, and that their "**disclaimers**" were fundamentally prospective. People *hedge*, they suggested, in prefacing their remarks to indicate a measure of uncertainty about what they are going to do ("I'm not sure this is going to work but . . . "). They use *credentializing* when they know their act will be discredited, but they are attempting to give a purpose or legitimacy to it ("I'm not prejudiced, in fact some of my best friends are XXX, but . . ."). Sometimes people invoke *sin licenses* when they know their behavior will be poorly received but want to suggest that this is a time where the ordinary rules might be suspended ("I realize you might think this is wrong but . . ."). *Cognitive disclaimers* try to make sense out of something that looks as if it might not be well understood ("This may seem strange to you . . ."). Finally, *appeals for the suspension of judgment* aim to deflect the negative consequences of acts or remarks that may be offensive or angering ("Hear me out before you explode . . ."). Disclaimers, then, are specifically conversational tactics that people invoke before they launch ahead into something commonly judged as inappropriate.

STIGMA MANAGEMENT

When people are labeled as deviant, it marks them with a stigma in the eyes of society. As we have seen, this label may lead to devaluation and exclusion. Consequently, people with deviant features learn how to "manage" their stigma so that they are not shamed or ostracized. This effort requires considerable social skills.

Goffman (1963) has suggested that people with potential deviant stigma fall into two categories: the "**discreditable**" and the "**discredited.**" The former are those with easily concealable deviant traits (ex-convicts, secret homosexuals) who may manage themselves so as to avoid the deviant stigma. The latter are either members of the former category who have revealed their deviance or those who cannot hide their deviance (the obese, racial minorities, the physically disabled). These people's lives are characterized by a constant focus on secrecy and information control. Goffman observed that most discreditables engage in "**passing**" as "normals" in their everyday lives, concealing their deviance and fitting in with regular people. They may do this by avoiding contacts with "stigma symbols," those objects or behaviors that would tip people off to their deviant condition (an anorectic avoiding family meals, mental patients surreptitiously taking their medications). Another technique for passing includes using "disidentifiers" such as props, actions, or verbal expressions to distract and fool people into thinking that they do not have the deviant stigma (homosexuals bragging about heterosexual conquests or taking a date to the company picnic, members of ethnic minorities who laugh at ethnic slurs about their group). Finally, they may "lead a double life," maintaining two different lifestyles with two distinct groups of people, one that knows about their deviance and one that does not.

In this endeavor, people may employ the aid of others to help conceal their deviance by "**covering**" for them. In these team performances, friends and family members may assist the deviants by concealing their identities, their whereabouts, their deficiencies, or their pasts. They may even coach the deviants on how to construct stories designed to hide their deviance.

Another form of stigma management, sometimes adopted when concealment fails, involves disclosing the deviance. People may do this for cathartic reasons (alleviating their burden of secrecy), therapeutic reasons (casting it in a positive light), or preventive reasons (so others don't find out in negative ways later). Although many people find that their disclosures lead to rejection, others are more fortunate when people may even sympathize with their condition, such as people with sexually transmitted diseases.

Disclosures of deviance can follow two courses. In observing the interactions between discredited deviants and normals, Davis (1961) noted that some nondeviant people go through a normalization process in their relationship with the deviant person largely through failing to acknowledge the deviant trait. Although this usually begins with a conspicuous and stilted ignoring of the individual's deviance (**deviance disavowal**), it progresses through stages where more relaxed interaction begins, interaction is directed at features of the person other than his or her deviant stigma, and finally gets to the point where the deviant stigma is overlooked and almost forgotten. This can be illustrated by the case of physically disabled people who, at first, are shunned by their co-workers or fellow students, but who gradually fit into the crowd when others realize that there is more to them than their disability (they root for the same teams, listen to the same music, share the same major).

In contrast, deviant people can strive to normalize their relationships with nondeviants through "**deviance avowal**" (Turner 1972), in which they openly acknowledge their stigma and try to present themselves in a positive light. This avowal often takes the form of humor, "breaking the ice" by joking about their deviant attribute. In this way they show others that they can take the perspective of the normal and see themselves as deviant too, thus forming a bridge to others. This action further asserts that they have nondeviant aspects and that they can see the world as others do. This can be the case when members of ethnic minority groups make self-deprecating comments about themselves based on stereotypes.

Thus far we have considered individual modes of adaptation to deviant stigma. Yet these stigma can also be managed through a group or collective effort. Many voluntary associations of stigmatized individuals exist, from the early organizations of prostitutes (COYOTE—Call Off Your Old Tired Ethics), to more recent ones such as the Gay Liberation Front, the Little People of America, the National Stuttering Project, and the Gray Panthers. Best known are the 12-step programs modeled after the tremendous success of Alcoholics Anonymous (AA), including such groups as Overeaters Anonymous, Narcotics Anonymous, and Gamblers Anonymous.

These groups vary in character. Some groups are organized as what Lyman (1970) calls an **expressive** dimension, whose primary function is to provide support for their members. This support can take the form of organizing social and recreational activities, dispersing legal or medical information, or offering services such as shopping, meals, or transportation. Expressive groups tend to be apolitical, helping their members adapt to their social stigma rather than evade it. They also serve as places where members can come together in the company of other deviants, avoid the censure of nonstigmatized normals, and seek collective solutions to their common problems. It is here that they can make disclosures to others without fear of rejection. For instance, the Little People of America have an annual conference which not only provides a social gathering for people of similar height but also offers support and advice for practical problems (setting up one's house) and social concerns (dating).

Lyman has also described groups with an **instrumental** dimension, where members gather together not only to accomplish the expressive functions but also to organize for political activism. This embodies Kitsuse's tertiary deviation, where individuals reject the societal conception and treatment of their stigma and organize to change social definitions. They fight to get others to modify their views of the status or behavior in question so that society, like them, will no longer regard it as deviant. Examples of such groups include ACT UP, an AIDS organization whose members have tried to change social attitudes toward AIDS patients, the National Organization for Women (NOW), and the Disabled in Action.

On another continuum, Lyman noted that groups may vary between **conformity** and **alienation.** Conformative groups fundamentally adhere to the norms and values of society. They accept most conventional views, with the occasional exception of their own deviance. They generally use their backstage arenas to counsel members on how to fit in with others who may neither accept nor understand them. Thus, support groups for bipolar people may offer advice to members on how to find the right doctor, various benefits and side effects of different drugs, and the risks and benefits of going drug-free. But they do not generally glorify either mania or depression. Where groups do break with society in the way they regard their own deviance and members instrumentally try to fight for the legitimation of their deviance, they use conventional means to attain their goals.

Groups may be alienative for one of two reasons: They are willing to step outside of conventional means to fight for changed definitions of their single form of deviance, or they have multiple conflicting values with society. Activists, such as the Black Panthers, a single-issue group, were willing to break the law to fight for improved social opportunities and status for African Americans. Radical feminists might not resort to violating laws, but their dissatisfaction with the social structures that disempower women are grounded in multiple dimensions of society. Modern-day descendants of the Ku Klux Klan such as the skinheads, Aryan Nation, and various militia groups may incorporate both elements, rejecting social attitudes of acceptance toward Blacks, Jews, immigrants, gays, and others. At the same time, they have resorted to violence to attain their ends, such as blowing up the Murrah Federal Building in Oklahoma City to avenge the Waco siege where members of the Branch Davidian cult and their leader David Koresh perished during an FBI assault. Members of other alienative groups, such as the Amish, nudists, and hippie communes, simply want to take their radically different values and form communities removed from conventional society.

23

The Adoption and Management of a "Fat" Identity

DOUGLAS DEGHER AND GERALD HUGHES

Degher and Hughes's selection on the way people come to think of themselves as fat is a study in identity transformation. They posit a model where individuals align their self-conception with cues that they derive from their external environment. Although the subjects in this study originally do not hold a view of themselves as obese, they receive active status cues (people say things) and passive status cues (their clothes no longer fit) that jar them away from their former self-conceptions. They follow a process of recognizing that they can no longer be considered to have a normal build and placing themselves with the new category that fits them more appropriately. This leads to their reconceptualizing themselves as fat. The fat status has a new, negative identity that they adopt, devaluing them and locating them within the deviant realm.

The interactionist perspective has come to play an important part in contemporary criminological and deviance theory. Within this approach, deviance is viewed as a subjectively problematic identity rather than an objective condition of behavior. At the core is the emphasis on "process" rather than on viewing deviance as a static entity. To paraphrase vintage Howard Becker, "... social groups create deviance by making the rules whose infraction constitutes deviance. Consequently, deviance is not a quality of the act ... but rather a consequence of the application by others of rules and sanctions to an offender" (Becker, 1963, p. 9). Attention is focused upon the *interaction* between those being labeled deviant and those promoting the deviant label. In the interactionist literature, emphases are in two major areas: (a) the conditions under which the label "deviant" comes to be applied to an individual and the consequences for the individual of having adopted that label (Tannenbaum, 1939; Lemert, 1951; Kitsuse, 1962,

p. 247; Goffman, 1963; Baum, 1987, p. 96; Greenberg, 1989, p. 79), and/or (b) the role of social control agents[1] in contributing to the application of deviant labels (Becker, 1963; Piliavin & Briar, 1964, p. 206; Cicourel, 1968; Schur, 1971; Conrad, 1975, p. 12).

Much of this literature frequently assumes that once an individual has been labeled, the promoted label and attendant identity is either internalized or rejected. As Lemert proposes, the shift from primary to secondary deviance is a categorical one, and is primarily a response to problems created by the societal reaction (Lemert, 1951, p. 40).

What is most often neglected is an examination of the mechanistic features of this identity shift. Our focus is on this "identity change process," which is what we have chosen to call this identity shift. Of interest is how individuals come to make some personal sense out of proffered labels and their attendant identities.

METHODOLOGY

The primary methodological tool employed to construct our identity change process model comes from "grounded" analysis.... The model presented in this paper emerged from comments and codes appearing in interviews with obese members of a weight reduction organization that had weekly meetings. The frequency of attendance allowed us to consider the members typical, and allowed us to suggest that major issues of obesity are transsituational and temporally durable. If obesity disappeared tomorrow, we would still be able to apply the generic concepts generated from our data to make statements about the process of "identity change." As suggested by Hadden, Degher, and Fernandez, our focus is on process rather than on unit characteristics of social phenomena (Hadden, Degher, & Fernandez, 1989, p. 9). This provides us with insights that have import for major issues in sociological theory.

SITE SELECTION

Because obese individuals suffer both internally (negative self-concepts) and externally (discrimination), they possess what Goffman refers to as a "spoiled identity" (Goffman, 1963). This seems to be the case particularly in contemporary America with what may be described as an almost pathological emphasis on fitness. The boom in health clubs, sales of videotapes on fitness, diet books, and so forth promote a definition of the "healthy" physical presence. As Kelly (1990) sees it, the boom in physical fitness in the mid-1980s is an attempt by many people to create a specific image of an ideal body. Thus, body build becomes a crucial element in self-appraisal. Consequently, fat people are an ideal strategic group within which to study the "identity change process."

Obese people are not only the subject of negative stereotypes, they are also actively discriminated against in college admissions (Canning & Mayers, 1966,

p. 1172), pay more for goods and services (Petit, 1974), receive prejudicial medical treatment (Maddox, Back, & Liederman, 1968, p. 287; Maddox & Liederman, 1969, p. 214), are treated less promptly by salespersons (Pauley, 1989, p. 713), have higher rates of unemployment (Laslett & Warren, 1975, p. 69), and receive lower wages (Register, 1990, p. 130). The obese label is one that seems to clearly fit Becker's description of a "master status," that is,

> Some statuses in our society, as in others, override all other statuses and have a certain priority . . . the status deviant (depending on the kind of deviance) is this kind of master status . . . one will be identified as a deviant first, before other identifications are made. (Becker, 1963, p. 33)

Obese people are "fat" first, and only secondarily are seen as possessing ancillary characteristics.

The site for the field observations had to meet two requirements: (a) it had to contain a high proportion of obese, or formerly obese individuals; and (b) these individuals had to be identifiable by the observer. The existence of a large number of national weight control organizations (a) whose membership is composed of individuals who have internalized an obese identity, and (b) who emphasize a radical program of identity change, make these organizations an excellent choice as strategic sites for study and analysis. The local franchise chapter of one of these national weight loss organizations satisfied both of our requirements, and was selected as the site for our study.

Attendance at the weekly meetings of this national weight control group is restricted to individuals who are current members of the organization. Since one requirement for membership is that the individual be at least 10 pounds over the maximum weight for his or her sex and height (according to New York Life tables), all of the people attending the meetings are, or were, overweight, and a high proportion of them are, or were, sufficiently overweight to be classified as obese."[2]

During the period of the initial field observations, the weekly membership of the group varied from 30 to 100 members, with an average attendance of around 60 members. Although there was a considerable turnover in membership, the greatest part of this turnover consisted of "rejoins" (individuals who had been members previously, and were joining again).

Although we have no quantitative data from which to generalize, the group membership appeared to represent a cross section of the larger community. The group included both male and female members, although females did constitute about three-fourths of the membership. Although the membership was predominantly white, a range of ethnicities, notably Hispanic and Native American, existed within the group. The majority of the members appeared to fall within the 30 to 50 age range, although there was a member as young as 11, and one over 70.

DATA COLLECTION

Two types of data were gathered for this study: field observations and in-depth interviews. The field observations were performed while attending meetings of a local weight control organization. The insights gained from these observations were used primarily to develop interview guides. There were two major sources of observation during this period: premeeting conversations; and exchanges during the meeting itself.[3] The observations were recorded in note form and served to provide an orientation for the subsequent interviews. The goal during this period of observation was to gain insight into the basic processes of obesity and the obese career.

The in-depth interviews were carried out with 29 members from the local group. The interviews were solicited on a voluntary basis, and each individual was assured anonymity. The interviewees were representative of the group membership. Although most were middle-aged, middle-income white females, various age groups, ethnicities, marital statuses, genders, and social classes were represented.

These interviews lasted in length from ½ to 2½ hours, with the average interview being about 1 hour and 15 minutes in duration. The interviews produced almost 40 hours of taped discussion, which yielded more than 600 pages of typed transcript for coding.

THE IDENTITY CHANGE PROCESS

In conceptualizing the "identity changes" process, the concept of "career" was employed. As Goffman notes, "career" refers ". . . to any social strand of any person's course through life" (Goffman, 1961, p. 127). In the present paper, our concern is the change process that takes place as individuals come to see definitions of self in light of specific transmitted information.

An important aspect of this career model is what Becker referred to as "career contingencies," or ". . . those factors on which mobility from one position to another depends. Career contingencies include both the objective facts of social structure, and changes in the perspectives, motivations, and desires of the individual" (Becker, 1963, p. 24).

Thus, the "identity change" process must be viewed on two levels: a public (external) and a private (internal) level. As Goffman has stated, "One value of the concept of career is its two-sidedness. One side is linked to internal matters held dearly and closely, such as image of self and felt identity; the other publicly acceptable institutional complex" (Goffman, 1961, p. 127).

On the public level, social status exists as part of the public domain; social status is socially defined and promoted. The social environment not only contains definitions and attendant stereotypes for each status, it also contains information, in the form of *status cues*, about the applicability of that status for the individual.

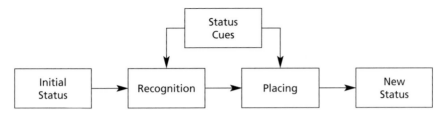

FIGURE 23.1 Visualization of the identity Change Process (ICP)

On the internal level, two distinct cognitive processes must take place for the identity change process to occur: first, the individual must come to recognize that the current status is inappropriate; and second, the individual must locate a new, more appropriate status. Thus, in response to the external status cues, the individual comes to recognize internally that the initial status is inappropriate; and then he or she uses the cues to locate a new, more appropriate status. The identity change occurs in response to, and is mediated through, the status cues that exist in the social environment (see Figure 23.1).

STATUS CUES: THE EXTERNAL COMPONENT

Status cues make up the public or external component of the identity change process. A status cue is some feature of the social environment that contains information about a particular status or status dimension. Because this paper is about obesity, the cues of interest are about "fatness." Such status cues provide information about whether or not the individual is "fat," and if so, how "fat."

"Recognizing" and "placing" comprise the internal component of the identity change process and occur in response to, and are mediated through, the status cues that exist in the social environment. In order to fully understand the identity change process, it is necessary to explain the interaction between outer and inner processes (Scheff, 1988, p. 396), or in our case, external and internal components of the process.

Status cues are transmitted in two ways: actively and passively. Active cues are communicated through interaction. For example, people are informed by peers, friends, spouses, etc., that they are overweight. The following are some typical comments that occurred repeatedly in the interviews in response to the question, "How did you know that you were fat?"

I was starting to be called chubby, and being teased in school.

When my mother would take me shopping, she'd get angry because the clothes that were supposed to be in my age group wouldn't fit me. She would yell at me.

Well, people would say, "When did you put on all your weight, Bob?" You know, something like that. You know, you kind of get the message, that, you know, I did put on weight.

A second category of cues might accurately be described as passive in form. The information in these cues exists within the environment, but the individual must in some way be sensitized to that information. For example, standing on a scale will provide an individual with information about weight. It is up to the individual to get on the scale, look at the numbers, and then make some sense out of them. Other passive cues might involve seeing one's reflection in a mirror, standing next to others, fitting in chairs, or, as frequently mentioned by respondents, the sizing of clothes. The comments below, all made in response to the question, "How did you know that you were fat?" are representative of passive cue statements.

> I think that it was not being able to wear the clothes that the other kids wore.
>
> How did I know? Because when we went to get weighed, I weighed more than my, uh, a girl my height should have weighed, according to the chart, according to all the charts that I used to read. That's when I first noticed that I was overweight.
>
> I would see all these ladies come in and they could wear size 11 and 12, and I thought, Why can't I do that? I should be able to do that.

Both active and passive cues serve as mechanisms for communicating information about a specific status. As can be seen from the data, events occur that force the individual to evaluate his or her conceptions of self.

RECOGNIZING

The term "recognizing" refers to the cognitive process by which an individual becomes aware that a particular status is no longer appropriate. As shown in the figure, the process assumes the individual's acceptance of some initial status. For obese individuals, the initial status is that of "normal body build."[4] This assumption is based on the observation that none of our interviewees assumed that they were "always fat." Even those who were fat as children could identify when they became aware that they were "fat." Through the perception of discrepant status cues, the individual comes to recognize that the initial status is inappropriate. It is possible that the person will perceive the discrepant cues and will either ignore or reject them, in which case the initial status is retained. The factors regulating such a failure to recognize are important, but are not dealt with in this paper. Further research on this point is called for.

Status cues are the external mechanisms through which the recognizing takes place, but it is paying attention to the information contained in these cues that triggers the internal cognitive process of recognizing.

An important point is that the acceptance (or rejection) of a particular status does not occur simply because the individual possesses a set of objective characteristics. For example, two people may have similar body builds, but one may have a self-definition of "fat" whereas the other may not. There appears to be a rather tenuous connection between objective condition and subjective definition. The following comments are supportive of this disjunction.

> I was really, as far as pounds go, very thin, but I had a feeling about myself that I was huge.
> Well, I don't remember ever thinking about it until I was about in eighth grade. But I was looking back at pictures when I was little. I was always chunky, chubby.

This lack of necessary connection between objective condition and subjective definition points up an important and frequently overlooked feature of social statuses: the extent to which they are *self-evident*. Self-evidentiality refers to the degree to which a person who possesses certain objective status characteristics is *aware* that a particular status label applies to them.[5]

Some statuses possess a high degree of evidentiality: gender identification is one of these.[6] On the other hand, being beautiful or intelligent is somewhat non-self-evident. This is not to imply that individuals are either ignorant of these statuses or of the characteristics upon which they are assigned. People may know that other people are intelligent, but they may be unaware that the label is equally applicable to them.

One idea that emerged quite early from the interviews was that being "fat" is a relatively non-self-evident status. Individuals do not recognize that "fat" is a description that applies to them.[7] The objective condition of being overweight is not sufficient, in itself, to promote the adoption of a "fat" identity. This non-self-evidentiality is demonstrated in the following excerpts.

> I think that I just thought that it was a little bit here and there. I didn't think of it, and I didn't think of myself as looking bad. But you know, I must have.
> I have pictures of me right after the baby was born. I had no idea that I was that fat.

The self-evidentiality of a status is important in the discussion of the identity change process. The less self-evident a status, the more difficult the recognizing process becomes. Further, because recognition occurs in response to status cues, the self-evidentiality of a status will influence the type of cues that play the most prominent role in identity change.

A somewhat speculative observation should be made about status cues in the recognizing process. For our subjects, recognizing occurred primarily through active cues. When passive cues were involved, they typically were highly visible and unambiguous. In general, active cues appear to be more potent in forcing the individual's attention to the information that the current status is inappropriate. The predominance of these active cues is possibly a consequence of the relatively non-self-evident character of the "fat" status. It is probable that the less self-evident a status, the more likely that the recognizing process will occur through active rather than passive cues.

Once the individual comes to recognize the inappropriateness of the initial status, it becomes necessary to locate a new, more appropriate status. This search for a more appropriate status is referred to as the "placing" process.[8]

PLACING

Placing refers to a cognitive process whereby an individual comes to identify an appropriate status from among those available. The number of status categories along a status dimension influences the placing process. A status dimension may contain any number of status categories. If there are only two status dimensional categories, such as in the case of gender, the placement process is more or less automatic. When individuals recognize that they do not belong in one category, the remaining category becomes the obvious alternative. The greater the number of status categories, the more difficult the placing process becomes.

The body build dimension contains an extremely large number of categories. When an individual recognizes that he or she does not possess a "normal" body build, there are innumerable alternatives open. The knowledge that one's status lies toward the "fat" rather than the "thin" end of the continuum still presents a wide range of choices. In everyday conversation, we hear terms that describe these alternatives: chubby, porky, plump, hefty, full-figured, beer belly, etc. All are informal descriptions reflecting the myriad categories along the body build dimension.

I wasn't real fat in my eyes. I don't think. I was just chunky.

Not fat. I didn't exactly classify it as fat. I just thought, I'm, you know, I am a pudgy lady.

I don't think that I have ever called myself fat. I have called myself heavy.

Even when individuals adopt a "fat" identity, they attempt to make distinctions about how fat they are. Because being fat is a devalued status, individuals attempt to escape the full weight of its negative attributes while still acknowledging the nonnormal status. The following responses exemplify this attempt to neutralize the pejorative connotations of having a "fat" status. The practice of differentiating one's status from others becomes vital in managing a fat identity.

Q. How did you know that you weren't *that* fat?

A. Well, comparing myself to others at the time, I didn't really feel that I was that fat. But I knew maybe because they didn't treat me the same way they treated people who were heavier than me. You know, I got teased lightly, but I was still liked by a lot of people, and the people that were heavy weren't.

As is apparent from this excerpt, the individual neutralized self-image by linking "fatness" with the level of teasing done by peers.

NEW STATUS

The final phase of the identity change process involves the acceptance of a new status. For our informants, it was the acceptance of a "fat" status, along with its previously mentioned pejorative characterizations.[9]

> I hate to look in mirrors. I hate that. It makes me feel so self-conscious. If I walk into a store, and I see my reflection in the glass, I just look away.
>
> We'd go somewhere and I would think, "I never look as good as everybody else." You know everybody always looks better. I'd cry before we'd go bowling because I'd think, "Oh, I just look awful."

As is clear, the final phase of the identity change process involves the internalization of a negative (deviant) definition of self. For many fat people accepting a new status means starting on the merry-go-round of weight reduction programs.[10] Many of these programs or organizations attempt to get members to accept a devalued status fully, and then work to change it. Consequently, individuals are forced to "admit" that they are fat and to "witness" in front of others.[11] The new identity becomes that of a "fat" person, which the weight reduction programs then attempt to transform. A further analysis of the impact of informal organizations on the identity change process will be attempted in another paper.

CONCLUSION

In this paper, we have attempted to fill a void within the interactionist literature by presenting an inductively generated model of the identity change process. The proposed model treats the change process from a career focus, and thus addresses both the external (public) and the internal (cognitive) features of the identity change.

We have suggested that the adoption of a new status takes place through two sequential cognitive processes, "recognizing" and "placing." First, the individual must come to recognize that a current status is no longer appropriate. Second, the individual must locate a new, more appropriate status from among those available. We have further suggested that these internal or cognitive processes are triggered by and mediated through status cues, which exist in the external environment. These cues can be either active or passive. Active cues are transmitted through interaction, whereas passive cues must be sought out by the individual.

We also found a relationship between the evidentiality of the status, that is, how obvious that status is to the individual, and the role of the different types of cues in the identity change process. Finally, we have suggested that the adoption of a new status is a trigger for further career changes.

Although the model presented in this paper was generated inductively from field data on obese individuals, we are confident that it may be fruitfully applied to the study of other deviant careers. It seems particularly appropriate where the identity involved has a low degree of self-evidentiality.

In addition, we feel that the focus upon the different types of status cues and their differing roles in the recognizing and placing processes can lead to a better understanding of how institutionally promoted identity change occur.

NOTES

1. Included here is research on both rule creators and rule enforcers. We have not made an attempt to analytically separate the two types of investigation.

2. Some of the members had successfully lost their excess weight. When these people were present at a meeting, a leader was careful to introduce them to the other members of the class and to tell how much weight they had lost. This was done to uphold their claim to acceptance by the other group members.

3. Access to this information was gained from an "insider" perspective because one of the researchers was well known among the membership, being an "off and on" member of the organization for three years. Thus, he was not confronted with the problem of gaining entry into a semiclosed social setting. Similarly, because the observer had "been an ongoing participant of the group," he did not have to desensitize the other members of the group to his presence.

4. It is important to note that this process can operate generically. That is, it is not only applicable to the "identity change" from a "normal" to a "deviant" identity, but can encompass the reverse process as well. In a forthcoming project, we will use the process to analyze how various rehabilitation programs attempt to get individuals back to the initial status.

5. This concept is different in an important way from what Goffman calls "visibility." He uses the term to refer to "... how well or how badly the stigma is adapted to provide means of communicating that the individual possesses it" (Goffman, 1963, p. 48). The focus of the concept is on how readily the social environment can identify that the individual possesses a stigmatized trait. The concept of self-evidentiality deals with how readily the individual can internalize

possession of the stigmatized trait. The focus is upon the actor's perceptions, not on the audience.

6. We are referring here to the physiological description of being male or female. We realize that sex roles are much less self-evident.

7. Conversely, a number of individuals thought of themselves as "fat" or "obese," and were objectively "normal." In this case, the existence of objective indicators was insufficient to prevent the individual from adopting a "fat" identity.

8. In some instances, recognizing and placing occur simultaneously. This is especially true when the cue involved is an active one, and contains information about both the initial and new statuses. For example, if peers call a child "fatty," this interaction informs the child that the "normal" status is inappropriate. At the same time, it informs the child that being "fat" is the appropriate status. Even here however, the individual must recognize before it is possible to place.

9. This phase corresponds closely to that presented in much of the "subcultural" research. (See Schur, 1971; Becker, 1963; Sykes & Matza, 1957, p. 664.)

10. Weight Watchers, TOPS, Overeaters Anonymous, Diet Center, and OptiFast are typical examples of this type of program.

11. By witnessing, we are referring to the process whereby individuals come to renounce, in front of others, a former self and former behaviors associated with that self. Some religious groups, Synanon, Alcoholics Anonymous, etc., seem to encourage this type of degradation of self.

REFERENCES

Baum, L. (1987, August 3). Extra pounds can weigh down your career. *Business Week,* p. 96.

Becker, H. S. (1963). *Outsiders: Studies in the sociology of deviance.* New York: Free Press.

Canning, H., & Mayers, J. (1966). Obesity: Its possible effects on college acceptance. *New England Journal of Medicine,* 275(24); November, 1172–1174.

Cicourel, A. (1968). *The social organization of juvenile justice.* New York: Wiley.

Conrad, P. (1975). The discovery of hyperkinesis: Notes on the medicalization of deviant behavior. *Social Problems,* 23(1); October, 12–21.

Goffman, E. (1961). *Asylums.* Garden City, NY: Anchor.

———. (1963). *Stigma: Notes on the management of spoiled identity.* Englewood Cliffs, NJ: Prentice Hall.

Greenberg, D. (1989). The antifat conspiracy. *New Scientist,* 22 (April 22): 79.

Hadden, S. C., Degher, D., & Fernandez, R. (1989). Sports as a strategic ethnographic arena. *Arena Review,* 13(1), 9–19.

Kelly, J. R. (1990). *Leisure* (2nd ed.). Englewood Cliffs, NJ: Prentice Hall.

Kitsuse, J. (1962). Societal reactions to deviant behavior: Problems of theory and method. *Social Problems* 9 (Winter): 247–256.

Laslett, B., & Warren, C. A. B. (1975). Losing weight: The organizational promotion of behavior change. *Social Problems,* 23(1), 69–80.

Lemert, E. (1951). *Social pathology,* New York: McGraw-Hill.

Maddox, G. L., Back, K. W., & Liederman, V. (1968). Overweight as social deviance and disability. *Journal of Health and Social Behavior,* 9(4): 287–298.

Maddox, G. L., & Liederman, V. (1969). Overweight as a social disability with medical implications. *Journal of Medical Education,* 9(4): 287–298.

Pauley, L. L. (1989). Customer weights as a variable in salespersons' response time. *Journal of Social Psychology,* 129: 713–714.

Petit, D. W. (1974). The ills of the obese. In G. A. Gray & J. E. Bethune (Eds.), *Treatment and management of obesity.* New York: Harper & Row.

Piliavin, I., & Briar, S. (1964). Police encounters with juveniles. *American Journal of Sociology,* (September): 206–214.

Register, C. A. (1990). Wage effects of obesity among young workers. *Social Science Quarterly,* 71 (March): 130–141.

Scheff, T. (1988). Shame and conformity: The deference emotion system. *American Journal of Sociology,* 53 (June): 395–406.

Schur, E. M. (1971). *Labeling deviant behavior: Its sociological implications.* New York: Harper & Row.

Sykes, G., & Matza, D. (1957). Techniques of neutralization: A theory of delinquency. *American Sociological Review,* (December): 664–670.

Tannenbaum, F. (1939). *Crime and the community.* New York: Columbia University Press.

24

Becoming Bisexual

MARTIN S. WEINBERG, COLIN J. WILLIAMS,
AND DOUGLAS W. PRYOR

Weinberg, Williams, and Pryor's study of the identity career followed by individuals who become bisexual illustrates a much more complex identity trajectory than Degher and Hughes's portrayal of their fat subjects. The attraction of individuals in this selection to members of the same gender leads them to question and reject their affiliation with the heterosexual identity. But their continuing attraction to members of the other gender leads them to question the appropriateness, for them, of the homosexual designation. Aware that they do not fit comfortably into either of these common sexual identity labels, they are confused and troubled. After struggling with their ambivalences about their dual sexual orientations, they discover the bisexual option and begin the process of reconceptualizing their selves. Although finding this label and discovering established bisexual others aid their process of self-acceptance, problems remain for people trying to establish identities outside of the more common sexual norms. They remain doubly stigmatized due to their rejection by both the heterosexual and homosexual communities and never feel completely comfortable with this liminal identity.

B ecoming bisexual involves the rejection of not one but two recognized categories of sexual identity: heterosexual and homosexual. Most people settle into the status of heterosexual without any struggle over the identity. There is not much concern with explaining how this occurs; that people are heterosexual is simply taken for granted. For those who find heterosexuality unfulfilling, however, developing a sexual identity is more difficult.

How is it then that some people come to identify themselves as "bisexuals"? As a point of departure we take the process through which people come to identify themselves as "homosexual." A number of models have been formulated that chart the development of a homosexual identity through a series of stages. While each model involves a different number of stages, the models all share three elements. The process begins with the person in a state of identity confusion—feeling different from others, struggling with the acknowledgment of same-sex attractions. Then there is a period of thinking about possibly being homosexual—involving associating with self-identified homosexuals, sexual experimentation, forays into the homosexual subculture. Last is the attempt to integrate one's self-concept and social identity as homosexual—acceptance of the label, disclosure about being homosexual, acculturation to a homosexual way of life, and the development of love relationships. Not every person follows through each stage. Some remain locked in at a certain point. Others move back and forth between stages.

To our knowledge, no previous model of *bisexual* identity formation exists. In this chapter we present such a model based on the following questions: To what extent is there overlap with the process involved in becoming homosexual? How far is the label "bisexual" clearly recognized, understood, and available to people as an identity? Does the absence of a bisexual subculture in most locales affect the information and support needed for sustaining a commitment to the identity? For our subjects, then, what are the problems in finding the "bisexual" label, understanding what the label means, dealing with social disapproval from two directions, and continuing to use the label once it is adopted? From our fieldwork and interviews, we found that four stages captured our respondents' most common experiences when dealing with questions of identity: initial confusion, finding and applying the label, settling into the identity, and continued uncertainty.

THE STAGES

Initial Confusion

Many of the people interviewed said that they had experienced a period of considerable confusion, doubt, and struggle regarding their sexual identity before defining themselves as bisexual. This was ordinarily the first step in the process of becoming bisexual.

They described a number of major sources of early confusion about their sexual identity. For some, it was the experience of having strong sexual feelings for both sexes that was unsettling, disorienting, and sometimes frightening. Often these were sexual feelings that they said they did not know how to easily handle or resolve.

> In the past, I couldn't reconcile different desires I had. I didn't understand them. I didn't know what I was. And I ended up feeling really mixed up, unsure, and kind of frightened. (F)

> I thought I was gay, and yet I was having these intense fantasies and feelings about fucking women. I went through a long period of confusion. (M)

Others were confused because they thought strong sexual feelings for, or sexual behavior with, the same sex meant an end to their long-standing heterosexuality.

> I was afraid of my sexual feelings for men and . . . that if I acted on them, that would negate my sexual feelings for women. I knew absolutely

no one else who had . . . sexual feelings for both men and women, and didn't realize that was an option. (M)

When I first had sexual feelings for females, I had the sense I should give up my feelings for men. I think it would have been easier to give up men. (F)

A third source of confusion in this initial stage stemmed from attempts by respondents trying to categorize their feelings for, and/or behaviors with, both sexes, yet not being able to do so. Unaware of the term "bisexual," some tried to organize their sexuality by using readily available labels of "heterosexual" or "homosexual"—but these did not seem to fit. No sense of sexual identity jelled; an aspect of themselves remained unclassifiable.

When I was young, I didn't know what I was. I knew there were people like Mom and Dad—heterosexual and married—and that there were "queens." I knew I wasn't like either one. (M)

I thought I had to be either gay or straight. That was the big lie. It was confusing. . . . That all began to change in the late 60s. It was a long and slow process. . . . (F)

Finally, others suggested they experienced a great deal of confusion because of their "homophobia"—their difficulty in facing up to the same-sex component of their sexuality. The consequence was often long-term denial. This was more common among the men than the women, but not exclusively so.

At age seventeen, I became close to a woman who was gay. She had sexual feelings for me. I had some . . . for her but I didn't respond. Between the ages of seventeen and twenty-six I met another gay woman. She also had sexual feelings towards me. I had the same for her but I didn't act on . . . or acknowledge them. . . . I was scared. . . . I was also attracted to men at the same time. . . . I denied that I was sexually attracted to women. I was afraid that if they knew the feelings were mutual they would act on them . . . and put pressure on me. (F)

I thought I might be able to get rid of my homosexual tendencies through religious means—prayer, belief, counseling—before I came to accept it as part of me. (M)

The intensity of the confusion and the extent to which it existed in the lives of the people we met at the Bisexual Center, whatever its particular source, was summed up by two men who spoke with us informally. As paraphrased in our field notes:

The identity issue for him was a very confusing one. At one point, he almost had a nervous breakdown, and when he finally entered college, he sought psychiatric help.

Bill said he thinks this sort of thing happens a lot at the Bi Center. People come in "very confused" and experience some really painful stress.

Finding and Applying the Label

Following this initial period of confusion, which often spanned years, was the experience of finding and applying the label. We asked the people we interviewed for specific factors or events in their lives that led them to define themselves as bisexual. There were a number of common experiences.

For many who were unfamiliar with the term bisexual, the discovery that the category in fact existed was a turning point. This happened by simply hearing the word, reading about it somewhere, or learning of a place called the Bisexual Center. The discovery provided a means of making sense of long-standing feelings for both sexes.

> Early on I thought I was just gay, because I was not aware there was another category, bisexual. I always knew I was interested in men and women. But I did not realize there was a name for these feelings and behaviors until I took Psychology 101 and read about it, heard about it there. That was in college. (F)

> The first time I heard the word, which was not until I was twenty-six, I realized that was what fit for me. What it fit was that I had sexual feelings for both men and women. Up until that point, the only way that I could define my sexual feelings was that I was either a latent homosexual or a confused heterosexual. (M)

> Going to a party at someone's house, and finding out there that the party was to benefit the Bisexual Center. I guess at that point I began to define myself as bisexual. I never knew there was such a word. If I had heard the word earlier on, for example as a kid, I might have been bisexual then. My feelings had always been bisexual. I just did not know how to define them. (F)

> Reading *The Bisexual Option* . . . I realized then that bisexuality really existed and that's what I was. (M)

In the case of others the turning point was their first homosexual or heterosexual experience coupled with the recognition that sex was pleasurable with both sexes. These were people who already seemed to have knowledge of the label "bisexual," yet without experiences with both men and women, could not label themselves accordingly.

> The first time I had actual intercourse, an orgasm with a woman, it led me to realize I was bisexual, because I enjoyed it as much as I did with a man, although the former occurred much later on in my sexual experiences. . . . I didn't have an orgasm with a woman until twenty-two, while with males, that had been going on since the age of thirteen. (M)

> Having homosexual fantasies and acting those out. . . . I would not identify as bi if I only had fantasies and they were mild. But since my fantasies were intensely erotic, and I acted them out, these two things led me to believe I was really bisexual. . . . (M)

After my first involved sexual affair with a woman, I also had feelings for a man, and I knew I did not fit the category dyke. I was also dating gay-identified males. So I began looking at gay/lesbian and heterosexual labels as not fitting my situation. (F)

Still others reported not so much a specific experience as a turning point, but emphasized the recognition that their sexual feelings for both sexes were simply too strong to deny. They eventually came to the conclusion that it was unnecessary to choose between them.

I found myself with men but couldn't completely ignore my feelings for women. When involved with a man I always had a close female relationship. When one or the other didn't exist at any given time, I felt I was really lacking something. I seem to like both. (F)

The last factor that was instrumental in leading people to initially adopt the label bisexual was the encouragement and support of others. Encouragement sometimes came from a partner who already defined himself or herself as bisexual.

Encouragement from a man I was in a relationship with. We had been together two or three years at the time—he began to define as bisexual.... [He] encouraged me to do so as well. He engineered a couple of threesomes with another woman. Seeing one other person who had bisexuality as an identity that fit them seemed to be a real encouragement. (F)

Encouragement from a partner seemed to matter more for women. Occasionally the "encouragement" bordered on coercion as the men in their lives wanted to engage in a *ménage à trois* or group sex.

I had a male lover for a year and a half who was familiar with bisexuality and pushed me towards it. My relationship with him brought it up in me. He wanted me to be bisexual because he wanted to be in a threesome. He was also insanely jealous of my attractions to men, and did everything in his power to suppress my opposite-sex attractions. He showed me a lot of pictures of naked women and played on my reactions. He could tell that I was aroused by pictures of women and would talk about my attractions while we were having sex.... He was twenty years older than me. He was very manipulative in a way. My feelings for females were there and [he was] almost forcing me to act on my attractions.... (F)

Encouragement also came from sex-positive organizations, primarily the Bisexual Center, but also places like San Francisco Sex Information (SFSI), the Pacific Center, and the Institute for Advanced Study of Human Sexuality,...

At the gay pride parade I had seen the brochures for the Bisexual Center. Two years later I went to a Tuesday night meeting. I immediately felt that I belonged and that if I had to define myself that this was what I would use. (M)

> Through SFSI and the Bi Center, I found a community of people . . . [who] were more comfortable for me than were the exclusive gay or heterosexual communities. . . . [It was] beneficial for myself to be . . . in a sex-positive community. I got more strokes and came to understand myself better. . . . I felt it was necessary to express my feelings for males and females without having to censor them, which is what the gay and straight communities pressured me to do. (F)

Thus our respondents became familiar with and came to the point of adopting the label bisexual in a variety of ways: through reading about it on their own, being in therapy, talking to friends, having experiences with sex partners, learning about the Bi Center, visiting SFSI or the Pacific Center, and coming to accept their sexual feelings.

Settling into the Identity

Usually it took years from the time of first sexual attractions to, or behaviors with, both sexes before people came to think of themselves as bisexual. The next stage then was one of settling into the identity, which was characterized by a more complete transition in self-labeling.

Most reported that this settling-in stage was the consequence of becoming more self-accepting. They became less concerned with the negative attitudes of others about their sexual preferences.

> I realized that the problem of bisexuality isn't mine. It's society's. They are having problems dealing with my bisexuality. So I was then thinking if they had a problem dealing with it, so should I. But I don't. (F)

> I learned to accept the fact that there are a lot of people out there who aren't accepting. They can be intolerant, selfish, shortsighted and so on. Finally, in growing up, I learned to say "So what, I don't care what others think." (M)

> I just decided I was bi. I trusted my own sense of self. I stopped listening to others tell me what I could or couldn't be. (F)

The increase in self-acceptance was often attributed to the continuing support from friends, counselors, and the Bi Center, through reading, and just being in San Francisco.

> Fred Klein's *The Bisexual Option* book and meeting more and more bisexual people . . . helped me feel more normal. . . . There were other human beings who felt like I did on a consistent basis. (M)

> I think going to the Bi Center really helped a lot. I think going to the gay baths and realizing there were a lot of men who sought the same outlet I did really helped. Talking about it with friends has been helpful and being validated by female lovers that approve of my bisexuality. Also the reaction of people who I've told, many of whom weren't even surprised. (M)

The most important thing was counseling. Having the support of a bisexual counselor. Someone who acted as somewhat of a mentor. [He] validated my frustration ..., helped me do problem solving, and guide[d] me to other supportive experiences like SFSI. Just engaging myself in a supportive social community. (M)

The majority of the people we came to know through the interviews seemed settled in their sexual identity. We tapped this through a variety of questions. . . . Ninety percent said that they did not think they were currently in transition from being homosexual to being heterosexual or from being heterosexual to being homosexual. However, when we probed further by asking this group "Is it possible, though, that someday you could define yourself as either lesbian/gay or heterosexual?" about 40 percent answered yes. About two-thirds of these indicated that the change could be in either direction, though almost 70 percent said that such a change was not probable.

We asked those who thought a change was possible what it might take to bring it about. The most common responses referred to becoming involved in a meaningful relationship that was monogamous or very intense. Often the sex of the hypothetical partner was not specified, underscoring that the overall quality of the relationship was what really mattered.

Love. I think if I feel insanely in love with some person, it could possibly happen. (M)

If I should meet a woman and want to get married, and if she was not open to my relating to men, I might become heterosexual again. (M)

Getting involved in a longer-term relationship like marriage where I wouldn't need a sexual involvement with anyone else. The sex of the ... partner wouldn't matter. It would have to be someone who I could commit my whole life to exclusively, a lifelong relationship. (F)

A few mentioned the breaking up of a relationship and how this would incline them to look toward the other sex.

Steve is one of the few men I feel completely comfortable with. If anything happened to him, I don't know if I'd want to try and build up a similar relationship with another man. I'd be more inclined to look towards women for support. (F)

Changes in sexual behavior seemed more likely for the people we interviewed ... than changes in how they defined themselves. We asked "Is it possible that someday you could behave either exclusively homosexual or exclusively heterosexual?" Over 80 percent answered yes. This is over twice as many as those who saw a possible change in how they defined themselves, again showing that a wide range of behaviors can be subsumed under the same label. Of this particular group, the majority (almost 60 percent) felt that there was nothing inevitable about how they might change, indicating that it could be in either a homosexual or a heterosexual direction. Around a quarter, though, said the change would be to exclusive heterosexual behavior and 15 percent to exclusive homosexual behavior. (Twice as many women noted the homosexual direction, while many more men than women said the heterosexual direction.) Just over 40 percent responded that a change to exclusive heterosexuality or homosexuality was not very probable, about a third somewhat probable, and about a quarter very probable.

Again, we asked what it would take to bring about such a change in behavior. Once more the answers centered on achieving long-term monogamous and involved relationship, often with no reference to a specific sex.

> For me to behave exclusively heterosexual or homosexual would require that I find a lifetime commitment from another person with a damn good argument of why I should not go to bed with somebody else. (F)

> I am a romantic. If I fell in love with a man, and our relationship was developing that way, I might become strictly homosexual. The same possibility exists with a woman. (M)

Thus "settling into the identity" must be seen in relative terms. Some of the people we interviewed did seem to accept the identity completely. When we compared our subjects' experiences with those characteristic of homosexuals, however, we were struck by the absence of closure that characterized our bisexual respondents—even those who appeared most committed to the identity. This led us to posit a final stage in the formation of sexual identity, one that seems unique to bisexuals.

Continued Uncertainty

The belief that bisexuals are confused about their sexual identity is quite common. The conception has been promoted especially by those lesbians and gays who see bisexuality as being in and of itself a pathological state. From their point of view, "confusion" is literally a built-in feature of "being" bisexual. As expressed in one study:

> While appearing to encompass a wider choice of love objects . . . [the bisexual] actually becomes a product of abject confusion; his self-image is that of an overgrown young adolescent whose ability to differentiate one form of sexuality from another has never developed. He lacks above all a sense of identity. . . . [He] cannot answer the question: What am I?

One evening a facilitator at a Bisexual Center rap group put this belief in a slightly different and more contemporary form:

> One of the myths about bisexuality is that you can't be bisexual without somehow being "schizoid." The lesbian and gay communities do not see being bisexual as a crystallized or complete sexual identity. The homosexual community believes there is no such thing as bisexuality. They think that bisexuals are people who are in transition [to becoming homosexual] or that they are people afraid of being stigmatized [as homosexual] by the heterosexual majority.

We addressed the issue directly in the interviews with two questions: "Do you *presently* feel confused about your bisexuality?" and "Have you ever felt confused ...?" ... For the men, a quarter and 84 percent answered "yes," respectively. For the women, it was about a quarter and 56 percent.

When asked to provide details about this uncertainty, the primary response was that *even after having discovered and applied the label "bisexual" to themselves, and having come to the point of apparent self-acceptance, they still experienced continued intermittent periods of doubt and uncertainty regarding their sexual identity.* One reason was the lack of social validation and support that came with being a self-identified bisexual. The social reaction people received made it difficult to sustain the identity over the long haul.

While the heterosexual world was said to be completely intolerant of any degree of homosexuality, the reaction of the homosexual world mattered more. Many bisexuals referred to the persistent pressures they experienced to relabel themselves "gay" or "lesbian" and to engage in sexual activity exclusively with the same sex. It was asserted that no one was *really* bisexual, and that calling oneself "bisexual" was a politically incorrect and unauthentic identity. Given that our respondents were living in San Francisco (which has such a large homosexual population) and that they frequently moved in and out of the homosexual world (to whom they often looked for support) this could be particularly distressing.

> Sometimes the repeated denial the gay community directs at us. Their negation of the concept and the term bisexual has sometimes made me wonder whether I was just imagining the whole thing. (M)

> My involvement with the gay community. There was extreme political pressure. The lesbians said bisexuals didn't exist. To them, I had to make up my mind and identify as lesbian.... I was really questioning my identity, that is, about defining myself as bisexual.... (F)

For the women, the invalidation carried over to their feminist identity (which most had). They sometimes felt that being with men meant they were selling out the world of women.

> I was involved with a woman for several years. She was straight when I met her but became a lesbian. She tried to "win me back" to lesbianism. She tried to tell me that if I really loved her, I would leave Bill. I did love her, but I could not deny how I felt about him either. So she left me and that hurt. I wondered if I was selling out my woman identity and if it [being bisexual] was worth it. (F)

A few wondered whether they were lying to themselves about their heterosexual side. One woman questioned whether her heterosexual desires were a result of "acculturation" rather than being her own choice. Another woman suggested a similar social dimension to her homosexual component:

> There was one period when I was trying to be gay because of the political thing of being totally woman-identified rather than being with

men. The Women's Culture Center in college had a women's studies minor, so I was totally immersed in women's culture.... (F)

Lack of support also came from the absence of bisexual role models, no real bisexual community aside from the Bisexual Center, and nothing in the way of public recognition of bisexuality which bred uncertainty and confusion.

I went through a period of dissociation, of being very alone and isolated. That was due to my bisexuality. People would ask, well, what was I? I wasn't gay and I wasn't straight. So I didn't fit. (F)

I don't feel like I belong in a lot of situations because society is so polarized as heterosexual or homosexual. There are not enough bi organizations or public places to go to like bars, restaurants, clubs.... (F)

For some, continuing uncertainty about their sexual identity was related to their inability to translate their sexual feelings into sexual behaviors. (Some of the women had *never* engaged in homosexual sex.)

Should I try to have a sexual relationship with a woman? ... Should I just back off and keep my distance, just try to maintain a friendship? I question whether I am really bisexual because I don't know if I will ever act on my physical attractions for females. (F)

I know I have strong sexual feelings towards men, but then I don't know how to get close to or be sexual with a man. I guess that what happens is I start wondering how genuine my feelings are.... (M)

For the men, confusion stemmed more from the practical concerns of implementing and managing multiple partners or from questions about how to find an involved homosexual relationship and what that might mean on a social and personal level.

I felt very confused about how I was going to manage my life in terms of developing relationships with both men and women. I still see it as a difficult lifestyle to create for myself because it involves a lot of hard work and understanding on my part and that of the men and women I'm involved with. (M)

I've thought about trying to have an actual relationship with a man. Some of my confusion revolves around how to find a satisfactory sexual relationship. I do not particularly like gay bars. I have stopped having anonymous sex.... (M)

Many men and women felt doubts about their bisexual identity because of being in an exclusive sexual relationship. After being exclusively involved with an opposite-sex partner for a period to time, some of the respondents questioned the homosexual side of their sexuality. Conversely, after being exclusively involved with a partner of the same sex, other respondents called into question the heterosexual component of their sexuality.

When I'm with a man or a woman sexually for a period of time, then I begin to wonder how attracted I really am to the other sex. (M)

In the last relationship I had with a woman, my heterosexual feelings were very diminished. Being involved in a lesbian lifestyle put stress on my self-identification as a bisexual. It seems confusing to me because I am monogamous for the most part, monogamy determines my lifestyle to the extremes of being heterosexual or homosexual. (F)

Others made reference to a lack of sexual activity with weaker sexual feelings and affections for one sex. Such learning did not fit with the perception that bisexuals should have balanced desires and behaviors. The consequence was doubt about "really" being bisexual.

On the level of sexual arousal and deep romantic feelings, I feel them much more strongly for women than for men. I've gone so far as questioning myself when this is involved. (M)

I definitely am attracted to and it is much easier to deal with males. Also, guilt for my attraction to females has led me to wonder if I am just really toying with the idea. Is the sexual attraction I have for females something I constructed to pass time or what? (F)

Just as "settling into the identity" is a relative phenomenon, so too is "continued uncertainty," which can involve a lack of closure as part and parcel of what it means to be bisexual.

We do not wish to claim too much for our model of bisexual identity formation. There are limits to its general application. The people we interviewed were unique in that not only did *all* the respondents define themselves as bisexual (a consequence of our selection criteria), but they were also all members of a bisexual social organization in a city that perhaps more than any other in the United States could be said to provide a bisexual subculture of some sort. Bisexuals in places other than San Francisco surely must move through the early phases of the identity process with a great deal more difficulty. Many probably never reach the later stages.

Finally, the phases of the model we present are very broad and somewhat simplified. While the particular problems we detail within different phases may be restricted to the type of bisexuals in this study, the broader phases can form the basis for the development of more sophisticated models of bisexual identity formation.

Still, not all bisexuals will follow these patterns. Indeed, given the relative weakness of the bisexual subculture compared with the social pressures toward conformity exhibited in the gay subculture, there may be more varied ways of acquiring a bisexual identity. Also, the involvement of bisexuals in the heterosexual world means that various changes in heterosexual lifestyles (e.g., a decrease in open marriages or swinging) will be a continuing, and as yet unexplored, influence on bisexual identity. Finally, wider societal changes, notably the existence of AIDS, may make for changes in the overall identity process. Being used to choice and being open to both sexes can give bisexuals a range of adaptations in their sexual life that are not available to others.

25

Anorexia Nervosa and Bulimia

PENELOPE A. McLORG AND DIANE E. TAUB

McLorg and Taub's study of eating disorders describes and analyzes women's progression along an identity career from their initial stage of hyperconformity through Lemert's stages of primary and secondary deviance. They illustrate how the intense societal preoccupation about weight leads women to the kind of deviant behavior that maintains (initially) their positive external status while they are deteriorating internally. Along the way, these women move through a progression of more common fixations about dieting, to frustration with dieting, and movement toward more radical solutions such as bingeing, purging, compulsive exercising, and lack of eating. These behaviors stand apart from their identities, McLorg and Taub argue, since they can avoid the deviant label and self-conception, remaining in the primary deviance stage, until they get caught and labeled as having an eating disorder. Once this occurs they move to secondary deviance, reconceptualized by others as anorectic or bulimic. With this label cast on them, they are forced to interact with others through the vehicle of their deviance, thus reinforcing their eating disorders.

Current appearance norms stipulate thinness for women and muscularity for men; these expectations, like any norms, entail rewards for compliance and negative sanctions for violations. Fear of being overweight—of being visually deviant—has led to a striving for thinness, especially among women. In the extreme, this avoidance of overweight engenders eating disorders, which themselves constitute deviance. Anorexia nervosa, or purposeful starvation, embodies visual as well as behavioral deviation; bulimia, binge-eating followed by vomiting and/or laxative abuse, is primarily behaviorally deviant.

Besides a fear of fatness, anorexics and bulimics exhibit distorted body images. In anorexia nervosa, a 20 to 25 percent loss of initial body weight occurs, resulting from self-starvation alone or in combination with excessive exercising, occasional binge-eating, vomiting and/or laxative abuse. Bulimia denotes cyclical (daily, weekly, for example) binge-eating followed by vomiting or laxative abuse; weight is normal or close to normal (Humphries et al., 1982). Common physical manifestations of these eating disorders include menstrual cessation or irregularities and electrolyte imbalances; among behavioral traits are depression, obsessions/compulsions, and anxiety (Russell, 1979; Thompson and Schwartz, 1982).

Increasingly prevalent in the past two decades, anorexia nervosa and bulimia have emerged as major health and social problems. Termed an epidemic on college campuses (Brody, as quoted in Schur, 1984: 76), bulimia affects 13 percent of college students (Halmi et al., 1981). Less prevalent, anorexia nervosa was diagnosed in 0.6 percent of students utilizing a university health center (Stangler and Printz, 1980). However, the overall mortality rate of anorexia nervosa is 6 percent (Schwartz and Thompson, 1981) to 20 percent (Humphries et al., 1982); bulimia appears to be less life-threatening (Russell, 1979).

Particularly affecting certain demographic groups, eating disorders are most prevalent among young, white, affluent (upper-middle to upper class) women in modern, industrialized countries (Crisp, 1977; Willi and Grossman, 1983). Combining all of these risk factors (female sex, youth, high socioeconomic status, and residence in an industrialized country), prevalence of anorexia nervosa in upper class English girls' schools is reported at 1 in 100 (Crisp et al., 1976). The age of onset for anorexia nervosa is bimodal at 14.5 and 18 years (Humphries et al., 1982); the most frequent age of onset for bulimia is 18 (Russell, 1979).

Eating disorders have primarily been studied from psychological and medical perspectives.[1] Theories of etiology have generally fallen into three categories: the ego psychological (involving an impaired child-maternal environment); the family systems (implicating enmeshed, rigid families); and the endocrinological (involving a precipitating hormonal defect). Although relatively ignored in previous studies, the sociocultural components of anorexia nervosa and bulimia (the slimness norm and its agents of reinforcement, such as role models) have been postulated as accounting for the recent, dramatic increases in these disorders (Schwartz et al., 1982; Boskind-White, 1985).[2]

Medical and psychological approaches to anorexia nervosa and bulimia obscure the social facets of the disorders and neglect the individuals' own definitions of their situations. Among the social processes involved in the development of an eating disorder is the sequence of conforming behavior, primary deviance, and secondary deviance. Societal reaction is the critical mediator affecting the movement through the deviant career (Becker, 1963). Within a framework of labeling theory, this study focuses on the emergence of anorexic and bulimic identities, as well as on the consequences of being career deviants.

METHODOLOGY

Sampling and Procedures

Most research on eating disorders has utilized clinical subjects or nonclinical respondents completing questionnaires. Such studies can be criticized for simply counting and describing behaviors and/or neglecting the social construction of the disorders. Moreover, the work of clinicians is often limited by therapeutic orientation. Previous research may also have included individuals who were not in therapy on their own volition and who resisted admitting that they had an eating disorder.

Past studies thus disregard the intersubjective meanings respondents attach to their behavior and emphasize researchers' criteria for definition as anorexic or bulimic. In order to supplement these sampling and procedural designs, the present study utilizes participant observation of a group of self-defined anorexics and bulimics.[3] As the individuals had acknowledged their eating disorders, frank discussion and disclosure were facilitated.

Data are derived from a self-help group, BANISH, Bulimics/Anorexics In Self-Help, which met at a university in an urban center of the mid-South. Founded by one of the researchers (D.E.T.), BANISH was advertised in local newspapers as offering a group experience for individuals who were anorexic or bulimic. Despite the local advertisements, the campus location of the meetings may have selectively encouraged university students to attend. Nonetheless, in view of the modal age of onset and socioeconomic status of individuals with eating disorders, college students have been considered target populations (Crisp et al., 1976; Halmi et al., 1981).

The group's weekly two-hour meetings were observed for two years. During the course of this study, 30 individuals attended at least one of the meetings. Attendance at meetings was varied: 10 individuals came nearly every Sunday; 5 attended approximately twice a month; and the remaining 15 participated once a month or less frequently, often when their eating problems were "more severe" or "bizarre." The modal number of members at meetings was 12. The diversity in attendance was to be expected in self-help groups of anorexics and bulimics.

[Most] people's involvement will not be forever or even a long time. Most people get the support they need and drop out. Some take the time to help others after they themselves have been helped but even they may withdraw after a time. It is a natural and in many cases *necessary* process. (Emphasis in original.) (American Anorexia/Bulimia Association, 1983)

Modeled after Alcoholics Anonymous, BANISH allowed participants to discuss their backgrounds and experiences with others who empathized. For many members, the group constituted their only source of help; these respondents were reluctant to contact health professionals because of shame, embarrassment, or financial difficulties.

In addition to field notes from group meetings, records of other encounters with all members were maintained. Participants visited the office of one of the researchers (D.E.T.), called both researchers by phone, and invited them to their homes or out for a cup of coffee. Such interaction facilitated genuine communication and mutual trust. Even among the 15 individuals who did not attend the meetings regularly, contact was maintained with 10 members on a monthly basis.

Supplementing field notes were informal interviews with 15 group members, lasting from two to four hours. Because they appeared to represent more extensive experience with eating disorders, these interviewees were chosen to amplify their comments about the labeling process, made during group meetings. Conducted near the end of the two-year observation period, the interviews focused on what the respondents thought antedated and maintained their eating disorders. In addition, participants described others' reactions to their behaviors as well as their own interpretations of these reactions. To protect the confidentiality of individuals quoted in the study, pseudonyms are employed.

Description of Members

The demographic composite of the sample typifies what has been found in other studies (Fox and James, 1976; Crisp, 1977; Herzog, 1982; Schlesier-Stropp, 1984). Group members' ages ranged from 19 to 36, with the modal age being 21. The respondents were white, and all but one were female. The sole male and three of the females were anorexic; the remaining females were bulimic.[4]

Primarily composed of college students, the group included four non-students, three of whom had college degrees. Nearly all members derived from upper-middle or lower-upper class households. Eighteen students and two non-students were never-marrieds and uninvolved in serious relationships; two non-students were married (one with two children); two students were divorced (one with two children); and six students were involved in serious relationships. The duration of eating disorders ranged from 3 to 15 years.

CONFORMING BEHAVIOR

In the backgrounds of most anorexics and bulimics, dieting figures prominently, beginning in the teen years (Crisp, 1977; Johnson et al., 1982; Lacey et al., 1986). As dieters, these individuals are conformist in their adherence to the cultural norms emphasizing thinness (Garner et al., 1980; Schwartz et al., 1982). In our society, slim bodies are regarded as the most worthy and attractive; overweight is viewed as physically and morally unhealthy—"obscene," "lazy," "slothful," and "gluttonous" (DeJong, 1980; Ritenbaugh, 1982; Schwartz et al., 1982).

Among the agents of socialization promoting the slimness norm is advertising. Female models in newspaper, magazine, and television advertisements are uniformly slender. In addition, product names and slogans exploit the thin orientation; examples include "Ultra Slim Lipstick," "Miller Lite," and "Virginia Slims." While retaining pressures toward thinness, an Ayds commercial attempts a compromise for those wanting to savor food: "Ayds... so you can taste, chew, and enjoy, while you lose weight." Appealing particularly to women, a nationwide fast-food restaurant chain offers low-calorie selections, so individuals can have a "license to eat." In the latter two examples, the notion of enjoying food is combined with the message to be slim. Food and restaurant advertisements overall convey the pleasures of eating, whereas advertisements for other products, such as fashions and diet aids, reinforce the idea that fatness is undesirable.

Emphasis on being slim affects everyone in our culture, but it influences women especially because of society's traditional emphasis on women's appearance. The slimness norm and its concomitant narrow beauty standards exacerbate the objectification of women (Schur, 1984). Women view themselves as visual entities and recognize that conforming to appearance expectations and "becoming attractive object[s] [are] role obligation[s]" (Laws, as quoted in Schur, 1984: 66). Demonstrating the beauty motivation behind dieting, a recent Nielsen survey indicated that of the 56 percent of all women aged 24 to 54 who dieted during the previous year, 76 percent did so for cosmetic, rather than health, reasons (Schwartz et al., 1982). For most female group members, dieting was viewed as a means of gaining attractiveness and appeal to the opposite sex. The male respondent, as well, indicated that "when I was fat, girls didn't look at me, but when I got thinner, I was suddenly popular."

In addition to responding to the specter of obesity, individuals who develop anorexia nervosa and bulimia are conformist in their strong commitment to other conventional norms and goals. They consistently excel at school and work (Russell, 1979; Bruch, 1981; Humphries et al., 1982), maintaining high aspirations in both areas (Theander, 1970; Lacey et al., 1986). Group members generally completed college-preparatory courses in high school, aware from an early age that they would strive for a college degree. Also, in college as well as high school, respondents joined honor societies and academic clubs.

Moreover, pre-anorexics and -bulimics display notable conventionality as "model children" (Humphries et al., 1982: 199), "the pride and joy" of their parents (Bruch, 1981: 215), accommodating themselves to the wishes of others. Parents of these individuals emphasize conformity and value achievement (Bruch, 1981). Respondents felt that perfect or near-perfect grades were expected of them; however, good grades were not rewarded by parents, because "A's" were common for these children. In addition, their parents suppressed conflicts, to preserve the image of the "all-American family" (Humphries et al., 1982). Group members reported that they seldom, if ever, heard their parents argue or raise their voices.

Also conformist in their affective ties, individuals who develop anorexia nervosa and bulimia are strongly, even excessively, attached to their parents. Respondents' families appeared close-knit, demonstrating palpable emotional ties. Several group members, for example, reported habitually calling home at prescribed times, whether or not they had any news. Such families have been termed "enmeshed" and "overprotective," displaying intense interaction and concern for members' welfare (Minuchin et al., 1978; Selvini-Palazzoli, 1978). These qualities could be viewed as marked conformity to the norm of familial closeness.[5]

Another element of notable conformity in the family milieu of pre-anorexics and -bulimics concerns eating, body weight/shape, and exercising (Kalucy et al., 1977; Humphries et al., 1982). Respondents reported their fathers' preoccupation with exercising and their mothers' engrossment in food preparation. When group members dieted and lost weight, they received an extraordinary amount of approval. Among the family, body size became a matter of "friendly rivalry." One bulimic informant recalled that she, her mother, and her coed sister all strived to wear a size five, regardless of their heights and body frames. Subsequent to this study, the researchers learned that both the mother and sister had become bulimic.

As preanorexics and -bulimics, group members thus exhibited marked conformity to cultural norms of thinness, achievement, compliance, and parental attachment. Their families reinforced their conformity by adherence to norms of family closeness and weight/body shape consciousness.

PRIMARY DEVIANCE

Even with familial encouragement, respondents, like nearly all dieters (Chernin, 1981), failed to maintain their lowered weights. Many cited their lack of willpower to eat only restricted foods. For the emerging anorexics and bulimics, extremes such as purposeful starvation or binging accompanied by vomiting and/or laxative abuse appeared as "obvious solutions" to the problem of retaining weight loss. Associated with these behaviors was a regained feeling of control in lives that had been disrupted by a major crisis. Group members' extreme weight-loss efforts operated as coping mechanisms for entering college, leaving home, or feeling rejected by the opposite sex.

The primary inducement for both eating adaptations was the drive for slimness: with slimness came more self-respect and a feeling of superiority over "unsuccessful dieters." Brian, for example, experienced a "power trip" upon consistent weight loss through starvation. Binges allowed the purging respondents to cope with stress through eating while maintaining a slim appearance. As former strict dieters, Teresa and Jennifer used binging/purging as an alternative to the constant self-denial of starvation. Acknowledging their parents' desires for them to be slim, most respondents still felt it was a conscious choice on their part to continue extreme weight-loss efforts. Being thin became the "most important thing" in their lives—their "greatest ambition."

In explaining the development of an anorexic or bulimic identity, Lemert's (1951; 1967) concept of primary deviance is salient. Primary deviance refers to a transitory period of norm violations which do not affect an individual's self-concept or performance of social roles. Although respondents were exhibiting anorexic or bulimic behavior, they did not consider themselves to be anorexic or bulimic.

At first, anorexics' significant others complimented their weight loss, expounding on their new "sleekness" and "good looks." Branch and Eurman (1980: 631) also found anorexics' families and friends describing them as "well-groomed," "neat," "fashionable," and "victorious." Not until the respondents approached emaciation did some parents or friends become concerned and withdraw their praise. Significant others also became increasingly aware of the anorexics' compulsive exercising, preoccupation with food preparation (but not consumption), and ritualistic eating patterns (such as cutting food into minute pieces and eating only certain foods at prescribed times).

For bulimics, friends or family members began to question how the respondents could eat such large amounts of food (often in excess of 10,000 calories a day) and stay slim. Significant others also noticed calluses across the bulimics' hands, which were caused by repeated inducement of vomiting. Several bulimics were "caught in the act," bent over commodes. Generally, friends and family required substantial evidence before believing that the respondents' binging or purging was no longer sporadic.

SECONDARY DEVIANCE

Heightened awareness of group members' eating behavior ultimately led others to label the respondents "anorexic" or "bulimic." Respondents differed in their histories of being labeled and accepting the labels. Generally first termed anorexic by friends, family, or medical personnel, the anorexics initially vigorously denied the label. They felt they were not "anorexic enough," not skinny enough; Robin did not regard herself as having the "skeletal" appearance she associated with anorexia nervosa. These group members found it difficult to differentiate between socially approved modes of weight loss—eating less and exercising more—and the extremes of those behaviors. In fact, many of their activities—cheerleading, modeling, gymnastics, aerobics—reinforced their pursuit of thinness. Like other anorexics, Chris felt she was being "ultra-healthy" with "total control" over her body.

For several respondents, admitting they were anorexic followed the realization that their lives were disrupted by their eating disorder. Anorexics' inflexible eating patterns unsettled family meals and holiday gatherings. Their regimented lifestyle of compulsively scheduled activities—exercising, school, and meals—precluded any spontaneous social interactions. Realization of their adverse behaviors preceded the anorexics' acknowledgment of their subnormal body weight and size.

Contrasting with anorexics, the binge/purgers, when confronted, more readily admitted that they were bulimic and that their means of weight loss was "abnormal." Teresa, for example, knew "very well" that her bulimic behavior was "wrong and unhealthy," although "worth the physical risks." While the bulimics initially maintained that their purging was only a temporary weight-loss method, they eventually realized that their disorder represented a "loss of control." Although these respondents regretted the self-indulgence, "shame," and "wasted time," they acknowledged their growing dependence on binging/purging for weight management and stress regulation.

The application of anorexic or bulimic labels precipitated secondary deviance, wherein group members internalized these identities. Secondary deviance refers to norm violations which are a response to society's labeling: "secondary deviation ... becomes a means of social defense, attack or adaptation to the overt and covert problems created by the societal reaction to primary deviance" (Lemert, 1967: 17). In contrast to primary deviance, secondary deviance is generally prolonged, alters the individual's self-concept, and affects the performance of his/her social roles.

As secondary deviants, respondents felt that their disorders "gave a purpose" to their lives. Nicole resisted attaining a normal weight because it was not "her"—she accepted her anorexic weight as her "true" weight. For Teresa, bulimia became a "companion"; and Julie felt "every aspect of her life," including time management and social activities, was affected by her bulimia. Group members' eating disorders became the salient element of their self-concepts, so that they related to familiar people and new acquaintances as anorexics or bulimics. For example, respondents regularly compared their body shapes and sizes with those of others. They also became sensitized to comments about their appearance, whether or not the remarks were made by someone aware of their eating disorder.

With their behavior increasingly attuned to their eating disorders, group members exhibited role engulfment (Schur, 1971). Through accepting anorexic or bulimic identities, individuals centered activities around their deviant role, downgrading other social roles. Their obligations as students, family members, and friends became subordinate to their eating and exercising rituals. Socializing, for example, was gradually curtailed because it interfered with compulsive exercising, binging, or purging.

Labeled anorexic or bulimic, respondents were ascribed a new status with a different set of role expectations. Regardless of other positions the individuals occupied, their deviant status, or master status (Hughes, 1958; Becker, 1963), was identified before all others. Among group members, Nicole, who was known as the "school's brain," became known as the "school's anorexic." No longer viewed as conforming model individuals, some respondents were termed "starving waifs" or "pigs."

Because of their identities as deviants, anorexics' and bulimics' interactions with others were altered. Group members' eating habits were scrutinized by friends and family and used as a "catchall" for everything negative that happened to them. Respondents felt self-conscious around individuals who knew of their disorders; for example, Robin imagined people "watching and whispering" behind her. In addition, group members believed others expected them to "act" anorexic or bulimic. Friends of some anorexic group members never offered them food or drink, assuming continued disinterest on the respondents' part. While being hospitalized, Denise felt she had to prove to others she was not still vomiting, by keeping her bathroom door open. Other bulimics, who lived in dormitories, were hesitant to use the restroom for normal purposes lest several friends be huddling at the door, listening for vomiting. In general, individuals interacted with the respondents largely on the basis of their eating disorder; in doing so, they reinforced anorexic and bulimic behaviors.

Bulimic respondents, whose weight-loss behavior was not generally detectable from their appearance, tried earnestly to hide their bulimia by binging and purging in secret. Their main purpose in concealment was to avoid the negative consequences of being known as a bulimic. For these individuals, bulimia connoted a "cop-out": like "weak anorexics," bulimics pursued thinness but yielded to urges to eat. Respondents felt other people regarded bulimia as "gross" and had little sympathy for the sufferer. To avoid these stigmas or "spoiled identities," the bulimics shrouded their behaviors.

Distinguishing types of stigma, Goffman (1963) describes discredited (visible) stigmas and discreditable (invisible) stigmas. Bulimics, whose weight was approximately normal or even slightly elevated, harbored discreditable stigmas. Anorexics, on the other hand, suffered both discreditable and discredited stigmas—the latter due to their emaciated appearance. Certain anorexics were more reconciled than the bulimics to their stigmas: for Brian, the "stigma of anorexia was better than the stigma of being fat." Common to the stigmatized individuals was an inability to interact spontaneously with others. Respondents were constantly on guard against topics of eating and body size.

Both anorexics and bulimics were held responsible by others for their behavior and presumed able to "get out of it if they tried." Many anorexics reported being told to "just eat more," while bulimics were enjoined to simply "stop eating so much." Such appeals were made without regard for the complexities of the problem. Ostracized by certain friends and family members, anorexics and bulimics felt increasingly isolated. For respondents, the self-help group presented a nonthreatening forum for discussing their disorders. Here, they found mutual understanding, empathy, and support. Many participants viewed BANISH as a haven from stigmatization by "others."

Group members, as secondary deviants, thus endured negative consequences, such as stigmatization, from being labeled. As they internalized the labels anorexic or bulimic, individuals' self-concepts were significantly influenced. When others interacted with the respondents on the basis of their eating disorders, anorexic or bulimic identities were encouraged. Moreover, group members' efforts to counteract the deviant labels were thwarted by their master statuses.

DISCUSSION

Previous research on eating disorders has dwelt almost exclusively on medical and psychological facets. Although necessary for a comprehensive understanding of anorexia nervosa and bulimia, these approaches neglect the social processes involved. The phenomena of eating disorders transcend concrete disease entities and clinical diagnoses. Multifaceted and complex, anorexia nervosa and bulimia require a holistic research design, in which sociological insights must be included.

A limitation of medical/psychiatric studies, in particular, is researchers' use of a priori criteria in establishing salient variables. Rather than utilizing predetermined standards of inclusion, the present study allows respondents to construct their own reality. Concomitant to this innovative approach to eating disorders is the selection of a sample of self-admitted anorexics and bulimics. Individuals' perceptions of what it means to become anorexic or bulimic are explored. Although based on a small sample, findings can be used to guide researchers in other settings.

With only 5 to 10 percent of reported cases appearing in males (Crisp, 1977; Stangler and Printz, 1980), eating disorders are primarily a women's aberrance. The deviance of anorexia nervosa and bulimia is rooted in the visual objectification of women and attendant slimness norm. Indeed, purposeful starvation and binging/

purging reinforce the notion that "a society gets the deviance it deserves" (Schur, 1979: 71). As recently noted (Schur, 1984), the sociology of deviance has generally bypassed systematic studies of women's norm violations. Like male deviants, females endure label applications, internalizations, and fulfillments.

The social processes involved in developing anorexic or bulimic identities comprise the sequence of conforming behavior, primary deviance, and secondary deviance. With a background of exceptional adherence to conventional norms, especially the striving for thinness, respondents subsequently exhibit the primary deviance of starving or binging/purging. Societal reaction to these behaviors leads to secondary deviance, wherein respondents' self-concepts and master statuses become anorexic or bulimic. Within this framework of labeling theory, the persistence of eating disorders, as well as the effects of stigmatization, is elucidated.

Although during the course of this research some respondents alleviated their symptoms through psychiatric help or hospital treatment programs, no one was labeled "cured." An anorexic is considered recovered when weight is normal for two years; a bulimic is termed recovered after being symptom-free for one and one-half years (American Anorexia/Bulimia Association Newsletter, 1985). Thus deviance disavowal (Schur, 1971), or efforts after normalization to counteract the deviant labels, remains a topic for future exploration.

NOTES

1. Although instructive, an integration of the medical, psychological, and socio-cultural perspectives on eating disorders is beyond the scope of this paper.

2. Exceptions to the neglect of socio-cultural factors are discussions of sex-role socialization in the development of eating disorders. Anorexics' girlish appearance has been interpreted as a rejection of femininity and womanhood (Orbach, 1979; Bruch, 1981; Orbach, 1985). In contrast, bulimics have been characterized as over-conforming to traditional female sex roles (Boskind-Lodahl, 1976).

3. Although a group experience for self-defined bulimics has been reported (Boskind-Lodahl, 1976), the researcher, from the outset, focused on Gestalt and behaviorist techniques within a feminist orientation.

4. One explanation for fewer anorexics than bulimics in the sample is that, in the general population, anorexics are outnumbered by bulimics at 8 or 10 to 1

(Lawson, as reprinted in American Anorexia/Bulimia Association Newsletter, 1985: 1). The proportion of bulimics to anorexics in the sample is 6.5 to 1. In addition, compared to bulimics, anorexics may be less likely to attend a self-help group as they have a greater tendency to deny the existence of an eating problem (Humphries et al., 1982). However, the four anorexics in the present study were among the members who attended the meetings most often.

5. Interactions in the families of anorexics and bulimics might seem deviant in being inordinately close. However, in the larger societal context, the family members epitomize the norms of family cohesiveness. Perhaps unusual in their occurrence, these families are still within the realm of conformity. Humphries and colleagues (1982: 202) refer to the "highly enmeshed and protective" family as part of the "idealized family myth."

REFERENCES

American Anorexia/Bulimia Association. 1983. Correspondence. April.

American Anorexia/Bulimia Association Newsletter. 1985. 8(3).

Becker, Howard S. 1963. *Outsiders*. New York: Free Press.

Boskind-Lodahl, Marlene. 1976. "Cinderella's stepsisters: A feminist perspective on anorexia nervosa and bulimia." *Signs, Journal of Women in Culture and Society* 2: 342–56.

Boskind-White, Marlene. 1985. "Bulimarexia: A socio-cultural perspective." In S. W. Emmett (ed.), ·

Theory and Treatment of Anorexia Nervosa and Bulimia: Biomedical, Sociocultural and Psychological Perspectives (pp. 113–26). New York: Brunner/Mazel.

Branch, C. H. Hardin, and Linda J. Eurman. 1980. "Social attitudes toward patients with anorexia nervosa." *American Journal of Psychiatry* 137: 631–32.

Bruch, Hilde. 1981. "Developmental considerations of anorexia nervosa and obesity." *Canadian Journal of Psychiatry* 26: 212–16.

Chernin, Kim. 1981. *The Obsession: Reflections on the Tyranny of Slenderness.* New York: Harper & Row.

Crisp, A. H. 1977. "The prevalence of anorexia nervosa and some of its associations in the general population." *Advances in Psychosomatic Medicine* 9: 38–47.

Crisp, A. H., R. L. Palmer, and R. S. Kalucy. 1976. "How common is anorexia nervosa? A prevalence study." *British Journal of Psychiatry* 128: 549–54.

Dejong, William. 1980. "The stigma of obesity: The consequences of naive assumptions concerning the causes of physical deviance." *Journal of Health and Social Behavior* 21: 75–87.

Fox, K. C., and N. Mel. James. 1976. "Anorexia nervosa: A study of 44 strictly defined cases." *New Zealand Medical Journal* 84: 309–12.

Garner, David M., Paul E. Garfinkel, Donald Schwartz, and Michael Thompson. 1980. "Cultural expectations of thinness in women." *Psychological Reports* 47: 483–91.

Goffman, Erving. 1963. *Stigma.* Englewood Cliffs, NJ: Prentice Hall.

Halmi, Katherine A., James R. Falk, and Estelle Schwartz. 1981. "Binge-eating and vomiting: A survey of a college population." *Psychological Medicine* 11: 697–706.

Herzog, David B. 1982. "Bulimia: The secretive syndrome." *Psychosomatics* 23: 481–83.

Hughes, Everett C. 1958. *Men and Their Work.* New York: Free Press.

Humphries, Laurie L., Sylvia Wrobel, and H. Thomas Wiegert. 1982. "Anorexia nervosa." *American Family Physician* 26: 199–204.

Johnson, Craig L., Marilyn K. Stuckey, Linda D. Lewis, and Donald M. Schwartz. 1982. "Bulimia: A descriptive survey of 316 cases." *International Journal of Eating Disorders* 2(1): 3–16.

Kalucy, R. S., A. H. Crisp, and Britta Harding. 1977. "A study of 56 families with anorexia nervosa." *British Journal of Medical Psychology* 50: 381–95.

Lacey, Hubert J., Sian Coker, and S. A. Birtchnell. 1986. "Bulimia: Factors associated with its etiology and maintenance." *International Journal of Eating Disorders* 5: 475–87.

Lemert, Edwin M. 1951. *Social Pathology.* New York: McGraw-Hill.

———. 1967. *Human Deviance, Social Problems and Social Control.* Englewood Cliffs, NJ: Prentice Hall.

Minuchin, Salvador, Bernice L. Rosman, and Lester Baker. 1978. *Psychosomatic Families: Anorexia Nervosa in Context.* Cambridge, MA: Harvard University Press.

Orbach, Susie. 1979. *Fat Is a Feminist Issue.* New York: Berkeley.

———. 1985. "Visibility/invisibility: Social considerations in anorexia nervosa—a feminist perspective." In S. W. Emmett (ed.), *Theory and Treatment of Anorexia Nervosa and Bulimia: Biomedical, Sociocultural, and Psychological Perspectives* (pp. 127–38). New York: Brunner/Mazel.

Ritenbaugh, Cheryl. 1982. "Obesity as a culture-bound syndrome." *Culture, Medicine and Psychiatry* 6: 347–61.

Russell, Gerald. 1979. "Bulimia nervosa: An ominous variant of anorexia nervosa." *Psychological Medicine* 9: 429–48.

Schlesier-Stropp, Barbara. 1984. "Bulimia: A review of the literature." *Psychological Bulletin* 95: 247–57.

Schur, Edwin M. 1971. *Labeling Deviant Behavior.* New York: Harper & Row.

———. 1979. *Interpreting Deviance: A Sociological Introduction.* New York: Harper & Row.

———. 1984. *Labeling Women Deviant: Gender, Stigma, and Social Control.* New York: Random House.

Schwartz, Donald M., and Michael G. Thompson. 1981. "Do anorectics get well? Current research and future needs." *American Journal of Psychiatry* 138: 319–23.

Schwartz, Donald M., Michael G. Thompson, and Craig L. Johnson. 1982. "Anorexia nervosa and bulimia: The socio-cultural context." *International Journal of Eating Disorders* 1(3): 20–36.

Selvini-Palazzoli, Mara. 1978. *Self-Starvation: From Individual to Family Therapy in the Treatment of Anorexia Nervosa.* New York: Jason Aronson.

Stangler, Ronnie S., and Adolph M. Printz. 1980. "DSM-III: Psychiatric diagnosis in a university population." *American Journal of Psychiatry* 137: 937–40.

Theander, Sten. 1970. "Anorexia nervosa." *Acta Psychiatrica Scandinavica Supplement* 214: 24–31.

Thompson, Michael G., and Donald M. Schwartz. 1982. "Life adjustment of women with anorexia nervosa and anorexic-like behavior." *International Journal of Eating Disorders* 1(2): 47–60.

Willi, Jurg, and Samuel Grossmann. 1983. "Epidemiology of anorexia nervosa in a defined region of Switzerland." *American Journal of Psychiatry* 140: 564–67.

26

Convicted Rapists' Vocabulary of Motive

DIANA SCULLY AND JOSEPH MAROLLA

Scully and Marolla's study of the way rapists rationalize their behavior offers a fascinating glimpse into the accounts offered by criminals. They interview the most hard-core segment of the rapist population, those sentenced to prison time. In analyzing these men's rationalizations, Scully and Marolla draw on Scott and Lyman's (1968) classic typology of accounts: excuses and justifications. In using excuses, men acknowledge the wrongfulness of the act but deny full responsibility. These, they find, are primarily used by those who admit to their deviant acts. Men who deny having committed rape (over 80 percent of the population) are more prone to use justifications, where they accept responsibility for their act but provide reasons that legitimate their behavior as not wrong. Scully and Marolla examine the various disavowal techniques, shed light on the repertoire of culturally available neutralizing accounts, and analyze the connection between types of accounts used and the way offenders locate blame.

P sychiatry has dominated the literature on rapists since "irresistible impulse" (Glueck, 1925: 243) and "disease of the mind" (Glueck, 1925: 243) were introduced as the causes of rape. Research has been based on small samples of men, frequently the clinicians' own patient population. Not surprisingly, the medical model has predominated: rape is viewed as an individualistic, idiosyncratic symptom of a disordered personality. That is, rape is assumed to be a psychopathologic problem and individual rapists are assumed to be "sick." However, advocates of this model have been unable to isolate a typical or even predictable pattern of symptoms that are causally linked to rape. Additionally, research has demonstrated that fewer than 5 percent of rapists were psychotic at the time of their rape (Abel et al., 1980).

From "Convicted Rapists' Vocabulary of Motive: Excuses and Justifications," by Diana Scully and Joseph Marolla, *Social Problems*, Vol. 31, No. 5, 1984. © 1984 Society for the Study of Social Problems. All rights reserved. Reprinted by permission of University of California Press Journals and Diana Scully.

We view rape as behavior learned socially through interaction with others; convicted rapists have learned the attitudes and actions consistent with sexual aggression against women. Learning also includes the acquisition of culturally derived vocabularies of motive, which can be used to diminish responsibility and to negotiate a nondeviant identity.

Sociologists have long noted that people can, and do, commit acts they define as wrong and, having done so, engage various techniques to disavow deviance and present themselves as normal. Through the concept of "vocabulary of motive," Mills (1940: 904) was among the first to shed light on this seemingly perplexing contradiction. Wrongdoers attempt to reinterpret their actions through the use of a linguistic device by which norm-breaking conduct is socially interpreted. That is, anticipating the negative consequences of their behavior, wrongdoers attempt to present the act in terms that are both culturally appropriate and acceptable.

Following Mills, a number of sociologists have focused on the types of techniques employed by actors in problematic situations (Hall and Hewitt, 1970; Hewitt and Hall, 1973; Hewitt and Stokes, 1975; Sykes and Matza, 1957). Scott and Lyman (1968) describe excuses and justifications, linguistic "accounts" that explain and remove culpability for an untoward act after it has been committed. *Excuses* admit the act was bad or inappropriate but deny full responsibility, often through appeals to accident, or biological drive, or through scapegoating. In contrast, *justifications* accept responsibility for the act but deny that it was wrong—that is, they show in this situation the act was appropriate. *Accounts* are socially approved vocabularies that neutralize an act or its consequences and are always a manifestation of an underlying negotiation of identity.

Stokes and Hewitt (1976: 837) use the term "aligning actions" to refer to those tactics and techniques used by actors when some feature of a situation is problematic. Stated simply, the concept refers to an actor's attempt, through various means, to bring his or her conduct into alignment with culture. Culture in this sense is conceptualized as a "set of cognitive constraints—objects—to which people must relate as they form lines of conduct" (1976: 837), and includes physical constraints, expectations and definitions of others, and personal biography. Carrying out aligning actions implies both awareness of those elements of normative culture that are applicable to the deviant act and, in addition, an actual effort to bring the act into line with this awareness. The result is that deviant behavior is legitimized.

This paper presents an analysis of interviews we conducted with a sample of 114 convicted, incarcerated rapists. We use the concept of accounts (Scott and Lyman, 1968) as a tool to organize and analyze the vocabularies of motive which this group of rapists used to explain themselves and their actions. An analysis of their accounts demonstrates how it was possible for 83 percent (n = 114)[1] of these convicted rapists to view themselves as nonrapists.

When rapists' accounts are examined, a typology emerges that consists of admitters and deniers. Admitters (n = 47) acknowledged that they had forced sexual acts on their victims and defined the behavior as rape. In contrast, deniers[2] either eschewed sexual contact or all association with the victim (n = 35),[3] or admitted to sexual acts but did not define their behavior as rape (n = 32).... By and large, the deniers used justifications while the admitters used excuses. In some cases, both groups relied on the same themes, stereotypes, and images: some admitters, like most deniers, claimed that women enjoyed being raped. Some deniers excused their behavior by referring to alcohol or drug use, although they did so quite differently than admitters. Through these narrative accounts, we explore convicted rapists' own perceptions of their crimes....

JUSTIFYING RAPE

Deniers attempted to justify their behavior by presenting the victim in a light that made her appear culpable, regardless of their own actions. Five themes run through rapists' attempts to justify their rapes: (1) women as seductresses; (2) women mean "yes" when they say "no"; (3) most women eventually relax and enjoy it; (4) nice girls don't get raped; and (5) guilty of a minor wrongdoing.

(1) Women as Seductresses

Men who rape need not search far for cultural language which supports the premise that women provoke or are responsible for rape. In addition to common cultural stereotypes, the fields of psychiatry and criminology (particularly the sub-field of victimology) have traditionally provided justifications for rape, often by portraying raped women as the victims of their own seduction (Albin, 1977; Marolla and Scully, 1979). For example, Hollander (1924: 130) argues:

> Considering the amount of illicit intercourse, rape of women is very rare indeed. Flirtation and provocative conduct, i.e. tacit (if not actual) consent is generally the prelude to intercourse.

Since women are supposed to be coy about their sexual availability, refusal to comply with a man's sexual demands lacks meaning and rape appears normal. The fact that violence and, often, a weapon are used to accomplish the rape is not considered. As an example, Abrahamsen (1960: 61) writes:

> The conscious or unconscious biological or psychological attraction between man and woman does not exist only on the part of the offender toward the woman but, also, on her part toward him, which in many instances may, to some extent, be the impetus for his sexual attack. Often a women [sic] unconsciously wishes to be taken by force—consider the theft of the bride in Peer Gynt.

Like Peer Gynt, the deniers we interviewed tried to demonstrate that their victims were willing and, in some cases, enthusiastic participants. In these accounts, the rape became more dependent upon the victim's behavior than upon their own actions.

Thirty-one percent (n = 10) of the deniers presented an extreme view of the victim. Not only willing, she was the aggressor, a seductress who lured them, unsuspecting, into sexual action. Typical was a denier convicted of his first rape and accompanying crimes of burglary, sodomy, and abduction. According to the presentence reports, he had broken into the victim's house and raped her at knife point. While he admitted to the breaking and entry, which he claimed was for altruistic purposes ("to pay for the prenatal care of a friend's girl-friend"), he also argued that when the victim discovered him, he had tried to leave but she had asked him to stay. Telling him that she cheated on her husband, she had voluntarily removed her clothes and seduced him. She was, according to him, an exemplary sex partner who "enjoyed it very much and asked for oral sex.[4] Can I have it now?" he reported her as saying. He claimed they had spent hours in bed, after which the victim had told him he was good looking and asked to see him again. "Who would believe I'd meet a fellow like this?" he reported her as saying.

In addition to this extreme group, 25 percent (n = 8) of the deniers said the victim was willing and had made some sexual advances. An additional 9 percent (n = 3) said the victim was willing to have sex for money or drugs. In two of these three cases, the victim had been either an acquaintance or picked up, which the rapists said led them to expect sex.

(2) Women Mean "Yes" When They Say "No"

Thirty-four percent (n = 11) of the deniers described their victim as unwilling, at least initially, indicating either that she had resisted or that she had said no. Despite this, and even though (according to presentence reports) a weapon had been present in 64 percent (n = 7) of these 11 cases, the rapists justified their behavior by arguing that either the victim had not resisted enough or that her "no" had really meant "yes." For example, one denier who was serving time for a previous rape was subsequently convicted of attempting to rape a prison hospital nurse. He insisted he had actually completed the second rape, and said of his victim: "She semistruggled but deep down inside I think she felt it was a fantasy come true." The nurse, according to him, had asked a question about his conviction for rape, which he interpreted as teasing. "It was like she was saying, 'rape me.'" Further, he stated that she had helped him along with oral sex and "from her actions, she was enjoying it." In another case, a 34-year-old man convicted of abducting and raping a 15-year-old teenager at knife point as she walked on the beach, claimed it was a pickup. This rapist said women like to be overpowered before sex, but to dominate after it begins.

> A man's body is like a Coke bottle, shake it up, put your thumb over the opening and feel the tension. When you take a woman out, woo her, then she says "no, I'm a nice girl," you have to use force. All men do this. She said "no" but it was a societal no, she wanted to be coaxed. All women say "no" when they mean "yes" but it's a societal no, so they won't have to feel responsible later.

Claims that the victim didn't resist or, if she did, didn't resist enough, were also used by 24 percent (n = 11) of admitters to explain why, during the incident, they believed the victim was willing and that they were not raping. These rapists didn't redefine their acts until some time after the crime. For example, an admitter who used a bayonet to threaten his victim, an employee of the store he had been robbing, stated:

> At the time I didn't think it was rape. I just asked her nicely and she didn't resist. I never considered prison. I just felt like I had met a friend. It took about five years of reading and going to school to change my mind about whether it was rape. I became familiar with the subtlety of violence. But at the time, I believed that as long as I didn't hurt anyone it wasn't wrong. At the time, I didn't think I would go to prison, I thought I would beat it.

Another typical case involved a gang rape in which the victim was abducted at knife point as she walked home about midnight. According to two of the rapists, both of whom were interviewed, at the time they had thought the victim had willingly accepted a ride from the third rapist (who was not interviewed). They claimed the victim didn't resist and one reported her as saying she would do anything if they would take her home. In this rapist's view, "She acted like she enjoyed it, but maybe she was just acting. She wasn't crying, she was engaging in it." He reported that she had been friendly to the rapist who abducted her and, claiming not to have a home phone, she gave him her office number—a tactic eventually used to catch the three. In retrospect, this young man had decided, "She was scared and just relaxed and enjoyed it to avoid getting hurt." Note, however, that while he had redefined the act as rape, he continued to believe she enjoyed it.

Men who claimed to have been unaware that they were raping viewed sexual aggression as a man's prerogative at the time of the rape. Thus they regarded their act as little more than a minor wrongdoing even though most possessed or used a weapon. As long as the victim survived without major physical injury, from their perspective, a rape had not taken place. Indeed, even U.S. courts have often taken the position that physical injury is a necessary ingredient for a rape conviction.

(3) Most Women Eventually Relax and Enjoy It

Many of the rapists expected us to accept the image, drawn from cultural stereotype, that once the rape began, the victim relaxed and enjoyed it.[5] Indeed, 69 percent (n = 22) of deniers justified their behavior by claiming not only that the victim was willing, but also that she enjoyed herself, in some cases to an immense degree. Several men suggested that they had fulfilled their victims' dreams. Additionally, while most admitters used adjectives such as "dirty," "humiliated," and "disgusted" to describe how they thought rape made women feel, 20 percent (n = 9) believed that their victim enjoyed herself. For example, one denier had posed as a salesman to gain entry to his victim's house. But he claimed he had had a previous sexual relationship with the victim, that she agreed to have sex for drugs, and that the opportunity to have sex with him produced "a glow, because she was really into oral stuff and fascinated by the idea of sex with a black man. She felt satisfied, fulfilled, wanted me to stay, but I didn't want her." In another case, a denier who had broken into his victim's house but who insisted the victim was his lover and let him in voluntarily, declared "She felt good, kept kissing me and wanted me to stay the night. She felt proud after sex with me." And another denier, who had hid in his victim's closet and later attacked her while she slept, argued that while she was scared at first, "once we got into it, she was OK." He continued to believe he hadn't committed rape because "she enjoyed it and it was like she consented."

(4) Nice Girls Don't Get Raped

The belief that "nice girls don't get raped" affects perception of fault. The victim's reputation, as well as characteristics or behavior which violate normative sex role expectations, are perceived as contributing to the commission of the crime. For example, Nelson and Amir Menachem (1975) defined hitchhike rape as a victim-precipitated offense.

In our study, 69 percent (n = 22) of deniers and 22 percent (n = 10) of admitters referred to their victims' sexual reputation, thereby evoking the stereotype that "nice girls don't get raped." They claimed that the victim was known to have been a prostitute, or a "loose" woman, or to have had a lot of affairs, or to have given birth to a child out of wedlock. For example, a denier who claimed he had picked up his victim while she was hitchhiking stated, "To be honest, we [his family] knew she was a damn whore and whether she screwed one or 50 guys didn't matter." According to presentence reports this victim didn't know her attacker and he abducted her at knife point from the street. In another case, a denier who claimed to have known his victim by reputation stated:

> If you wanted drugs or a quick piece of ass, she would do it. In court she said she was a virgin, but I could tell during sex [rape] that she was very experienced.

When other types of discrediting biographical information were added to these sexual slurs, a total of 78 percent (n = 25) of the deniers used the victim's reputation to substantiate their accounts. Most frequently, they referred to the victim's emotional state or drug use. For example, one denier claimed his victim had been known to be loose and, additionally, had turned state's evidence against her husband to put him in prison and save herself from a burglary conviction. Further, he asserted that she had met her current boyfriend, who was himself in and out of prison, in a drug rehabilitation center where they were both clients.

Evoking the stereotype that women provoke rape by the way they dress, a description of the victim as seductively attired appeared in the accounts of 22 percent (n = 7) of deniers and 17 percent (n = 8) of admitters. Typically, these descriptions were used to substantiate their claims about the victim's reputation. Some men went to extremes to paint a tarnished picture of the victim, describing her as dressed in tight black clothes and without a bra; in one case, the victim was portrayed as sexually provocative in dress and carriage. Not only did she wear short skirts, but she was observed to "spread her legs while getting out of cars." Not all of the men attempted to assassinate their victim's reputation with equal vengeance. Numerous times they made subtle and offhand remarks like, "She was a waitress and you know how they are."

The intent of these discrediting statements is clear. Deniers argued that the woman was a "legitimate" victim who got what she deserved. For example, one denier stated that all of his victims had been prostitutes; presentence reports indicated they were not. Several times during his interview, he referred to them as "dirty sluts," and argued "anything I did to them was justified." Deniers also claimed their victim had wrongly accused them and was the type of woman who would perjure herself in court.

(5) Only a Minor Wrongdoing

The majority of deniers did not claim to be completely innocent and they also accepted some accountability for their actions. Only 16 percent (n = 5) of deniers argued that they were totally free of blame. Instead, the majority of deniers pleaded guilty to a lesser charge. That is, they obfuscated the rape by pleading guilty to a less serious, more acceptable charge. They accepted being oversexed, accused of poor judgment or trickery, even some violence, or guilty of adultery or contributing to the delinquency of a minor, charges that are hardly the equivalent of rape.

Typical of this reasoning is a denier who met his victim in a bar when the bartender asked him if he would try to repair her stalled car. After attempting unsuccessfully, he claimed the victim drank with him and later accepted a ride. Out riding, he pulled into a deserted area "to see how my luck would go." When the victim resisted his advances, he beat her and he stated:

> I did something stupid. I pulled a knife on her and I hit her as hard as I would hit a man. But I shouldn't be in prison for what I did. I shouldn't have all this time [sentence] for going to bed with a broad.

This rapist continued to believe that while the knife was wrong, his sexual behavior was justified.

In another case, the denier claimed he picked up his under-age victim at a party and that she voluntarily went with him to a motel. According to presentence reports, the victim had been abducted at knife point from a party. He explained:

> After I paid for a motel, she would have to have sex but I wouldn't use a weapon. I would have explained. I spent money and, if she still said no, I would have forced her. If it had happened that way, it would have been rape to some people but not to my way of thinking. I've done that kind of thing before. I'm guilty of sex and contributing to the delinquency of a minor, but not rape.

In sum, deniers argued that, while their behavior may not have been completely proper, it should not have been considered rape. To accomplish this, they attempted to discredit and blame the victim while presenting their own actions as justified in the context. Not surprisingly, none of the deniers thought of himself as a rapist. A minority of the admitters attempted to lessen the impact of their crime by claiming the victim enjoyed being raped. But despite this similarity, the nature and tone of admitters' and deniers' accounts were essentially different.

EXCUSING RAPE

In stark contrast to deniers, admitters regarded their behavior as morally wrong and beyond justification. They blamed themselves rather than the victim, although some continued to cling to the belief that the victim had contributed to the crime somewhat, for example, by not resisting enough.

Several of the admitters expressed the view that rape was an act of such moral outrage that it was unforgivable. Several admitters broke into tears at intervals during their interviews. A typical sentiment was,

> I equate rape with someone throwing you up against a wall and tearing your liver and guts out of you.... Rape is worse than murder ... and I'm disgusting.

Another young admitter frequently referred to himself as repulsive and confided:

> I'm in here for rape and in my own mind, it's the most disgusting crime, sickening. When people see me and know, I get sick.

Admitters tried to explain their crime in a way that allowed them to retain a semblance of moral integrity. Thus, in contrast to deniers' justifications, admitters used excuses to explain how they were compelled to rape. These excuses appealed to the existence of forces outside of the rapists' control. Through the use of excuses, they attempted to demonstrate that either intent was absent or responsibility was diminished. This allowed them to admit rape while reducing the threat to their identity as a moral person. Excuses also permitted them to view their behavior as idiosyncratic rather than typical and, thus, to believe they were not "really" rapists. Three themes run through these accounts: (1) the use of alcohol and drugs; (2) emotional problems; and (3) nice guy image.

(1) The Use of Alcohol and Drugs

A number of studies have noted a high incidence of alcohol and drug consumption by convicted rapists prior to their crime (Groth, 1979; Queen's Bench Foundation, 1976). However, more recent research has tentatively concluded that the connection between substance use and crime is not as direct as previously thought (Ladouceur, 1983). Another facet of alcohol and drug use mentioned in the literature is its utility in disavowing deviance. McCaghy (1968) found that child molesters used alcohol as a technique for neutralizing their deviant identity. Marolla and Scully (1979), in a review of psychiatric literature, demonstrated how alcohol consumption is applied differently as a vocabulary of motive. Rapists can use alcohol both as an excuse for their behavior and to discredit the victim and make her more responsible. We found the former common among admitters and the latter common among deniers.

Alcohol and/or drugs were mentioned in the accounts of 77 percent (n = 30) of admitters and 84 percent (n = 21) of deniers and both groups were equally likely to have acknowledged consuming a substance—admitters, 77 percent (n = 30); deniers, 72 percent (n = 18). However, admitters said they had been affected by the substance; if not the cause of their behavior, it was at least a contributing factor. For example, an admitter who estimated his consumption to have been eight beers and four "hits of acid" reported:

Rapists' Accounts of Own and Victims' Alcohol and/or Drug (A/D) Use and Effect

	Admitters n = 39%	Deniers n = 25%
Neither self nor victim used A/D	23	16
Self used A/D	77	72
Of self used, no victim use	51	12
Self affected by A/D	69	40
Of self affected, no victim use or effect	54	24
Self A/D users who were affected	90	56
Victim used A/D	26	72
Of victim used, no self use	0	0
Victim affected by A/D	15	56
Of victim affected, no self use or effect	0	40
Victim A/D users who were affected	60	78
Both self and victim used and affected by A/D	15	16

> Straight, I don't have the guts to rape. I could fight a man but not that. To say, "I'm going to do it to a woman," knowing it will scare and hurt her, takes guts or you have to be sick.

Another admitter believed that his alcohol and drug use,

> ... brought out what was already there but in such intensity it was uncontrollable. Feelings of being dominant, powerful, using someone for my own gratification, all rose to the surface.

In contrast, deniers' justifications required that they not be substantially impaired. To say that they had been drunk or high would cast doubt on their ability to control themselves or to remember events as they actually happened. Consistent with this, when we asked if the alcohol and/or drugs had had an effect on their behavior, 69 percent (n = 27) of admitters, but only 40 percent (n = 10) of deniers, said they had been affected.

Even more interesting were references to the victim's alcohol and/or drug use. Since admitters had already relieved themselves of responsibility through claims of being drunk or high, they had nothing to gain from the assertion that the victim had used or been affected by alcohol and/or drugs. On the other hand, it was very much in the interest of deniers to declare that their victim had been intoxicated or high: that fact lessened her credibility and made her more responsible for the act. Reflecting these observations, 72 percent (n = 18) of deniers and 26 percent (n = 10) of admitters maintained that alcohol or drugs had been consumed by the victim. Further, while 56 percent (n = 14) of deniers declared she had been affected by this use, only 15 percent (n = 6) of admitters made a similar claim. Typically deniers argued that the alcohol and drugs had sexually aroused their victim or rendered her out of control. For example, one denier insisted that his victim had become hysterical from drugs, not from being raped, and it was because of the drugs that she had reported him to the police. In addition, 40 percent (n = 10) of deniers argued that while the victim had been drunk or high, they themselves either hadn't ingested or weren't affected by alcohol and/or drugs. None of the admitters made this claim. In fact, in all of the 15 percent (n = 6) of cases where an admitter said the victim was drunk or high, he also admitted to being similarly affected.

These data strongly suggest that whatever role alcohol and drugs play in sexual and other types of violent crime, rapists have learned the advantage to be gained from using alcohol and drugs as an account. Our sample were aware that their victim would be discredited and their own behavior excused or justified by referring to alcohol and/or drugs.

(2) Emotional Problems

Admitters frequently attributed their acts to emotional problems. Forty percent (n = 19) of admitters said they believed an emotional problem had been at the root of their rape behavior, and 33 percent (n = 15) specifically related the problem to an unhappy, unstable childhood or a marital-domestic situation. Still others claimed to have been in a general state of unease. For example, one admitter said that at the time of the rape he had been depressed, feeling he couldn't do anything right, and that something had been missing from his life. But he also added, "being a rapist is not part of my personality." Even admitters who could locate no source for an emotional problem evoked the popular image of rapists as the product of disordered personalities to argue they also must have problems:

> The fact that I'm a rapist makes me different. Rapists aren't all there. They have problems. It was wrong so there must be a reason why I did it. I must have a problem.

Our data do indicate that a precipitating event, involving an upsetting problem of everyday living, appeared in the accounts of 80 percent (n = 38) of admitters and 25 percent (n = 8) of deniers. Of those experiencing a precipitating event, including deniers, 76 percent (n = 35) involved a wife or girlfriend. Over and over, these men described themselves as having been in a rage because of an incident involving a woman with whom they believed they were in love.

Frequently, the upsetting event was related to a rigid and unrealistic double standard for sexual conduct and virtue which they applied to "their" woman but which they didn't expect from men, didn't apply to themselves, and, obviously, didn't honor in other women. To discover that the "pedestal" didn't apply to their wife or girlfriend sent them into a fury. One especially articulate and typical admitter described his feeling as follows. After serving a short prison term for auto theft, he married his "childhood sweetheart" and secured a well-paying job. Between his job and the volunteer work he was doing with an ex-offender group, he was spending long hours away from home, a situation that had bothered his wife. In response to her request, he gave up his volunteer work, though it was clearly meaningful to him. Then, one day, he discovered his wife with her former boyfriend "and my life fell apart." During the next several days, he said his anger had made him withdraw into himself and, after three days of drinking in a motel room, he abducted and raped a stranger. He stated:

> My parents have been married for many years and I had high expectations
> about marriage. I put my wife on a pedestal. When I walked in on her,
> I felt like my life had been destroyed, it was such a shock. I was bitter and
> angry about the fact that I hadn't done anything to my wife for cheating.
> I didn't want to hurt her [victim], only to scare and degrade her.

It is clear that many admitters, and a minority of deniers, were under stress at the time of their rapes. However, their problems were ordinary—the types of upsetting events that everyone experiences at some point in life. The overwhelming majority of the men were not clinically defined as mentally ill in court-ordered psychiatric examinations prior to their trials. Indeed, our sample is consistent with Abel et al. (1980) who found fewer than 5 percent of rapists were psychotic at the time of their offense.

As with alcohol and drug intoxication, a claim of emotional problems works differently depending upon whether the behavior in question is being justified or excused. It would have been counter-productive for deniers to have claimed to have had emotional problems at the time of the rape. Admitters used psychological explanations to portray themselves as having been temporarily "sick" at the time of the rape. Sick people are usually blamed for neither the cause of their illness nor for acts committed while in that state of diminished capacity. Thus, adopting the sick role removed responsibility by excusing the behavior as having been beyond the ability of the individual to control. Since the rapists were not "themselves," the rape was idiosyncratic rather than typical behavior. Admitters asserted a non-deviant identity despite their self-proclaimed disgust with what they had done. Although admitters were willing to assume the sick role, they did not view their problem as a chronic condition, nor did they believe themselves to be insane or permanently impaired. Said one admitter, who believed that he needed psychological counseling: "I have a mental disorder, but I'm not crazy." Instead, admitters viewed their "problem" as mild, transient, and curable. Indeed, part of the appeal of this excuse was that not only did it relieve responsibility, but, as with alcohol and drug addiction, it allowed the rapist to "recover." Thus, at the time of their interviews, only 31 percent (n = 14) of admitters indicated that "being a rapist" was part of their self-concept. Twenty-eight percent (n = 13) of admitters stated they had never thought of themselves as rapists, 8 percent (n = 4) said they were unsure, and 33 percent (n = 16) asserted they had been a rapist at one time but now were recovered. A multiple "ex-rapist," who believed his "problem" was due to "something buried in my subconscious" that was triggered when his girlfriend broke up with him, expressed a typical opinion:

> I was a rapist, but not now. I've grown up, had to live with it. I've hit the bottom of the well and it can't get worse. I feel born again to deal with my problems.

(3) Nice Guy Image

Admitters attempted to further neutralize their crime and negotiate a non-rapist identity by painting an image of themselves as a "nice guy." Admitters projected the image of someone who had made a serious mistake but, in every other respect, was a decent person. Fifty-seven percent (n = 27) expressed regret and sorrow for their victim indicating that they wished there were a way to apologize for or amend their behavior. For example, a participant in a rape-murder, who insisted his partner did the murder, confided, "I wish there was something I could do besides saying 'I'm sorry, I'm sorry.' I live with it 24 hours a day and, sometimes, I wake up crying in the middle of the night because of it."

Schlenker and Darby (1981) explain the significance of apologies beyond the obvious expression of regret. An apology allows a person to admit guilt while at the same time seeking a pardon by signaling that the event should not be considered a fair representation of what the person is really like. An apology separates the bad self from the good self, and promises more acceptable behavior in the future. When apologizing, an individual is attempting to say: "I have repented and should be forgiven," thus making it appear that no further rehabilitation is required.

The "nice guy" statements of the admitters reflected an attempt to communicate a message consistent with Schlenker's and Darby's analysis of apologies. It was an attempt to convey that rape was not a representation of their "true" self. For example,

> It's different from anything else I've ever done. I feel more guilt about this. It's not consistent with me. When I talk about it, it's like being assaulted myself. I don't know why I did it, but once I started, I got into it. Armed robbery was a way of life for me, but not rape. I feel like I wasn't being myself.

Admitters also used "nice guy" statements to register their moral opposition to violence and harming women, even though, in some cases, they had seriously injured their victims. Such was the case of an admitter convicted of gang rape:

> I'm against hurting women. She should have resisted. None of us were the type of person that would use force on a woman. I never positioned myself on a woman unless she showed an interest in me. They would play to me, not me to them. My weakness is to follow. I never would have stopped, let alone pick her up without the others. I never would have let anyone beat her. I never bothered women who didn't want sex; never had a problem with sex or getting it. I loved her—like all women.

Finally, a number of admitters attempted to improve their self-image by demonstrating that, while they had raped, it could have been worse if they had not been a "nice guy." For example, one admitter professed to being especially gentle with his victim after she told him she had just had a baby. Others claimed to have given the victim money to get home or make a phone call, or to have made sure the victim's children were not in the room. A multiple rapist, whose pattern was to break in and attack sleeping victims in their homes, stated:

> I never beat any of my victims and I told them I wouldn't hurt them if they cooperated. I'm a professional thief. But I never robbed the women I raped because I felt so bad about what I had already done to them.

Even a young man, who raped his five victims at gun point and then stabbed them to death, attempted to improve his image by stating:

> Physically they enjoyed the sex [rape]. Once they got involved, it would be difficult to resist. I was always gentle and kind until I started to kill them. And the killing was always sudden, so they wouldn't know it was coming.

SUMMARY AND CONCLUSIONS

Convicted rapists' accounts of their crimes include both excuses and justifications. Those who deny what they did was rape justify their actions; those who admit it was rape attempt to excuse it or themselves. This study does not address why some men admit while others deny, but future research might address this question. This paper does provide insight on how men who are sexually aggressive or violent construct reality, describing the different strategies of admitters and deniers.

Admitters expressed the belief that rape was morally reprehensible. But they explained themselves and their acts by appealing to forces beyond their control, forces which reduced their capacity to act rationally and thus compelled them to rape. Two types of excuses predominated: alcohol/drug intoxication and emotional problems. Admitters used these excuses to negotiate a moral identity for themselves by viewing rape as idiosyncratic rather than typical behavior. This allowed them to reconceptualize themselves as recovered or "ex-rapists," someone who had made a serious mistake which did not represent their "true" self.

In contrast, deniers' accounts indicate that these men raped because their value system provided no compelling reason not to do so. When sex is viewed as a male entitlement, rape is no longer seen as criminal. However, the deniers had been convicted of rape, and like the admitters, they attempted to negotiate an identity. Through justifications, they constructed a "controversial" rape and attempted to demonstrate how their behavior, even if not quite right, was appropriate in the situation. Their denials, drawn from common cultural rape stereotypes, took two forms, both of which ultimately denied the existence of a victim.

The first form of denial was buttressed by the cultural view of men as sexually masterful and women as coy but seductive. Injury was denied by portraying the victim as willing, even enthusiastic, or as politely resistant at first but eventually yielding to "relax and enjoy it." In these accounts, force appeared merely as a seductive technique. Rape was disclaimed: rather than harm the woman, the rapist had fulfilled her dreams. In the second form of denial, the victim was portrayed as the type of woman who "got what she deserved." Through attacks on the victim's sexual reputation and, to a lesser degree, her emotional state, deniers attempted to demonstrate that since the victim wasn't a "nice girl," they were not rapists. Consistent with both forms of denial was the self-interested use of alcohol and drugs as a justification. Thus, in contrast to admitters, who accentuated their own use as an excuse, deniers emphasized the victim's consumption in an effort to both discredit her and make her appear more responsible for the rape. It is important to remember that deniers did not invent these justifications. Rather, they reflect a belief system which has historically victimized women by promulgating the myth that women both enjoy and are responsible for their own rape.

While admitters and deniers present an essentially contrasting view of men who rape, there were some shared characteristics. Justifications particularly, but also excuses, are buttressed by the cultural view of women as sexual commodities, dehumanized and devoid of autonomy and dignity. In this sense, the sexual objectification of women must be understood as an important factor contributing to an environment that trivializes, neutralizes, and, perhaps, facilitates rape.

Finally, we must comment on the consequences of allowing one perspective to dominate thought on a social problem. Rape, like any complex continuum of behavior, has multiple causes and is influenced by a number of social factors. Yet, dominated by psychiatry and the medical model, the underlying assumption that rapists are "sick" has pervaded research. Although methodologically unsound, conclusions have been based almost exclusively on small clinical populations of rapists—that extreme group of rapists who seek counseling in prison and are the most likely to exhibit psychopathology. From this small, atypical group of men, psychiatric findings have been generalized to all men who rape. Our research, however, based on volunteers from the entire prison population, indicates that some rapists, like deniers, viewed and understood their behavior from a popular cultural perspective. This strongly suggests that cultural perspectives, and not an idiosyncratic illness, motivated their behavior. Indeed, we can argue that the psychiatric perspective has contributed to the vocabulary of motive that rapists use to excuse and justify their behavior (Scully and Marolla, 1984).

Efforts to arrive at a general explanation for rape have been retarded by the narrow focus of the medical model and the preoccupation with clinical populations. The continued reduction of such complex behavior to a singular cause hinders, rather than enhances, our understanding of rape.

NOTES

1. These numbers include pretest interviews. When the analysis involves either questions that were not asked in the pretest or that were changed, they are excluded and thus the number changes.

2. There is, of course, the possibility that some of these men really were innocent of rape. However, while the U.S. criminal justice system is not without flaw, we assume that it is highly unlikely that this many men could have been unjustly convicted of rape, especially since rape is a crime with traditionally low conviction rates. Instead, for purposes of this research, we assume that these men were guilty as charged and that their attempt to maintain an image of nonrapist springs from some psychologically or sociologically interpretable mechanism.

3. Because of their outright denial, interviews with this group of rapists did not contain the data being analyzed here and, consequently, they are not included in this paper.

4. It is worth noting that a number of deniers specifically mentioned the victim's alleged interest in oral sex. Since our interview questions about sexual history

indicated that the rapists themselves found oral sex marginally acceptable, the frequent mention is probably another attempt to discredit the victim. However, since a tape recorder could not be used for the interviews and the importance of these claims didn't emerge until the data was being coded and analyzed, it is possible that it was mentioned even more frequently but not recorded.

5. Research shows clearly that women do not enjoy rape. Holmstrom and Burgess (1978) asked 93 adult rape victims, "How did it feel sexually?" Not one said they enjoyed it. Further, the trauma of rape is so great that it disrupts sexual functioning (both frequency and satisfaction) for the overwhelming majority of victims, at least during the period immediately following the rape and, in fewer cases, for an extended period of time (Burgess and Holmstrom, 1979; Feldman-Summers et al., 1979). In addition, a number of studies have shown that rape victims experience adverse consequences prompting some to move, change jobs, or drop out of school (Burgess and Holmstrom, 1974; Kilpatrick et al., 1979; Ruch et al., 1980; Shore, 1979).

REFERENCES

Abel, Gene, Judith Becker, and Linda Skinner. 1980. "Aggressive behavior and sex." *Psychiatric Clinics of North America* 3(2): 133–151.

Abrahamsen, David. 1960. *The Psychology of Crime.* New York: John Wiley.

Albin, Rochelle. 1977. "Psychological studies of rape." *Signs* 3(2): 423–435.

Burgess, Ann Wolbert, and Lynda Lytle Holmstrom. 1974. *Rape: Victims of Crisis.* Bowie: Robert J. Brady.

———. 1979. "Rape: Sexual disruption and recovery." *American Journal of Orthopsychiatry* 49(4): 648–657.

Feldman-Summers, Shirley, Patricia E. Gordon, and Jeanette R. Meagher. 1979. "The impact of rape on sexual satisfaction." *Journal of Abnormal Psychology* 88(1): 101–105.

Glueck, Sheldon. 1925. *Mental Disorders and the Criminal Law.* New York: Little, Brown.

Groth, Nicholas A. 1979. *Men Who Rape.* New York: Plenum Press.

Hall, Peter M., and John P. Hewitt. 1970. "The quasi-theory of communication and the management of dissent." *Social Problems* 18(1): 17–27.

Hewitt, John P., and Peter M. Hall. 1973. "Social problems, problematic situations, and quasi-theories." *American Journal of Sociology* 38(3): 367–374.

Hewitt, John P., and Randall Stokes. 1975. "Disclaimers." *American Sociological Review* 40(1): 1–11.

Hollander, Bernard. 1924. *The Psychology of Misconduct, Vice and Crime.* New York: Macmillan.

Holmstrom, Lynda Lytle, and Ann Wolbert Burgess. 1978. "Sexual behavior of assailant and victim during rape." Paper presented at the annual meetings of the American Sociological Association, San Francisco, September 2–8.

Kilpatrick, Dean G., Lois Veronen, and Patricia A. Resnick. 1979. "The aftermath of rape: Recent empirical findings." *American Journal of Orthopsychiatry* 49(4): 658–669.

Ladouceur, Patricia. 1983. "The relative impact of drugs and alcohol on serious felons." Paper presented at the annual meetings of the American Society of Criminology, Denver, November 9–12.

Marolla, Joseph, and Diana Scully. 1979. "Rape and psychiatric vocabularies of motive." In Edith S. Gomberg and Violet Franks (eds.), *Gender and Disordered Behavior: Sex Differences in Psychopathology* (pp. 301–318). New York: Brunner/Mazel.

McCaghy, Charles. 1968. "Drinking and deviance disavowal: The case of child molesters." *Social Problems* 16(1): 43–49.

Mills, C. Wright. 1940. "Situated actions and vocabularies of motive." *American Sociological Review* 5(6): 904–913.

Nelson, Steve, and Amir Menachem. 1975. "The hitchhike victim of rape: A research report." In Israel Drapkin and Emilio Viano (eds.), *Victimology: A New Focus* (pp. 47–65). Lexington, KY: Lexington Books.

Queen's Bench Foundation. 1976. *Rape: Prevention and Resistance.* San Francisco: Queen's Bench Foundation.

Ruch, Libby O., Susan Meyers Chandler, and Richard A. Harter. 1980. "Life change and rape impact." *Journal of Health and Social Behavior* 21(3): 248–260.

Schlenker, Barry R., and Bruce W. Darby. 1981. "The use of apologies in social predicaments." *Social Psychology Quarterly* 44(3): 271–278.

Scott, Marvin, and Stanford Lyman. 1968. "Accounts." *American Sociological Review* 33(1): 46–62.

Scully, Diana, and Joseph Marolla. 1984. "Rape and psychiatric vocabularies of motive: Alternative perspectives." In Ann Wolbert Burgess (ed.), *Handbook on Rape and Sexual Assault.* New York: Garland Publishing.

Shore, Barbara K. 1979. "An examination of critical process and outcome factors in rape." Rockville, MD: National Institute of Mental Health.

Stokes, Randall, and John P. Hewitt. 1976. "Aligning actions." *American Sociological Review* 41(5): 837–849.

Sykes, Gresham M., and David Matza. 1957. "Techniques of neutralization." *American Sociological Review* 22(6): 664–670.

27

The Devil Made Me Do It: Use of Neutralizations by Shoplifters

PAUL CROMWELL AND QUINT THURMAN

Cromwell and Thurman offer a discussion of shoplifters' rationalizations that many people will find familiar. Stealing from stores has long been widespread among American youth, and practitioners have found it convenient and easy to rationalize pilfering from large companies that do not have a visible or identifiable local owner. This chapter complements the Scully and Marolla analysis by drawing on Sykes and Matza's (1957) techniques of neutralizations, a conceptualization of accounts that predates the distinction between excuses and justifications. Cromwell and Thurman show how shoplifters make ample use of these existing categories and invent a few new ones of their own. These accounts, like all others, help individuals deflect the labeling process and the deviant identity.

"You know that cartoon where the guy has a little devil sitting on one shoulder and a little angel on the other? And one is telling him 'Go ahead on, do it,' and the angel is saying 'No, don't do it.' You know? . . . Sometimes when I'm thinking about boosting something, my angel don't show up." (30-year-old male shoplifter)

Nearly five decades ago Gresham Sykes and David Matza (1957) introduced neutralization theory as an explanation for juvenile delinquency. Sykes and Matza's (1957) theory is an elaboration of Edwin Sutherland's (1947) proposition that individuals can learn criminal techniques, and the "motives, drives, rationalizations and attitudes favorable to violations of the law." Sykes and Matza argued that these justifications or rationalizations protect the individual from self-blame and the blame of others. Thus, the individual may remain committed to the value system of the dominant culture while committing criminal acts without experiencing the cognitive dissonance that might be otherwise expected. He or she deflects or "neutralizes" guilt in advance, clearing the way to blame-free crime. These neutralizations also protect the individual from any residual guilt following the crime. It is this ability to use neutralizations that differentiates delinquents from non-delinquents (Thurman 1984).

While Sykes and Matza (1957) do not specifically maintain that only offenders who are committed to the dominant value system make use of these techniques of neutralizations, they appear to contend that delinquents maintain a commitment to the moral order and are able to drift into delinquency through the use of techniques of neutralization. This approach assumes that should delinquents fail to internalize conventional morality, neutralization would be unnecessary since there would be no guilt to neutralize.

One issue that has not been satisfactorily settled is when neutralization occurs. Sykes and Matza (1957) contend that deviants must neutralize moral prescriptions prior to committing a crime. However, most research is incapable of determining whether the stated neutralization is a before-the-fact neutralization or an after-the-fact rationalization.

TECHNIQUES OF NEUTRALIZATION

Sykes and Matza (1957) identified five techniques of neutralization commonly offered to justify deviant behavior—denial of responsibility, denial of injury, denial of the victim, condemning the condemners, and appeal to higher loyalties.

Five additional neutralization techniques have since been identified. These include defense of necessity (Klockars 1974), metaphor of the ledger (Minor 1981), denial of the necessity of the law, the claim that everybody else is doing it, and the claim of entitlement (Coleman 1994).

This article examines neutralization theory as it might apply to a specific form of criminal activity that is highly prevalent across a wide range of the population. The purpose of this study is to determine the extent to which adult shoplifters use techniques of neutralizations and to analyze the various neutralizations available to them. We examine offenders who shoplift and explore the justifications that they say they rely upon to excuse behavior they also acknowledge as morally wrong.

STATEMENT OF THE PROBLEM

Shoplifting may be the most serious crime with which the most people have some personal familiarity. Research has shown that one in every 10–15 persons who shops has shoplifted at one time or another. Further, losses attributable to shoplifting are considerable, with estimates ranging from 12 to 30 billion dollars lost annually. Shoplifting also represents one of the most prevalent forms of larceny, accounting for approximately 15 percent of all larcenies according to data maintained by the Federal Bureau of Investigation (1996).

Unlike many other forms of crime, people who shoplift do not ordinarily require any special expertise or tools to engage in this crime. Consequently, those persons who shoplift do not necessarily conform to most people's perception of what a criminal offender is like. Instead, shoplifters tend to be demographically similar to the "average person." In a large study of non-delinquents,

Klemke (1982) reported that as many of 63 percent of those persons he interviewed had shoplifted at some point in their lives. Students, housewives, business
and professional persons, as well as professional thieves constitute the population
of shoplifters. Loss prevention experts routinely counsel retail merchants that
there is no particular profile of a potential shoplifter. Turner and Cashdan
(1988) conclude, "While clearly a criminal activity, shoplifting borders on what
might be considered a 'folkcrime.'" In her classic study, Mary Cameron
(1964:xi) wrote:

> Most people have been tempted to steal from stores, and many have been
> guilty (at least as children) of "snitching" an item or two from counter
> tops. With merchandise so attractively displayed in department stores
> and supermarkets, and much of it apparently there for the taking, one
> may ask why everyone isn't a thief.

Neutralization theory argues that ordinary individuals who engage in deviant
or criminal behavior may use techniques that permit them to recognize extenua-
ting circumstances that enable them to explain away delinquent behavior. Without worrying about guilty feelings that would stand in their way of committing a
criminal act, the theory asserts that those persons are free to participate in delinquent acts that they would otherwise believe to be wrong.

METHOD

The data presented here were obtained in 1997 and 1998 in Wichita, Kansas. We
obtained access to a court-ordered diversion program for adult "first-offenders"
charged with theft. Of these, the majority of offenders were charged with misdemeanor shoplifting and required to attend an eight-hour therapeutic/education
program as a condition of having their record expunged. A new group met each
Saturday. The average group size was 18–20 participants. Participants were encouraged by the program facilitator to discuss with the group the offense that brought
them to the diversion program, why they did what they did, and how they felt
about it. We obtained interviews with 137 subjects from approximately 350 subjects who were approached. Ethnicity and gender of the sample are shown in
Table 27.1. The mean age of the sample was 26. The age range was 18 to 66
years of age. Although the diversion program was designed for first offenders,
over one-half of the participants had been apprehended for shoplifting in the past.

TABLE 27.1 Neutralizations by Shoplifters Respondents by Gender and
Ethnicity (N = 137)

	White	Hispanic	Black	Total
Male	48	11	29	88
Female	30	6	13	49
Total	78	17	42	137

FINDINGS

The informants appeared to readily use neutralization techniques. We identified nine categories of neutralizations; the **five Sykes and Matza (1957) categories,** the **Defense of Necessity** and **Everybody Does It,** identified by Coleman (1994) and two additional, which we labeled **Justification by Comparison** and **Postponement.** Only 5 of the 137 informants failed to express a rationalization or neutralization when asked how they felt about their illegal behavior. Three of these subjects responded by admitting their guilt and expressing remorse. Two others simply stated that they had nothing to say on that issue. In many cases, the respondents offered more than one neutralization for the same offense. For example, one female respondent stated, "I don't know what comes over me. It's like, you know, is somebody else doing it, not me" (Denial of Responsibility). "I'm really a good person. I wouldn't ever do something like that, stealing, you know, but I have to take things sometimes for my kids. They need stuff and I don't have any money to get it" (Defense of Necessity). They frequently responded with a motivation ("I wanted the item and could not afford it") followed by a neutralization ("Stores charge too much for stuff. They could sell things for half what they do and still make a profit. They're just too greedy"). Thus, in many cases, the motivation was linked to excuse in such a way as to make the excuse a part of the motivation. The subjects were in effect explaining the reason the deviant act occurred and justifying it at the same time. The following section illustrates the neutralizations we discovered in use by the informants.

Denial of Responsibility ("I Didn't Mean It")

Denial of responsibility frees the subject from experiencing culpability for deviance by allowing him or her to perceive themselves as victims of their environment. The offender views him- or herself as being acted upon rather than acting. Thus, attributing behavior to poor parenting, bad companions or internal forces (the devil made me do it) allows the offender to avoid disapproval of self or others, which in turn, diminishes those influences as mechanisms of social control. Sykes and Matza (1957: 666) describe the individual resorting to this neutralization as having a "billiard ball conception of himself in which he see himself as helplessly propelled into new situations."

> I admit that I lift. I do. But, you know, it's not really me—I mean, I don't believe in stealing. I'm a church-going person. It's just that sometimes something takes over me and I can't seem to not do it. Its like those TV shows where the person is dying and he goes out of his body and watches them trying to save him. That's sorta how I feel sometimes when I'm lifting. (26-year-old female)

> I wasn't raised right. You know what I mean? Wasn't nobody to teach me right from wrong. I just ran with a bad group and my mamma didn't ever say nothin' about it. That's how I turned out this way—stealin' and stuff. (22-year-old female)

> If it wasn't for the bunch I ran with at school I never would have started taking things. We used to go the mall after school and everybody would have to steal something. If you didn't get anything, everybody called you names—chicken-shit and stuff like that. (20-year-old male)

Many of the shoplifter informants neutralized their activities citing loss of self-control due to alcohol or drug use. This is a common form of denial of responsibility. If not for the loss of inhibition due to drug or alcohol use, they argue, they would not commit criminal acts.

> I was drinking with my buddies and we decided to go across the street to the [convenience store] and steal some beer. I was pretty wasted or I wouldn't done it. (19-year-old White male)

> I never boost when I'm straight. It's the pills, you know? (30-year-old White female)

Denial of Injury ("I Didn't Really Hurt Anybody")

Denial of injury allows the offender to perceive of his or her behavior as having no direct harmful consequences to the victim. The victim may be seen as easily able to afford the loss (big store, insurance company, wealthy person) or the crime may be semantically recast, as when auto theft is referred to as joyriding, or vandalism as a prank.

> They [stores] big. Make lotsa money. They don't even miss the little bit I get. (19-year-old male)

> They write it off their taxes. Probably make a profit off it. So, nobody gets hurt. I get what I need and they come out O.K. too. (28-year-old male)

> Them stores make billions. Did you ever hear of Sears going out of business from boosters? (34-year-old female)

Denial of the Victim ("They Had It Coming")

Denial of the victim facilitates deviance when it can be justified as retaliation upon a deserving victim. In the present study, informants frequently reported that the large stores from which they stole were deserving victims because of high prices and the perception that they made excessive profits at the expense of ordinary people. The shoplifters frequently asserted that the business establishments from which they stole overcharged consumers and thus deserved the pay back from shoplifting losses.

> Stores deserve it. It don't matter if I boost $10,000 from one, they've made 10,000 times that much ripping off people. You could never steal enough to get even … I don't really think I'm doing anything wrong. Just getting my share. (48-year-old female)

> Dillons [food store chain] are totally bogus. A little plastic bag of groceries is $30, $25. Probably cost them $5. … Whatta they care about me? Why

should I care about them? I take what I want. Don't feel guilty a bit. No sir. Not a bit. (29-year-old female)

I have a lot of anger about stores and the way they rip people off. Sometimes I think the consumer has to take things into their own hands. (49-year-old female)

Condemning the Condemners ("The System Is Corrupt")

Condemning the condemners projects blame on law-makers and law-enforcers. It shifts the focus from the offender to those who disapprove of his or her acts. This neutralization views the "system" as crooked and thus unable to justify making and enforcing rules it does not itself live by. Those who condemn their behavior are viewed as hypocritical since many of them engage in deviant behavior themselves.

I've heard of cops and lawyers and judges and all kind of rich dudes boosting. They no better than me. You know what I'm saying. (18-year-old male)

Big stores like J.C. Penneys—when they catch me with something—like two pairs of pants, they tell the police you had like 5 pairs of pants and 2 shirts or something like that. You know what I'm saying? What they do with the other 3 pairs of pants and shirts? Insurance company pays them off and they get richer—they's bigger crooks than me. (35-year-old female)

They thieves too. Just take it a different way. They may be smarter than me—use a computer or something like that—but they just as much a thief as me. Fuck'em. Cops too. They all thieves. Least, I'm honest about it. (22-year-old male)

Appeal to Higher Loyalties ("I Didn't Do It For Myself")

Appeal to higher loyalties functions to legitimize deviant behavior when a non-conventional social bond creates more immediate and pressing demands than one consistent with conventional society. The most common use of this technique among the shoplifters was pressure from delinquent peers to shoplift and the perceived needs of one's family for items that the informant could not afford to buy. This was especially common with mothers shoplifting for items for their children.

I never do it 'cept when I'm with my friends. Everybody be taking stuff and so I do too. You know—to be part of the group. Not to seem like I'm too good for 'em. (17-year-old female)

I like to get nice stuff for my kids, you know. I know it's not O.K., you know what I mean? But, I want my kids to dress nice and stuff. (28-year-old female)

The Defense of Necessity ("I Had No Other Choice")

The defense of necessity (Coleman 1998) serves to reduce guilt through the argument that the offender had no choice under the circumstances but to engage in a criminal act. In the case of shoplifting, the defense of necessity is most often used when the offender states that the crime was necessary to help one's family.

> I had to take care of three children without help. I'd be willing to steal to give them what they wanted. (32-year-old female)

> I got laid off at Boeing last year and got behind on all my bills and couldn't get credit anywhere. My kids needed school clothes and money for supplies and stuff. We didn't have anything and I don't believe in going on welfare, you know. The first time I took some lunch meat at Dillons (grocery chain) so we'd have supper one night. After that I just started to take whatever we needed that day. I knew it was wrong, but I just didn't have any other choice. My family comes first. (42-year-old male)

Everybody Does It

Here the individual attempts to reduce his or her guilt feelings or to justify his or her behavior by arguing that the behavior in question is common (Coleman 1998). A better label for this neutralization might be "diffusion of guilt." The behavior is justified or the guilt is diffused because of widespread similar acts.

> Everybody I know do it. All my friends. My mother and her boyfriend are boosters and my sister is a big-time booster. (19-year-old female)

> All my friends do it. When I'm with them it seems crazy not to take something too. (17-year-old male) I bet you done did it too . . . when you was coming up. Like 12–13 years old. Everybody boosts. (35-year-old female)

Justification By Comparison ("If I Wasn't Shoplifting I Would Be Doing Something More Serious")

This newly identified neutralization involves offenders justifying their actions by comparing their crimes to more serious offenses. While it might be argued that Justification by Comparison is not a neutralization in the strict Sykes and Matza (1957) sense in that these offenders are not committed to conventional norms, they are nonetheless attempting to maintain their sense of self-worth by arguing that they could be worse or are not as bad as some others. Even persons with deviant lifestyles may experience guilt over their behavior and/or feel the necessity to justify their actions to others. The gist of the argument is that "I may be bad, but I could be worse."

> I gotta have $200 every day—day in and day out. I gotta boost a thousand, fifteen-hundred dollars worth to get it. I just do what I gotta

do.... Do I feel bad about what I do? Not really. If I wasn't boosting, I'd be robbing people and maybe somebody would get hurt or killed. (40-year-old male)

Looka here. Shoplifting be a little thing. Not a crime really. I do it 'stead of robbing folks or breaking in they house. [Society] oughta be glad I boost, stead of them other things. (37-year-old male)

It's no thing. Not like its "jacking" people or something. It's just a little lifting. (19-year-old male)

Postponement ("I Just Don't Think About It")

In a previous study one of the present authors (Thurman 1984) suggested that further research should consider the excuse strategy of Postponement, by which the offender suppresses his or her guilt feelings—momentarily putting them out of mind to be dealt with at a later time. We found this strategy to be a common occurrence among our informants. They made frequent statements that indicated that they simply put the incident out of their mind. Some stated that they would deal with it later when they were not under so much stress.

I just don't think about it. I mean, if you think about it, it seems wrong, but you can ignore that feeling sometimes. Put it aside and go on about what you gotta do. (18-year old male)

Dude, I just don't deal with those kinda things when I'm boosting. I might feel bad about it later, you know, but by then it's already over and I can't do anything about it then, you know? (18-year-old male)

I worry about things like that later. (30-year-old female)

DISCUSSION AND CONCLUSION

We found widespread use of neutralizations among the shoplifters in our study. We identified two new neutralizations: *Justification by Comparison* and *Postponement*. Even those who did not appear to be committed to the conventional moral order used neutralizations to justify or excuse their behavior. Their use of neutralizations was not so much to assuage guilt but to provide them with the necessary justifications for their acts to others. Simply because one is not committed to conventional norms does not preclude their understanding that most members of society do accept those values and expect others to do so as well. They may also use neutralizations and rationalizations to provide them with a convincing defense for their crimes that they can tell to more conventionally oriented others if the need arises.

As stated earlier, our research approach could not determine whether the informants neutralized before committing the crime or rationalized afterwards. Pogrebin, Poole, and Martinez (1992: 233) suggest that post-event reasons given for deviant behavior are not neutralizations but accounts, or "socially approved vocabularies that serve as explanatory mechanisms for deviance." No one, however, has yet been able to empirically verify the existence of pre-event neutralizations. In fact, neutralization theory depends upon analysis of post-event accounts by the offender. We suggest that accounts, neutralizations, and rationalizations are essentially the same behavior at different stages in the criminal event. We argue that Hirschi (1969) was correct in stating that a post-crime rationalization may serve as a pre-crime neutralization the next time a crime is contemplated. Whether neutralization allows the offender to mitigate guilt feelings before the crime is committed or afterward, the process still occurs. Once an actor has reduced his or her guilt feelings through the use of techniques of neutralization, he or she can continue to offend, assuaging guilt feelings and cognitive dissonance both before and after each offense. It would follow that continued utilization of neutralization and rationalization habitually over time might serve to weaken the social bond, reducing the need to neutralize at all.

Our exploratory study of shoplifters' use of neutralization techniques also suggests that neutralization (theory) may not be a theory of crime but rather a description of a process that represents an adaptation to morality that leads to criminal persistence. Neutralization focuses on how crime is possible, rather than why people might choose to engage in it in the first place. In a sense, neutralization serves as a form of situational morality. While the offender knows an act is morally wrong (either in his or her eyes or in the eyes of society), he or she makes an adaptation to convention that permits deviation under certain circumstances (the various neutralizations discussed). Whether the adaptation is truly neutralizing (before the act) or rationalizing (after the act) the result is same—crime without guilt.

REFERENCES

Cameron, Mary. 1964. *The Booster and the Snitch.* New York: Free Press.

Coleman, James W. 1994. "Neutralization Theory: An Empirical Application and Assessment." Ph.D. Dissertation, Oklahoma State University, Department of Sociology, Stillwater.

Coleman, James W. 1998. *Criminal Elite: Understanding White Collar Crime.* New York: St. Martin's Press.

Federal Bureau of Investigation 1996. Crime in the United States—1995. Washington, DC: U.S. Department of Justice.

Hirschi, Travis. 1969. *Causes of Delinquency.* Berkeley, CA: University of California Press.

Klemke, Lloyd. 1982. "Exploring Juvenile Shoplifting." *Sociology and Social Research* 67: 59–75.

Klockars, Carl B. 1974. *The Professional Fence.* New York: Free Press.

Minor, William W. 1981. "Techniques of Neutralization: A Reconceptualization and Empirical Examination." *Journal of Research in Crime and Delinquency* (July): 295–318.

Pogrebrin, M., E. Poole, and A. Martinez. 1992. "Accounts of Professional Misdeeds: The Sexual Exploitation of Clients by Psychotherapists." *Deviant Behavior* 13: 229–52.

Sutherland, Edwin H. 1947. *Principles of Criminology.* Philadelphia: Lippincott.

Sykes, Gresham M., and David Matza. 1957. "Techniques of Neutralization: A Theory of Delinquency." *American Sociological Review* 22(6): 664–70.

Thurman, Quint C. 1984. "Deviance and Neutralization of Moral Commitment: An Empirical Analysis." *Deviant Behavior* 5: 291–304.

Turner, C. T., and S. Cashdan. 1988. "Perceptions of College Students' Motivations for Shoplifting." *Psychological Reports* 62: 855–62.

28

Managing the Stigma
of Personal Bankruptcy

DEBORAH THORNE AND LEON ANDERSON

Twenty-first-century developments in American law have made it more difficult and less beneficial to file for personal bankruptcy. In changing the bankruptcy laws, politicians were swayed by a moral campaign from the banking industry to conceptualize debtors as irresponsible spenders who wantonly rode lines of credit into mountains of unnecessary debt and then irresponsibly shrugged off their obligations, leaving solid financial institutions to shoulder their burden. Previous competing, though less successful, moral campaigns presented information that portrayed most bankruptcy filings as arising from divorce, loss of employment, or catastrophic illness. In winning the moral entrepreneurial battle, bankers have successfully increased the stigma of personal bankruptcy, branding those unable to pay their debts as morally irresponsible (the "sin" definition of deviance).

In this chapter, Thorne and Anderson portray the shame felt by 28 middle-class couples they interviewed who had to declare bankruptcy. They discuss a three-part stigma management strategy whereby people first tried to conceal their deviant act from various parties, then avoided situations where their discredited status would be revealed, and finally avowed their deviance while offering accounts to help explain or to neutralize their behavior and thereby diminish the damage to their reputations and self-conceptions.

Issues of stigma have long been of concern in the study of bankruptcy. While documenting shifts in moral perceptions of bankrupt debtors, historical research ... has consistently acknowledged the long-enduring and robust nature of bankruptcy stigma. Academic attention to the topic has increased in recent years, particularly as some scholars have questioned contemporary levels of such stigma. From a sociological vantage point, one significant weakness of academic

"Managing the Stigma of Personal Bankruptcy," Deborah Thorne and Leon Anderson,
Sociological Focus, the Journal of the North Central Sociological Association, Vol. 39, No. 2,
May 2006. Reprinted by permission of the North Central Sociological Association.

discussions of bankruptcy stigma has been that virtually no scholars have examined the experiences of bankrupt debtors themselves. Perhaps this omission is due, in part, to the fact that research on bankruptcy in general has tended to focus on legal and financial issues at a macro level of analysis. . . .

But as McIntyre (1989) observed in her call for a sociological perspective on bankruptcy, "knowledge about not only the legal and economic, but the social and social psychological underpinnings of these relations is essential" (124). If this is the case regarding bankruptcy in general, it is even more relevant for understanding such a quintessentially social and symbolic phenomenon as bankruptcy stigma.

In contrast to previous research, this paper provides in-depth empirical data gathered directly from bankrupt debtors to assess their experiences of stigmatization. Further, to accurately represent the concerns expressed by our informants, we center our analysis on a discussion of the stigma management strategies they invoked to mitigate the shame and social disapprobation they experienced as a result of their bankruptcies.

Analytically, our focus on stigma management has two advantages. First, it provides a vantage point from which to view bankrupt debtors as social agents, rather than as passive victims of stigma. This concern has guided symbolic interactionist studies of stigma for over four decades, beginning with Davis' (1961) study of techniques used by the visibly handicapped to manage social interaction and, particularly, with Goffman's seminal essay, Stigma (1963), which was subtitled, "Notes on the Management of Spoiled Identity."

The second advantage of a focus on stigma management is that it enables us to compare the experiences and actions of bankrupt debtors to members of a host of other stigmatized groups. As Anderson and Snow (2001) conclude in their review of symbolic interactionist studies of inequality, "Humans routinely and creatively take measures to reduce the prospect of status affronts and degradation or to moderate the force of their impact" (410). Despite the nearly universal urge to mitigate stigma, not all stigmatized individuals and groups avail themselves of the same strategies. A comparison of stigma management techniques used by members of dissimilar groups can enhance our understanding of the specific group under study as well as refine our knowledge of stigma management at a broader generic level.

The analysis of stigma management among bankrupt debtors offers a unique case study that expands the sociological understanding of stigmatization and inequality more generally, providing deeper knowledge about the strategies and struggles for social agency among stressed and marginalized populations in U.S. society.

CONTEXT AND METHODS

Context

Over the past 20 years, despite an oft-touted strong economy, there has been a fourfold increase in the number of petitions filed for personal bankruptcy. In

1984, 291,532 petitions were filed; by 2004, that number had ballooned to 1,567,846. . . .

Many explanations for the increase have been asserted, the most frequent of which is that bankruptcy has shed its stigma and become little more than a financial planning tool for opportunistic filers. Representative Asa Hutchinson, an avid supporter of the recent bankruptcy reform, claimed that, "Having lost its social stigma, bankruptcy 'convenience' filings have become a tool to avoid financial obligations rather than a measure of last resort" (1999). This sentiment was echoed by Alan Greenspan, then Chairman of the Federal Reserve: "Personal bankruptcies are soaring because Americans have lost their sense of shame" (quoted in Zywicki 2005, p. 30). The message from advocates of bankruptcy reform was loud and clear: yesterday's morally responsible consumer had transformed into today's Homo oeconomicus. Prodded by lobbying efforts and large campaign contributions from the credit card industry, Congress and President Bush concluded that the most efficient way to simultaneously restore fiscal integrity and reduce the number of filings was to restrict access to the bankruptcy process altogether. Thus, in spring of 2005, Congress passed and President Bush signed the Bankruptcy Abuse Prevention and Consumer Protection Act. As he signed the bill into law, Mr. Bush stated: "[T]oo many people have abused the bankruptcy laws" and with a stroke of his pen, he was "restoring integrity to the bankruptcy process" (2005). Interestingly, debtors were labeled deadbeats and reforms to the bankruptcy laws were made without input from debtors or valid empirical evidence about the stigma associated with filing.

For the past two decades, the most recognized and respected scholarship on consumer bankruptcy has been conducted by Sullivan, Warren, and Westbrook, lead investigators on the multi-state Consumer Bankruptcy Project. Some of their most important findings include the demographics of filers and debtors' chief reasons for filing . . . Although these scholars recognize and stress the stigma associated with personal bankruptcy, none of their extensive research has measured it directly. . . .

To address the kinds of issues we have just described, we analyze direct testimony from debtors who filed for Chapter 7 personal bankruptcy in 1999 in the Eastern District of Washington state. The first author conducted face-to-face interviews with 37 individuals from 19 married couples who had filed joint petitions for personal bankruptcy. (A total of 38 interviews should have been completed. However, just before the interview was to begin, the husband of one couple hid upstairs because, as his wife said, he was ashamed of their bankruptcy.) Originally, the names of ninety debtor couples, whose Chapter 7 cases had closed between April and July 1999, were randomly selected from the website of the Eastern District of Washington State, United States Bankruptcy Court, Each couple received a letter that described the research and explained that they would be contacted to request an interview. Phones were disconnected or letters were undeliverable for 48 (53 percent) of the potential respondent couples. Stanley and Girth (1971) reported similar difficulties when they attempted to contact bankrupt debtors; over two-thirds of their letters requesting interviews were returned and marked as "undeliverable." The remaining 42 couples were contacted, and 15 (36 percent) declined to participate.

In total, 27 couples agreed to interviews. Eight of these couples were excluded because they had either moved, were too ill to participate within the research time frame, or had separated or divorced.

The decision to interview both members of marital couples was based on two considerations. First, marital couples are jointly responsible for most financial obligations. Consequently, the legal act of filing is almost always made as a couple. Second, many of the core research questions for this study concerned the effects of bankruptcy on marital relationships and were thus addressed most appropriately by interviewing both partners. One objective of the original research was to explore the effects of debt and bankruptcy on couples' relationships. To insure that husbands and wives were forthcoming with their assessments and that they described their experiences without bias, men and women were interviewed separately.

The interviews were conducted within three months of the couples' bankruptcy filings, and, with only one exception, took place in couples' homes. Whenever possible, husbands and wives were interviewed separately—in only two instances did a spouse refuse to leave while the other was interviewed. On average, the interviews lasted an hour and a half; a few lasted over three hours. The semi-structured interviews loosely followed an interview guide . . . that consisted of a list of general topics and questions covering a range of bankruptcy-related experiences. These topics included the nature of the respondents' debts and the conditions under which they had accrued; how debtors reached the decision to declare bankruptcy; how their insolvency and ultimate bankruptcy influenced their personal, marital, and family lives; how they felt about their decision to file; and their experiences with the legal and bureaucratic processes associated with bankruptcy. Beyond these general themes, the interviewer encouraged debtors to explore other topics that emerged during the interview process, thus maximizing receptiveness to "developing meanings" that might either render the original questions irrelevant or reveal the importance of new ones. . . .

The conversational format of the interviews allowed comments about stigma and stigma management to surface in a variety of contexts, normally without specific prompting. For example, debtors voiced feelings of stigma or embarrassment when they were asked to describe how they felt about their decision to file, their first visit to the bankruptcy attorney, or their attendance at the Section 341 hearing. For others, their assertions of stigma came as unexpected interjections, statements that were outside the immediate topic of conversation, as in the case of a woman who abruptly shifted from a description of the repossession of their vehicles to her generalized feelings of shame about filing: "It was the most embarrassing experience, to file bankruptcy, because [we believed that] we've got to pay our debts, even if it takes forever." Interestingly, the two debtors who did not spontaneously offer comments regarding stigma, and who were therefore prompted to discuss the issue, were the only respondents in the study who said they felt little or no shame or embarrassment. Thus, all 35 debtors who reported feelings of shame broached the subject independently in the course of more general conversation and without any direction from the interviewer. . . .

Sample Characteristics

Like most intensive interview studies, the sample for this research was relatively small. Nonetheless, it was demographically quite diverse and consistent in many respects with other bankruptcy studies. The debtors ranged in age from mid-20s to early-80s. The modal level of education among primary petitioners (husbands) was as follows: 58 percent graduated high school, 16 percent had some college, another 16 percent earned a bachelor's degree, and 10.5 percent had either a graduate or professional degree. Family size ranged from two members (couples with no children in the home) to six—a couple whose two adult daughters and their two children were living with them. Mean family size was 3.6. Annual income at time of filing ranged from $13,000 to $58,560; median income was $29,832. Unsecured debt at time of filing ranged from $9,800 to $66,500; median debt was $28,900. The most frequently acknowledged cause of bankruptcy was employment problems: 68 percent of debtors claimed either a job loss or a general decline in income as a major reason for filing. Medical expenses and credit card debt were also frequently mentioned. For families with young children, the cost of raising them and providing daycare was also a major factor in their bankruptcies. . . .

STIGMA AND STIGMA MANAGEMENT AMONG BANKRUPT DEBTORS

The Existence of Stigma

As the earlier quote by Congressmen Hutchinson illustrates, there is a common assumption that bankruptcy, which was once inextricably linked with stigma, is now embraced by debtors as a purely rational economic decision, nothing more than a "convenient financial planning tool." In glaring contrast to this assumption, debtors who had been through the process insisted that filing for personal bankruptcy was, more than anything else, especially stigmatizing. Specifically, 95 percent of the debtors (all but two) in the study stated that they felt shame and stigmatization as they negotiated the bankruptcy process. For example, when a young mother of two was asked to discuss the disadvantages of filing, she responded: "The disadvantage was just the humiliation." Another respondent, a man in his early 60s who had retired from the U.S. Postal Service, told of how he had kept their bankruptcy from family members because, "I thought of it as a mark against my name . . . I just felt it had to be private. It was too embarrassing. . . . I feel like I failed. You know, to go bankrupt, that's a sign of failure." When asked if he would ever recommend bankruptcy to a friend who was struggling with debt, he said, "I guess I would if there was no way [to pay their bills]. But it isn't that easy. I still feel ashamed about the whole thing." . . .

To avoid the stigma of bankruptcy altogether, the majority of debtors reported that they postponed filing for months and even years after recognizing that their debts were unmanageable. Several couples unsuccessfully attempted to solve their financial problems by making large withdrawals from retirement accounts or borrowing against insurance annuities. Others turned their homes and vehicles back to lenders in last-ditch, and eventually futile, attempts to manage their debt loads. As these efforts met with failure and financial pressures mounted, some marriages were strained to the breaking point. Several debtors sank into serious self-described depression, and two female respondents reported that they considered suicide as a means to escape the pressures and worries associated with insolvency. Both expressed hope that their life insurance policies would be adequate to cover their family's debts.

Despite their best efforts and reluctance, however, the couples in our study did file for bankruptcy. While this certainly relieved them from major financial obligations and harassment by bill collectors, it simultaneously placed them in the awkward and uncomfortable position of having engaged in an act that they, their families, and American society at large, disparage. This dilemma and their efforts to manage it served as a focal point of attention for almost all of the bankrupt debtors we interviewed. Consistent with the broader literature on stigma, we found that stigma management among bankrupt debtors can be categorized into three broad, strategies which provide the organizational structure for our analysis: (1) concealment, (2) avoidance, and (3) deviance avowal.

Concealment

Declaration of bankruptcy is a public act, and petitioners cannot restrict the many venues through which their bankruptcies are publicly declared. The legal records section of most city newspapers reports the names of people who have filed for personal bankruptcy. Further, anyone can visit the courthouse and request a copy of any bankruptcy petition that was filed in that district. For those with access to the internet, debtors' petitions from around the nation are available on-line for a minimal charge. Finally, there are companies, such as Bankruptcy News, that notify residents about neighbors who have filed for bankruptcy. For a price, the company will provide the resident with the bankrupt neighbor's name, Social Security number, list of creditors and amount owed to each—all information that was provided by the debtor to the court.

Such public disclosure notwithstanding, roughly 80 percent of debtors made a concerted effort to keep the news of their bankruptcy contained and particularly to minimize its disclosure to individuals or groups who they felt presented potential risks. Similar to the formerly hospitalized mental patients studied by Nancy Herman (1993), bankrupt debtors engaged in "selective concealment" (307–11), most frequently from parents, co-workers, and employers.

Parents   Debtors perceived parents and in-laws as the most critical individuals from whom to hide information. Consequently, three-quarters of those debtors who reported feelings of stigma and whose parents were living worked to keep the bankruptcy from them. . . . For example, a man whose mother had described those who file as "bastards who rip off the system" was adamant that she would never learn of his bankruptcy. When asked if he had mentioned either his financial problems or his bankruptcy to her, he virtually erupted:

> OH HELL NO! No, no, no way, no way! Nope. And she won't ever know. Never. Never. . . . She'd be like, "Argh, you piece of shit. Why did you do that?" . . . I was afraid, you know, and I, my mom has no clue and I think it would just crush her. Because she has a view of me. . . . I'm her success. I am such a success to her because, despite having really been in trouble and all these awful things that went on, here I am. I'm married and I have two kids and I'm making it. Yeah, I'm getting my Ph.D.

This debtor went on to say that he and his wife had also been very careful to keep his siblings in the dark—lest they accidentally or injudiciously share the news with their mother. . . .

Employers and Co-Workers   In addition to concealing the bankruptcy from parents, well over half of debtors who were employed said that they had tried to conceal the bankruptcy from their employers and/or co-workers. While fear of disappointment dominated debtors' concerns with parents finding out, a fear of scorn and criticism prevailed in relation to those with whom they worked. Such fear of disdain was captured by one woman who explained that she had done everything possible to keep the bankruptcy from her employer:

> My boss is terribly critical about welfare mothers, people who file bankruptcy, people that have the subsidized medical program, any kind of anybody that doesn't do things on their own. Very vocal, very critical. . . . I just didn't need it. . . . And he's very, I wouldn't say he'd get in your face and say, "You're a loser," but he's very, very judgmental.

Another female debtor worried that a specific co-worker, the accounts receivable clerk, would find out. When customers were late with payment or filed for bankruptcy, it was the job of the co-worker to place a lien against any property, and the debtor witnessed her contempt for people who were behind on their bills and feared the same derisive comments would be directed toward her:

> She says things like "Loser," and she comes up with [derogatory] names [for the clients with delinquent accounts]. She's just, she doesn't call it to their faces, but when she hangs up the phone with this person, or if she gets a note, "Oh, great, another bankruptcy hearing I've got to go to and try and cut some more money out of these people." And you just think to yourself, "Shit, I'm not like that, though. Really." You know, it wasn't like I set out and said, "OK, in six years I'm going to file bankruptcy.". . . And the people that take care of credit and stuff like that [at work], they're always harsh on those people. But I'm thinking, "Shit, I don't want them to know about this."

Despite their best efforts at concealment, it was inevitable that some debtors' parents and coworkers would read about the bankruptcies in the local newspapers. Being confronted by someone from whom they had tried to withhold the information was particularly mortifying for debtors, since they then had to face interrogation about not only their bankruptcies, but about their attempts to hide the information as well. A debtor in her late-30s who worked as bookkeeper at an auto parts store, for instance, related her most difficult experience:

> The most horrible thing in the whole world is, we didn't tell our parents, both sets of parents. . . . But my mom read it in the paper and called me. I just about died. They happened to be out of town when it was printed. It just happened that way. . . . I was glad that they were going to be out of town when it came out. But then they read the papers, all of the back ones [back issues] and she called me and said, "Why didn't you tell me?" Yeah, like I'm going to call and brag! . . .

Avoidance

The preceding section documents bankrupt debtors' efforts to conceal information from intimates. Fear of stigmatization also led debtors to avoid situations that might lead to embarrassing or degrading interactions with non-intimates who would have particular reason to uncover their economic troubles and failures. In fact, bankrupt debtors had typically learned techniques of avoidance earlier in their debt histories. As they slid into bankruptcy and became delinquent with payments, they often found themselves pursued by bill collectors. Contact with bill collectors often represented many debtors' first direct experience with the stigmatization of insolvency. In her classic work on emotional labor, Hochschild (1983) studied how flight attendants and bill collectors perform their jobs. She concluded that the emotional labor performed by flight attendants generally served to make customers feel good about themselves and "enhance the customer's status." In contrast, during the final stages of collection, bill collectors worked to humiliate debtors by suggesting that they were "lazy and of low moral character" which served to "deflate the customer's status" (139). Moreover, to insure that collectors did not sympathize with debtors, and thereby reduce the likelihood of collection, the landscape of the collectors' workplace was peppered with descriptions of the debtor as "loafer" and "cheat" (139).

Only four of the nineteen couples escaped collection calls—they filed what credit card lenders describe as "no advanced warning" or "surprise" bankruptcies. Essentially, these couples remained current with all their bills until the time at which they concluded they were fighting a losing battle. At this point, rather than become delinquent on payments, something they described as irresponsible, they filed. Debtors who did receive collection calls attested to collectors' abilities to successfully shame and intimidate them. One debtor, a woman in her late-thirties who had managed her family's bills, said:

> It was to the point where I couldn't stand to be in the house. I was scared to come home because I didn't want to be faced with the phone calls

and the letters. You don't want to go to the mailbox and you don't want to answer the phone. Because all I can say is, "I sent you this much on this date and I'll send you more when I can".... I was scared the phone was going to ring. To hear it ring was just horrible.... You feel trapped in your own home, you feel trapped....

Deviance Avowal

The stigma management strategies discussed so far are based on attempts to evade the stigmatizing label of bankruptcy. Debtors postponed filing, concealed the information, and/or avoided interactions with those whom they worried might confront them with their negative status. But it was not always possible to hide the bankruptcy. As debtors' status evolved from "discreditable" to "discredited," the stigma management strategies they employed also changed. The most prevalent efforts to manage discredited identity involved deviance avowal (Turner 1972): while debtors acknowledged the general wrongfulness of bankruptcy, they endeavored to separate the stigmatized act from the deviant role with which it was generally associated. Deviance avowal enabled individuals to cope with stigma by arguing in one way or another that their particular cases were emphatically not examples of typical deviant role enactment. In this respect, deviance avowal exemplifies Stokes and Hewitt's concept of "aligning actions" or communications that individuals use to ward off potential problems in inter-subjective understandings of conduct (1976). Specifically, deviance avowal involves, "making a separation between condition and role, [in such a way that] it is impossible to overlook, forget, or redefine prior 'failing'" (Pfuhl and Henry 1993, pp. 202–3). Three varieties of deviance avowal were particularly evident among debtors in this study: (1) distancing strategies, (2) "accounts," and (3) post-bankruptcy actions and statements directed toward transcending their stigmatized status.

Distancing Discourse about poverty has long been framed in terms of conceptions of the deserving and undeserving poor. The non-poor often employ this distinction to differentiate the poor who are worthy of charity and state support from those who should be denied such support. The distinction has also been used by the poor themselves to manage the stigma of poverty by distancing themselves from images of the unworthy poor and presenting themselves as among those whose poverty is not a reflection of their moral character. So, for instance, Snow and Anderson (1993, pp. 215–16) found that a significant number of the homeless engaged in "associational distancing" as a form of identity talk to distinguish themselves from other homeless individuals who they felt rightly deserved to be stigmatized. Schwalbe et al. have described such distinctions among the worthy and unworthy, in which potentially stigmatized individuals sharply differentiate themselves from images of unworthy peers, as "defensive othering" (2000, p. 425).

Associational distancing was extremely common among the debtors in this study. They went to considerable lengths to distinguish their "legitimate" reasons for declaring bankruptcy from the otherwise illegitimate and morally objectionable actions and rationales of other bankrupt debtors. Interviewees often used the term "deadbeats" to describe illegitimate debtors, whom they depicted as having various negative characteristics, including frivolous spending habits, intentionally accrued insurmountable debts, and an inability to learn from their financial mistakes. Three particular stereotypes were drawn upon consistently for this purpose: the extravagant bankrupt, the credit card bankrupt, and the repeat filer.

Distancing from the "Extravagant Bankrupt"   The image of the extravagant bankrupt created by interviewees described debtors who reveled in inappropriately luxurious lifestyles and engaged in profligate spending patterns, only to jettison their debt when it caught up with them. Twenty-nine of the debtors (78 percent) asserted that they knew of someone who fit this negative stereotype, from which they were quick to distance themselves. Fancy cars and expensive vacations were repeatedly used to symbolize a bankruptcy of extravagance. One debtor, who had grown up in a family of Nascar drivers, said he knew someone who "had over $200,000 in credit, went hog-wild, Caribbean cruise.... And then field for bankruptcy and dumped it all.... I mean he just used and abused everything" In contrast, he described himself as a "poor slob just trying to get by, just trying to live, versus work the system, you know, file this fancy and extravagant stuff and bail out on it."

Another debtor, a stay-at-home mother of three, juxtaposed her "bankruptcy of necessity" with that of the extravagant bankrupt who charged lavish vacations and home furnishings.

> We haven't gotten carried away.... most people I talked to, you know, they've charged stereos, they've went on vacations. You know, we weren't afforded that luxury. This [bankruptcy] was all accumulated debt out of necessity.... I guess I felt better because, like I said, we didn't go to Tahiti and we didn't drive up major credit cards. It was all debt of necessity....

Distancing from the "Credit Card Bankrupt"   While the stereotype of the extravagant bankrupt emphasized opulent spending, the image of the credit card bankrupt focused attention primarily on fiscal immaturity and a lack of self-control. Faced with easily available "plastic money," credit card bankrupts were portrayed by those we interviewed as having a juvenile tendency to deny the trouble they were getting into until it was too late. Credit card bankrupts' inability to manage their spending provided many debtors with a handy stereotype of what they themselves were not. Over 70 percent of the debtors asserted that they had tried to be conscientious in their credit card charges—even though many of these debtors also discharged considerable credit card balances in their bankruptcies.

Debtors who distanced themselves from credit card bankrupts were quick to point out what they believed was a clear distinction between the kinds of legitimate debt that forced them into bankruptcy, especially medical bills, and the credit card bankrupt's irresponsible debts. One couple in their early thirties, for instances, was adamant that they felt no shame whatsoever from their bankruptcy. The husband had fought stomach cancer and their daughter had been born with severe medical problems. They were without health insurance, and their medical bills were exorbitant. Consequently, they filed what they both described as a "medical bankruptcy." They contrasted their bankruptcy with a "credit card bankruptcy" and argued that the two should even be treated differently by the courts. "I think there should be rules for different circumstances," the husband said. "Like medical conditions, something you can't avoid. That's different than going out and maxing out your credit card, and 'I don't want to pay for it.' . . . I think people should be responsible for their debt."

Even debtors who had amassed significant consumer debt of other varieties argued that they were unlike the much-disparaged credit card bankrupts. A debtor for whom the catalysts for bankruptcy were two large car loans and her husband's unstable employment insisted that their bankruptcy was unique in that it did not include any credit card debt, an observation that, she emphasized, their attorney corroborated.

> I didn't mind losing all my cars [two of their vehicles were repossessed], but what I was filing bankruptcy on was NOT credit cards. . . . It was NOT a credit card bankruptcy. . . . I NEVER turned a credit card over to bankruptcy, ever. We've never done that, So, we look at it as that wasn't our problem. . . . And even Linda [their bankruptcy attorney] said, "This is not a normal bankruptcy. Normally you see a lot of credit card [debt]." . . .

Distancing from the "Repeat Filer"   The final stereotype from which debtors worked to distance themselves was that of the "repeat filer." While research reveals that only a small minority of those who declare Chapter 7 bankruptcy do so more than once, . . . debtors frequently mentioned "repeat filers," whom they believed were prevalent among bankrupts. One woman, who worked as a baker, put the matter succinctly: "You know, I'm sure there is a lot of people out there who make a career out of racking up bills every six years and filing bankruptcy and starting all over again." Discussions of repeat filers were often laced with resentment, in part because these, "deadbeats" increased the potential stigma for everyone, legitimate and illegitimate filers alike. As a woman in her mid-20s, who had sold her home, a vehicle, and a boat in a futile attempt to avoid bankruptcy, stated:

> Those people shouldn't get the third and fourth chance. . . . It pisses me off because, you know, we're playing by the rules. We've given some things up, we've made some sacrifices and we are paying for it. . . . And there's people out there that are just raping it [the bankruptcy system] and make it hard for everyone else and that's what's happening. The government is going to make it hard for everybody [via bankruptcy law reform]. . . . If people don't accept it as legitimate, then no one's going to respect it.

To distance themselves from the repeat filer stereotype, three-quarters of the debtors asserted that they would never set foot in bankruptcy court again. Some stoically stated this as fact, while others became quite heated when discussing the possibility. A father of two young boys shouted in defiance when the issue was broached:

> HELL NO! I'd kill myself first. Damned right. I ain't doing it. I told my wife, "We will not do it again." No way, because that would mean that we've learned nothing. . . . Damn it, man, we're not gonna go through that again. No way. No way. I don't care if we lose a job or whatever. We will live poor.

Not only did the majority of debtors insist that they would never file again, but they also shared an explanation for why they would not do so that clearly separated them from repeat filers: bankruptcy had taught them a critical, albeit painful, life lesson. For example, a woman who was forced into early retirement because of multiple chemical sensitivities that began when her workplace was sprayed with pesticides, insisted: "We would do anything in our power to not do it again. It isn't stigma. It's more personal worth. Because if you do this, you should learn and examine why in the hell you got in this position when you didn't believe in it."

In all their distancing activities, the debtors suggested that illegitimacy and irresponsibility were prevalent among those who declare bankruptcy, but they asserted that they themselves were relatively exceptional in not fitting the stereotypical profiles. They were neither extravagant in their spending nor profligate in racking up credit card debt, and they had learned from their mistakes. In the case of "learning from their mistakes," they not only distanced themselves from less legitimate bankruptcy filers, but also imbued their struggles with a deeper purpose of personal growth and hard-won maturity. . . .

Transcendence The final form of deviance avowal observed among the debtors is what Warren (1980) has referred to as "transcendence." In her discussion of ways in which individuals may seek "destigmatization," Warren notes that some stigmatized individuals attempt to "transcend" or "rise above their condition by means of some persistent type of action [that] display[s] a 'better' self"(64–5). Consistent with Warren's conceptualization, several bankrupt debtors attempted transcendence by engaging in actions in the present, or proclaiming intended future actions, that would lift them above their stigmatized identity.

One notable means of transcendence involved a commitment to repay some of the debts from which debtors had been legally relieved. According to court records examined by researchers with the 2001 wave of the Consumer Bankruptcy Project, "One in four families signs on to pay off debts they no longer owe after filing for bankruptcy" (Warren and Tyagi 2003, p. 170). Consistent with Zelizer's (1994) analysis of the "social meaning of money," the debts earmarked as critical for repayment were those that reflected valued social relationships. Specifically, debtors in our study focused on repaying creditors with whom they had close personal or professional relationships. So, for example, the previously quoted woman who ran a horse boarding facility explained, "I had some veterinary bills also. I see that as a personal service and a friendship there, so you really hate to file on something like that. . . . [So] I'm still paying them, because, you know, there's a personal stigma that goes along with it too." Another woman, a mother of three who worked as a waitress, also mentioned that she felt an obligation to repay her veterinarian. "I didn't want to include my vet [in the bankruptcy], and I didn't want to have to include my dentist. That was personal and that definitely, I thought, was so wrong that eventually I want to pay those people back, when we can. I will not pay back a credit card company, but I will pay my vet and my dentist."

Yet other debtors sought to transcend the "bankrupt" identity by using the knowledge they had gained through their financial travails to help others avoid similar problems. In their efforts to teach others, debtors demonstrated that they themselves had "learned from their mistakes" and desired to use their hard-won understanding to benefit others. One debtor, whose two granddaughters were living with her and her husband, described how she was teaching her oldest granddaughter to manage credit cards and the associated debt: "She [the granddaughter] will pay it [the credit card] off, then she'll run it up again. And I say to her, 'uh uh, no. That's not the way to go. If you can't pay cash, then...don't buy.' She doesn't like to pay the interest, because, see, this is what I've been teaching them. The interest will eat you up." Another debtor, a young woman in her late twenties, recounted that she and her husband had offered to help their friends learn how to work from a budget: "I mean, my husband and I are advocates of budgets. Some of our friends are having problems, and we're like, "This is how you do a budget. Come on over and we'll teach you how to do it; it's real simple.' "

Finally, a number of debtors made proclamations of intended transcendence. Many insisted that despite futures that might well be plagued by unstable income and new debts, they would do almost anything to avoid a subsequent bankruptcy. As a man in his early forties who laid cable for a living put it, "I would just give up everything. I would sell the house. Everything." Others said they would sacrifice their marriages before they would file again. And in the most extreme case, quoted previously, a man asserted, "I'd kill myself first. ... We're not gonna go through that again. No way."

CONCLUSIONS AND IMPLICATIONS

Drawing upon data from a study of married couples who filed for bankruptcy, we have explored the stigma of personal bankruptcy. In doing so, we have accomplished two things. First, we have challenged the popular argument that contemporary Americans do not experience bankruptcy as stigmatizing. Second, we have demonstrated the means through which bankrupt debtors manage this stigma. While our limited sample size constrains our ability to broadly generalize our findings, we have nonetheless presented the only in-depth U.S. data reported to date that directly address the issue of bankruptcy stigma and bankruptcy stigma management.

Our findings are clear. Feelings of stigmatization were a pervasive feature of our informants' bankruptcy experiences. Thirty-five of the thirty-seven individuals in our study stated—frequently, often emphatically, and in a variety of ways—that they felt shame and stigmatization as they negotiated the bankruptcy process. Significantly, these comments emerged. without specific prodding from the interviewer regarding stigmatization. Further, our informants exhibited many classic techniques documented in previous sociological literature for attempting to manage stigma. They strived to conceal their spoiled identities, especially from particularly significant others. Fearing embarrassment, they avoided interactions with those who might know of their recent failings. And when put in a position of having to openly face their spoiled identities in the context of the research interview, they distanced themselves from stereotypical images of illegitimate bankruptcy filers, provided excuses and justifications for their own bankruptcies, and described their attempts at activities that would enable them, at least partially, to transcend their stigmatized identities....

Our data inform sociological analysis more broadly by speaking to issues related to the achievement of social agency among stressed and marginalized sectors of the population. As we observed in our introduction, members of stigmatized social categories routinely endeavor to alleviate the status affronts and degradation associated with stigma. Yet it is clear that not all individuals or groups have similar opportunities to mitigate their stigmatization and inequality. The bankrupt debtors in our study provide a case in point. While they drew upon many of the most widely recognized forms of stigma management, two types were notably absent: in-group strategies and collective action. The absence of these techniques among debtors seems largely due to the social context within which bankruptcy is experienced. Unlike many kinds of stigmatized individuals, who either seek each other out (e.g., gays and lesbians) or are institutionally brought into close contact (e.g., as the homeless are brought together at soup kitchens and homeless shelters), those who experience the stigma of bankruptcy have neither a clear reason nor an institutional venue through which to come into sustained contact with each other. As a result, bankrupt debtors do not develop the "subcultural adaptations" to their subordinate status that characterize many other subordinate groups (Schwalbe et al. 2000:427–29). While bankrupt debtors are not alone among marginal groups in lacking access to in-group and collective action opportunities to mitigate stigma, the preceding observation highlights the importance of attending to contextual factors when seeking to analyze stigmatization processes.

REFERENCES

Anderson, Leon, and David A. Snow. 2001. "Inequality and the Self: Exploring Connections from an Interactionist Perspective." *Symbolic Interaction* 24: 395–406.

Bush, George W. 2005. "Remarks by the President in Signing the Bankruptcy Abuse Prevention and Consumer Protection Act of 2005." U.S. Newswire. Retrieved September 9, 2005 (http://releases.usnewswire.com/GetRelease. asp?id=46148).

Davis, Fred. 1961. "Deviance Disavowal: The Management of Strained Interaction by the Visibly Handicapped." *Social Problems* 9: 120–132.

Goffman, Erving. 1963. *Stigma: Notes on the Management of Spoiled Identity.* Englewood Cliffs, NJ: Prentice Hall.

Herman, Nancy. 1993. "Return to Sender: Reintegrative Stigma-Management Strategies of Ex-Psychiatric Patients." *Journal of Contemporary Ethnography* 22: 295–330.

Hochschild, Arlie Russell. 1983. *The Managed Heart: Commercialization of Human Feeling*. Los Angeles, CA: University of California Press.

Hutchinson, Asa. 1999. "Bankruptcy Reform." Federal Document Clearing House, Inc., Congressional Press Releases. April 20.

McIntyre, Lisa J. 1989. "A Sociological Perspective on Bankruptcy." *Indiana Law Journal* 65: 123–139.

Pfuhl, Erdwin H., and Stuart Henry. 1993. *The Deviance Press*. Hawthorne, NY: Aldine de Gruyter.

Schwalbe, Michael, Sandra Godwin, Daphne Holden, Douglas Schrock, Shealy Thompson, and Michele Wolkomir. 2000. "Generic Processes in the Reproduction of Inequality: An Interactionist Analysis." *Social Forces* 79: 419–452.

Snow, David A., and Leon Anderson. 1993. *Down on Their Luck: A Study of Homeless Street People*. Berkeley, CA: University of California Press.

Stanley, David T., and Marjorie Girth. 1971. Bankruptcy: Problem, Process, Reform. Washington, DC: The Brookings Institution.

Stokes, Bandall, and John P. Hewitt. 1976. "Aligning Actions." *American Sociological Review* 41: 838–49.

Turner, Ralph H. 1972. "Deviance Avowal as Neutralization of Commitment." *Social Problems* 19: 308–21.

Warren, Carol A. B. 1980. "Destigmatization of Identity: From Deviant to Charismatic." *Qualitative Sociology* 3: 59–72.

Warren, Elizabeth, and Amelia Warren Tyagi. 2003. *The Two-Income Trap: Why Middle-Class Mothers and Fathers Are Going Broke*. New York: Basic Books.

Zelizer, Viviana A. 1994. *The Social Meaning of Money: Pin Money, Paychecks, Poor Relief, and Other Currencies*. New York: Basic Books.

Zywicki, Todd J. 2005. "Institutions, Incentives, and Consumer Bankruptcy Reform." George Mason University School of Law, Working Paper 05–07. Retrieved July 15, 2005 (http://www.gmu.edu/departments/law/faculty/papers/wpDetail php?wpID=285).

29

Men Who Cheer

MICHELLE BEMILLER

Just as the lesbian athletes discussed by Blinde and Taub violated gender roles in being intercollegiate athletes, Bemiller's male cheerleaders encounter gender stigma from venturing into a female-dominated activity. Female labels of lesbian in sport correspond to male labels of homosexuality in cheerleading. The men Bemiller studied know that their descent into a girls' realm will lead to masculinity challenges of various sorts from the people they encounter, and they take a variety of measures intended to forestall their deviant labeling. Their face-saving strategies are fairly aggressive, as they attempt to invoke hypermasculine demeanors to counter their taint of female association. This chapter, like many others in the book, reveals the hierarchy of gender stratification that positions hypermasculine men at the top, soft/gentle/androgynous men in a lower position, gay men below them, and all women at the bottom. In attempting to elevate themselves on this ladder, the male cheerleaders demean the role of their female squad mates to distance themselves and to enhance their position. We earlier noted this behavior in Tuggle and Holmes' article on the antismoking campaign, where nonsmokers gained status and power by stigmatizing and diminishing smokers. Male cheerleaders also draw on other high-status attributes of their gender role by emphasizing hypermasculine features such as toughness and the sexual objectification of women. Demeaning women is revealed as more than an innocuous form of male jocularity; it is a powerful strategy for maintaining differential access to status, opportunity, and power in society.

The institution of sports has long been associated with the construction and maintenance of masculinity among boys and men (Anderson 1999; Connell 1995; Koivula 2001; Lantz and Schroeder 1999; Messner 2002). As an institution, sports reinforce the patriarehal superstructure where masculinity is valued over femininity. The devaluation of femininity is reflected in the subordination of women, as well as men who participate in non-masculine activities or exhibit non-masculine characteristics or mannerisms. Participation in competitive sports that emphasize physical size, strength, and power reinforces and reaffirms the masculinity of men who participate as viewers or players (Connell 2002; Messner

"Men Who Cheer," by Michelle Bemiller, *Sociological Focus,* the Journal of the North Central Sociological Association, Vol. 38, No. 3, August 2005. Reprinted by permission of the North Central Sociological Association.

1992; Suitor and Reavis 1995). Thus, sports such as football, basketball, ice hockey, and baseball that emphasize mental toughness, competitiveness, and domination are viewed as the domain of men (Griffin 1995).

Athletics provide young men with status among their peers, increasing their popularity and acceptance (Griffin 1995), assuming, of course, that these men are participating in appropriate "male" sports. Men who do not participate in masculine sports are stigmatized, leading to negative appraisals regarding their gender and sexuality (Suitor and Reavis 1995).

Despite the possibility of negative appraisals, men have become more visible within female-dominated sports such as cheerleading, leading to a unique opportunity for research on gender presentation, relations, and identity. Yet, this area has not received a lot of attention to date. In contrast, notable work has been done on men's entry into female-dominated occupations. Williams (1989, 1995), for example, indicates that when men do women's work, gender differences are reproduced. Men are viewed as highly competent at their work and rise quickly through the ranks. The same is not true for women in male-identified occupations.

Men in female-dominated occupations do encounter questions regarding their sexuality. This questioning, however, does little to impede their progress in the organization. To reaffirm their masculinity, these men seek out male-identified specialties, emphasize masculine aspects of the job, and pursue administrative positions (Williams 1989, 1995). Similarly, in her work on men working in "safe" and "embattled" organizations, Dellinger (2004) found that when men work in an organization that is dominated by feminist ideals, they construct their masculinity by separating themselves from women in the work context and aligning with other males. In contrast, when men work in an environment that is supportive of masculinity, they have better relationships with their female co-workers.

While these findings are useful in helping us understand men who do women's work, we still know little about men who do women's sports. Do these same patterns and outcomes emerge when men participate in women's sports? To further our understanding of men in female-dominated arenas, this paper will use cheerleaders at one northeastern Ohio university to examine the maintenance of gender and sexuality in a female-dominated sport. . . .

METHODS

Sample and Procedures

Participants for this study were selected from a cheerleading squad at a northeastern Ohio public university. To recruit participants, I attended a cheerleading practice and asked for volunteers to participate in a study about men who cheer. I was introduced by the cheerleading coach as a sociologist interested in doing research on cheerleaders. The coach's introduction provided me with status as a legitimate researcher, which may have influenced the cheerleaders' decision to participate in the study. Upon introduction, I simply told the cheer-leaders that I was interested in learning about men's participation as cheerleaders and that I would appreciate their assistance in learning more about the men who cheer. At the time of recruitment, the men and women had just finished practice and were talking among themselves about their plans after practice. Most of them were planning on going out to the local bars. A few men and women agreed to participate. Once these individuals agreed, their friends decided to participate as well. My assumption was that they agreed to participate together, because they were all going out after the practice.

Convenience sampling—sampling cases that were available at the time of the study—was used (Singleton, Straits, and Straits 1993). Out of 25 possible partici-pants, 17 volunteered: 8 men and 9 women between the ages of 18 and 25. Four men and four women declined to participate. Because both men and women were equally willing to participate, there does not seem to be anything unique about the individuals who chose to participate. The individuals who chose not to participate did not provide me with a reason for declining. Of the 17 partici-pants in the study, no racial minorities were present. Had any minorities participated, some of the information provided by the respondents might have been different, since experiences may have differed not only based on gender, but also based on race.

The data provided are meant to present a glimpse into the lives of men who cheer at one university, not to be generalized to all cheerleaders at all universities. University and regional culture may certainly play a role in an individual's expe-riences, with the size of the university and region and the cultural diversity of the campus and community affecting individual attitudes regarding men's and women's participation in gendered sports. The community where the university is located is a large urban area with over 200,000 residents. The residents of the city are predominantly white, middle class individuals with high school educa-tions. The university is a large, urban campus with roughly 20,000 undergraduate students; roughly 11,000 are women and 9,000 are men. Non-whites make up roughly 20% of the overall student population.

Data collection consisted of two stages. Focus group discussions were held in the summer of 1999. Reviews of earlier versions of the focus group research indicated that men may not be willing to discuss issues related to sexuality in focus group settings so in-depth; one-on-one interviews were conducted in 2001 to probe for a deeper understanding of issues related to sexuality and stigmatization. One-on-one interviews were conducted with both new partici-pants (n = 7) as well as participants from the 1999 focus groups (n = 4)....

FINDINGS

Two main themes emerged from the cheerleaders' narratives. The first was the stigma that coincides with being a male cheerleader. Stigmatization included the negative experiences men had because they are cheerleaders as well as accusations regarding their sexuality. The second theme analyzes face-saving strategies that men use to protect their masculinity in a female-dominated sport. These strategies include emphasizing the masculine qualities of cheerleading (e.g., injury and risk) and acting in hyperheterosexual ways (e.g., claiming ownership over the sport and objectifying the female cheerleaders).

Stigma

Goffman (1963) argued that an individual is stigmatized when they possess an undesired differentness from what is anticipated. In the present study, the differentness under investigation is men participating in cheerleading. Goffman (1963) pointed to the need to determine if one's stigma is evident (e.g., race) or less visible and difficult to discern. Men who cheer possess a potentially discreditable identity due to their deviation from gendered proscriptions regarding participation in sports. This identity is "discreditable" because it is not a status that everyone necessarily knows about, yet if known or found out, it can be stigmatizing.

Participation in a Feminine Sport   Throughout the focus group discussions and one-on-one interviews, the male and female cheerleaders talked about the stigmatization and resulting undesirable outcomes associated with being a male cheerleader. This is consistent with prior research showing that men are stigmatized because of their participation in what is considered a feminine sport (Connell 1987, 1992; Goffman 1963; Griffin 1995, Kimmel 1996; Messner 1992; Messner and Sabo 1996; Suitor and Reavis 1995). Thus, male cheerleaders are labeled as non-masculine and homosexual for "crossing over" into a female domain. The following quotes capture some of these comments. According to Betty in her one-on-one interview:

> Cheerleading is known more as a girl's sport than a guy's sport, so the guys get a lot of teasing. I give them a lot of credit for putting up with people talking about them. The comments bother them, but they really like doing it, so they put up with them.

Annie elucidates this point further, "I think it is more acceptable to be a girl cheerleader than a guy cheerleader on campus. It is hard for people to adjust to the idea that men are cheerleaders too." In the all-male focus group discussion, one participant asserted that his fraternity brothers give him a hard time about being a cheerleader, teasing and taunting him about participation in a female-dominated sport. Another stated that he works in construction and that the guys at work "rib him pretty good" about being a male cheerleader.

The stigma of participation is further magnified because cheerleading is in a category of activities that many people would not define as a sport at all. John discusses other men's views of cheerleading as a sport by specifically focusing on the teasing that goes along with cheerleading. He claims that, "men from other sports say 'you can't participate in a real sport.'" The suggestion that cheerleading is not a "real" sport implies that men should be playing football or some other contact sport that demonstrates their manliness and that the majority of society labels as a sport (Suitor and Reavis 1995).

Brett provides an interesting look at the importance of playing a "real" sport as opposed to a feminine sport. While Brett participates in cheerleading practices, he has not yet performed at a football or basketball game. In other words, he has not made a "public" appearance as a male cheerleader. Brett tells me that he does not intend to cheer at the games. Although he enjoys cheerleading, he fully understands and attempts to avoid the stigma that he faces as a male cheerleader. "I'd rather be playing football than cheering for the football team. It would hurt my self-esteem, because guys don't cheer. I mean what percentage of the male population cheers? I mean, I like cheerleading, but it's a girl's sport." By participating in the cheerleading practices but not the public events, Brett attempts to have his cake and eat it too. In other words, Brett enjoys participating in the practices, but has avoided being publicly labeled as "a man who cheers." Thus, he has attempted to avoid the stigma associated with participation in a female activity. The emphasis on "real" sports not only stigmatizes the males who choose to be cheerleaders but also marginalizes cheerleading as a sport altogether.

Sexuality   Besides the stigma that occurs with participation in a female-dominated sport, men's participation in cheerleading calls into question their sexual identity, interpreted through the lens of gender. Men who cheer are perceived and labeled as homosexual. As a case in point, prior to becoming a male cheerleader, Adam assumed that men who cheered were homosexual. "I was hesitant to become a cheerleader because of the idea that you have to be gay to be a cheerleader." After discussing this issue with several of the men and women who cheer, Adam decided to participate in the sport despite these stereotypes. Even after discussing this issue with his teammates, however, Adam still faces stereotyping regarding his and his teammate's sexuality. "A lot of people think that the men are gay and ask me if they are. People make fun of them. People say they're all a bunch of fags." Similarly, Brett is teased by his close friends. "People definitely perceive that the male cheerleaders are fags. All of my friends give me a hard time about being a cheerleader." Brett's words demonstrate that not only do strangers react to male cheerleaders, but also the people who know them. Since significant others (e.g., family and friends) often have a strong impact, Brett's friends' reactions affected how he saw himself as a cheerleader. Thus, Brett's friends' reactions probably had a stronger impact on how he saw himself as a cheerleader than strangers' comments would.

Perceptions about male cheerleaders' sexuality, however, do not just come from external forces. As in the cases of Adam and Brett, men who cheer bring internalized stereotypes with them into the sport. Specifically, many of the respondents claimed that they thought that male cheerleaders were gay prior to their own participation in the sport, and some current participants indicated that some of the men on the squad might be gay. Brett provides an example of that belief:

> The men here have more feminine characteristics. I would assume that
> they are gay. I would say they seem to be gay because they are cheerleaders
> hut then I put myself in that group. I think that I perceive them as gay
> because they gossip a lot and have more girl friends than guy friends.
> Something is weird about a guy who can have all girls for friends.

By claiming that men who exhibit these qualities are gay, Brett buys into the
stereotypes associated with nonmasculine men or men who participate in
nonmasculine activities. Brett acknowledges that he is a male cheerleader, but
he does not belong "in that group" (i.e., gay cheerleaders). Both Adam and
Brett distance themselves from the homosexual stereotype of male cheerleaders
by referring to other male cheerleaders as "the men" and "they." By using the
third person, they verbally and mentally remove themselves from an association
with men whom they perceive as gay. In other words, they might be cheerleaders, but they are certainly not gay.

Brett and Adam articulate the gendered belief system that limits men and
women to certain activities and claims that deviation must relate to sexuality and
a departure from heterosexual normativeness (Connell 1987; Lorber 1994; Pharr
1997), For example, Brett's comment regarding the strangeness of males with
female friends coincides with perceptions that all male-female relationships are
inherently sexual. In other words, if a male has a female friend, he must have a sexual interest in her, and if he does not, he must be gay. Brett's comments reaffirm
what is normatively assumed and accepted, that women and men cannot be friends
unless a sexual relationship exists. Brett's assumptions regarding cross-gender
friendships is not necessarily supported by research on this topic. While some
research on cross-gender friendships does indicate that male-female friends may
experience sexual feelings toward one another (Egland, Spitzberg, and Zormeier
1996), other research finds the opposite to be true. It is possible for men and
women to be friends without a sexual interest or dynamic (Monsour 2002).

In Annie's discussion of male cheerleaders, she draws attention to the male
cheerleaders' tendency to believe in the stereotypical image of men who cheer:

> We had a guy come in the last week of tryouts and he fit the media's
> image of a gay person and the guys made fun of him behind his back.
> It's almost like the guy cheerleaders believe the perception that male
> cheerleaders are gay. So, they participate but they still believe the
> stereotype.

This adherence to stereotypical images of male cheerleaders, however, does
not rest solely with men who cheer. The female cheerleaders also indicated that
they had thought all male cheerleaders were gay until they participated in college
cheerleading. In her one-on-one interview, Cindy asserted:

> Before I started cheering, I thought it was different for guys to cheer
> because when I was in high school we only had girls on the squad.
> People think the guys are gay and they say they wouldn't want to cheer.
> They say 'they're gay' and I tell them that I know all of them and
> they're not gay.

Cindy's views are not far from the other women's perceptions. In the focus group discussions, one woman said, "I thought they were gay" and another woman stated, "Yeah, I thought they were freaky."

Throughout both the focus group discussions and the one-on-one interviews, the male and female cheerleaders acknowledged that it is more acceptable for females to cheer based on perceptions about sexuality. One woman explains why participation is less acceptable for the men. "The people that don't know the men still have the idea that the guys are gay and the girls are okay." This response was followed by nods of agreement by the rest of the females in the focus group. In his one-on-one interview, Adam states, "I've been made fun of by people, they've said like, 'What are you doing? You're in a girl's sport, you're gay, blah, blah, blah.'" These responses capture a central issue: if the image of male cheerleading can be heterosexually validated, then any man should be able to cheer without the assumption that he is homosexual. Unfortunately, the link that was established by the female participants between being gay and being "freaky" reflects the power of the labeling and stigmatization process of people believed to be sexually deviant (Connell 1987; Lorber 1994; Pharr 1997; Schur 1984). Homophobia limits choices by labeling anyone who deviates from gender-appropriate norms as sexually deviant or gay (Johnson 1997).

SAVING FACE

In response to stigma and stereotyping, the male cheerleaders participate in strategies to help them save face as they participate in a female-dominated activity. According to Goffman (1959), face-saving behavior involves attempts to salvage an interactional performance that hasn't gone as planned. By participating in a female-dominated sport, men who cheer fail to "do gender" appropriately because of their failure to participate in masculine sports. They, therefore, must use face-saving techniques to be viewed as acceptable in interactions. The men in this study saved face by acting in hypermasculine ways. According to Connell (1992) hypermasculinity becomes manifest when hostility exists toward homosexual men, and heterosexual men attempt to create social distance from homosexuality by emphasizing their heterosexuality. For the men who cheer this hypermasculinity was demonstrated as the men claimed territoriality over the sport, stressed the masculine nature of the sport, and sexually objectified the female cheerleaders. Through these actions, the men attempted to deflect accusations of homosexuality.

Territoriality   Both the male and female cheerleaders discussed the fact that cheerleading is viewed as a female-dominated sport. However, many of the male cheerleaders insisted that cheerleading is becoming a male-dominated sport and that the female cheerleaders would not be able to participate in this sport without the help of the men. Female cheerleaders have been cheering without men at the high school and professional level for quite some time (Hanson 1995). Yet, these statements coincide with attitudes of entitlement, superiority, and solidarity that exist in male athletics, which encourage homophobia and sexism (Griffin 1995).

This sense of entitlement, superiority, and solidarity was apparent during the focus group discussions. The men were adamant about the importance of men in cheerleading, marginalizing female participation. The tenor of the conversation became tense and defensive as the men proclaimed their dominance in the sport. One male stated, "Males started the sport of cheerleading in the first place." This comment was followed by agreement from other respondents, "Yeah, they wouldn't let women do it. So, we started it, we're just coming back." Males' reclaiming the sport requires marginalizing the women's status. Women's importance in the sport has been relegated to the margins because "men started the sport." Once this territoriality was established, the atmosphere became much more comfortable. They had asserted their authority and defended their masculinity (McGuffey and Rich 1999).

In the all-female focus group, the female cheerleaders acknowledged that cheerleading started as a male-dominated sport. One female said, "Cheerleading started with men. I think the media has turned it into a sexualized female sport. That's what people like to see." In so saying, the women recognize the second-ary, sexualized status of women in cheering. At the same time, the women are invested in helping the men to protect and maintain their masculine guise. The women's willingness to elevate men into a superior position within the sport demonstrates that gender maintenance occurs in social interactions with both men and women cooperating. By subordinating their role as cheerleaders, the women "do femininity" as they demonstrate submissiveness to the men who cheer (West and Zimmerman 1987). The women also help the males to "do masculinity" by assisting in the construction of cheerleading as a masculine sport through the emphasis placed on aggression and injury.

Masculine Aspects of the Sport: Toughness and Aggression   The male cheerleaders used imagery and examples from their cheerleading lives to demon-strate their heterosexuality and establish their masculinity. Through these mascu-line examples, the men saved face, or protected their masculine identities. Specifically, some of the men referred to fighting, or the possibility of fighting, with individuals who made derogatory comments to them about participation in cheerleading. These men said that no one would have the nerve to say any-thing offensive for fear of being injured for their remarks. One male stated, "I don't think that anyone's ever had the balls to say anything negative to me," implying that there would be consequences for anyone who voiced negative comments. Another respondent demonstrated his masculinity by stating, "Me and Bob fought the football team one year. I don't think that had anything to do with cheerleading though. We kicked their asses." Demonstrating physical prowess and the ability to defeat the most masculine of sports teams, the two cheerleaders not only saved face, but also were elevated in status. Adam, in a one-on-one interview, alluded to this incident with the football team, saying: "I've heard of cheerleaders taking on the football team and winning before. I wasn't a part of that but I did hear about it." Again, he uses this incident to demonstrate that male cheerleaders should not be messed with and that they are able to protect themselves through violence if necessary. Since he had "heard of this incident, it is apparent that the story had been talked about within the cheerleading group, serving to reinforce the idea that the men who cheer can take care of themselves physically. This storytelling demonstrates how the men collectively maintain their masculinity in a female-dominated activity.

During the female focus group discussion, this masculine facade was also discussed. One female commented that, "They always say, 'They won't mess with me, I'm a cheerleader.'" This comment and the comments stated above illustrate how the men are creating and exuding a collective identity in which male cheerleaders are defined as dominant and aggressive. These statements demonstrate how the male cheerleaders maintain their masculinity by emphasizing their ability to defend themselves in physical altercations with other men. The use of violence and aggression are common characteristics utilized in maintaining masculinity for men (Connell 1995, 2002; Griffin 1995; Messner 2005).

The emphasis on toughness and aggression is also apparent when the male and female cheerleaders discuss the men's role as cheerleaders. Male-dominated sports such as football, wrestling, and hockey are often labeled as "real" sports because they contain elements of violence and an increased likelihood of injury. In order to equate cheerleading with these "real" sports, Leeann talks about the men's injuries and the need for strength to masculinize cheerleading, showing that cheerleading is not just a "girl's sport."

> You're always going to have people that think it's just a girl's sport.
> But, as people start to see more men doing it and once people see what
> they do and see all of these bloody noses and broken mouths, then
> they realize that it's not a girl's sport. It's tough for them. They do a lot
> of lifting and stuff that requires a lot of strength.

Similarly, Mary states, "I think guys like cheerleading because they get the satisfaction of knowing that they can lift two girls at once. It takes a lot of strength." Cindy also acknowledges the difficulty of the sport, "It's a lot more physical than you would think." Adam agrees with the women as he states, "I don't think that people realize how hard it is." Brett also emphasizes the difficulty of cheerleading for guys. He says, "It's physical. I'm tired [when I cheer]."

The male and female cheerleaders define cheerleading as a sport because it requires strength, manual labor, physical exertion, and involves competition, much like sports defined as masculine (e.g., football, wrestling). The male cheerleaders act tough and aggressive to prove their masculinity, which then confirms their heterosexuality. The link between toughness, masculinity, and heterosexuality for these men is woven throughout their face-saving strategies.

Sexual Objectification of Women   Connell (1992) asserts that hypermasculinity becomes manifest when hostility exists toward homosexual men, and heterosexual men attempt to create social distance. In cheerleading, the men do this in large part by sexualizing the female cheerleaders. The women indicated this sexual objectification in both the focus group discussions and in the one-on-one inter-views. One participant stated, "People say the guys are girly. Not at all ... they are probably more manly than guys who don't participate in the sport. They're perverted and sexual sometimes." In agreement with this female, another partici-pant stated, "The male cheerleaders are the most heterosexual males I have ever met, so people have the wrong perception when they say they are gay. It makes me laugh. They are the most perverted guys that I have ever met." By "perverted," the women mean that the men talk about the women's bodies among themselves and to the women. For example, one female focus group respondent stated, "They say, like, 'Oh people think I'm gay but I get to grab your butt.' They're perverts." Another cheerleader in the female focus group said that, "They always think all the girls want them. They'll say, 'She wants me.'"

The female focus group participants agreed that when a group of men and women work closely, sexual innuendos occur, and sexual tensions build. Cindy provides an explanation why the male cheerleaders sexualize the females:

> When the guys are accused of being gay, I think they try to over-compensate for these accusations by being all about the girls. They act very sexual toward girls. I think that the guys look at us as sex objects, but all guys would have that attraction to the women. But they've never made me or any of the other girls feel uncomfortable. I'm sure they find us attractive, but it's kind of more like a sisterly-brotherly thing.

Cindy's statement is contradictory as she asserts that the men are sexually attracted to the women, but that the relationship is sisterly-brotherly. This confusion continues as another participant also comments on the attraction between the men and women as well as the family-like relationship between the male and female cheerleaders:

> At first, I think that the girls and guys are attracted to each other. It's only natural when you have guys and girls together in a sport. You are constantly touching and it is sexual at first, but you get over that. It becomes like kissing your grandma.

The contradiction between sexual and familial relationships demonstrated by the females reveals the limited frame of reference available for mixed gender settings. The women construct the men as both "brotherly" and "perverted." Both instances provide defenses for the men's sexualization of the women. In the first instance, the women argue that the men's comments are often miscon-strued as perverted when in actuality they are simply jokes and that their touching is more like that of a brother than a boyfriend. By claiming this, the women become accepting of the men's antics. In the second instance, the women do claim that the men are perverted. However, the women defend the men's com-ments and actions as "boys being boys." In essence, the women normalize the perverse comments and actions by inadvertently stating that this sexual banter is part of the men's attempts to demonstrate their masculinity. By defending the men's actions, the women allow the men to display their gender (Lorber

1994; Lucal 1999; West and Zimmerman 1987), which serves to negate perceptions that male cheerleaders are gay.

Both the men and the women see the objectification of the women as a component of being a man or being masculine within society (Connell 1987; Johnson 1997; Pharr 1997). While the women demonstrate contradictions regarding their relationships with the male cheerleaders, the men emphasize a sexualized relationship with the women as opposed to a familial relationship. When asked why they chose to cheer, many of the men asserted that it was because of the presence of the beautiful girls and because they get to have intimate physical contact with the women. Brett said,

> I participate in cheerleading because of the girls. Most of the cheerleaders are hot all around the board. I would love to go to a cheerleading competition. There are tons of girls there. That's the only reason that I'm a cheerleader.

George sees the female cheerleaders as sex objects because all of the women are in good physical shape. George states, "If they were ugly cheerleaders, I probably wouldn't cheer."

In the male focus group discussion the men continuously talked about the fact that they get to touch the female cheerleaders. The males asserted that one of the best perks of being a male cheerleader is the closeness to the females' bodies. The "hot chicks" that stand beside them are seen as prizes and a major reason for participating in the sport of cheerleading. One male even stated that people who might make allegations of homosexuality do not realize that they get to "touch the girls' butts and stuff." Another respondent stated, "I worked at a trucking company for three years and all the guys loved it (that I was a male cheerleader). They would come over and ask me stories and stuff. Like, what do you do with these girls?"

The construction of masculinity through the objectification of women is applied when men are in the minority, engaged in a female-dominated activity. The route to power in this situation is the use of sex and sexuality to realize gender relations. By talking about these women as sex objects, they are subordinating women within the sport. The men were so excited about touching the women that they even suggested that more men might be interested in cheering if they were aware of how beautiful the women are and how much time they would get to spend with them, talking to them, and touching them. One male specifically suggested that the athletic department needs to "sell the female cheerleaders" in order to recruit more men.

The objectification of women provides the male cheerleaders with a mechanism for asserting their heterosexuality and masculinity in the context of cheering to the non-cheering world. In so doing, the male cheerleaders are maintaining a socially constructed ideal of manliness through the objectification of women, which confirms their heterosexuality and masculinity to themselves, while also reinforcing hegemonic, dominating, sexualized, masculinity for public consumption.

CONCLUSION: CROSSING THE GREAT GENDER DIVIDE

The aim of this research was to explore male participation in a female-dominated sport by focusing on men who cheer. The findings highlight the methods used by both male and female cheerleaders to redefine men's roles and activities within cheerleading in ways that stabilize and reassert hegemonic masculinity. Both the male and female cheerleaders articulated the stigma that goes hand in hand with participation as a male cheerleader.

In an attempt to masculinize cheerleading, the male and female cheerleaders focused attention on the men's strength and toughness. The male cheerleaders asserted their dominance through claims that cheerleading started as a male sport and that the women "need" the men in order to perform stunts. In addition, the males used objectification of the women to pronounce their heterosexuality and counter the stigma attached to men engaged in "women's" activities.

Claiming territoriality over a female-domain happens in other female-dominated areas besides sports, because in a patriarchal society maleness is valued over femaleness, allowing men to take over organizations that were previously female-identified. For example, Williams (1989, 1995) found that men in female-dominated occupations were paid better and promoted more quickly than females in the organizations. Some of these men, however, were perceived as homosexuals when outsiders observed their participation in a female-dominated occupation. Because of these perceptions, the men adhered to a macho image and distanced themselves from the women in the organization by participating in male-identified specialties, emphasizing the masculine nature of their jobs, and pursuing higher administration jobs. Similar to Williams' findings, men who cheer also experience accusations of homosexuality because of their participation in a female-dominated sport. Contrary to Williams' findings, however, the men who cheer did not distance themselves from their female counterparts. Instead, to prove their masculinity, these men physically objectified the female cheerleaders. This sexual objectification may be unique to the sport of cheerleading because of the close physical contact that is essential within the sport. Men and women are expected to perform stunts that require physical contact, and this physical contact is used by the men to demonstrate their heterosexuality.

These findings demonstrate that both male and female cheerleaders redefine cheerleading to emphasize masculinity and subordinate femininity. The tactics used to re-create and maintain masculinity for the male cheerleaders demonstrate the importance of maintaining male dominance over women and gay men. While the negotiation of masculinities is discussed in the masculinity literature (Anderson 1999; Bordo 1993; Connell 1987), the cheerleaders' adherence to male dominance and male power within cheerleading reinforces the maintenance of a masculinity that emphasizes strength, skill, aggression and competition (Anderson 1999), as well as heterosexuality.

REFERENCES

Anderson, Kristin. 1999. "Snowboarding: The Construction of Gender in an Emerging Sport," *Journal of Sport and Social Issues* 23: 55–79.

Bordo, Susan. 1993. *Unbearable Weight*. Berkeley: University of California Press.

Connell, Robert W. 2002. *Gender*. Cambridge: Blackwell Publishers.

———. 1995. *Masculinities*. Berkeley: University of California Press.

———. 1992. "A Very Straight Gay: Masculinity, Homosexual Experience, and the Dynamics of Gender." *American Sociological Review* 57: 735–751.

———. 1987. *Gender and Power: Society, the Person, and Sexual Politics*. Stanford, CA: Stanford University Press.

Dellinger, Kirsten. 2004. "Masculinities in 'Safe' and 'Embattled' Organizations: Accounting for Pornographic and Feminist Magazines." *Gender & Society* 18: 545–566.

Egland, Kori; Brian H. Spitzberg, and Michelle M. Zormeier. 1996. "Flirtation and Conversational Competence in Cross-Sex Platonic and Romantic Relationships." *Communication Reports* 9: 105–117.

Goffman, Erving. 1963. *Stigma: Notes on the Management of Spoiled Identity*. Englewood Cliffs, NJ: Prentice Hall.

———. 1959. *The Presentation of Self in Everyday Life*. Garden City: Doubleday.

Griffin, Pat. 1995. "Homophobia in Sport: Addressing the Needs of Lesbian and Gay High School Athletes." Pp. 53–65 in *The Gay Teen: Educational Practice and Theory for Lesbian, Gay, and Bisexual Adolescents*, edited by G. Unks. New York: Routledge.

Hanson, Mary E. 1995. *Go, Fight, Win! Cheerleading in American Culture*. Bowling Green, Ohio: Bowling Green State University Popular Press.

Johnson, Allan G. 1997. *The Gender Knot: Unraveling our Patriarchal Legacy*. Philadelphia: Temple University Press.

Kimmel, Michael. 1996. *Manhood in America*. New York: The Free Press.

Koivula, Nathalie. 2001. "Perceived Characteristics of Sports Categorized as Gender-Neutral, Feminine, and Masculine." *Journal of Sport Behavior* 24: 377–393.

Lantz, Christopher D., and Peter J. Schroeder. 1999. "Endorsement of Masculine and Feminine Gender Roles: Differences between Participation in and Identification with the Athletic Role." *Journal of Sport Behavior* 22: 545–557.

Lorber, Judith. 1994. *Paradoxes of Gender*. New Haven: Yale University Press.

Lucal, Betsy. 1999. "What It Means to Be Gendered Me: Life on the Boundaries of a Dichotomous Gender System." *Gender and Society* 13: 781–797.

McGuffey, C. Shawn, and B. Lindsay Rich. 1999. "Playing in the Gender Transgression Zone: Race, Class and Hegemonic Masculinity in Middle Childhood." *Gender & Society* 13: 608–627.

Messner, Michael. 2005. "Still a Man's World?" Pp. 313–326 in *Handbook of Studies of Men & Masculinities* edited by M.S. Kimmel, J. Hearn, and R.W. Connell. Thousand Oaks, CA: Sage.

———. 2002. *Taking the Field*. Minneapolis: University of Minnesota Press.

———. 1992. *Power at Play: Sports and the Problem of Masculinity*. Boston: Beacon Press.

Messner, Michael and Donald Sabo. 1996. *Sex, Violence and Power in Sports: Rethinking Masculinity*. Thousand Oaks, CA: Sage Publications.

Monsour, Michael. 2002. *Women and Men as Friends: Relationships across the Life Span in the 21st Century*. Mahwah, NJ: Laurence Erlbaum Associates, Inc.

Pharr, Suzanne. 1997: *Homophobia: A Weapon of Sexism*. Berkeley, CA: Chardon Press.

Schur, E. 1984. *Labeling Women Deviant: Gender, Stigma and Social Control*. New York: Random House.

Singleton, Royce A., Bruce C. Straits, and Margaret Miller Straits. 1993. *Approaches to Social Research*. New York: Oxford University Press.

Suitor, Jill J., and Rebel Reavis. 1995. "Football, Fast Cars, and Cheerleading: Adolescent Gender Norms, 1978–1989." *Adolescence* 30: 265–270.

West, Candace, and Don Zimmerman. 1987. "Doing Gender." *Gender & Society* 1–2: 125–151.

Williams, Christine. 1995. *Still a Man's World: Men Who Do 'Women's Work.'* Berkeley, CA: University of California Press.

———. 1989. *Gender Differences at Work: Women and Men in Nontraditional Occupations*. Berkeley, CA: University of California Press.

30

Fitting In and Fighting Back: Homeless Kids' Stigma Management Strategies

ANNE R. ROSCHELLE AND PETER KAUFMAN

The homeless occupy a particularly stigmatized position in society, serving as general pariahs. People go out of their way to avoid seeing them or interacting with them, and as Reinarman (Chapter 15) notes, the American "vocabulary of attribution" assigns blame to them based on their individual failures rather than to the unequal opportunity structure of society. Nowhere is this blame less worthy than it is for homeless kids, who have inherited their homeless status from parents unable to house them. Based on research conducted in moderate-term housing shelters for (usually) single mothers with children, Roschelle and Kaufman articulate some of the strategies these youngsters use to disguise or manage their deviant status.

They divide their techniques into strategies of inclusion and exclusion. In the former, children attempt to fit into society and attain the appearance, friendships, and acceptance they see among their "homed" peers. Their attempts to pass, cover, and forge relationships are poignant, revealing their pain and insecurity. When these fail, however, and they cannot escape their stigma, they muster a bravado designed to achieve a tough and intimidating persona. Like the cheerleaders and the bankrupt, they seek to raise their status by comparing themselves to others whom they deem lower than themselves, although in so doing they denigrate others in similarly unfortunate situations whom they can construct as more socially and morally object than themselves.

B eginning in the 1980s the gap between the rich and the poor widened significantly, and there was a concomitant increase in homelessness. Social scientists responded to this crisis by studying the rates and causes of homelessness. However, with the exception of a few notable studies (Liebow 1993; Snow and Anderson 1993), there has been little ethnographic research on the daily struggles this population encounters. Moreover, research has been slow to address the growing feminization of the homeless. The proportion of women and children constituting the urban homeless population increased from 9 percent in 1987 to 34 percent in

1997 to a staggering 40 percent in 2001. Although there has been some recent work on familial homelessness, most accounts have disproportionately examined street homelessness among single men. More research is needed on women and children who are homeless and, in particular, on the strategies they use to construct meaning, participate in social interactions, and negotiate the boundaries with the nonhomeless world. . . .

In our analysis, stigma and homelessness are both construed as structural locations. Like stigma, homelessness must be recognized as a structural component characterizing the individual's relationship to the social world. Individuals may find themselves delegitimized socially, politically, or economically if they cannot accumulate normative resources, are unable to receive positive appraisals, and are incapable of following expected rules of behavior. Such is the case with homeless individuals and other stigmatized populations. In daily life, these populations often find themselves disadvantaged because they do not have the means to engage in successful social interactions. Implicitly then, stigma, like homelessness, is about power—or the lack thereof. As Link and Phelan (2001, p. 367) argue, "[S]tigmatization is entirely contingent on access to social, economic, and political power that allows the identification of differentness, the construction of stereotypes, the separation of labeled persons into distinct categories, and the full execution of disapproval, rejection, exclusion, and discrimination."

Recognizing stigma as a condition deriving from macro-level forces does not negate the micro-level processes that characterize an individual's stigma management. Although stigma should be understood as a consequence of structural relationships, the ways in which individuals manage their stigma is micro-interactional. Acknowledging the macro-origins of stigma forces researchers to incorporate an analysis of the structural while studying the ways in which individuals manage their stigma interpersonally. Because stigmatized individuals lack power, their ability to protect their sense of self may be severely compromised. As a result, their actions may unintentionally augment their stigmatization and perpetuate the structural relationships that generate their stigmatized status (Link and Phelan 2001). Much like impoverished individuals, therefore, acquiring the capital (social, cultural, economic, political, etc.) that is needed to overcome stigma may be difficult because of the spoiled identity. In short, stigma may be a chronic status until the individual accumulates the resources necessary to break the cycle.

Much of the existing literature concentrates on strategies that presumably lessen stigma and neglects to consider the extent to which stigma management strategies may perpetuate the individual's stigmatized status. Since Goffman's 1963 work, the assumption that stigma management strategies result in positive outcomes has prevailed. Researchers generally explore the ways in which stigmatized individuals protect their sense of self and attempt to gain social acceptance. The underlying assumption is that through these strategies, deviant individuals will achieve some degree of normalcy. Clearly, some actions benefit the individual's social standing. However, more attention should be given to the process through which stigma management strategies might result in the acquisition of characteristics that further spoil one's identity. Recognizing stigma management strategies as micro-level reactions to macro-level conditions elucidates how such strategies may lead to unintended and potentially negative consequences.

In the case of kids who are homeless, the structural disadvantages are significant. They come from poor families and are also marginalized because of their age, race, ethnicity, and gender. Unlike some of their privileged peers who may go to great lengths to distinguish themselves through dress, taste, and behavior, kids who are homeless experience what Goffman (1963, p. 5) calls "undesired differentness." Homeless kids are social outcasts because of the situation into which they were born or because of their parents' current predicament. Consequently, when we attempt to understand the homeless kids' strategies to manage their stigma, we must examine how structural disparities influence these strategies and how the nonhomeless interpret them. As we illustrate, although some of their stigma management strategies produced positive results with regard to their identities and social standing, others led to unanticipated and negative consequences. In effect, it is through the use of these strategies that kids who are homeless, like some stigmatized populations, unwittingly reproduce their stigmatization and cement their status loss.

RESEARCH SITE

The research site is an organization in San Francisco called A Home Away From Homelessness. Home Away serves homeless families living in shelters, residential motels, foster homes, halfway houses, transitional housing facilities, and low-income housing. The program operates a house in Marin County (called the Beach House), provides shelter support services, a crisis hotline, a family drop-in center (the Club House), a mentorship program, and, in conjunction with the San Francisco Unified School District, an afterschool educational program (the School House). Home Away is neither a typical homeless service agency nor a shelter.

Home Away serves a population of approximately one hundred five- to eighteen-year-old children annually. Fifty percent of the participants are boys and 50 percent are girls. The racial-ethnic breakdown of homeless families participating in Home Away programs is 40 percent African American, 30 percent white, 20 percent Latino, and 10 percent multiracial. All the kids participating in Home Away programs lived with at least one parent (none are runaways or homeless youth living independently), although some were essentially on their own as a result of parental neglect. Many of the kids suffered from a variety of physical, emotional, and developmental deficits, which is not surprising given the harsh conditions under which they live (see, e.g., Bassuk and Gallagher 1990; Rafferty 1991).

A majority of Home Away's families come from northern California; some families are from southern California and the Pacific Northwest. These families, often mother-only, typically come from chronically poor communities and usually become homeless after losing a job or stable housing. When extended kinship networks deteriorate as a result of poverty (Menjivar 2000; Roschelle 1997) and violence, these transient families are forced to rely on institutional forms of social support. There are primarily two types of families using the services of Home Away: short-term participants who experience brief spells of homelessness and the chronically homeless who use programs over several years. . . .

Chronically homeless families who cycled in and out of Home Away were the most frequent users of Home Away programs. Because of the severe housing shortage and exorbitant rents in San Francisco, many homeless families had their shelter stays extended past the three-month deadline. Some families eventually moved to motels, others doubled up with family members, and some shuffled between sub-standard apartments in violent neighborhoods. Many of the chronically homeless kids went to the Beach House on a weekly basis over the course of several years. As they got older, some of the teenagers who no longer wanted to go to the Beach House went to the Club House a few times a week. Some of these kids attended the afterschool educational program several times a week, but their transience made it difficult for many to stay in the program. During the past six years, nearly one thousand children and their parents have participated in various programs provided by Home Away.

METHODS AND DATA

The ethnographic component of our research included participant observation at the Beach House, the School House, and the family drop-in center where Anne collected the data and did volunteer work for four years (1995–99). During this time, she also attended meetings of relevant social service agencies at which she took copious notes and conducted observational research at residential motels, transitional housing facilities, and homeless shelters throughout the San Francisco Bay area. In addition, she conducted thousands of hours of informal interviews, sometimes asking leading questions (Gubrium and Holstein 1997) and sometimes just listening. Many of the informal interviews were taped and transcribed. In addition, she conducted and transcribed verbatim formal taped interviews with ninety-seven kids and parents who were currently or previously homeless. While in the field, when something that seemed profound about the homeless experience was uttered or occurred, Anne would go to the bathroom and quickly record it. This strategy was necessary because sitting in a room full of kids and simultaneously taking notes creates an artificial atmosphere that inspires skepticism and mistrust among kids who may already be wary of adults. Anne typically spoke her field notes and critical interpretations of the days' happenings into a tape recorder immediately after working with the kids. These tapes were eventually transcribed. . . .

Gathering ethnographic data among kids who are homeless presents a number of methodological challenges. The first is the difficulty of gaining entry, although this proved easier than expected. Anne developed rapport with the kids by spending considerable time with them at a variety of locations. Entry was further facilitated by the House Mother (a formerly homeless woman) who regularly conveyed to the kids her support for and fondness of Anne. This legitimacy was crucial because it allowed kids to be interviewed and observed in the context of their social group interactions. . . .

The second, related challenge is being attentive to their devalued social position. This awareness is particularly necessary because of our interest in how homeless kids manage their stigma. As noted, because these kids are poor, young, and predominantly racial-ethnic minorities, their behavior is often interpreted pejoratively even when it mirrors that of their nonhomeless peers. For this reason, it is important to record and respect the ways in which kids construct meanings and to not assume "the existence of a unidimensional external reality" (Charmaz 2000, p. 522). . . .

The third strategy, and challenge, was participating in impromptu conversations, which captured the ways kids managed their devalued social status. These gatherings, which Anne tape recorded and later transcribed verbatim, were not focus groups because they had neither an identifiable agenda nor a formal mediator. Rather, these were freewheeling conversations that allowed Anne to "remain as close as possible to accounts of everyday life while trying to minimize the distance between [herself] and [the] research participants" (Madriz 2000, p. 838). These conversations were particularly appropriate because they allowed for the expression of ideas in a forum where individuals felt comfortable speaking up. Furthermore, talking with kids in relaxed group settings in which the kids outnumber the adults, minimizes the inherent power differential between the adult researcher and the respondents and gives voice to those who have been subjugated.

STIGMA MANAGEMENT STRATEGIES

Goffman (1963, p. 3) suggests that stigma should be understood as "a language of relationships" as opposed to the attributes a person possesses. Moreover, he notes that "the stigmatized are not persons but rather perspectives" (p. 138). These ideas are particularly relevant for understanding the stigmatization of kids who are homeless. While these kids do not necessarily exhibit physical manifestations of their stigma, they are stigmatized because of their marginalized status in society. The discourse surrounding poverty and homelessness has a long history of blaming individuals for their predicament. Kids who may have been defined in the past as deserving of societal sympathy have more recently been recast as part of the legions of the undeserving poor.

Home Away kids were aware of their relationship to normative society and understood that others generally saw them as undesirable. Because of the large numbers of homeless in San Francisco and the visibility of the street homeless, there were numerous negative public discussions about the issue. The local newspapers constantly featured articles that were contemptuous of the homeless, and mayoral races often focused on ridding the city of its unsightly homeless population. The kids in the sample were attuned to the negative discourse surrounding homelessness and often expressed their anxiety about being demonized by the public. In fact, the kids spent many hours sitting in front of the fireplace discussing how hard it was for them to be routinely disparaged. As Dominic, an articulate sixteen-year-old, said, "Everyone hates the homeless because we represent what sucks in society. If this country was really so great there wouldn't be kids like us.". . .

Kids who are homeless manage their stigma in a variety of ways. One common typology of stigma management strategies assumes an in-group/out-group dichotomy (Anderson, Snow, and Cress 1994; Blum 1991). Although this typology is useful for understanding some stigmatized populations, the stigma management strategies of Home Away kids were more fluid and could not be so neatly categorized. As kids who are homeless make their way through the world, they transgress—yet simultaneously create—boundaries and often use similar stigma management strategies with both peers and strangers.

Although the in-group/out-group dichotomy did not fit our data, we developed an alternative schema that enabled us to better categorize the stigma management strategies Home Away kids use. We observed that the kids displayed two sets of strategies. The first set conformed to societal norms of appropriate behavior and aimed at creating a harmonious environment with both peers and strangers. In other words, these strategies represented attempts at establishing self-legitimation in both hostile and supportive environments. We refer to these as *strategies of inclusion* because they reflected the kids' desire to eradicate the boundary between a homeless and a nonhomeless identity. Through these inclusive strategies, Home Away kids hoped to be recognized simply as kids—without the stigmatizing label and discredited status of being homeless. The most common strategies of inclusion among Home Away kids were forging friendships, passing, and covering.

The second set of strategies were also attempts at gaining social acceptance but were not necessarily aimed at creating a harmonious atmosphere. We refer to these as *strategies of exclusion* because Home Away kids use them to distinguish themselves from both peers and strangers. With these strategies, kids who are homeless attempted to redress their spoiled identity by declaring themselves tougher, more mature, and better than others. These exclusive stigma management strategies included verbal denigration and physical and sexual posturing. Unlike strategies of inclusion in which kids used conciliatory tactics to be accepted, strategies of exclusion were aggressive and forceful attempts by the kids to blend in. These strategies largely reflected the kids' interpretations of socially acceptable behavior. However, given their disenfranchised social position, these behaviors were often perceived as maladaptive and threatening. As a result, when kids engaged in strategies of exclusion, they provided members of the dominant culture with a seemingly legitimate justification to further disqualify and disparage them . . .

STRATEGIES OF INCLUSION

Forging Friendships

The language of relationships that identifies kids who are homeless as a stigmatized group is so embedded in society that it is implicit in all their social relations. Even when these kids are in situations in which their homelessness is unknown or unimportant, they still feel the need to manage their spoiled identities. For example, one would expect that in the consonant social environment of the Beach House they would not have to engage in stigma management strategies. However, because their stigmatization is defined by their relationship to mainstream society, their stigma is always part of their identity....

At the Beach House, the kids attempted to construct a positive identity. Children were treated with dignity and respect by volunteers and were included in important decision-making processes. The services provided by Home Away gave kids an opportunity to experience childhood in a way most individuals take for granted. In addition to providing a break from the difficulties of living in shelters, transitional housing, residential motels, and so on, Home Away gave the kids a place to forge friendships with caring adults. Many of these relationships offered the only opportunity for homeless kids to obtain positive self-appraisals from nondisenfranchised adults. This supportive environment allowed for greater self-legitimation, gave the kids a respite from their stigmatization, and provided them with a sense of belonging (Brooks 1994). As a result of this inclusive strategy, many of the kids developed more favorable self-images. Silvia's discussion of her friendship with Tami, a volunteer, illustrates the importance of forging friendships as a way to manage stigma.

SILVIA: I call her whenever I feel really bad. She is so nice. She makes me feel better when I'm depressed and she never makes me feel like a freak because I'm homeless.

ANNE: Is that important to you?

SILVIA: Yeah. Most of the time I feel pretty bad about my life. I mean my mom has a new boyfriend every few weeks, we live in this nasty ass hotel, and I feel like everyone knows I'm a loser. Tami is always trying to make me feel better about my life. She tells me how smart and pretty I am and that I should feel good about being such a great older sister. Tami really cares about me and makes me feel better about myself. I don't know what I'd do without her.

Home Away also provided these kids with the opportunity to befriend other impoverished youths. By embracing other disenfranchised children, Home Away kids created a safe space where they could construct a positive identity and manage their stigmatized status. For example, when new kids were incorporated into Beach House activities, instead of rejecting them, the "old-timers" served as mentors and attempted to create a harmonious climate. Old-timers initiated the new kids into the program by teaching them the rules, showing them around, and introducing them to secret hiding places. Much like the traditional African American role of "old heads" and "other mothers," this mentor relationship allowed the old-timers to gain self-esteem. By initiating newcomers, they became experts at something and shared their wisdom with other impoverished kids. In addition, as the following conversation suggests, old-timers took great pride in their knowledge of the surrounding environment.

CARLOS: Hey, Hernán, don't go in the water without your life jacket.

HERNÁN: *No hay problema*—I'm a great swimmer.

CARLOS: It doesn't matter, man, the undertow is really strong and you can get swept under really easily.

HERNÁN: Don't be such a pussy.

CARLOS: Seriously man, people drown here all the time. Last week some kid almost died.

HERNÁN: Really?

CARLOS: Yeah, man. Listen I been coming here for two years and I've seen a lot of shit—you gotta believe me.

HERNÁN: *Gracias*, Carlos—thanks for watching my back. You really are a cool dude. I guess I better to stick close to you so I don't get into any trouble.

CARLOS: Hey, after we finish swimming Marcus and I can show you the secret hiding place.

HERNÁN: Cool.

Forging friendships with newcomers is an important stigma management strategy because it enables kids who are homeless an opportunity to transcend their discredited status and assume a role invested with interactional legitimacy. Through forging friendships, Carlos is able to negate feelings of worthlessness bestowed on him by society and feel like a valuable member of the kids' community. Similarly, Hernán's stigma is mitigated because he now feels like a legitimate member of Home Away. This strategy of inclusion anchors kids in a particular social group and subsequently provides them with a desired social identity that is conferred by significant others.

Passing

Goffman (1963) suggests that visibility is a crucial factor in attempts at passing. To pass successfully, an individual must make his or her stigma invisible so that it is known only to himself or herself and to other similarly situated individuals. Unlike their nonhomeless peers who have legitimate access to the public domain, kids who are homeless must often pass as nonhomeless as a means of appropriating heretofore unavailable and legitimate public space. One "passing" strategy kids in our study used entailed adopting the dress and demeanor of nonhomeless kids. Whenever clothing was donated to Home Away, the kids selected outfits based on style rather than function. It was extremely important that clothing and shoes looked new and were "hip." On numerous occasions, kids refused to take donated coats during the winter because they were ugly and out of style. The importance of fitting in is evidenced by the following conversation between Anne, Jamie, and Cynthia (Jamie's mom) while they were hanging out at Stonestown Mall.

JAMIE: Hey, do you think these people can tell we are homeless?

ANNE: No, how could they possibly know?

JAMIE: I don't know. I always feel like people are looking at me because they know I am poor and they think I am a loser.

CYNTHIA: I feel like that a lot too—it makes me feel so bad—like I'm a bad mother and somehow being homeless is my fault. I feel so ashamed

JAMIE: Me too.

ANNE: Jamie, what are some of the ways you keep people from knowing you are homeless?

JAMIE: I try and dress like the other kids in my school. When we get clothes from Home Away, I always pick stuff that is stylin' and keep it clean so kids won't know I'm poor. Sometimes it's hard though because all the kids try and get the cool stuff and there isn't always enough for every one. I really like it when we get donations from people who shop at the Gap and Old Navy. I got one of those cool vests and it made me feel really great.

ANNE: Is it important for you to keep your homelessness a secret?

JAMIE: Yeah, I would die if the kids at school knew.

Home Away kids also attempt to pass by using code words. Their use of code words illustrates the importance of language as a symbolic indicator of membership in a social group. As Herman (1993) argues, individuals selectively withhold or disclose information as a way to maintain secrecy about their negative attributes. Knowing the rules of social interaction and the symbolic importance of words, Home Away kids resist language that reveals their homelessness to others. They purposefully choose words they associate with middle-class life to articulate their social reality and simultaneously conceal their marginalized social existence. The following impromptu conversation illustrates this point.

ROSINA: I hate that kid Jamal that lives with us in Hamilton [Family Shelter], you know his cot is three over from mine.

SHELLEY: Sssh, be quiet, someone will hear you and then people will know we are homeless.

ROSINA: I don't care.

SHELLEY: I do.

LINDA: So do I. You should say that you don't like Jamal who lives three houses down—that way people will think you are talking about a kid in your neighborhood.

ANNE: Is that how you guys talk in school?

SHELLEY: Yeah, we say things like that and we make up other stuff so people don't know we live in the shelter.

LINDA: Or in the motels.

SHELLEY: When we talk about our caseworkers we say our aunts and when we talk about shelter staff we call them our friends.

ANNE: That is a really clever way of keeping people from knowing you're homeless.

ROSINA: It totally is, but then you can't invite friends over after school because they think you live in a house.

LINDA: We can't hang out with kids who are not poor. There is no way I would invite a kid from school over to my room in that skanky motel we live in, I'd be so embarrassed.

SHELLEY: Yeah, we can't make friends with a lot of kids because how can we bring them to the shelter after school?

This example illustrates how homeless kids use words to construct a positive identity, protect their sense of self, and feel integrated with the larger society. Moreover, it shows that language reflects and reproduces power relations in society. Interestingly, although these kids use code words to mask their stigmatized status and construct a positive identity, their language choices disqualify them from participating in normative social interactions. These verbal gestures reflect the extent to which homeless kids are in danger of having their homelessness revealed. When they do not use code words, their stigmatized identity will be evident, and they will be unlikely to befriend middle-class children. Alternatively, when they do use code words, they reduce their chances of befriending nonhomeless kids because they risk exposing their stigma. In managing stigma through passing, the threat of discovery is ever present. Although the kids desire acceptance in mainstream society, the threat of discovery leads them to monitor and curtail their social involvement.

Covering

Individuals engage in covering when they attempt to minimize the prominence of their spoiled identity. Covering allows individuals to participate in more normative social interactions by reducing the effects their stigma elicits (Goffman 1963). Unlike passing, the point of covering is not to deny one's stigma but rather to make it less obtrusive and thereby reduce social tension (Anderson, Snow, and Cress 1994). Home Away kids often engaged in this stigma management strategy when they became friends with nonhomeless kids who knew about their predicament. Home Away kids would especially use this strategy around the parents of their non-homeless friends, and they would often ask the staff to advise them on appropriate dress or behavior when they were going out with non-homeless kids and their families.

One Home Away kid, Ellie, became a close friend of an affluent schoolmate, Carol, who lived in the Marina district of San Francisco. Ellie was honest with Carol and her family about where she lived. Still, when Ellie went to Carol's house for dinner, she wore the least tattered, cleanest clothes she owned and would eliminate her ghetto swagger and jargon. As she said to Anne, "They know I'm homeless and all, but I don't want to act like I'm homeless. I don't want to embarrass them. You know my clothes aren't that fancy and I'm used to eating at the shelter where everyone is talking really loud and eating with their hands and being kinda sloppy." To fit in and be "normal," Ellie decided to minimize the obvious manifestations of her homelessness. . . . For the most part, these strategies had the intended effect of protecting the kids' identities by offering them a degree of social legitimacy and by aligning them with nonhomeless individuals. In spite of the chaotic nature of their lives and their awareness of their stigmatization, it is noteworthy that these kids attempt to conform to society through socially accept-able means. In this sense, Home Away kids do not fit the traditional conception of the homeless as disaffiliated and socially isolated individuals. They seem surprisingly resilient in their attempts to develop relationships with nonhomeless people in spite of being denigrated by them. . . .

STRATEGIES OF EXCLUSION

Verbal Denigration

When individuals face a social world that labels them deviant, they are likely to fight back by maligning others as a way to augment their self-esteem. Termed "defensive othering" by Schwalbe et al. (2000), this type of stigma management strategy was common among Home Away kids. Many kids in the sample protected their sense of self by verbally denigrating other stigmatized groups such as homosexuals and homeless street people. This was a form of identity work that allowed homeless kids to distance themselves from the stigmatized "other" (Snow and Anderson 1987) and proclaim their superiority over these similarly disparaged groups. Because homeless kids recognize that they are prob-lematic in the eyes of society, their denigration of other stigmatized groups is a "largely verbal effor[t] to restore or assure meaningful interaction" and align themselves with the dominant culture (Stokes and Hewitt 1976, p. 838).

Along these lines, many of the kids were homophobic and freely expressed their disgust for homosexuality. This attitude can be attributed to the kids' (arguably accurate) interpretation of societal norms and values with regard to homosexuality. By portraying gays and lesbians as "freakin' faggots" and "child molesters," the kids were placing others lower than themselves in the social hier-archy. The kids' stigmatized status was deflected onto others, thereby bolstering their own sense of self. This exclusionary strategy was especially interesting in the context of San Francisco—a city in which there is a large population of gay and lesbian activists and citizens, many of whom have achieved positions of power and prestige and who themselves often denigrate the homeless. The following conversation illustrates the pejorative discourse Home Away kids often used when talking about gays and lesbians.

ANTOINE: Look at those homos, they make me sick.

FRED: Yeah, they be all trying to grab your ass when you walk by.

ANTOINE: I know man, they'd love a piece of us but I'd make 'em suck on a bullet before I'd let 'em suck on my thang.

FRED: People think we are nasty because we live at the Franciscan, but at least we don't do little boys. I mean those guys are freaks.

ANTOINE: I hear ya, I'd rather be a dope fiend livin' in the Tenderloin than a fuckin' faggot! [Loud laughter]

FRED: I'd rather be a dope fiend, livin' in the Tenderloin, selling crack to ho's than be a fuckin' faggot! [Louder laughter]

Homeless kids also spoke disparagingly about homeless street people. Homeless kids in our sample lived in shelters, residential motels, transitional housing, foster homes, and other institutional settings. Though many of them have spent a night or two sleeping in parked cars, in a park, or on the street with their parents, their homeless experience has taken place primarily in some type of sheltered environment. The main reason they have not spent the majority of their lives on the streets is because there are many more programs for homeless families in San Francisco than there are for childless homeless adults. Essentially, the only factor preventing their parents from becoming part of the homeless street population is them. Despite this, Home Away kids often made fun of homeless street people.

ROSITA: Man look at those smelly street people, they are so disgusting, why don't they take a shower.

JALESA: Yeah, I'm glad they don't let them into Hamilton with us.

ROSITA: Really, they would steal our stuff and stink up the place! [Laughter]

JALESA: Probably be drunk all the time too.

ROSITA: Yeah, and smokin' crack all night long!

Ironically, the mothers of both these girls have struggled with drug and alcohol addiction and have had several episodes of homelessness. By distancing themselves from the "true" social pariahs of San Francisco, Jalesa and Rosita mitigate their own stigmatized status and maintain some semblance of a positive self. Further, identifying a group as more discredited than themselves allows the girls to feel superior to "those losers on the street."...

Physical Posturing

Physical posturing is another form of identity work that grants homeless and non-homeless kids a momentary degree of empowerment. Kids who are homeless use physical posturing to feel powerful in their interaction with nondisenfranchised kids, a feeling that is rarely available to them. It has long been recognized that many adolescents are filled with insecurity as they attempt to establish their identities. However, middle- and upper-class kids have more socially acceptable means than low-income ones to demonstrate to others (and to themselves) that they are important. As Anderson (1999, p. 68) notes, privileged kids "tend to be more verbal in a way unlike those of more limited resources.

In poor inner-city neighborhoods verbal prowess is important for establishing identity, but physicality is a fairly common way of asserting oneself. Physical assertiveness is also unambiguous." In this sense, Home Away kids' use of physical demeanor is a stigma management technique that emerges from their social structural location and protects them from a hostile world.

One common manifestation of this physical posturing was the use of body language. When encountering nonhomeless children, kids in our study often adopted threatening postures by altering their walk, speech, and clothing to mimic "gangsta" bravado. In short, these kids adopted what Majors (1986) refers to as the "cool pose." For example, to get from the parking lot to the beach, we had to walk over a very small, narrow wooden bridge. As we crossed the bridge, we often passed kids coming from the opposite direction. As soon as the Beach House kids spotted other kids coming toward them, they would immediately change their demeanor. They metamorphosed from sweet cuddly kids to ghetto gangstas. The kids grabbed their pants and pulled them down several inches to mimic the large baggy low-riding pants of the gangbanger. They turned their baseball caps around so that the brim was in the back and swaggered in exaggerated ways. Their speech became filled with ghetto jargon, and they spoke louder than usual. When encountering other kids, the topic of conversation on the bridge would also suddenly shift from how much fun we just had to the fate of a gang member they knew who had just been arrested.

These self-presentations mirror what Anderson (1990, p. 175) calls "going for bad." Anderson argues that this intimidation tactic is clearly intended to "keep other youths at bay" and allow disempowered kids to feel tougher, stronger, and superior. Like verbal denigration, this strategy of exclusion also helped Home Away kids to lessen their stigma. Interestingly, these behaviors clearly imitate images of adolescence that pervade popular culture. In music videos, movies, television shows, video games, and so on, images of baggy pants, baseball caps, and swaggering walks abound. Although this strategy had the intended effect of empowering Home Away kids by intimidating their nonhomeless peers, it also reinforced their stigmatization. As Nonn (1998, p. 322) notes, "While the coping mechanism of cool pose weakens the stigma of failure, it undermines identity ... and contribute[s] to their own alienation from other groups in society." Though some kids recognized that their behavior may stigmatize them further, they still engaged in such posturing because it was one of the few areas in their lives over which they had some control. ...

Sexual Posturing

Some Home Away kids used sexuality to validate themselves. Although nonhomeless youth also engage in sexual posturing to establish a sense of self in their social groups, Home Away kids engage in it more explicitly and overtly than their nonhomeless peers. For Home Away kids, sexual posturing and promiscuity articulate the sexual exploitation and violence they experienced both within and outside the family. As researchers have documented, victims of child sexual abuse often resort to promiscuity in adolescence and adulthood.

Molestation by older men was not uncommon, and many of these kids learned at an early age to use their sexuality to gain status among their peers. For example, several young women in the sample dressed and behaved in overtly sexualized ways that surpassed what we expect from the budding sexuality of "normal" adolescents. In one poignant incident, a fourteen-year-old girl was teaching a younger girl how to perform oral sex on an eleven-year-old boy. By bragging to his friends about his newfound sexual prowess and achieving prominence among his peers, this boy exhibited an exaggerated masculine alternative commonly found among lower-income racial-ethnic males (Oliver 1984). The girls involved increased their status by demonstrating a level of sexual maturity that is typically associated with older women.

In another example, Patti, a thirteen-year-old, discussed the self-esteem she gained through her sexual posturing.

> I mean not everyone is pretty enough to get a man to pay for a room for the night. At least I know that I can get anything I need from a man because I am totally hot. I get lots of attention and some of my friends are jealous of me because they know I got it going on. Some girls are so butt ugly no man would ever even want to fuck them anyway—but not me—they think I'm a hotty—you know what they say—if you got it flaunt it!

As these examples illustrate, sexual posturing is an exclusionary tactic kids use to distinguish themselves from their peers and to lessen their stigmatization. Research has shown that some poor racial-ethnic girls invoke a discourse of sexuality in order to negotiate and construct a more empowering identity (Emihovich 1998), and others may attempt to achieve similar results through motherhood (Luker 1996). In a society bombarded with images of hyper-sexuality, it is not surprising that Home Away kids imitate these sexual scripts in an attempt to strengthen their social standing. . . .

CONCLUSION

In this article, we examine the stigma management strategies of kids who are homeless. Although researchers of stigma management have studied a variety of populations, our work contributes to existing knowledge by including this previously neglected group. By examining the stigma management strategies homeless kids use, we address a number of gaps in the literature. First, our work emphasizes the need to acknowledge stigma and homelessness as structural locations. Much like homelessness, we must posit stigma not as an individual attribute but as a relationship to the social structure (Goffman 1963; Link and Phelan 2001). This perspective is noteworthy because it attends to how individuals' behaviors are both informed by and interpreted through their social structural location. This insight is particularly true of homeless kids who are oppressed socially because of their age, race, ethnicity, and social class. Kids who are homeless encounter and interpret a world that is characterized by hunger, uncertainty, chaos, pain, drug abuse, violence, sexual abuse, degradation, and social rejection, among other things. This social reality epitomizes the violence of poverty and ultimately results in a lack of consistency, stability, and safety in their lives. Furthermore, these kids exist in a constrained public domain and are forced to carve out their own space in a limited urban environment that is generally hostile to them. It is in this environment that kids who are homeless engage in social interactions that aim to construct positive identities and overcome their discredited status.

By positing stigma and homelessness as social structural phenomena, our work also illustrates the processes whereby stigma may become a chronic status. In contrast to much of the literature on stigma management, we found that the tactics used by Home Away kids sometimes had the unintended effect of perpetuating their spoiled identities. Although their actions were attempts at protecting their sense of self and gaining a degree of social acceptance, strategies such as sexual posturing and verbal denigration substantiated societal stereotypes of kids who are homeless as violent, disrespectful, and dangerous delinquents to be avoided. Obviously, it is not these kids' intention to engage in behaviors that perpetuate their stigmatized identities and their disenfranchised positions. Rather, it is their social structural location that offers them limited opportunities to exert their agency in a socially acceptable way.

In attempting to negate their stigmatization, Home Away kids engaged in strategies that often imitated the behaviors of mainstream youth. By interpreting popular culture and the social interactions of their nonhomeless peers, homeless kids behaved in ways that are often exhibited, condoned, and even rewarded when enacted by their more privileged peers. In mimicking this behavior, Home Away kids naively expected to obtain a modicum of social acceptance. Unfortunately, they failed to understand that the parameters of acceptable and unacceptable behavior are mediated by one's social location. In other words, the deviant behaviors middleclass and affluent peers engage in have different consequences than when they are perpetrated by homeless kids (Chambliss 1973).

REFERENCES

Anderson, Elijah. 1990. *Streewise: Race, Class, and Change in an Urban Community*. Chicago: Chicago University Press.

———. 1999. *Code of the Street: Decency, Violence and the Moral Life of the Inner City*. New York: Norton.

Anderson, Leon, David A. Snow, and Daniel Cress. 1994. "Negotiating the Public Realm: Stigma Management and Collective Action among the Homeless." *Research in Community Sociology* (Supplement 1): 121–43.

Bassuk, Ellen L., and Ellen M. Gallagher. 1990. "The Impact of Homelessness on Children." *Child and Youth Services* 14(1): 19–33.

Blum, Nancy. 1991. "The Management of Stigma by Alzheimer Family Caregivers." *Journal of Contemporary Ethnography* 20(3): 263–84.

Brooks, Robert B. 1994. "Children at Risk: Fostering Resilience and Hope." *American Journal of Orthopsychiatry* 64(4): 545–53.

Chambliss, William J. 1973. "The Saints and the Roughnecks." *Society* 2(1): 24–31.

Charmaz, Kathy. 2000. "Grounded Theory: Objectivist and Constructivist Methods." Pp. 509–35 in *Handbook of Qualitative Research*, edited by N. K. Denzin and Y. S. Lincoln. Thousand Oaks, CA: Sage.

Emihovich, Catherine. 1998. "BodyTalk: Discourses of Sexuality among Adolescent African American Girls." Pp. 113–33 in *Kids Talk: Strategic Language Use in Later Childhood*, edited by S. M. Hoyle and C. Temple Adger. Oxford: Oxford University Press.

Goffman, Erving. 1963. *Stigma: Notes on the Management of Spoiled Identity*. New York: Simon and Schuster.

Gubrium, Jaber F., and James A. Holstein. 1997. *The New Language of Qualitative Method*. New York: Oxford University Press.

Herman, Nancy J. 1993. "Return to Sender: Reintegrative Stigma-Management Strategies of Ex-Psychiatric Patients." *Journal of Contemporary Ethnography* 22(3): 295–330.

Liebow, Elliot. 1993. *Tell Them Who I Am: The Lives of Homeless Women*. New York: Penguin.

Link, Bruce G., and Jo C. Phelan. 2001. "Conceptualizing Stigma." *Annual Review of Sociology* 27: 363–85.

Luker, Kristine. 1996. *Dubious Conceptions: The Politics of Teenage Pregnancy*. Cambridge, MA: Harvard University Press.

Madriz, Esther. 2000. "Focus Groups in Feminist Research." Pp. 835–50 in *Handbook of Qualitative Research,* edited by N. K. Denzin and Y. S. Lincoln. Thousand Oaks, CA: Sage.

Majors, Richard. 1986. "Cool Pose: The Proud Signature of Black Survival." *Changing Men: Issues in Gender, Sex and Politics* 17: 5–6.

Menjivar, Cecilia. 2000. *Fragmented Ties: Salvadoran Immigrant Networks in America*. Berkeley: University of California Press.

Nonn, Timothy. 1998. "Hitting Bottom: Homelessness, Poverty, and Masculinity." Pp. 318–27 in *Men's Lives*, edited by M. S. Kimmel and M. S. Messner. Boston: Allyn and Bacon.

Oliver, W. 1984. "Black Males and the Tough Guy Image: A Dysfunctional Compensatory Adaptation." *Western Journal of Black Studies* 8: 201–2.

Rafferty, Yvonne. 1991. "Developmental and Educational Consequences of Homelessness on Children and Youth." Pp. 105–39 in *Homeless Children and Youth*, edited by J. H. Kryder-Coe, L. M. Salamon, and J. M. Molnar. New Brunswick, NJ: Transaction Publishers.

Roschelle, Anne R. 1997. *No More Kin: Exploring Race, Class, and Gender in Family Networks*. Thousand Oaks, CA: Sage.

Schwalbe, Michael, Sandra Godwin, Daphne Holden, Douglas Schrock, Shealy Thompson, and Michele Wolkomir. 2000. "Generic Processes in the Reproduction of Inequality: An Interactionist Analysis." *Social Forces* 79(2): 419–52.

Snow, David A., and Leon Anderson. 1987. "Identity Work among the Homeless: The Verbal Construction and Avowal of Personal Identities." *American Journal of Sociology* 92: 1336–71.

———. 1993. *Down on Their Luck: A Study of Homeless Street People*. Berkeley: University of California Press.

Stokes, Randall, and John R. Hewitt. 1976. "Aligning Actions." *American Sociological Review* 41: 838–49.

31

Collective Stigma Management and Shame: Avowal, Management, and Contestation

DANIEL D. MARTIN

The way organizations help people collectively manage their stigma is the focus of Martin's study of three stigma management groups associated with obesity: Overeaters Anonymous (OA), Weight Watchers (WW), and the National Association to Advance Fat Acceptance (NAAFA). Martin compares the way these groups frame their purpose and set their goals. While OA and WW help members navigate their way through the world and conform to the prevailing body norms, NAAFA fights these norms and seeks to redefine them. Martin ties the characteristics of these organizations to the way they deal with shame, avowing it, managing it, and contesting it. The selection offers an excellent comparative view of the varied rationales and strategies associated with differently structured collective stigma management groups. You may want to think about the breakdown of participants into each of these groups by gender and form your own analysis of why certain groups attract higher rates of men versus women.

. . . . Both cross-cultural research and feminist studies have observed the connection between shame and gender identity, demonstrating that, while women are more susceptible to body shame, for men the emotion is associated with failure to live up to culturally prescribed norms of masculinity (Gilmore 1987; Horowitz 1983).

Feminist scholars investigating the U.S. beauty culture have observed its emergence in the 1870s (Banner 1983; Schwartz 1986) as new appearance norms were created and distributed by industries such as cinema (Featherstone 1982), advertising (Ewen 1976), and scale and corset companies (Schwartz 1986). As a result, the social formation that exists at the end of the twentieth century is a "cult of thinness" (Hesse-Biber 1996) that promotes and normalizes slenderness and its attendant anxieties (Bordo 1993). This cult has far-reaching effects. Barrie Thorne and Zella Luria's (1986) research on children's games involving cross-sex interaction reveal that girls' bodies are more sexually defined and accrue more penalties for an overweight appearance than do boys'. This pattern is consistent with findings of cross-cultural research on mate selection that demonstrate that men evaluate the social value of women more heavily in terms of appearance and that women evaluate the social value of men in terms of their occupational status and earning potential (Buss 1989). In the United States, where fat is stigmatizing (Allon 1982; Cahnman 1968), feelings of shame may result in debilitating eating disorders. Research on adolescent girls has demonstrated that the internalization of beauty norms is responsible for producing high degrees of body dissatisfaction and a profound sense of body shame (McFarland and Baker-Baumann 1990) leading to bulimia and anorexia nervosa (Rodin, Striegel-Moore, and Silberstein 1990, p. 362). Of course, the fat content of bodies, along with the degree to which it might be stigmatized and accompanied by body shame, is variable. A slightly overweight body may be experienced as an object of great shame and the focus of disordered eating, or it may be experienced as only a "minor bodily stigma" (Ellis 1998).

However, ... the research on ... feminist studies of body shame or sociological studies on shaming as a form of "strategic interaction" (Goffman 1969) have [not] addressed how organizations attempt to assist their members in managing shame....

Drawing upon Erving Goffman's (1986) frame analysis, I assess how shame is managed within three different "appearance organizations" through the discursive and bodily strategies that they supply their members. By appearance organizations, I mean organizations for whom the physical appearance of members is a primary concern. These organizations include Overeaters Anonymous (OA), Weight Watchers, Inc. (WW), and the National Association to Advance Fat Acceptance (NAAFA). The theoretical and substantive questions concerning these organizations are twofold. First, what kinds of organizational frames do these organizations offer their members to make sense of shame? Second, what strategies are used within the organizations in managing or contesting experiences of shame?

THE ORGANIZATIONS

Given the sex ratio of the organizations, all three of the appearance organizations I studied might properly be recognized as "women's organizations." Therefore, the present analysis of shame is an analysis of women's shame. During interviews and participant observation, I found that men rarely talked about shame or embarrassment over the body while women seemed to be continually, acutely aware of it. Based upon information reported by the organizations, participation rates reveal significant sex differences in membership (Table 31.1). These differences were also revealed in the accounts that women and men constructed about joining and participating in the organizations and about the gender relations that were negotiated within them—a topic beyond the scope of the present article (Martin 1995).

TABLE 31.1 Organizational Membership

	Sex Differences in Participation		
	WW	**NAAFA**	**OA**
Women	95%	78%	86%
Men	5%	22%	14%
Total*	100%	100%	100%

*These figures represent national membership rather than participation rates in local chapters.

At the beginning of data collection, Weight Watchers Inc. (WW) was the leading national weight loss organization in the United States, with annual revenues of approximately $1.3 billion (Weber 1990, p. 86). Of its 15 million members worldwide, approximately 14.3 million are women. In contrast, the exact size and composition of Overeaters Anonymous (OA), a program for compulsive overeaters, remains undetermined, even though the organization estimates that 86 percent of its U.S. members are female (OA 1992, p. 1). Because anonymity in the twelve-step program is strictly enforced, there is no comprehensive roster of OA members. Currently, the number of OA groups is estimated at over seven thousand worldwide (OA 1987, p. 10), but group size varies from five to twenty-five people, making estimates of total membership difficult.

The National Association to Advance Fat Acceptance (NAAFA) describes itself as a civil rights organization in the "size rights movement." It also defines itself as a social support organization and as a self-help group (NAAFA 1990, pp. 1–6). Unlike members of OA and WW, the NAAFA members in the local chapter I studied engaged in a variety of social and political activities. These ranged from informal parties and organized dances, to speaking at community conferences, to holding protests and conducting write-in campaigns. The national organization, founded in 1969, has headquarters in Sacramento, California, and a membership of 2,500 to 3,000 members. NAAFA consists of a loosely federated set of local chapters and special interest groups, including gay and lesbian groups, fat admirers groups, couples groups, singles, teen and youth groups, groups for super-sized and mid-sized individuals, and feminist groups.

METHODS

Over the course of approximately two years, I conducted participant observation and in-depth interviews at Weight Watchers and NAAFA. In attempting to gain access to OA, I sought out an acquaintance and long-standing group leader in OA who was centrally located in local OA networks. She was helpful in providing me with extensive information about the organization, its program, and its local chapters. Believing that her interest in my project, her sensitivity to research needs, and the fact that we had worked together indicated an openness to

TABLE 31.2 Interviews

	Women	Men
WW	20	5
NAAFA*	8	4
OA**	41	8
Total	69	17

*Defines the entire population of active chapter members during the period of my observation.

**Includes recorded personal stories of members from the OA audiotape library.

facilitating entree, I directly broached the topic. My request for an introduction into the closed meetings and access to OA members was met, instead, by recalcitrance motivated by her desire to ensure the anonymity of local members. I was frozen out of the local chapters and was unable to conduct participant observation in groups other than the publicly available "open meetings." A couple of months later, I learned of a local OA audiotape library. I selectively sampled personal stories and presentations that had been taped at national and local conferences and then were produced for OA members as well as members of Alcoholics Anonymous. I utilized the strategy of "theoretical saturation" recommended by Barney G. Glaser and Anselm L. Strauss (1967), selecting audiotapes for analysis until the additional data yielded few additional insights. I was assisted by the proprietor of the service, who was instrumental in locating tapes containing life stories that were as complete as possible.[1] I conducted interviews with all of the active members of NAAFA after I had established myself with the group and had been conducting participant observation and with members of Weight Watchers who responded to ads that I had run in local newspapers. The total number of interviews for all three organizations is in Table 31.2. Because the organizations themselves constituted the sampling frames, little variation in race, class, or ethnicity was provided within the samples. All of the informants in this study were white and most were middle income.

DATA COLLECTION, EMBARRASSMENT, AND SHAME

Because my own bodily appearance serves as a "dis-identifier" (Goffman 1967), a symbol belying a legitimate, potential, future claim for group membership in WW, OA, and NAAFA, I intentionally gained about twenty-five pounds before participation. I hoped to learn as much as I could about members' own meanings and feelings of the body and participation as well as how the organizational "frames" that were established shaped emotional experience and participation. Having gained my undergraduate degree on a wrestling scholarship attuned me to the issues of extreme dieting, binge-eating, and weigh-ins, which could similarly be found in the lived experience of Weight Watchers, NAAFA, and OA members. While my own weight gain facilitated somatic insight into the lives and meanings of members of OA and Weight Watchers, my burgeoning body size was clearly not comparable to the experience of members of NAAFA. Because I was not subjected to the stigmatization or public inconvenience suffered daily by NAAFA members (some of whom weighed in excess of three hundred pounds) my own "fat identity" never fully crystalized. Indeed, if anything, I was considered a novelty among friends, peers, and colleagues who issued positive sanctions for my "sociological commitment."

At Weight Watchers, I regularly attended weekly meetings, which entailed a logical stream of activities once members entered the meeting place. These included paying weekly dues, being weighed on physician's scales by WW personnel, receiving organizational literature concerning meal plans along with myriad other promotional materials distributed by the organization (such as the monthly newspaper), and finally selecting a seat in the meeting area. As a participating observer, I experienced the anxiety that members later recounted in interviews about "facing the scale," that is, weigh-ins. Because weigh-ins take place in semipublic space, it is possible that queuing members will learn of one's weight, increasing the anxiety that is already present for some members. Having failed weigh-in several times by gaining weight, I was struck by the capacity for the ritual to evoke, simultaneously, feelings of dependency and embarrassment. WW personnel are aware of these feelings, and they commonly query deviating members about the possible causes of weight gain. They also join in the construction of accounts and remedial work directed at either exonerating deviating members or diminishing their presumed culpability.

ORGANIZATIONAL FRAMES

Experiences of shame, humiliation, and embarrassment can be found in the life stories of members of all three appearance organizations. What establishes the meaning of these experiences is the organizational "frame" (Goffman 1986) within which the experiences of members are socially and discursively organized Table 31.3. A "frame," according to Goffman (1986, pp. 10–11), is a set of "definitions of situations" that are "built up in accordance with the principles of organization which govern events—at least social ones—and our subjective involvement in them." *Organizational* frames are those definitions constructed and maintained by organizational actors within which experience, interaction, and communication are structured and rendered both personally and organizationally meaningful. For the programs of organizations to be subjectively meaningful, they must facilitate a linkage of individual interpretations to organizational meanings and definitions, that is, "frame alignment" (Snow, Rochford, Worden, and Benford 1986).

In the case of these appearance organizations, the meanings, as well as the strategies that are dispensed to members for dealing with shame, reflect the structure, agenda, and ideology of the organization, its "frame." The organizational frames of Weight Watchers, OA, and NAAFA represent three unique sets of social and discursive practices developed in accordance with different historical

T A B L E 31.3 Shame Work: A Comparison of Organizational Frames

	Overeaters Anonymous	Weight Watchers	NAAFA
Type of Organization	Twelve-step program	Multinational corporation	Civil rights organization
Organizational Frame	Redemption	Rationality	Activism sociality
Organizational Goal	Support members in abstinence from overeating	Help clients lose weight/profit	Organize against "size discrimination"
Source of Shame	Compulsive overeating/fat body	Fat body/body image	Societal definitions of beauty, cultural appearance norms
Consequence of Shame	Inhibits recovery	Leads clients to exit program	Inhibits activism
Shame Strategy	Shame avowal/ expiation	Shame management	Contestation of shame
Approach to Shame	Self-transformation	Bodily transformation	Societal transformation
Ritual for Shame Removal	Shame avowal	Dieting	Identity announcement/ public confrontation
Meaning of the Body	Body as symptomatic display of spiritual deficit; thinner body as a symbol of regaining control over personal, spiritual, and emotional life	Body as a site of contestation over appetite: locus of self-control and self-indulgence	Body as symbol of self-acceptance; political transformation
Body as Medium of Communication	"Body announcement" used as evidence of program efficiency; thin bodies as part of personal testimonies	"Body announcement" via fashion shows; display and glorification of thin bodies	"Body announcement" via fashion shows; display and glorification of fat bodies
Vocabulary of Motive	Overeating as disease	Overeating due to irrational management and lack of education; program as a skilling process for a total lifestyle change	Genetic accounts: large body a product of nature; program for empowerment; stories as "Oppression Tales"

circumstances and audiences. Thus, while the autobiographical content of participants' shame experiences seemed to bear great similarity across the organizations, the "organizational" meaning of these experiences and the strategies employed to manage or alleviate them differed vastly.

The Organizational Frame of OA: Redemption

The Overeaters Anonymous program is a redemptive model of treatment directed at developing a spiritual consciousness though a therapeutic group process. OA meetings, whether open to the public ("open meetings") or open to OA members only ("closed meeting"), are typically held in public facilities, often churches. Meetings begin with a recitation of OA's preamble and the evening's speaker testifying to the effectiveness of the OA program in her or his life. Most meetings end with members joining in the "serenity prayer."[2] OA explicitly acknowledges its reliance upon the Alcoholics Anonymous program of recovery, including the Twelve Steps and the Twelve Traditions developed by Bill W. and other AA founders. Reproduced in most literature published by the organization, the twelve-step program serves as a blueprint for achieving abstinence from compulsive overeating (OA 1990, p. 114).

The Twelve Steps of Overeaters Anonymous

1. We admitted we were powerless over food—that our lives had become unmanageable.
2. Came to believe that a power higher than ourselves could restore us to sanity.
3. Made a decision to turn our will and our lives over to the care of God *as we understood him.*
4. Made a searching and fearless moral inventory of ourselves.
5. Admitted to God, to ourselves and to another human being the exact nature of our wrongs.
6. Were entirely ready to have God remove all these defects of character.
7. Humbly ask Him to remove our shortcomings.
8. Made a list of all persons we had harmed, and became willing to make amends to them all.
9. Made direct amends to such people wherever possible, except when to do so would injure them or others.
10. Continued to take personal inventory and when we were wrong, promptly admitted it.
11. Sought through prayer and meditation to improve our conscious contact with God *as we understood him,* praying only for knowledge of his will for us and the power to carry that out.
12. Having had a spiritual awakening as the result of these steps, we tried to carry this message to compulsive overeaters and to practice these principles in all our affairs.

Essentially the organization prescribes a spiritual program of recovery for what it considers to be an "incurable disease"—compulsive overeating. The concept of disease represents OA's primary "vocabulary of motive" (Mills 1940) that alleviates members of responsibility for their unhealthy overeating: "OA believes that compulsive overeating is an illness—a progressive illness—which cannot be cured but which, like many other illnesses, can be arrested.... Once compulsive overeating as an illness has taken hold will power is no longer involved because the suffering overeater has lost the power of choice over food" (OA 1988, p. 2). In defining "compulsive overeating," OA tells the prospective member that "in OA, compulsive overeaters are described as people whose eating habits have caused growing and continuing problems in their lives. It must be emphasized that only the individuals involved can say whether food has become an unmanageable problem" (OA 1988, p. 2).

The redemptive aspect of OA's program is both ideological and structural, as well as observable in relationships that members have with their sponsors. The alignment of personal with collective definitions of compulsive overeating is accomplished, in part, with the assistance of an OA sponsor, a person with substantial "program experience" who serves as the member's confessor, friend, advocate, and spiritual leader as well as a purveyor and mediator of organizational ideology. Sponsors, along with other OA members, are part of the emotion management team that members may use in expressing and policing emotions and securing social support. Redemption from compulsive overeating and an obsessive-compulsive self, according to OA members, is facilitated by sponsors and fellow group members as one "works through denial"—that is, gives up one's idiopathic justifications for compulsive overeating and the avoidance of unpleasant emotions in favor of collective definitions and meanings found in OA's twelve steps and twelve traditions.

The Organizational Frame of NAAFA: Activism

Throughout its history, NAAFA has mobilized lobbying efforts for antidiscrimination bills with state legislatures,[3] held rallies, conducted protests,[4] and brought cases of discrimination to the attention of the mass media. The objective of such action is the creation and enforcement of human rights provisions under which fat people might be identified as a protected group. Yet a bifurcation of NAAFA's primary objectives can be found in the orientation literature for new members: NAAFA refers to itself as both a "human rights organization" (NAAFA 1990, p. ii) and "a self-help group" (NAAFA 1990, p. 2). For NAAFA members both intensive involvement in social activities and activist participation constitute the dominant frame of meaning or, in the words of NAAFA members, "a way of life." Activism, as defined by the organization, includes a broad array of personal and political activities that are marked by a displayed willingness to engage in confrontation. Such activism, as described by the organization (NAAFA 1990, p. 3–4), may include:

Personal Activism: Educating people around you, not letting negative comments about your weight or preference go unchallenged, and your life as a "role model" are all forms of personal activism.

Legislative Activism: is specific activist work that leads to changes in the law. Considering the amount of time, effort, and expertise required in such undertakings, it deserves a category all its own.

Advocacy Activism: includes letter writing and other forms of communication with the "powers that be" regarding your opinions. Did you like how the fat person was portrayed on the TV show? Are you happy with the article in the magazine? Did the salesperson treat you with disrespect because of your fat? Writing letters of praise or protest to the station, the magazine, or the store owner are examples of advocacy activism.

The organizational frame of activism in NAAFA represents a way in which participants situate themselves, emotionally and practically,[5] within the ideological contours of the organization. It is a "relevance structure" through which not only social or political events but also interpersonal lines of action are defined, evaluated, and constructed.

The Organizational Frame of WW: Rationality

The weekly Weight Watchers meetings were held in weight loss centers located in shopping malls, high-rise business complexes, and basements leased from other companies. In contrast to OA, where developing a meal plan and taking inventory of daily consumption is strongly encouraged though largely voluntary, these activities are preestablished, compulsory features of WW's program. The official purpose of the program is to help clients achieve bodily reduction. Hence, emotional and spiritual "recovery" is not pertinent to WW clients as it is for OA members, and bodily reduction for clients of WW is largely a matter of technocratic administration or "technique,"[6] not spiritual practice.

The materials distributed to new clients during the first week of WW membership include a basic program guidebook (containing a multi-item questionnaire assessing eating patterns, levels of desire to succeed in Weight Watchers, and the degree of stress in one's life), a "food diary" that includes daily menus, and a list of items from various food groups. In each subsequent week, members are given additional diaries to keep along with new daily menus and comprehensive guides on various topics including exercise, eating out, socializing, tips for better eating, and dealing with obstacles to weight loss. "Portion control" (the allotment of measured meals items for which WW markets food scales, frozen entrees, and cookbooks) is the fundamental principle underlying its program.

While the topic of compulsive overeating may receive a modicum of attention within meetings, the term cannot be found within WW's literature. Instead, the term "volume eating" is used, and what is emphasized is the type rather than the amount of food consumed. "Portion control" is to be applied within those food groups that are seen as leading to the creation of fat bodies but can be ignored within food groups whose caloric content is negligible.

Thus, according to WW, there are no good or bad foods, only those necessitating more or less portion control. . . .

SHAME WORK

In *The Managed Heart,* Arlie Russell Hochschild (1983, p. 7) described "emotion work" as labor that "requires one to induce or suppress feeling in order to sustain the outward countenance that produces the proper state of mind in others." Hochschild's definition pertains to the self-monitoring, management, and modeling of emotion as a strategy used by service personnel to evoke certain feelings. But other activities in which people engage (to elicit, constrain, and manage the emotions of others or encourage their self-management) might also be classified as emotion work, given a less restrictive definition of the term.[7] For example, confronting other people about the discrepancy between their verbal actions and nonverbal cues, such as the tone in their voice, or pointing out apparent discrepancies between seemingly frustrating situations and a lack of emotional response (as family therapists can attest) are very much work and can be emotionally exhausting. Hence, evoking emotional expression within WW's men's meetings might be considered emotion work, while displacing or channeling such expression constitutes work within women's meetings. Within OA, "working through" denial is a form of emotional labor, while for members of NAAFA, emotion work includes attempting to impassion members over issues of weight-based discrimination.

Within all three organizations, a common form of emotion work is "shame work," that is, emotional labor aimed at evoking, removing, or managing shame—though in the contemporary contexts of OA, WW, and NAAFA only the latter two objectives are sought. Within OA and WW, shame work involves activities directed at mitigating the internalized experience of shame that emanated from the personal and social practices of consuming food. Within NAAFA shame work was directed at removing shame that had been acquired through the stigmatization of a "fat identity." Like other forms of emotional labor, shame work includes communicative and expressive action that may take both discursive and bodily forms. In the following sections, I discuss the use of shame avowal (OA), shame management (WW), and shame contestation (NAAFA) and how these are accompanied by use of "body announcements" in managing and systematizing members' shame.

Shame Work in Overeaters Anonymous

For members of OA, patterns of compulsive overeating and resultant body size are seen as a source of shame. Yet this is "only secondary shame" insofar as these patterns have developed as a result of some earlier "primary shame" experience invoked by others. Both female and male members commonly located the development of primary shame experiences in early childhood, citing failures at living up to adult expectations as the reason for their present "shame-based" state.

Primary sources of shame, as OA members define them, are profoundly social, most often residing in interaction episodes with significant others who have knowingly or unintentionally evoked it. One strategy used by OA members engaged in shame work might best be termed "shame avowal," an open acknowledgement of past and present shame experiences that serves to expiate the shame experience. Such avowals are commonly woven into the public testimony of OA members:

> Jason (open OA meeting): I also need to say I've never had one perfect
> day of abstinence, there is no such thing for me. You know there is
> no perfect abstinence. I've lost weight in this program. I've gained
> weight in this program. But I have had enough shame over my food
> and my weight to last me the rest of my life. I don't need to put more on
> me. You know my goal weight . . . somebody asked me at a convention
> what my goal weight was. And I said, "My goal weight, if I'm abstinent,
> it's exactly what I weigh today."

The acknowledgment of shame is cited by OA members as problematic because it is premised on a self-awareness that, if not already present, must be organizationally manufactured. As indicated by one member:

> Rob (OA conference workshop): The second part [of dealing with
> shame] is to recognize it. To even find out that we have it. And as far
> as I'm concerned that is the most difficult part of it. To recognize that
> we have it is, is the thing that's so easy to cover up with denial. I've been
> in the OA program for about five years and AA about six years. And
> it took a long time for me to discover that I had any shame at all.

Determining whether the self exists in a deep shame-based state is a subjective evaluation, but the evaluation is also linked to a collective assessment as OA sponsors and other group members attempt to facilitate the awareness of shame-based actions on the part of nascent members. While such an awareness involves the acquisition and application of an organizational vocabulary of motive that is acquired through socialization in OA, it also necessitates the presence of viable experiences and circumstances to which it can be applied. As in other twelve-step programs (Denzin 1987; Rudy 1986), shame experiences are recognized and formulated as members of Overeaters Anonymous pragmatically invoke the organizational vocabulary for understanding and overcoming patterns of compulsive overeating. As it is applied to shame, the motive of denial is used in retrospectively interpreting compulsive overeating as behavior connected to a shame-based state. Working through denial is facilitated by "hitting bottom." Incidences of binging, stealing food, or lying about consumption, which lead to acute intrapersonal states of shame, degradation, and humiliation, may all be defined subjectively as "hitting bottom," that is, reaching the lowest possible point in one's life spiritually, emotionally, and socially.[8] According to the redemptive frame of the OA, salvation from a shame-based state of compulsive overeating is only possible after one acknowledges both to self and others the causes and consequences of compulsive overeating and then turns one's life over to a "higher power." As in Alcoholics Anonymous, the concept of a "higher power" is one that is idiosyncratically constructed,[9] yet members are encouraged to let it intervene in their lives and assist them in working through their denial of compulsive overeating. . . .

Shame Work in Weight Watchers

Within Weight Watchers, NAAFA, and OA, body shame is commonly discussed by women. Female members in OA also cite shame experiences in episodes of emotional, physical, or sexual abuse. By contrast, most male members of Weight Watchers and NAAFA appear to be exempt from experiences of body shame: neither did male members in OA seem to share experiences of body shame though they mentioned shame experiences rooted in a general failure to live up to a multiplicity of social expectations. Weight Watchers group leaders claim body shame as exclusive to the experience of female members:

> Robin: Men are not as shameful about their bodies. They don't care
> if they are thirty pounds overweight, they go ahead and take off their
> shirts at the beach anyway. Women just aren't able to do that—we're
> ashamed to wear bathing suits and show off our cellulite or flabby legs or
> varicose veins at the beach. We are overwhelmingly told we can't do
> that, that we must be slim.

As a former longtime member of Weight Watchers and now group leader, Robin claims that body shame is not only central to the experience of women vis-à-vis men but that being female makes one susceptible to body shame in ways that men are not. Some men in WW, however, did discuss similar experiences:

> Carl: For phys-ed class we had to go swimming. And I always got the largest
> swim suit. And people would just laugh at me and call me "fatso" and,
> you know, names. And then if I went to lunch no one would sit with me.

Carl does not specifically identify a personal sense of shame as a residual of this experience but he cites feelings of profound differences induced by episodes of stigmatization. While it seems reasonable to expect the presence of shame in episodes such as the one above, it remained liminal in the experiences of the men I interviewed. In contrast to the OA strategy of shame avowal, the work of WW personnel consisted of shame management that is, attempts to neither deny, contest, or specifically avow the experience but to contain it in hopes of a future, natural dissipation through dieting and weight loss. The management of shame represents a change in WW's treatment ideology over the course of its thirty-year history. Until the 1970s the organization had relied heavily upon the "shame-aversion" strategy (Serber 1970) of "card calling" in modifying the eating behaviors of clients.[10] Card calling involved placing a client's name along with the number of pounds that the client had either gained or lost during the week on an index card. The cards of all clients would then be read publicly at the beginning of each meeting. Similar instances of shaming appear to have been used in other weight loss organizations during this same period. Marcia Millman's *Such a Pretty Face* (1980) reveals that the shaming ritual used to discipline candy-sneaking "kids at fat camp" (children at summer weight loss camps) was to extract public confessions from these children on stage, humiliating them in front of their peers. Earlier research on the organization Take Off Pounds Sensibly (TOPS) reported the use of an equally drastic strategy for motivating clients to lose weight:

> There is a public announcement of each member's success or failure during the previous week. When a client has lost weight, there is applause and much vocal behavior; if she has gained weight, there is obvious diapproval with booing and derision.... In some groups the member who has gained the most weight is given the title of "Pig of the Week" and may be made to wear a bib with a picture of a pig on it or may have to sit in a specified area called the "pig pen." (Wagonfeld and Wolowitz 1968)

The loss of Weight Watchers revenues, which resulted when clients terminated their membership after such experiences, proved to be a compelling reason for abandoning shame aversion strategies in favor of strategies directed at its containment. One former client and now group leader of Weight Watchers recounted the outcome: "Those were the old days. Jeez, can you imagine how uncomfortable everyone was? We found that half the group had left before we even got seated." However, the absence of shame talk in WW meetings also reflects corporate concerns for liability. The training provided group leaders does not equip them to conduct psychotherapy or deal intensively with revelation of traumatic events such as sexual, emotional, or physical abuse that might produce shame. In light of episodes where women may make such disclosures, WW developed a strategy that allows organizational personnel to maneuver efficiently through the most troublesome situations. When I asked a group leader what happens if people divulge very personal or very traumatic information that is related to weight loss, she replied:

> The line is drawn that you never touch anything that is not directly affecting the members' weight. If they can't verbalize it you move on.... You don't touch it, you let them tell you. And they'll take care of it. It's just amazing. I've never been as scared as when that woman remembered in the meeting having been raped. Because I didn't know what to do with her. And the question that we tell leaders and that we practice in any group—men, women, any kind of group that we're doing—is, "And how does that affect your weight control?" That's your appropriate follow up, no matter what they say, if you think you're into something that you're not qualified for, you let them answer and then you let the group take care of that person.... You know, there's always someone that's gonna say, "Well, gee, you really need counseling for rape." So

that's the point, once you've thrown it back to them, to back off. . . . [Y]ou don't have anything to offer them. You're not qualified. If we could just know what we're doing we could do so much more. And we do have an awful lot of psychology background given to us by the company but we can't use it; we're not really qualified.

Here, the emotion work of leaders must strike a precarious balance between the emotional resolution of members' personal problems and corporate exposure, which writes large for group leaders as "uncertified" professionals. This is accomplished by displacing the responsibility for resolving emotionally charged situations onto members themselves. Thus, group leaders do shame work as well as other kinds of emotion work but always within limits rationally guided by concerns for organizational liability.

The emotion work now done by WW personnel in women's meetings consists of assisting in the management of client's shame experiences. Shame management in WW largely included remedial work done during weight loss meetings as well as teaching clients intrapersonal strategies for minimizing body shame. Remedial work within the social context of group meetings primarily involved supplying clients with viable disclaimers of excuses (Hewitt and Stokes 1975; Lyman and Scott 1970) for program violation. An example of the latter is demonstrated in the following interaction:

LEADER: How did everyone's week go this week?

CLIENT 2: [female, approximately age twenty-four]: I almost didn't come in tonight because I was afraid to face the scale.

LEADER: Ohhhh . . . what happened?

CLIENT 2: I was really stressed this week and I ate everything in sight. I went off the plan, I'm afraid. I kept saying to myself [laughs] I wonder if they take your ribbon back [a red ribbon is given to clients after they lose their first ten pounds].

LEADER: Okay . . . but the positive thing is that you came back this week, see . . . you could have done a lot of other unhealthy things to yourself than just eat. Does anyone have anything to say to [name]?

CLIENT 3: (Woman looks at client 2) Well, tell her why you were so stressed.

CLIENT 2: Well, I went to visit my mother and there was food everywhere. My future brother-in-law made it to the finals at the state wrestling tournament, and a cousin who I was very close to died this last week.

CLIENT 3: I think you need to give yourself a break.

CLIENT 4: I think you are being way too hard on yourself, you shouldn't worry about the program [WW] with all that you're going through.

LEADER: Like I said, I think it's very positive that you *did* decide to come back this week, knowing that you may have not stuck with the program. I think that's just positive in itself. [To other clients] Don't you agree? [Everyone nods their heads yes.]

Here, the group leader enlists the support of other group members in ratifying the excuse. Because bodily monitoring means heightened awareness of body size, it evokes body shame for some members. In such cases, group leaders also recommended nondiscursive strategies for dealing with shame:

> Margaret [group leader]: I had one woman who was ashamed to find out how many inches she had lost every week and so instead of using a measuring tape would use different colored yarns to measure her bust, hips, waist, thighs, calves, and upper arms. Each week, this member would cut the tails off these pieces of yarn according to how many inches she had reduced and therein came to amass a pile of different colored yarn. She could then hold and feel the cut off yarn which represented her weight loss.

In suggesting that clients objectify their weight loss, the group leader offers a technique by which clients might manage shame and gain a sense of control over body size and weight. The tactile strategy offered above is designed to contain or manage rather than avow shame. . . .

Shame Work in NAAFA

In contrast to avowing or managing shame, NAAFA members contest it. The objective of activism as presented in NAAFA's national newsletter is to "Change the World, Not Your Body" (Wolfe 1992, p. 5). Such contestation can readily be seen in the communicative but nondiscursive practice of making a "body announcement," where the fat body is displayed as shameless. NAAFA members make a public avowal of a "fat identity" not only through these displays but also through forms of public and private confrontation. Letter-writing campaigns and confrontations with public officials, as well as proprietors of businesses and services who engage in size discrimination, comprise part of the protest activities of NAAFA members. No social context is considered exempt from "fat activism." What is formulated in these activities, according to NAAFA members, is a way of life, an organized set of attitudes, and a mode of responding to myriad situations that members face daily.

> Beth: Like one time they were making cracks in the lounge about eating, well it was at lunchtime and, oh God, this one woman said, who's always dieting, "If we eat this we'll all have to shop at Women's World [clothing store for large women]." And everybody laughed. And I said, "Oh pardon me." I said, "Don't make fun of the places where I buy my clothes." I just said it in a nice, lighthearted tone. There was just dead silence. God, everybody at the table was embarrassed to death. I thought, "Hey Beth, good for you for saying that."
> Deborah: So I was shopping and all of a sudden I found myself in front of this huge section of Slimfast products. And all of a sudden the idea came to me, [shouts] gee, I don't have to use the entire card [NAAFA business card]. So I just tore the bottom half of the card off, the

part that had the NAAFA address and left the part that said, "Do something about your weight, accept it." And I just stuck it on the shelf, sort of like behind one of those little plastic place cards and said, "There."

Contestation emanates from the lived experiences of members who organize their lives around their organizational "fat"—identity. The identity is viewed as relevant in all of the daily activities of NAAFA members, whether confronting fellow employees like Beth or grocery shopping like Deborah. According to NAAFA members, shame militates against the active initiation of individual and, hence, collective, political practice. NAAFA sees social stigmas as directly linked to the internalized oppression of shame. Shame contestation is thus a requisite component of identity politics where the aggrieved attempts to transform themselves by transforming both societal definitions of beauty and human value and the feeling rules that govern fat bodies....

CONCLUSION

Sociologists studying formal organizations contend that organizations are boundary-maintaining social units where society and culture are both experienced by individuals and reproduced in various sets of social practices (Perrow 1986). Society, it is argued, is now absorbed by organizations that mediate culture and cultural themes for the individuals who participate within them. Rudolph Bell's (1985) analysis of "Holy Anorexia" has delineated the historic role of the Catholic church in the creation and social reproduction of themes of shame and obesity, purity and thinness, as they came to be interwoven with the phenomenon anorexia from the thirteenth century on. Feminist scholars (Bordo 1993: Hesse-Biber 1996; Schwartz 1986; Wolf 1991) have argued that the increasing commercialization of the body and bodily needs for the sake of expanding markets has ensured that themes of shame, slothfulness, loss of control, and embarrassment will continue to be associated with fat bodies, particularly fat female bodies. The present study indicates that the cultural meanings of shame associated with the body are not simply residues that exist as historical abstractions.[11] Rather, it is in interaction that takes into consideration the organizational frames that a shame experience is formulated as a social object and given meaning. Organizational frames may be quite variable depending upon the nature, objectives, and ideology within which the organizations attempts to align members' experiences, yet they provide the template in which shame experiences may be systematized in organizational routines.

NOTES

1. Many of the tapes were truncated or contained several short testimonies by members of how OA had helped restore their lives. Yet the tapes yielded little information about the lived experience of the person or how she or he came to OA.

2. The serenity prayer of OA: "God grant me the serenity to accept the things I cannot change; the courage to change the things I can; and the wisdom to know the difference."

3. One of the more recent efforts was bill A 3484 New York State, sponsored by assemblyman Daniel Feldman. The bill was not supported by Governor Mario Cuomo. The event was reported in NAAFA's national newsletter, which commented, "Asked by his interviewer whether questions on the bill had him skating on thin ice with voters who are fat, Cuomo quipped, 'If you are overweight, you shouldn't do that—you're liable to fall through the ice' " (NAAFA 1993, p.1).

4. On May 23, 1992, NAAFA chapters in five cities launched demonstrations against Southwest Airlines for ejecting a very large man from a half-full flight and for disallowing another large passenger to board after she had already completed one leg of her flight. (NAAFA, 1992, p. 1)

5. I use the term "practically" in reference to the "investment model" of participation, discussed by Rudy and Greil (1987), observing that all organizational members mobilize personal resources to be used in and by the organizations. This is clear in the case of WW where both time and money represent "sunken costs." Yet it is also the case in OA and NAAFA where members invest time, money, and intellectual and emotional labor to keep the organization operating.

6. The collection of all shame management strategies may be labeled "technique." I have in mind Jacques Ellul's (1964, p. vi) meaning: a "complex of standardized means for attaining a predetermined result. Thus it converts spontaneous and unreflective behavior into behavior that is deliberate and rationalized." Here, the emphasis must be on the idea of a "complex," that is, not only the measurement of food and calories but the restructuring of social relations and development of adaptive strategies for dealing with unexpected situations.

7. Goffman's notion of "cooling the mark out" (1952) would appear to be a rather pervasive form of emotion management found among all types of service personnel. Yet, curiously, in light of a rather extensive citation of Goffman's works, Hochschild (1983) does not mention it. Beyond the management of emotion, the concept of emotion work suggests a range of interactional strategies that might be used in managing the emotions of others. Lying, for example, might be used to "cool out" airline passengers who might otherwise become disruptive.

8. Rudy (1986), in his analysis of Alcoholics Anonymous, observes that "hitting bottom" is ultimately a subjective perception.

9. While some OA members refer to their higher power as "God," others adopt a Durkheimian (Durkheim 1965) conception of this power, defining it as the social force that exists outside of themselves and abides in the group.

10. Shaming appears to have been a treatment modality that was gaining limited popularity within quite diverse fields. Serber (1970), in his research on behavioral modification among transvestites, referred to the treatment as "shame aversion therapy." Braithwaite (1989) revisited the idea in the concept of "reintegrative shaming" within the context of criminal justice.

11. Pareto observed that "residues" (nonscientific belief systems) rarely mobilize people into action even though they may historically express deep collective sentiments (Coser 1977). My point is that cultural themes such as shame are socially reproduced within concrete episodes of interaction.

REFERENCES

Allon, Natalie. 1982. "The Stigma of Overweight in Everyday Life." Pp. 130–174 in *Aspects of Obesity*, edited by Benjamin B. Wolman. New York: Van Nostrand Reinhold.

Banner, Lois W. 1983. *American Beauty*. New York: Knopf.

Bell, Rudolph. 1985. *Holy Anorexia*. Chicago: University of Chicago Press.

Bordo, Susan. 1993. *Unbearable Weight*. Berkeley: University of California Press.

Braithwaite, John. 1989. *Crime Shame and Reintegration*. New York: Cambridge University Press.

Buss, David M. 1989. "Sex Differences in Human Mate Preferences: Evolutionary Hypotheses Tested in 37 Cultures." *Behavioral and Brain Sciences* 12: 1–49.

Cahnman, Werner. 1968. "The Stigma of Obesity," *The Sociological Quarterly* 9: 283–299.

Coser, Lewis A. 1977. *Masters of Sociological Thought: Ideas in Historical and Social Content*. 2d ed. New York: Harcourt, Brace, Jovanovich.

Denzin, Norman K. 1987. *The Recovering Alcoholic*. Beverly Hills, CA: Sage.

Durkheim, Emile. 1965. *The Elementary Forms of Religious Life*. New York: The Free Press.

Ellis, Carolyn. 1998. " 'I Hate My Voice': Coming to Terms with Minor Bodily Stigmas." *The Sociological Quarterly* 39: 517–538.

Ellul, Jacques. 1964. *The Technological Society*. New York: Vintage.

Ewen, Stuart. 1976. *Captains of Consciousness: Advertising and the Roots of the Consumer Culture*. New York: McGraw-Hill.

Featherstone, Mike. 1982. "The Body in Consumer Culture." *Theory Culture & Society* 1: 18–33.

Gilmore, David B. 1987. *Honor and Shame and the Unity of the Mediterranean*. Washington, DC: American Anthropological Association.

Glaser, Barney G., and Anselm L. Strauss. 1967. *The Discovery of Grounded Theory*. Chicago: Aldine de Gruyter.

Goffman, Erving. 1952. "On Cooling the Mark Out: Some Aspects of Adaptation to Failure." *Psychiatry* 15: 451–463.

———. 1967. *Interaction Ritual.* Garden City, NY: Anchor.

———. 1969. *Strategic Interaction.* Philadelphia: University of Pennsylvania Press.

———. 1986. *Frame Analysis.* Boston, MA: Northeastern University Press.

Hesse-Biber, Sharlene. 1996. *Am I Thin Enough Yet? The Cult of Thinness and the Commercialization of Beauty.* New York: Oxford University Press.

Hewitt, John P., and Randall Stokes. 1975. "Disclaimers." *American Sociological Review,* 60: 1–11.

Hochschild, Arlie Russell. 1983. *The Managed Heart: Commercialization of Human Feeling.* Berkeley: University of California Press.

Horowitz, Ruth. 1983. *Honor and the American Dream: Culture and Identity in a Chicano Community.* New Brunswick, NJ: Rutgers University Press.

Lyman, Stanford, and Marvin B. Scott. 1970. *A Sociology of the Absurd.* Pacific Palisades, CA: Goodyear.

Martin, Daniel D. 1995. "The Politics of Appearance: Managing Meanings of the Body, Organizationally." Ph.D. dissertation, University of Minnesota.

McFarland, Barbara, and Tyeis L. Baker-Baumann. 1990. *Shame and the Body: Culture and the Compulsive Eater.* Deerfield Beach, FL: Health Communications.

Millman, Marcia. 1980. *Such a Pretty Face: Being Fat in America.* New York: W. W. Norton.

Mills, C. Wright. 1940. "Situated Actions and Vocabularies of Motive." *American Sociological Review* 5: 904–913.

National Association to Advance Fat Acceptance (NAAFA). 1990. *NAAFA Workbook: A Complete Study Guide.* Sacramento, CA: NAAFA.

———. 1992. "Southwest Protest: NAAFAns Demonstrate in Five Cities." *NAAFA Newsletter* XXII: 1.

———. 1993. "Cuomo: One Law Too Many!" *NAAFA Newsletter* XXIII: 6 (May): 1.

Overeaters Anonymous (OA). 1987. *To the Newcomer: You're Not Alone Anymore.* Torrance, CA: OA.

———. 1988. *Questions & Answers About Compulsive Overeating and the OA Program.* Torrance, CA: OA.

———. 1990. *The Twelve Steps of Overeaters Anonymous.* Torrance, CA: OA.

———. 1992. *Overeaters Anonymous: Membership Survey Summary.* Torrance, CA: OA.

Perrow, Charles. 1986. *Complex Organizations.* 3d ed. New York: McGraw-Hill.

Rudy, David R. 1986. *Becoming Alcoholic: Alcoholics Anonymous and the Reality of Alcoholism.* Carbondale: Southern Illinois University Press.

Rudy, David R., and Arthur L. Greil. 1987. "Taking the Pledge: The Commitment Process in Alcoholics Anonymous." *Sociological Focus* 20: 45–59.

Schwartz, Hillel. 1986. *Never Satisfied: A Cultural History of Diets, Fantasies and Fat.* New York: Free Press.

Serber, Michael. 1970. "Shame Aversion Therapy." *Experimental Psychiatry* 1: 213–215.

Snow, David A., E. Burke Rochford, Jr., Steven K. Worden, and Robert D. Benford. 1986. "Frame Alignment Processes, Micromobilization, and Movement Participation." *American Sociological Review* 51: 461–481.

Thorne, Barrie, and Zella Luria. 1986. "Sexuality and Gender in Children's Daily Worlds." *Social Problems* 33: 176–190.

Wagonfeld Samuel, and Howard M. Wollowitz. 1968. "Obesity and the SelfHelp Group: A Look at TOPS." *American Journal of Psychiatry* 125: 249–252.

Weber, Joseph. 1990. "The Diet Business Takes It on the Chin." *Business Week,* April 16, pp. 86–88.

Wolf, Naomi. 1991. *The Beauty Myth: How Images of Beauty Are Used against Women.* New York: William Morrow.

Wolfe, Louise. 1992. "Boston 'Free' Party: The Revolution Within; the Revolution Without." *NAAFA Newsletter* 23: 5.

PART VI

The Social Organization
of Deviance

In Part VI we turn to a closer examination of the lives and activities of deviants. Once they get past dealing with outsiders, they must deal with other members of their deviant communities and with the specifics of accomplishing their deviance. There are several ways of looking at how deviants organize their lives. We start by looking at the relationships among groups of deviants, focusing on the character, structure, and consequences of different types of organizations. These encompass the structure or patterns of relationships in which individuals engage when they enter the pursuit of deviance.

As Best and Luckenbill (1980) have noted in their analysis of the social organization of deviants, relationships among deviants can follow many models. These vary along a dimension of sophistication, involving complexity, coordination, and purpose. Deviant associations vary in their numbers of members, the task specialization among members, the stratification within the group, and the amount of authority concentrated in the hands of a leader or leaders. Some groups of deviants are loose and flexible, with members entering or leaving at their own will, uncounted or monitored by anybody. Others maintain more rigid boundaries, with access granted only by the consent of one or more insiders. Membership rituals may vary from none to highly specific acts that must be performed by prospective inductees, thereby granting them not only membership but also a place in the pecking order once they are inside. In some ways, rigidity inside deviant groups is related to its insulation from conventional society: The more its members withdraw into a social and economic world of their own, the more they will develop norms and rules to guide them, replacing those of the outside order.

Groups of deviants also vary in their organizational sophistication, with the more organized groups capable of more complex activities. Such organized groups provide greater resources and services to their members; they pass on the norms, values, and lore of their deviant subculture; they teach novices specific skills and techniques where necessary; and they help one another out when they get in trouble. As a result, individuals who join more tightly knit deviant scenes tend to be better protected from the efforts of social control agents and more deeply committed to a deviant identity.

Best and Luckenbill (1980) outlined a range of ways by which deviants may socially organize. **Loners** are the most solitary, interacting with people but keeping their deviant attitudes, behaviors, or conditions secret. They lack the company of other similar deviants with whom they can share their interests, troubles, and strategies. Serial rapists or murderers and embezzlers commit their acts without the benefit of the camaraderie of like others. Many individuals who hold themselves as loner deviants in the real world, however, have been able to find that they are not alone due to the rise of Internet communities. Without jeopardizing their identities and social relations with conventional people, they have found ways to connect anonymously with people who share their deviance, seeking their company and advice. People involved in deviance, such as sexual asphyxiates, self-injurers, anorectics and bulimics, computer hackers, depressives, pedophiles, and others, now have the opportunity to go online and find international cybercommunities populated 24/7 by a host of like others. These websites offer chatrooms, Usenet groups, e-mail discussion lists, and bulletin/message boards for individuals to post where they can seek the advice and cybercompany of others. Some, such as the "proana" (anorexia) and "promia" (bulimia) sites explicitly state that they reinforce and support the deviant behavior, regarding this as a lifestyle choice (Force 2005). Others, such as many self-injury sites, purport to help users desist from their deviance but may actually end up reinforcing it by providing a supportive and accepting community where individuals can go when they feel misunderstood and rejected by the outside world (Adler and Adler 2007).

Whether the sites aim to reinforce or discourage the deviance, nearly all tend to serve several unintended functions that have significant consequences for participants. First, they transmit knowledge of a practical and ideological sort among people, enabling them to more effectively engage in and legitimate the behavior. This helps people learn new variants of their activities, how to carry them out, how to obtain medical or legal services, and how to deal with outsiders. Second, they tend to be leveling, bringing people together into a common discourse regardless of their age, gender, marital status, ethnicity, or socioeconomic status (although users do need a computer, and most have high-speed Internet access). Third, they bridge huge spans of geographic distance, putting Americans in contact with English-speaking people from the UK, Australia, New Zealand, Canada, and all over the world.

These interactions, regularly conducted among a range of regular and moderate users as well as periodic posters and "lurkers" (those who read but do not post), forge deviant communities. Participants develop ties to them by virtue of the support and acceptance they offer, especially to individuals who are lonely or semi-isolated. People unable to find "real" friends "FTF" (face-to-face) may come to rely on these cybercommunities and cyberrelationships, interacting with members for years and even traveling large distances to meet each other. They may, then, take the place of core friendships. In this way, if in no other, they reinforce continuing participation in the deviance as a way of maintaining membership. The stronger and more frequent these bonds, the greater effect they have on strengthening members' deviant identities. Deviant cybercommunities thus provide a space and mechanism for deviance to grow and thrive in a way that it has not previously had.

Colleagues represent the next most organizationally sophisticated associational form. Participants have face-to-face relationships with other deviants like themselves but do not need the cooperation of fellow deviants to perform their deviant acts. The jump from loners to colleagues is the greatest leap in the spectrum of deviant social organization, as mutual association brings the possibilities of membership in a deviant subculture or counterculture from which people can learn specific norms, values, rationalizations, helpful information, specialized terms or vocabulary, and gossip about people like them. From others they can gain social support, as we see with the expressive groups discussed earlier, and a sense of their position in the status hierarchy of their kind. Deviants organizing as colleagues include the homeless, recreational drug users, and con artists. Colleagues may interact and perform their deviance with nondeviants, such as the lowest level drug dealers and their customers.

Deviants socially organized as **peers** engage in their deviance with others like themselves but have no more than a minimal division of labor. Neighborhood gangs who congregate with their friends generally engage in all of the same types of activities and only see role specialization when it comes to the leader versus the followers. Most peers traffic in a black market of illegal goods and services such as drugs, guns, endangered species, stolen art, and exotic forms of sex.

Especially fascinating to the media and the public is the **crew** form, where groups of anywhere from three to a dozen individuals band together to engage in more sophisticated deviant capers than less organized deviants can accomplish. Crews fascinate observers because their more sophisticated division of labor usually requires specialized training and socialization, giving them a more professional edge. Bounded by their lack of affiliation with other crews, they are dependent on a leader who organizes and recruits them, sets and enforces the group rules, plans their activities, and organizes travel and lodging if they go on the road. Crews usually commit intricate forms of theft, but they may also engage in smuggling and hustling at cards and dice.

The top of this continuum is deviant **formal organizations,** which are much larger than crews and extend over time and space. They may stand alone or be connected to other similar organizations domestically or even internationally, as we see with the Cosa Nostra Mafia families and the Colombian drug cartels. Their affiliations may take the form of what Godson and Olson (1995) call "transnational links," where they have regular connections to do business or exchange services with other criminal organizations, or they may have a "global scope," where they conduct extensive operations in various continents through franchised branches of their own organization located in different places, akin to multinational corporations. Much larger in size than crews, deviant formal organizations may have 100 members or more so that when their leaders are killed the group endures. Ethnically homogeneous, these organizations trade in a currency of violence, are vertically and horizontally stratified, and have the resources to corrupt law enforcement. They are the most organizationally sophisticated of the deviant associations formed for purely deviant purposes.

Legitimate individuals and organizations also engage in deviant activities, although these may be their side rather than primary purpose. It is worth noting that while most studies of crime and deviance focus primarily on crime in the streets, a more socially injurious amount of deviance occurs at the top, in the suites. In the Introduction to Part III, we noted Cloward and Ohlin's belief, advanced in their differential opportunity theory, that access to illegitimate opportunity is unequal. They talked about people from distinct neighborhoods, ethnic groups, and criminal ladders having better criminal opportunities, but their concept of differential opportunity can apply to the privileged as well. **White-collar crime** is directly related to opportunities to abuse positions of financial, organizational, and political power. We often tend to associate white-collar crimes with strictly financial activities, but they extend to bodily injury and death as well.

White-collar crime can be divided into two main subsections: occupational and organizational crime. **Occupational crime** is pursued by individuals acting on their own behalf. Employees at all levels of organizations may steal from their companies, and we have also seen the rise of embezzlement (recently, from the Red Cross) and computer crime. Corporate executives at firms such as Enron, Tyco, and WorldCom looted their companies, shareholders, and employee retirement plans of millions through fraudulent accounting, offshore and dummy corporations, and the manipulation of information to live in high style. Individuals in charge of purchasing for their firms also frequently accept bribes to give business to sales vendors.

In the government sector we see people evading taxes through offshore companies and fraudulent tax shelters, often sold to them by top accounting and brokerage firms who charged millions for these services. Politicians have been caught selling power, especially those in charge of awarding government contracts, such as jailed U.S. Republican Congressman Randy "Duke" Cunningham, the former Air Force flying ace (whom Tom Cruise portrayed in the film *Top Gun*), who had a price list in the tens of thousands of dollars for military appropriations. Politicians also sell business to companies (sometimes in no-bid contracts) whose products and services are inferior or if they have a financial stake in their operations, such as we saw with Vice-President Dick Cheney, former CEO of Halliburton, who continued to receive "deferred compensation" from the firm throughout his term in the White House, and U.S. Democratic Congressman William Jefferson, who accepted $100,000 to steer contracts to a broadband telecommunications company in Nigeria and hid it in his freezer. Politicians and police officers may also receive individual remuneration for selling immunity from prosecution to criminals and companies, either in direct cash payments or campaign contributions. People connected to the government may also sell their influence, such as we saw in the scandals surrounding former U.S. Republican Congressman Tom DeLay and lobbyist Jack Abramoff, who collected millions from Indian tribes to secure their gambling interests.

Professionals are not above collecting money for their individual benefit, as we see with doctors who accept gifts from pharmaceutical companies to steer their patients toward certain drugs, who overcharge and overservice their patients, and who commit Medicare and Medicaid fraud. Top stockbrokers and their clients make money through insider trading, such as we saw in the Martha Stewart scandal.

Organizational crime, committed with the support and encouragement of a legitimate formal organization, is intended to advance the goals of the firm or agency. Looking at the corporate world, we see many instances of false advertising, where products are misleadingly alleged to do one thing and either fail to do so or have the reverse effect (air and water purifiers, fire retardants) and fraud, where companies misrepresent themselves to investors and the general public. Antitrust violations are common, where companies engage in monopolistic practices to control the market (Microsoft), artificially subsidizing or cheapening their products or services to drive competitors out of business (microchips, airlines), or conspiring with other companies to set minimum threshold prices for consumers (Samsung, General Electric). Corruption among companies with government service contracts is rife, with $5 to $7 billion lost annually during the Iraq war.

Injury and loss of life may result from unsafe working conditions, such as we saw in the numerous mining deaths of 2005, where governmental regulators let enforcement slide among several drilling organizations to strengthen their business. Widespread illness and fatalities have also arisen due to the working conditions found in nuclear power plants, oil and chemical companies, and pesticide manufacturers. Unsafe products represent another area where companies put their balance sheet above the lives of consumers, figuring that it is cheaper to settle lawsuits against them than to fix their goods. Notable offenders are the pharmaceutical companies, the automobile and tire industry, and medical manufacturers.

A disturbing amount of government activity also falls into this category, with politicians abusing the public trust, manipulating information, and breaking laws to advance their administrations. American international policy is often clearly tied to the interests of the corporate sector, most notably recently with the oil industry in Middle Eastern diplomatic and military activities. The K Street project was designed by Republican strategist Karl Rove to engineer a Republican takeover of the lobbying industry, bringing corporate and governmental interests and financial obligations closer together to enrich politicians and favor their contributors. Numerous domestic governmental scandals have erupted in violation of the law, such as the Watergate breakin, the Iran-Contra scandal, and the NSA warrantless surveillance controversy. The intervention of the CIA, the military, the FBI, and private burglars into the telephone records, bank accounts, Internet logs, library records, and credit card transactions of alleged terrorists may be acceptable to the American public, but when these things are done in violation of law or directed for political parties against reporters, political opponents, chaplains or lawyers counseling political prisoners, or antiwar activists, they are regarded as very grave indeed. International violations have also been common, with the secret CIA "black site" torture prisons in unknown Eastern European countries, Iraq's Abu Ghraib prisoner abuse violations, and the numerous political dirty tricks and secret assassination attempts. All these white-collar crimes have led to greater cost, loss of life, forfeit of international prestige, and violation of conventional norms and values than the sum total of conventionally recognized crime and deviance.

32

Self-Injurers

PATRICIA A. ADLER AND PETER ADLER

The year 1996 marked a turning point as awareness about the existence of self-injury started to seep into the culture. Based on 35 interviews with people who self-injure, mostly through cutting and burning. Adler and Adler discuss the social organization of this practice. Self-injurers, they find, are loners who cut and burn themselves in private to relieve the pain and pressure of loneliness, sadness, and depression. Usually practiced by adolescents, self-injury conforms to sociological descriptions of loner behavior: Participants hide their behavior, subscribe to conventional norms and often feel shame about their activities, withdraw into private and personal space to conduct their acts, and generally avoid social intercourse with other similar deviants. This behavior compares to other loner forms of deviance, including sexual asphyxia, anorexia and bulimia, embezzling, rape, and physician and pharmacist drug addiction.

Research in the sociology of deviance has characterized different types of deviant acts by the social organization of participants. Best and Luckenbill (1982) have suggested that some deviants organize and commit their acts as loners, without the support of fellow deviants. According to this definition, loners such as sexual asphyxiates (O'Halloran and Dietz 1993; Lowery and Wetli 1982), anorectics and bulimics (Gordon 1990; McLorg and Taub 1987; Way 1995), embezzlers (Cressey 1971), rapists (Scully and Marolla 1984; Stevens 1999), and physician (Winick 1964) and pharmacist drug addicts (Dabney and Hollinger 1999) do not know other individuals who participate in their form of deviance, or if they do, they generally do not congregate with these people and do not discuss their deviance together. This relative isolation requires loner deviants to move into their norm violations on their own, without the help and support of others. They must decide to do their deviance themselves and figure out on their own how to do it. Without the company of others, they lack the benefits of a deviant subculture from which to draw rationalizations and justifications that might help them neutralize their acts. Of all forms of deviants, loners are characterized as those most entrenched in the normative subculture and are most likely, then, to view their deviant acts through a conventional value system.

In this paper we examine people who self-injure, either by cutting, burning, or branding, the majority in the former category. The research draws on 25 in-depth interviews with self-injurers conducted between 2001 and 2004. Partici-pants ranged in age from 16 to 35 and had mostly, but not entirely, given up the behavior. Most self-injury occurred when people were in middle and high school, with only a smattering of individuals continuing past that age. Nearly three-quarters of the people we interviewed were women, and all were white. Subjects were gathered through a convenience sample of individuals who heard, usually on one of our campuses, that we were interested in talking with people about their self-injury. Those who were interested came forward and contacted us by email, asking for an interview. All of the interviews were con-ducted on campus in our faculty offices, many with college students, friends of students, university employees, or local high school students. If there is a bias in this self-selected sample, it may be that these people do not represent the most severe segment of self-injurers. Conversations with friends and acquaintan-ces who worked in hospital emergency rooms as well as articles and books about self-injury (Conterio and Lader 1998; Favazza 1996; Favazza and Conterio 1989; Harris 2000; Strong 1998) indicate that some people perform more damaging acts of mutilation on themselves than the people with whom we spoke. Our sample, however, likely represents the majority of self-injurers.

Not all of the self-injurers we interviewed were loners. Some cut, branded, burned, or electro-shocked themselves in the company of others. The social dynamics and meanings of this kind of self-injury were dramatically different from those in the loner group. In this paper we focus only on the loners, who comprised nearly 80 percent of our sample. We outline the characteristics of loners and describe the way our self-injurers correspond to or differ from Best and Luckenbill's (1982) ideal typical model.

THE SOCIAL ORGANIZATION
OF SOLITARY DEVIANCE

People became involved in self-injury as loners for a variety of reasons, including depression, malaise, alienation, and rebellion. For these young people, self-injury provided a form of comfort that assisted them during a stressful period of their lives.

Formulating Deviant Ideology

As loners, people who self-injure were on their own in formulating the meanings and set of rationalizations legitimating their deviance. They often drew on their respectable training and experiences, not only to develop their techniques but to develop their rationales as well. We see this in the rationalizations of convicted rapists (Stevens 1999; Scully and Marolla 1984), for instance, where the men commonly denied the violent and forced nature of their acts and suggested instead that their victims precipitated or desired the incidents. Rapists drew on cultural myths, learned-hypermasculinity, and their sense of righteous entitlement to justify their behavior. Self-injurers had a much more difficult time giving social meanings and legitimacy to their acts, which were often, especially initially, unclear and undefined. Natalie, a 19-year-old college sophomore, discussed how she viewed her self-injury:

> I guess at first I didn't think much about it. I knew that it was a source of relief for me, and that was all that mattered to me. And, of course, over time when it becomes more of an issue, and people start noticing it, and other people maybe start commenting on it. And then I had to start thinking more about it and what I was really doing and the consequences and what it meant. And I guess just the way that I thought about it was, at first, I was glad that I was able to do it because it made me feel better.

After some time they developed personally acceptable views of their deviance. Some people focused on their neatness, that they were able to do it without making a mess. For others, control was the issue: they could control where their hurt would be. A common feature that many self-injurers shared was their relationship to the pain. Dana, a 19-year-old college sophomore, talked about her pain:

> The thing with emotional pain—you can't see it! It's all inside. I keep it bottled up inside 90 percent of the time, and people can never, like, quantify emotional pain because it's all inside. And so I try to put a picture to that. That's how much I hurt; I did that to myself. Those cuts, that pain, came from inside here. It's kind of a representation of . . . I mean, no cut on my body will ever embody what I feel inside, but it's a start.

Kyle, a 20-year-old junior, described how he would say to himself when he was ready to begin an episode of branding, "It's time for some pain."

Social Isolation

For loners, self-injury deviance was personal. Part of the reason they stayed to themselves was that they viewed their behavior as private, not to be shared with others. Unlike people such as embezzlers or pharmacist drug addicts, who had no special feeling of personal intensity about committing their deviant acts, self-injurers, like many sexual asphyxiates (O'Halloran and Dietz 1993), needed the focus and concentration of being alone while they were engaged in their deviance. It was all about them, and they were focused so completely on themselves while in the act that it would detract from what they could get out of the experience if they did it in the company of others. Dana discussed the feeling she had about her inner-directedness:

When I hurt like that, I get really self-involved. I get my blinders on. I'm
all about me, and don't disturb me... so that if someone was cutting
themselves in my house, even though I do it, I'd be uncomfortable. It's
my thing, you know? I'm in control.

When presented with the opportunity to meet or interact with other people
who self-injured, many cutters withdrew from or avoided the interactions. While
some knew others who self-injured or were introduced to the idea by knowing
others who did it, they generally did not want to form a subculture of self-
injurers and did not want to know or be responsible for others. When
approached in high school by a classmate who tried to bond with her over
their cutting, Mandy, an 18-year-old college freshman, rejected these overtures:

It scared me, I think, more than anything, because I didn't want to
be the person she depended on because I didn't feel ready for that.
I still wouldn't, even if she came to me today and said that she needed
something. I would be like, "I'm sorry."

One of the primary reasons people self-injured was that they were lonely and
depressed. Episodes of self-injury tended to occur when people were away from
the company of friends and family, often in the afternoons after school or at
night. They had time to sit around and reflect, and they felt bad. In thinking
back on high school, many people noted this as a period in their lives where
they spent a lot of time alone. It is ironic that while people self-injured to
avert their feelings of loneliness and depression, their self-injury could exacerbate
this condition. This often occurred when they held themselves back from being
with people in situations where their scars would be noticeable. Janice flunked
physical education, and endured the ignorant jeers of her classmates for it,
because she did not want to let the scars on her legs show. Alice described a negative consequence she incurred as a result of her cutting:

Sometimes I would just, I would not go with my friends to a hot tub at a
hotel or something. We were all having a good time, but I would stay
up in the room by myself. And then I'd end up in tears up there because I
was so frustrated with myself and with the fact that I have to live with that.

Practical Problems

Lacking a deviant subculture, self-injurers often found themselves on their own
in coping with the practical problems posed by their deviance. Many of them
worried about how to deal with these issues. Dana recalled a conversation she
had with a therapist who pointed out some of the auxiliary issues she might
have to deal with in the future if she continued with her cutting:

The practical side to it—how am I going to explain these cuts all
over myself in the summer, to my friends? Am I going to go to a
job interview with a big scar or a cut on my arm? And that really freaked
me out—I don't want anyone to know; no one can know, no!

Many were unable to anticipate what people might say to them or ask them, and they had no ready response when confronted by others. Prior to 1996 questions about scars were not as great a problem, because people could easily explain them away with almost any ridiculous answer, and nobody questioned them. Alice talked about the kind of carte blanche she had as a 12- and 13-year-old to cut without being challenged:

> You know, sometimes stuff that didn't even need to appear very realistic, but people believed it because why wouldn't they? I don't think that many people thought about it when they saw cuts, what if that's self-inflicted?

This all changed, however, when self-injury began to be more widely recognized and people became more suspicious. Longer-term cutters, whose parents had become aware of their behavior, were watched very carefully and often quizzed about their scars. This led them to move their cutting away from their hands and arms to less visible bodily locations such as their stomachs and the inside of their thighs.

In contrast to other forms of loner behavior such as sexual asphyxia (Lowery and Wetli 1982), where participants often shared their knowledge with a nonparticipant who assisted them in their deviance and also helped them keep it secret, self-injurers lacked this type of support. Cutters and burners sometimes told their closest friends or boyfriends about their behavior, but these others usually tried to get them to desist, rather than aiding them. Amy, a 19-year-old young woman working part-time in a hospital while attending community college, recalled that as a burner during high school, she had a boyfriend who cut himself and twice did it when he called her on the phone. Trying to convince him to stop, she said that she would burn herself if he cut himself. Although he denied cutting during the conversations, she felt she could tell from the quivering tone of his voice and the lags in their talk that he was incurring pain. Following through on her threat, she burned herself with a lighter (her preferred method), but this proved ineffective in restraining him. This dual self-injury happened once again, when they both were interrupted by a girlfriend of Amy's who came in, figured out what was going on, grabbed the phone, somehow got in the middle, and stopped them both. Maggie, a 19-year-old college sophomore, described the way a friend of hers who self-injured relied on her for help in restraining herself:

> She'd call me once in a while and be like, "I'm gonna." I told her not to do it, I told her to call me or do something, so she'd call me sometimes and be like, "I'm in the bath," and I'd be like, "I'm coming to get you," or we'd talk for a long time if I wasn't in a position where I could leave.

Normative Socialization

Another ironic juxtaposition about loners is that according to Best and Luckenbill (1982), they are socialized by society, not by fellow deviants, yet they choose deviance. They choose deviance not because they want to contradict their socialization but because they face situations where respectable courses of action are unattractive or unsatisfactory. Dana described the nature of her inner ambivalence:

> It's not, and I'm not, like, and I never really argue with myself, like, I really want to cut myself but I shouldn't, you know? If I really want to, I do, because it's not all the time, so when it does, I just do it because I'm desperate to feel better, so it's never like, I never, I'm never at war with how I should deal with it, like "Oh, I really want to do it, where's the knife," and my other side being like, "Dana, don't do that." It's always like, one or the other. If I have the urge, I will.

As a result, people who self-injured often condemned their behavior and felt ashamed of themselves. Lisa, a 20-year-old college sophomore, talked about how she viewed her deviance:

> Oh yeah. I was ashamed of myself because it's disgusting and it's not normal, which is OK, but it's just bad to do, I think, to yourself, especially if it's because someone else is hurting you or some other reason. People that do it because someone else is doing it and get a high from it, I think that's sick, too.

Strain

Without the support of fellow deviants, loners are unstable in their deviance, as Best and Luckenbill (1982) have suggested, and they have more difficulty sustaining their deviance over extended periods of time. Ellen compared the isolation of her cutting with other forms of individual deviance such as eating disorders:

> Women hear it all the time. Not so much with cutting, but with bulimia—you hear about it all the time happening, and percentages, and how many women are bulimic and how many women are anorexic. So therefore you, you not necessarily find a common bond with other people, but more like, you know, you know that other people do it. You know that you're not weird doing it. It's not like that with cutters.

For people who self-injured, there was often a structural strain between their normative expectations and their deviant behavior. Some wanted to rebel, yet they did not want to rebel too far. One young woman found an intermediate point for balancing this strain where she could burn herself occasionally but never so much that it endangered her access to the benefits of conventional society. Like many self-injurers, she "drifted" (Matza, 1964) in deviance, keeping one foot firmly anchored in legitimacy while she dabbled in deviance, able to return to a normative lifestyle without unduly damaging her identity.

Another form of strain experienced by loners arose from their lack of fellow deviants who might share their perspectives and reaffirm the meaning of their deviance. A few people we met did have contact with other self-injurers.

None cut together, but they had some limited contact. Maggie described the sense of contact she got from two other cutters she knew, one whom she described as a leader and the other as a copycat:

> I don't think they ever sat down and said, "Let's cut together," but I know they discussed it, and they were like, "Oh, I use a razor; what do you use?" you know? And I know they had conversations—while I was never, you know, directly involved in the conversations, I know they happened because of what they said. If I was with one or the other, they would say something that would make me think that they discuss this with each other.

When self-injurers moved away from these relationships, no matter how superficial, they noted how strongly they felt their absence. Lilly, a 21-year-old college senior, traced the demise of her high school cutting to moving to another school where she was away from the presence of a cutting acquaintance. By her own admission, they had not talked about it much. At most they occasionally said things to each other such as, "Had a bad night last night," and this would signify a cutting incident. But in the new environment, she really noted the absence of this support:

> It wasn't that we relied on each other so much. Not at all. But somehow when I was at the new school and she wasn't there, I suddenly felt her absence a lot. It was like it had been propping me up more than I realized. Without her there, I moved away from those thoughts and those feelings and got into a new life easier.

Most loners, then, lacking this kind of support, had difficulty sustaining their behavior over extended periods of time.

DISCUSSION

As loners, self-injurers represent the least organizationally sophisticated form of deviant association. They are on their own without the benefits of fellow deviants to either assist them in their deviant acts or to keep them company in their private backstage moments when they are not engaged in deviance. They thus lack not only the human resources to mount intricate deviant capers characterized by multiple members and differential role specialization but also the basic rudiments of deviant camaraderie to provide social support; a guide to core deviant norms and values; an ideology that offers rationalizations and neutralizations legitimating their deviant acts; information diffusal providing practical, legal, and medical advice; subcultural jargon and stories to enrich the deviant experience; and a system of status stratification to help them measure participants against each other. Although they become absorbed in their deviance, they often find themselves lonely.

REFERENCES

Best, Joel, and David F. Luckenbill. 1982. *Organizing Deviance*. Englewood Cliffs, NJ: Prentice Hall.

Conterio, Karen, and Wendy Lader. 1998. *Bodily Harm: The Breakthrough Treatment Program for Self-Injurers*. New York: Hyperion.

Cressey, Donald R. 1971. *Other People's Money: A Study in the Social Psychology of Embezzlement*. Belmont, CA: Wadsworth.

Dabney, Dean A., and Richard C. Hollinger. 1999. "Illicit Prescription Drug Use Among Pharmacists: Evidence of a Paradox of Familiarity." *Work and Occupations* 26(1): 77–107.

Favazza, Armand R. 1996. *Bodies Under Siege: Self-Mutilation and Body Modification in Culture and Psychiatry*, 2nd ed. Baltimore: Johns Hopkins University Press.

Favazza, Armand R., and Karen Conterio. 1989. "Female Habitual Self-Mutilators." *Acta Psychiatrica Scandinavica* 79(3): 283–89.

Gordon, Richard. 1990. *Anorexia and Bulimia: Anatomy of a Social Epidemic*. Cambridge, MA: Basil Blackwell.

Harris, Jennifer. 2000. "Self Harm: Cutting the Bad Out of Me." *Qualitative Health Research*. March 10(2): 164–73.

Lowery, Shearon A., and Charles V. Wetli. 1982. "Sexual Asphyxia: A Neglected Area of Study." *Deviant Behavior* 4: 19–39.

Matza, David. 1964. *Delinquency and Drift*. New York: Wiley.

McLorg, Penelope A., and Diane E. Taub. 1987. "Anorexia and Bulimia: The Development of Deviant Identities." *Deviant Behavior* (8): 177–89.

O'Halloran, Ronald L., and Park Elliott Dietz. 1993. "Autoerotic Fatalities with Power Hydraulics." *Journal of Forensic Sciences*, JFSCA 38(2): 359–64.

Scully, Diana, and Joseph Marolla. 1984. "Convicted Rapists' Vocabulary of Motive: Excuses and Justifications." *Social Problems* 31(5): 530–44.

Stevens, Dennis J. 1999. *Inside the Mind of a Serial Rapist*. San Francisco: Austin & Winfield.

Strong, Marilee. 1998. *A Bright Red Scream: Self-Mutilation and the Language of Pain*. New York: Penguin Putnam.

Way, Karen. 1995. "Never Too Rich . . . or Too Thin: The Role of Stigma in the Social Construction of Anorexia Nervosa." Pp. 91–113 in Donna Maurer and Jeffrey Sobal (eds.), *Eating Agendas: Food and Nutrition as Social Problems*. Hawthorne, NY: Aldine de Gruyter.

Winick, Charles. 1964. "Physician Narcotic Addicts." Pp. 261–80 in Howard S. Becker (ed.), *The Other Side: Perspectives on Deviance*. New York: Free Press.

33

Recreational Ecstasy Users

MICHELLE GOURLEY

Deviants who have the benefit of congregating with others like themselves differ in significant ways from loners. Not only can they count on the company and support of colleagues, but they also derive valuable knowledge from their friends. Gourley draws on Becker's (1953) classic study of becoming a marijuana user to examine the way people who share the enjoyment of the drug "ecstasy" learn the attitudes, values, and behavior surrounding the drug. She notes that their word-of-mouth transmission of subcultural knowledge effectively overcomes publicly promoted antidrug messages, teaching participants how to use the drug safely and to get the most out of their experiences. Ecstasy users' behaviors, like those in any sub-culture, are guided by group norms that promote socially shared expectations and understanding while sanctioning deviations.

E arly research on youth drug use suggested that it occurred predominantly in deviant subcultures. Subcultural theories of deviance have therefore been the dominant theoretical framework within which youth drug use has been examined and understood since the 1950s. Subcultural theories focus on the importance of deviant subcultures to the initiation and maintenance of deviant acts. They argue that deviance is the result of a learned acquisition of deviant values and norms within the context of a subculture. . . .

THE STUDY

Ecstasy was chosen as the drug of interest in this study for two main reasons. First, because of the relative newness of the drug, few studies have looked at ecstasy use in any qualitative way (Pedersen and Skrondal, 1999: 1696). Second, ecstasy has become a common recreational drug for middle-class youth. Australian statistics from 2001 show that the lifetime use of the drug ecstasy is 19.7 percent for

From *Journal of Sociology*, 40(1), pp. 59–73, copyright © 2004 by Sage Publications.
Reprinted by permission of Sage Publications, Inc.

people aged 20 to 29 years old, which makes it the third most popular drug for young people after marijuana and amphetamines (AIHW, 2002). The growing popularity of ecstasy is strongly associated with the development of the rave/dance culture in the late 1980s. It is important to note that the ecstasy culture and the dance culture are not one and the same phenomenon, although they are strongly linked (Gourley, 2002).

Observation and semi-structured interviewing were used to investigate ecstasy use in the young population. The observations were drawn from approximately six venues in Canberra and Sydney at which recreational drugs were commonly used, such as dance events, nightclubs and music concerts. These direct observations were accumulated over approximately three years. In addition, semi-structured interviews were conducted with 12 recreational ecstasy users (six female and six male) aged between 20 and 22. They were all of middle-class background and were either university students or in full-time work. The interviewing sessions aimed to uncover subcultural characteristics such as norms of behaviour, values and shared understandings surrounding ecstasy use, and the subcultural learning and socialization processes involved in ecstasy use.

INITIATION INTO ECSTASY USE

Subcultural theories argue that drug users acquire motivations to use a drug in interaction with other users. Howard Becker argued that people's initiation, maintenance and discontinuance of drug use are the consequence of changes in their conception of the drug and that membership in a drug subculture can vary over time as interaction with other users and acceptance of drug lore change (1963, 1967). Individuals' socialization into ecstasy use can thus be looked at as a function of their degree of interaction with users and their attitude change over time towards the use of illicit drugs.

Interviewees reported that at some stage in the past their attitudes towards drugs were fairly negative and that they considered all drugs to be destructive. A couple of interviewees explained that in high school they were fairly distanced from drug use and there was a general perception among students that only the 'bad' kids were associated with drugs. However, they stated that as they grew older they became increasingly exposed to certain drugs such as ecstasy and discovered that these drugs were being used by normal, conventional youth. This led to a change in their attitude towards drug use. They were now of the view that maybe all drugs were not such a bad thing, and could in fact be used to have fun. Such an attitude change had not taken place for other drugs such as heroin. As one interviewee said: "smackees [heroin users] give drug users a bad name as they go past the point of recreation."...

Peer involvement with ecstasy was a key factor in an individual beginning and maintaining ecstasy use. The interviewees in this study shared similar experiences in terms of entry into drug networks and patterns of consumption through involvement in friendship networks that provided them with access to ecstasy and to a social setting for its consumption. The majority of the interviewees' first experiences with ecstasy occurred with friends from their immediate social networks who had used ecstasy before and who played a major role in their decision to try the drug. The importance of being with friends who were accepting of ecstasy use, and being in a setting that was conducive to ecstasy use was stressed by the interviewees for all occasions of use. The need for these conditions to be met when using ecstasy highlights the subcultural underpinnings of ecstasy use. . . .

Initial fears of dangerous experiences from taking ecstasy were challenged by the reassurances of other users who had engaged in ecstasy use with little trouble. It was stressed that friends who were experienced users gave advice and information on safety issues and on the effects of ecstasy before the interviewees made the decision to take ecstasy for the first time. Interaction with friends who use drugs and, in turn, participation in a drugusing subculture, persuade the beginner that drug use can be safe. This supports Zinberg's work which suggests that the drugusing group reinforces an individual's discovery that use of a particular drug is not a bad activity and is worthwhile engaging in (1984: 16). . . .

LEARNING TO PERCEIVE
THE EFFECTS OF ECSTASY

According to Becker's subcultural theory of deviance, users learn to perceive a drug's effects through participation in a drug-using subculture. Becker stressed that the novice drug user acquires the concepts that make an awareness of being high possible through direct communication with other users or via picking up cues from other users within a subcultural context (1963: 50).

Being high not only involves the presence of symptoms, but the recognition and perception of those symptoms and their connection with use of the drug by the user (Becker, 1963: 49). In the present study, ecstasy users were asked about their own experiences of being high from ecstasy use and their observations of others. All those interviewed said that they had seen people who were high on ecstasy but did not believe they were high themselves. This was especially the case with first-timers whom the respondents believed were in denial about the drug's effectiveness and needed to be coached through the experience to be able to recognize the effects of ecstasy. A respondent explained that sometimes people need someone to point out their abnormal behaviours to alert them to the fact that they are high as they are not always conscious of their behaviour. For example, a female respondent remarked that she was unaware that she was high for the first time until other users told her that she was unusually talkative, honest and had huge pupils. Such an example suggests that drug effects alone do not automatically provide the experience of being high, but a user must recognize them (usually through interaction with other users) and consciously connect them with having taken ecstasy. . . .

LEARNING TO DEFINE THE EFFECTS
OF ECSTASY AS PLEASURABLE

Becker argued that drug users learn the pleasures to be derived from an act in the course of interaction with more experienced deviants (1963: 30). He stated that before engaging in the activity of drug use, people have little notion of the pleasures to be derived from it. It is only through experience with the drug coupled with subcultural support that an individual can conceive of the drug experience as pleasurable (Becker, 1963: 52).

All sensations of the drug ecstasy are not necessarily pleasurable such as the potential hallucinations, high alertness and altered sense of touch, which can be perceived as physically unpleasant or ambiguous. Users must learn to define these effects as pleasurable. For example, one interviewee described how she felt paranoid (she believed everyone was staring at her) the first time she took ecstasy, and thought this was abnormal and unpleasant until other users told her that such a feeling was common with the first use of ecstasy and would pass. She was able to relax after this advice and ended up defining her experience as highly pleasurable. This supports Becker's assertion that "how a person experiences the effects of a drug depends greatly on the way others define those effects for him [or her]" (1967: 105)....

Paul Willis argued that drug experiences are socially constructed, culturally learned and defined in the context of subcultural use rather than being based on the properties of different types of drugs (cited in Hall and Jefferson, 1975: 119). Most of the respondents in this study attributed their best experiences with ecstasy to the atmosphere, their frame of mind and presence of close friends. The purity of the pill was considered only a secondary influence. For example, one respondent said that her best experiences with ecstasy were not the result of a good pill but had been when she was around a group of people who were all in a good mood and in a setting conducive to ecstasy use. Another interviewee stated "It's not where you go but who you're with" in reference to what makes an enjoyable ecstasy experience. These comments attest to the importance of atmosphere, companions and state of mind in determining the outcome of the drug experience. It is through involvement in a subculture in which drug use is normalized that a user can get the most out of the drug experience by having access to a drug-conducive environment and interaction with other like-minded users.

TRANSMISSION OF KNOWLEDGE
WITHIN THE ECSTASY SUBCULTURE

Becker's studies of marijuana use along with more recent drug research have emphasized that communication and transmission of knowledge about drugs through stories and folklore represent important aspects of user subcultures (Becker, 1967; McElrath and McEvoy, 2001). Cultural information and options for behaviour are spread through interacting social networks, which leads to a common world of discourse for the members of a subculture (Fine and Kleinman, 1988 cited in Moore, 1990: 338).

The importance of expert knowledge, which is accumulated within subcultures, is emphasized in Grebenc's study of drug subcultures in which prohibited knowledge was shown to circulate among drug users through storytelling (2001: 105). A picture emerged from observations and interviews that much expert knowledge is accumulated inside the ecstasy culture. For example, it is common knowledge among ecstasy users that if you chew rather than swallow an ecstasy pill it will work faster, that smoking marijuana during the come-down phase helps to alleviate the negative effects of ecstasy and sucking a lolly-pop prevents the jaw grinding associated with ecstasy use.

Becker illustrated the importance of an informed user subculture as a way to create a safe context for drug use. He found that a continually evolving folklore educates users which results in individuals being able to gain experience with the drug and communicate their experiences to other users (Becker, 1967; cited in Beck and Rosenbaum, 1994: 143). Within the ecstasy culture stories are transmitted from user to user about how to use the drug safely. For example, stories involving instances of people getting caught with ecstasy or having an unpleasant experience with the drug provide users with the appropriate information and knowledge to guide their own drug-related behaviour. . . .

Deviant activities tend to generate a specialized language to describe the events, people and objects involved, which is learnt through participation in the deviant subculture and through interaction with other members (Becker, 1967). Within the culture surrounding ecstasy use, drug lingo is common among users. There is a vast array of specialized terms used for taking ecstasy, being high, coming down, and names for both the drug and ecstasy users themselves. For example, "dropping," "racking up," "pilling," "peaking," "scattered," "lollies," "disco biscuits" and "eccy-heads" are all terms that have little meaning to those outside the ecstasy subculture.

SHARED UNDERSTANDINGS/VALUES

Subcultural theories maintain that membership in a drug subculture requires evidence that drug users share similar cultural ideas about drugs. Part of the underlying symbolism of drug use is the shared understandings which organize and make sense out of the reality of drug use. The information obtained from interviews with, and observations of, ecstasy users reveals that there are shared ideas, values and beliefs that govern the use of ecstasy among groups of users, which suggests the presence of a drug subculture.

When the effects of ecstasy were discussed with those interviewed, it became clear that users share similar cultural ideas about the positive effects of ecstasy. All those interviewed believed that the drug reduces people's inhibitions by making them more honest and open, and enhances communication and closeness with others. . . .

Becker stressed that most deviant subcultures have a self-justifying rationale to neutralize conventional attitudes and legitimize the continuance of the deviant act (1963: 38). Justifications for ecstasy use were very similar among respondents and can be placed into three main categories. The first type of justification is to do with the perceived safety of the drug. All interviewees were aware that there may be long-term dangers associated with ecstasy use such as brain damage. They showed little concern however and focused on the fact that they hadn't personally seen or heard of many bad experiences with the drug and believed that they did not find it addictive. One interviewee stated that "ecstasy is very safe when you know what you are doing and when you are with others who know what they are doing." Ecstasy's reputation as easily controllable and user-friendly relieves the fears of many people (Beck and Rosenbaum, 1994: 58).

The second self-justifying rationale that came to light when interviewing ecstasy users was that ecstasy use is thought to be so commonplace among young people that the respondents had difficulty in perceiving ecstasy users as committing a criminal act. One respondent said that people see other people using ecstasy who aren't risk takers or the type of people typically associated with drug use and therefore come to see ecstasy use as legitimate behaviour. None of the users interviewed regarded ecstasy use as deviant or as a rejection of social convention. It would seem that the drug ecstasy has had a huge impact on young people's perceptions of drug taking in that it has given ordinary youth a positive experience of illegal drug use.

The third main justification that users gave for their drug use was that ecstasy, like alcohol and tobacco, is used as part of people's leisure. Many interviewees considered ecstasy use to be a good alternative to drinking alcohol. This is because a night out can be cheaper when using ecstasy than drinking alcohol and enables users to drive home without the fear of being caught for drink driving. The belief that ecstasy use can be more convenient than drinking alcohol is further illustrated by this response from a male interviewee: "Taking a pill can be so much easier than drinking 10 beers in order to feel a high." Some respondents even saw the drug ecstasy as less of a danger than alcohol, believing it allowed people to stay more in control of their body.

NORMS OF BEHAVIOUR/SANCTIONS

Subcultures, like cultures in general, prescribe norms which regulate conduct and are ordered socially (Rubington and Weinberg, 1968: 207). This study identified norms of conduct and social sanctions surrounding the use of ecstasy that define moderate and acceptable use, condemn compulsive use, limit use to settings conducive to the drug experience and routinize use.

A picture emerged from interviews and observations that characteristic occasions when ecstasy was used were planned activities, and that a cycle of socializing, dancing and drinking is the norm when the drug is used. Ecstasy can be both a social and an antisocial drug. There are times when users can't stop talking and want to be with others, and there are times when they just want to dance for hours by themselves, get immersed in the music and block out those around them. Part of learning the effects of ecstasy and the norms that go with it is that there are times to be sociable, times to dance and times for self-contemplation, especially when the comedown is experienced.

When asked about where ecstasy is taken, interviewees reported that they had used ecstasy most often at raves or dance events because of the drug-conducive atmosphere of such settings. They said that they had only used ecstasy at specific clubs, bars and house parties where their drug-using behaviours were accepted by those around them. It was clear from observations that social sanctions limit ecstasy use to settings conducive to the drug experience. For example, disapproving looks are often directed at people who are clearly high on drugs in mainstream bars, pubs and clubs where alcohol is the substance of choice. Social interaction with drug users in such venues is typically avoided, especially on the dance floor. Such subtle sanctions are used to enforce conformity.

There are also social sanctions that condemn compulsive use in the ecstasy culture. People who use ecstasy compulsively often find themselves excluded from their drug-using peer group. For example, many of the interviewees said that they no longer socialized with friends who they believed had taken ecstasy use to an extreme. Mechanisms such as social exclusion therefore help to ensure that individuals will conform to convention. Unspoken rules of the drug subculture exert a powerful influence in regulating individual drug-taking behaviour and conduct. . . .

CONCLUSION

First, the findings indicate that ecstasy use occurs in an ecstasy-using subculture. Specific norms of behaviour, social sanctions, shared understandings and values can be identified that are particular to ecstasy users as well as clear rules concerning why, where and how much it is considered legitimate to use ecstasy by groups of users. The research suggests that there is a great deal of normality in user perceptions of their drug use and in the way ecstasy is used by groups in specific social contexts. This supports an underlying assumption of subcultural theories that deviance needs to be understood as normal behaviour in specific social circumstances. It is evident that knowledge and behaviour concerning ecstasy use are transmitted through an interlocking social network, which further suggests the presence of an ecstasy subculture.

REFERENCES

AIHW (Australian Institute of Health and Welfare) (2002) *2001 National Drug Strategy Household Survey: First Results*. AIHW Cat. No. PHE35, Drug Statistics Series No. 9. Canberra.

Beck, J. and M. Rosenbaum (1994) *Pursuit of Ecstasy: The MDMA Experience*. Albany: State University of New York Press.

Becker, H.S. (1963) *Outsiders: Studies in the Sociology of Deviance*. New York: Free Press.

Becker, H.S. (1967) "History, Culture and Subjective Experience: An Exploration of the Social Bases of Drug-induced Experiences," *Journal of Health and Social Behaviour* 8(3): 163–77.

Gourley, M. (2002) "A Subcultural Study of Recreational Ecstasy Use," ch. 5, unpublished Honours Thesis, Australian National University.

Grebenc, V. (2001) "Expert Reports on Basis of Experience and the Forbidden Knowledge in the Narrative of Drug Users," *Socialno-Delo* 40(4): 105–17.

Hall, J. and T. Jefferson (eds) (1975) *Resistance through Rituals: Youth Subcultures in Post-war Britain.* London: Routledge.

McElrath, K. and K. McEvoy (2001) "Fact, Fiction and Function: Mythmaking and the Social Construction of Ecstasy Use," *Substance Use and Misuse* 36(1–2): 1–22.

Moore, D. (1990) "Anthropological Reflections on Youth Drug Use Research in Australia: What We Don't Know and How We Should Find Out," *Drug and Alcohol Review* 9(4): 333–42.

Pedersen, W. and A. Skrondal (1999) "Ecstasy and New Patterns of Drug Use: A Normal Population Study," *Addiction* 94(11): 1695–706.

Rubington, E. and M.J. Weinberg (1968) *The Interactionist Perspective: Text and Readings in the Sociology of Deviance.* New York: Macmillan.

Zinberg, N.E. (1984) *Drug, Set and Setting: The Basis for Controlled Intoxicant Use.* New Haven, CT: Yale University Press.

34

Real Punks and Pretenders: The Social Organization of a Counterculture

KATHRYN J. FOX

Fox's study of a Midwestern city's punk scene, or subculture, examines the social relations found in this group of people who associate together but do not need each other to be punk. She presents four types of punks in the scene and examines the relations among them. Starting at the center, she looks at the hardcore punks, gradually working outward to the softcore punks, the preppie punks, and finally to the spectators. For each group, she examines their immersion in the punk lifestyle and commitment to punk ideals, activities, style of dress, and mode of survival. Membership in the hardcore inner group requires a more serious dedication than does participation in the transitory outer fringe, yet these style leaders feed on the presence and adulation of the more peripheral groups for sustenance. Each group fills a distinctly different role in the subculture and its members identify themselves largely in relation to each other.

Modern Western society has been characterized by a variety of anti-establishment style countercultures following in succession (i.e., the Teddy Boys of 1953–1957, the Mods and Rockers of 1964–1966, the Skinheads of 1967–1970, and the Punk Rockers of the late 1970's [Taylor, 1982]). The punk culture is but the latest in this series. Since most studies of youth- and style-oriented groups are British (Frith, 1982), very little has been written from a sociological perspective about punks in the United States. This study is an attempt to fill that void.

Whether or not the punk scene in the United States could be legitimately classified as a social movement is debatable. Most writers on this contemporary phenomenon agree that American punks have a more amorphous, less articulated ideological agenda than punks elsewhere (Brake, 1985; Street, 1986). While the

From "Real Punks and Pretenders: The Social Organization of a Counterculture," Kathryn J. Fox, *Journal of Contemporary Ethnography*, Vol. 16, No. 3, 1987. Reprinted by permission of Sage Publications, Inc.

punk scene in England responded to youth unemployment and working-class problems, the phenomenon in the United States was more closely connected to style than to politics. Street (1986: 175) notes that even for English punks, "politics was part of the style." I would argue that this was even more the case for American punks. The consciousness of the youth in the United States did not parallel the identification with the plight of youth found in Europe. Nonetheless, the "style" code for punks in the United States contained an insistent element of conflict with the dominant value system. The consensual values among the punks, as ambiguous as they were, could best be understood by their contradictory quality with reference to mainstream society. In this respect, punk in America fit the definition of a "counterculture" offered by Yinger (1982: 22–23). According to this definition, the salient feature of a counterculture is its contrariness. Further, as opposed to individual deviant behavior, punks constituted a counterculture in that they shared a specific normative system. Certain behaviors were considered punk, while others were not. Indeed, style was the message and the means of expression. Observation of behaviors that were consistent with punk sensibilities were viewed as indicative of punk "beliefs." These behaviors, along with verbal pronouncements, verified commitment. Within the groups of punks I studied, the degree of commitment to the counterculture lifestyle was the variable that determined placement within the hierarchy of the local scene.

Previous portrayals of youthful, antiestablishment style cultures have discussed their norms and values (Berger, 1967; Davis, 1970; Hebdige, 1981; Yablonsky, 1968), their relationship to conventional society (Cohen, 1972; Douglas, 1970; Flacks, 1971), their focal concerns and ideology (Flacks, 1967; Miller, 1958), and their relationship to social class (Brake, 1980; Hall and Jefferson, 1975; Mungham and Pearson, 1976). With the exception of Davis and Munoz (1968), Kinsey (1982), and Yablonsky (1959), few of the studies of antiestablishment, countercultural groups discuss their implicit stratification. In this essay I will describe and analyze the various categories of membership in the punk scene and show how members of these strata differ with regard to their ideology, appearance, taste, lifestyle, and commitment.

I begin by discussing how I became interested in the topic and the methods I employed to gain access to the group and to gather data. I then offer a description of the setting and the people who frequented this scene. Next, I offer a structural portrayal of the social organization of this punk scene, showing how the layers of membership form. I then examine each of the three membership categories (hardcore punks, softcore punks, and preppie punks), as well as the spectator category, focusing on the differences in their attitudes, behavior, and involvement with this antiestablishment style culture. I conclude by outlining the contributions each of these types of members makes to the continuing existence of the punk movement and, more broadly, by describing the relation between the punk counterculture and conventional society.

METHODS

My interest in the punk movement dates back to 1978. At that time, I attended a local punk bar fairly regularly and wore my hair and clothing in "punk" style, albeit not the radical version. I also visited a major northeastern city at about that same time, when the punk scene was in full flower, and spent several nights visiting what are now famous punk hangouts. My early interest and involvement in this scene laid the groundwork for this study, as it permitted me to gain knowledge of the punk vernacular, styles, and motives. The research continued, with active, weekly participation, through the middle of 1986. I have continued to keep a close watch on national trends and developments in punk culture. Further, I continually frequent local punk bars in an effort to deepen my understanding and to observe the decline of this counterculture. However, I conducted the bulk of my interviews in the fall of 1983 over a period of about two months. During that time, I attended "punk night" at a local bar once a week. In addition, I was invited to other punk functions, such as parties, midnight jam sessions, and public property destruction events. I thereby observed approximately 30 members of this movement with some degree of regularity. I used mainly observational techniques, along with some participation. Although I frequented numerous punk gatherings, my participatory role was constrained by the limited time I spent there. I also had to tread a line between covert and overt roles. While some people knew I was researching this setting, I could not reveal this to others because they might have denied me further access to the group. This created a problem, much the same as that experienced by Henslin (1972) and Adler (1985), in that I had to be careful about what I said and to whom I confided my research interests. This "tightrope effect" severely limited my active participation in the scene.

Nevertheless, by following the investigative research techniques advocated by Douglas (1976), I was able to gain the trust of some key members. I tried to establish friendly relations by running errands for them, buying them drinks and food, and driving them to pick up their welfare checks and food stamps. After several weeks, when I began to be recognized, I was able to broach the topic of doing interviews with several people. I formally interviewed nine people at locations outside the bar. These tape-recorded interviews were unstructured and open-ended. Additionally, I conducted 15 informal interviews at the bar. Finally, I had countless conversations with members, nonmembers, interested bystanders, and social scientists who had an interest in the punk counterculture. In all, my somewhat punk appearance, similar age, regularity at the scene, and apparent acceptance of their lifestyle allowed me to move within the scene freely and easily.

SETTING

The research took place in a small cowboy bar, "The Glass Gun," which was transformed into a punk bar one night a week. The bar was situated in a southwestern city with a population of about 500,000. The city itself is located in the "Bible Belt," characterized by conservative religious and political views. The bar was a small, dark, and dilapidated place. There was a stage area where the bands played, surrounded by a wooden rail. Wobbly tables and torn chairs formed a U-shape around the stage. There was a pool table in the corner, which the punks rarely used. Most of the patrons of the club stood at or near the bar.

For the punks in this city, the Glass Gun was the only place to congregate regularly at that time. On these designated nights, local punk bands played to an audience of about 20 people; some were punks, others were not. The typical audience ranged in age from about 16 to 30, although a few were younger or older. Basically, the punk counterculture was a youth phenomenon. It seemed to attract young, single, mobile people. Snow et al. (1980) have suggested that these characteristics make a person more "structurally available" for movement recruitment. Within the punk scene the number of men and women was fairly equal.

The punk style codes were somewhat diverse. Different styles existed for different kinds of punk. Pfohl (1985: 381) has referred to Hebdige's description of punk style as "the outrageous disfigurement of commonsensical images of aesthetics and beauty and the abrasive, destructive codes of punk style. These are aesthetic inversions of the normal, or consensus-producing, rituals of the dominant culture's style." The basic identifiable element was a subculturally accepted punk hairstyle. These ranged from a very short, uneven haircut, sticking straight up in front, to an American Indian mohawk style, to a shaven head. Along with the haircut, a punk fashion prevailed. The two were inextricably associated. The fashion ranged from torn, faded jeans, T-shirts, and army boots to expensive leather outfits.

The punk dress code was also fairly androgynous. There was no real distinction between male and female fashions. Both men and women wore faded jeans, although leather pants and miniskirts were also quite common among the women. The middle-class punk women, who tended to be students, dressed in a more traditionally feminine manner, glorifying and exaggerating the "glamour girl" image reminiscent of the sixties. This included tight skirts, teased hair, and dark, heavy makeup. The other punk women identified with a more masculine, working-class image, deemphasizing their feminine attributes. Both sexes also wore and admired leather jackets. It was also quite common to see both men and women with multiple pierced earrings all the way around the outside of their ears. Men sometimes wore eye makeup as well. (One man wore miniskirts, makeup, and rhinestones. However, this type of behavior occurred infrequently.) Basically, punk style ran counter to what the dominant culture would deem aesthetically pleasing. One major reason punks dressed as they did was to set themselves apart and to make themselves recognizable. The image consisted of dark, drab clothing, short, spiky, "homemade" haircuts, and blank, bored, expressionless faces reminiscent of those of concentration camp prisoners.[1] The punks created a new aesthetic that revealed their lack of hope, cynicism, and rejection of societal norms.

THE SOCIAL ORGANIZATION
OF THE PUNK SCENE

Like the youth gangs that Yablonsky (1962) studied, members of this local punk scene constituted a "near-group." The membership was impermanent and shifting, members' expectations were not always clearly defined, consensus within the group was problematic, and the leadership was vague. Yet out of this uncertainty surfaced an apparent consensus about the stratification of the local community and the roles of the three types of members and peripheral hangers-on who participated in this scene. These four typologies can be hierarchically arranged by the presence (or absence) and intensity of their commitment to the punk counterculture and their consequent display of the punk affectations and belief system. They thus formed a series of outwardly expanding concentric circles, with the most committed members occupying the core, inner roles, and the least involved participants falling around the periphery.

Starting from the center, the number of members occupying each stratum progressively increased as the commitment level of the participants diminished. The categories to which I refer come from the terms used by the participants themselves.[2] The *hardcore punks* were the most involved in the scene, and derived the greatest amount of prestige from their association with it. They set the trends and standards for the rest of the members. The *softcore punks* were less dedicated to the antiestablishment lifestyle and to a permanent association with this counterculture, yet their degree of involvement was still high. They were greater in number and, while highly respected by the less committed participants, did not occupy the same social status within the group as the hardcores. Their roles were, in a sense, dictated by the hardcores, whom they admired, and who defined the acceptable norms and values. The *preppie punks* were only minimally committed, constituting the largest portion of the actual membership. They were held in low esteem by the two core groups, following their lead but lacking the inner conviction and degree of participation necessary to be considered socially desirable within the scene. Finally, the *spectators* made up the largest part of the crowd at any public setting where a punk event transpired. They were not truly members of the group, and therefore did not necessarily revere the actions and dedication of the hardcores as did the two intermediary groups. They did not attempt to follow the standards of those committed to this near-group. They were merely outsiders with an interest in the punk scene.

These four groups constituted the range of participants who attended and were involved with, to varying degrees, the punk counterculture. I will now examine in greater detail their styles, beliefs, practices, intentions, and roles in the scene.

PUNKS AND COMMITMENT

At the time of this study, the group of punks was small and disorganized. The number of people at any given punk event had steadily declined since my first encounters with the scene in 1978. Punk was no longer a new phenomenon, and this particularly conservative community did not provide a very conducive atmosphere for a large countercultural group to flourish. Every member of the group expressed dissatisfaction and boredom with the events (or rather, lack of events) within the scene. Even within the limits necessitated by the relatively small size of the group, there was a great deal of variation in terms of punk roles and characteristics. The qualities attributed to the different roles were based upon commitment to the scene. The punks' perceptions of levels of commitment were based principally on their evaluation of physical appearances and lifestyles. The punks categorized members of the scene on these bases and invented terms to describe them. The four types of participants, described above, varied according to their level of commitment to the scene.

Hardcore Punks

Hardcore punks made up the smallest portion of the scene's membership. In the eyes of the other punks, though, they were the essence of the local movement. The hardcores expressed the greatest loyalty to the punk scene as a whole. Although the hardcores embodied punk fashion and lifestyle codes to the highest degree, their commitment to the counterculture went much deeper than that. As one hardcore punk said:

> There's been so much pure bullshit written about punks. Everyone is shown with a safety pin in their ear or blue hair. The public image is too locked into the fashion. That has nothin' to do with punk, really. . . . For me, it is just my way of life.

The feature that distinguished hardcore punks from other punks was their belief in, and concern for, the punk counterculture. In this sense, the hardcore punks had gone beyond commitment; they had undergone the process of conversion (Snow et al., 1986). In other words, not only did they have membership status, but they believed in and espoused the virtues and ideology of the counterculture. Although many hardcores differed on what the counterculture's core values were, they all expressed some concern with punk ideology. These values were ambiguous at best, but included a distinctly antiestablishment, anarchistic sentiment. Street (1986: 175) has described punk as celebrating chaos and "a life lived only for the moment." The associated value system of punk was understood by the incorporation of cynicism and a distrust of authority. In keeping with other subcultures that intentionally distinguish themselves from the dominant culture, the punk aesthetic, lifestyle, and worldview directly confronted those of the larger society and its traditions. While the other types of punks made no reference to group beliefs or values, the hardcores revered the counterculture. For them, being punk had a profound effect on all aspects of their lives. As one hardcore said:

> There are a lot of punks around, even real punks, who don't mean it. At least, not all the way, like I do. Sometimes I feel so good about punk that I cry. And when I see people getting into some band with real punk lyrics, it's like a religious experience.

It was precisely this belief in "punk" as an external reality, like a higher good, that set the hardcores apart. Similar to Sykes and Matza's (1957) "appeal to higher loyalties," hardcore punks based their rejection of conventional society on their commitment to their antiestablishment lifestyles and beliefs. This imbued their self-identity with a sense of seriousness and purpose. Unlike other punks, they did not view their punk identity as a temporary role or a transitory fashion, but as a permanent way of life. As one hardcore member said:

> Punk didn't influence me to be the way I am much. I was always
> this way inside. When I came into punk, it was what I needed all my life.
> I could finally be myself.

Without exception, the hardcores reported having always held the values or qualities associated with the punk counterculture. The local scene, in fact, was just a convenient way of expressing these ideas collectively. Perhaps the most essential value professed by the punks was a genuine disdain for the conventional system. Their use of the term *system* here referred to a general concept of the way the material world works: bureaucracies, power structures, and competition for scarce goods. This "system" further referred to the ethic of deferred gratification, conventional hard work for profit, and the concept of private property. While this bears some similarity to Flacks's (1967) discussion of the student movement and Davis's (1970) portrayal of hippies, hardcore punks generally had a disdain for these earlier youth subcultures. There was a general attitude among punks of the need to create and maintain their own distinctive style. Kinsey found this same feature in the antibourgeois "killum and eatum" subculture. According to Kinsey (1982: 316), "K and E offered an attractive setting as its ideology presented an excellent vehicle for expressing hostility toward conventional society." This contempt for authority and the conventional culture was, in fact, such an essential value for the punks that if one expressed prosystem sentiments or support for the present administration, one could not be considered a committed member, no matter how well one looked the part. Overt behavioral and physical attributes, though, were major ways hardcore punks showed disdain for the system. Particular characteristics were essential for consideration as hardcores. Most fundamentally, a verbal commitment to punk values and the punk scene, in general, was required. For example, John, the epitome of a hardcore punk, claimed to hate the system. He talked about the inequality of the system quite often. In John's words:

> Punk set me free. It let me out of the system. I can walk the streets
> now and do what I want and not live by the demands of the system.
> When I walk the streets, I am a punk, not a bum.

However, this verbal pronouncement had to be backed by a certain lifestyle that further indicated commitment to the group. This lifestyle consisted of escaping the system in some way. Almost all of the hardcores were unemployed and lived in old, abandoned houses or moved into the homes of friends for periods of time. Some survived from the charity of sycophantic, less committed punks. Others worked in jobs that they considered to be outside the system, such as musicians in rock bands or artists.

Another central feature of the lifestyle was the hardcores' use of dangerous drugs. Many hardcores indulged heavily in sniffing glue. Glue was inexpensive and readily available to the punks. Its use also symbolized the self-destructive, nihilistic attitude of hardcores and their desire to live outside of society's norms. As one member said:

> It is kind of like a competition, a show-off thing. . . . See who has the most guts by seeing who can burn his brain up first. It is like a total lack of care about anything, really.

This closely corresponds to Davis and Munoz's (1968) description of "freaks." Both punks and freaks were "in search of drug kicks as such, especially if [their] craving carries [them] to the point of drug abuse where [their] health, sanity and relations with intimates are jeopardized" (Davis and Munoz, 1968: 306). Again, here we see a rejection of anything the larger society sees as "sensible."

However, the most salient feature of the hardcore lifestyle was the radical physical appearance. In every case, people who were labeled as hardcore had drastically altered some aspect of their bodies. For example, in addition to the hairstyles discussed earlier, they often had tattoos, such as swastikas, on their arms or faces. Brake (1985: 78) has referred to the use of the swastika as a symbol for punks that was actually devoid of any political significance. Rather, the swastika was a "symbol of contempt" employed as a means of offending the traditional culture. The hardcore punks did their best to alienate themselves from the larger society.

According to Kanter (1972), the first requisite in the principle of a gestalt sociology is that a group forms maximum commitment to this higher ideal by sharply differentiating itself from the larger society. The hardcore members did this by going through the initiation rite of passage: semipermanently altering their appearance. As one hardcore said:

> Did you see Russell's mohawk? I'm so glad for him. He finally decided to go for it. Now he is a punk everywhere . . . no way he can hide it now.

This was similar to certain religious cults, such as the Hare Krishnas, where a drastic change in appearance was required for consideration as a total convert (Rochford, 1985). The punk counterculture informally imposed the same prerequisite. By doing something so out of the ordinary to their appearance, the punks voluntarily deprived themselves of some of the larger society's coveted goods. For example, many of the hardcores were desperately poor. They said that they knew all they would have to do to obtain a job would be to grow their hair into a conventional style; yet they refused. This kind of action based on commitment was what Becker (1960) has called "side bets," where committed people act in such ways that affect their other interests separate and apart from their commitment interests. By making specific choices, people who are committed sacrificethe possible benefits of their other roles. An important characteristic of Becker's notion of side bets is that people are fully aware of the potential ramifications of their actions. This point was illustrated by one punk:

> Some of my friends that aren't punk say, "Why don't you get a job? All you'd have to do is grow your hair out or get a wig and you could get a job." I mean, I know I could. Don't they think I knew that when I did it? It was a big step when I finally cut my hair in a mohawk.

For the hardcore punk, being punk was worth the sacrifices; it was perceived as an inherently good quality. In this respect, the hardcores differed from other types of punks. They held the larger punk scene in esteem. As one loyal member said:

> It really pisses me off when people act like ours is the only punk group in the world. They don't even care what bigger and better groups exist. If this whole thing ended tomorrow, they wouldn't care what happened to the whole punk scene.

The hardcores continually expressed their disgust with the local scene. Much like the hippies studied by Davis (1970), "the scene," in itself, was the message. While not enough people joined the group to satisfy the punks, they were, nonetheless, grateful that they had any kind of group environment to which they could attach themselves. The Glass Gun, with its regularly scheduled "punk night," nonhostile attitude toward them, and coterie of interested bystanders, at least gave them a place to express their values collectively. It was essential in maintaining the group's solidarity and social organization.

Softcore Punks

The softcore punks made up a larger portion of the local scene than the hardcores. There were around 15 softcore punks. There was one fundamental difference between the hardcore and softcore punks. For the hardcores, it was not sufficient just to be antiestablishment or to wear one's hair in a certain way. Rather, one had to embody the punk lifestyle and ideology in all possible ways. As one hardcore punk put it:

> Everybody thinks she is hardcore because she looks so hardcore. I mean, yeah, she has a mohawk, and she won't get a job and she says she's for anarchy, but she doesn't care that much about being punk. She likes all these different kinds of music and stuff. She seems sometimes like she is just in it for fun. She even says she'll be whatever's in when punk goes out!

The softcore punks lived similar lifestyles to the hardcores. However, the element of "seriousness" about the scene, so pervasive among the hardcores, was absent among the softcores. Visually, the two types were basically indistinguishable. They were different only in their level of commitment. The commitment for softcores was to the lifestyle and the image only, not to "punk" as an ideology or an intrinsically valuable good. The softcores made no pretense of concern for either the larger counterculture or the feeling of permanence about their punk roles. As one softcore, Beth, said:

> Everyone thinks I am so serious about it because I have a mohawk. Some people just can't get past it. Sometimes I get tired of it. Other times, I like to play jokes on people; like another friend of mine who has a mohawk, we'll walk down the street and point at someone with regular hair and say, "Wow! Look at him, he's weird, he doesn't have a mohawk." The fact is, if everyone did have one, I'd do something different to my hair.

The softcores identified with the punk image only temporarily. This distinguished their level of commitment from the *conversion* of the hardcores. The softcores' interest in the scene had only to do with what it could offer them at the present time. While participating, they did what was considered a good job of being "punk." However, if a new cultural trend surfaced, it would be just as likely that they would use their energy effectively to create that particular image. As one softcore punk said:

> I've spent time identifying myself as a hippie, then as a women's libber, then an ecologist, and now as a punk. I'm punk now, but I am in the process of changing into something else. I don't know what. I'm getting bored with this scene. But for now, if I'm gonna do it, I'll do it right.

The softcore punks were somewhat committed in that they participated in some of the more drastic elements of punk lifestyle. For example, softcores had their hair cut in severe ways, just like the hardcores. They were, at least temporarily, committed to being punk (or playing punk) in that they "cut" themselves off from some of society's goods as well. However, the softcores did not share the self-destructive bent of the hardcores. The drugs that they consumed, such as marijuana, alcohol, and amphetamines, were not so potentially dangerous. Yet, because of their apparent visual commitment, and because of the lip service they gave to punk values, softcores were viewed as members in good standing. The hardcores liked and respected the softcores; the two groups associated freely. Some hardcores considered softcores to be simply members in transition. Lofland and Stark (1965) have suggested that movements themselves play a role in promoting the ideology in the new members, rather than the members coming to the movement because its ideology coincides with their own established beliefs. This was the case for the softcore punks. They did not claim to have held punk values before becoming punk. As Anne, a softcore, recalled:

> It was scary to me at first. The hype from the magazines and stuff—all this weird shit, y'know. Then I went there and just hung out. The reason it was frightening is that a lot of people had different ideas about life than me. And I had to change myself to be with them. I had to be more intense, be an outcast. It was exciting because there seemed to be an element of danger in it—like living on the edge.

A process of simply happening onto the scene was typical of softcore members. Many recounted the feelings of purposelessness that preceded the drift into the punk scene. This drift is similar to the drifts that occur in other deviant lifestyles (Matza, 1964). What Matza called the "mood of desperation" often caused people to drift into delinquency or deviant lifestyles. As Joanie said:

> I was really doing nothing with my life and I just kinda accidentally came into the punk scene. I gradually got involved in it that way. The music, and the people to an extent, really raised my consciousness about the system.

Softcores' verbal recognition of punk attitudes, such as awareness of the system, helped to validate their punk performances. Hardcores felt that verbal commitment was an essential first step to further commitment. For this reason, the hardcores accepted the softcores and considered them to be genuine and authentic in their punk identity. Such identification with punk values, along with a typical punk lifestyle, made the distinction between hardcores and softcores difficult. Again, the distinction became clear only with regard to the level of commitment, or seriousness, of the two types. One softcore made this qualification more apparent:

> They get mad at me and think I'm insincere or whatever 'cause I like to have fun. I take my politics serious, too, but I feel if you are here, you might as well enjoy it. They think being punk is so serious, they are depressed or stoned all the time.

This statement indicates that the hardcores defined the situation for the local scene. The hardcores decided what differentiated real punks (or committed punks) from pretenders. The hardcores considered only themselves and the softcores to be real. The "realness" of a punk was based on the level of commitment. The level was judged on the basis of willingness to sacrifice other identities for the punk identity. To prove this, a member would have to make permanent his or her punk image. What the punk identity offered was status within its own subculture for those who could not or would not achieve it in conventional society (Cohen, 1955). However, commitment to the deviant identity did not stem from a forced label. On the contrary, commitment to the punk identity was a "self-enhancing attachment" (Goffman, in Stebbins, 1971). The punks' self-esteem was enhanced by the approval they received. It would follow, then, that the more consistent one's behavior was with the superficial signs of commitment, the more prestige one would be able to obtain. Doug, a softcore, commented on this aspect of subcultural prestige among the punks:

> In their own way, they're elitist. It's kind of like because they're not part of the general run of things, because they've actually *chosen* to be rejected in a lot of cases, they've kind of set up their own little social order. It seems to me like it's based on, like a contest, who can be more cool than who. With the really hardcore punks, it's who can self-destruct first; in the name of punk, I guess.

The hardcores and the softcores used the same criteria to judge commitment. Both types agreed that the difference between them was their levels of commitment. Both types fully realized that the softcores did not share the same loyalty to, and identification with, the punk counterculture as a whole. Although both types expressed some commitment to the punk identity and lifestyle, they both realized that the hardcores viewed their own identities as permanent and the softcores' as temporary.

Preppie Punks

The preppie punks made up an even larger portion of the crowd at punk events. The preppies frequented the scene, but approached it similarly to a costume party. They were concerned with the novelty and the fashion. The preppies bore some resemblance to Yablonsky's (1968) "plastic hippies" in that they were drawn to the excitement of the scene. Whereas the core members acted nonchalant and natural about being punk, the preppies could not hide their enthusiasm about being part of the scene. This feature contributed to the core members' perceptions of preppies as "not real" punks. As one core member said:

> It really kills me when these preppie girls come up to me and say "Oh, wow, you're so punk; you're so new wave," like I'm really trying or something.

Preppie punks did not lead the lifestyle of the core members. The preppie punks tended to be from middle-class families, whereas the core punks were generally from lower- or working-class backgrounds.[3] Preppie punks often lived with their parents; they tended to be younger, and were often in school or in respectable, system-sanctioned jobs. This quasi-commitment meant that preppies had to be able to turn the punk image on and off at will. For example, a preppie punk hairstyle, although short, was styled in such a versatile way that it could be manipulated to look punk sometimes and conventional at other times. Preppie fashion was much the same way. Mary, a typical preppie punk, put her regular clothes together in a way she thought would look punk. She ripped up her sorority T-shirt. She bought outfits that were advertised as having the "punk look." Her traditional bangs transformed into "punk" bangs, standing straight up using hair spray or setting gel. The distinguishing feature of preppie punks was the manufactured quality of their punk look. This obvious ability to change roles kept the preppies from being considered real or committed. The preppies were not willing to give anything up for a punk identity. As one softcore said:

> They come in with their little punk outfits from Ms. Jordan's [an exclusive clothing store] and it's written all over 'em: money. They think they can have their nice little jobs and their semipunk hairdo and live with mom and dad and be a real punk, too. Well, they can't.

Another said of preppies:

> It's a little hard to take when you have nothing and they try to have everything. Having all that goes against punk. They gotta choose to not have it. Otherwise, they're just playing a game.

The preppies liked to disavow their punk association in situations that would sanction them negatively for such associations. This state of "dual commitments," in which they never had to reject the conventional world in order to be marginally a part of the group, was characteristic of preppie punks (Cohen, 1955). Kanter (1972) has described a process of conversion and commitment that is commonly found in communes. The first step in the process was the renunciation of previous identities. According to this model, the preppie punks would not be considered committed at all. Thus they could not have been categorized as punks in any meaningful sense. As one core member said, "Being 'punk' to them is like playing cowboys and Indians."

Criticizing and joking about the preppies made up a large portion of core members' conversations. Some truly disliked the preppies and others were flattered by their feeble attempts at imitation of core behavior. For example, when a preppie punk approached one core member, he rolled his eyes and said, "Here comes my fan club," with a half-embarrassed smile and a distinct look of pleasure on his face.

Also, the financial function that the preppies served to core members made them more tolerable. Preppies almost always had jobs or survived by their parents' support. Many of the core members subsisted on the continued generosity of their devout fans. The preppies were more than willing to help the other punks. Preppies sometimes offered hardcores financial help in the form of buying them groceries, driving them places, and providing them with cigarettes, alcohol, and other drugs. Because of this, many punks felt that they could not afford to reject outwardly those who were less committed. As Anne said, "One of these days, this kindness is going to dry up."

Yet the joking and poking fun at preppies was a constant activity. It served to separate, for the committed punks, "us" from "them." It reinforced their sense of being the only real punks. Again the distinction made by core members was grounded in the preppies' attempts to play numerous roles. Haircut and clothing were the decisive clues. The real punks could spot a preppie from a distance; they never had to say a word. As one core member said,

> Oh look, she's punked out her hair. Yes, we're impressed. Tomorrow she'll look just like a Barbie doll again.

Perhaps the most definitive statement separating the real punks from the preppies referred to lifestyle:

> All I know is that I live this seven days a week, and they just do it on weekends.

The preppies, though, while definitely removed from core members, still played an important role in the scene.

Spectators

The category of spectators referred to everyone who observed the scene fairly regularly, but were not punks themselves. This type consisted of, literally, "everyone else." They made up the largest portion of the crowd at the Glass Gun on any given night. They were different from the preppie punks in that they did not try to look punk. They made no pretense of commitment to the scene at all. They did not identify themselves as punks; they had no stake in the scene. Spectators consisted of all different types of people and varied in their occupations, clothes, and reasons for being there. The only common denominator this group shared was the desire to stand back and watch, rather than to participate actively in punk activities. One spectator said of his involvement in this scene:

> People on the fringe are usually voyeurs of a sort. They like to be on the receiver's end of what's happening. Maybe punk is really their alter ego. And maybe that need is satisfied just by watching and pretending. That's how it is with me, anyway.

The spectators liked to observe the fashion, to listen to the music, and to be "in the know" about the scene. They were, in other words, punk appreciators.

For the most part, spectators on the fringe were ignored by the core members. They never received the attention that preppies did because they made no attempts to "play punk." However, if a spectator appeared on the scene looking completely antithetical to punk, core members would simply laugh or say something derogatory about them and drop the subject. For example, one time a hippie-looking character came in and one punk said, "Oh my God, I think we're in a time warp," to which another punk responded, "Maybe we should tell him that Woodstock's over and that it is 1983."

Following such statements, the punks would watch the spectator's reaction to the scene. For the most part, except as a diversion, the punks were uninterested in the spectators. They did not generally associate with them or talk about them much. Presumably this was the case because of the tremendous turnover in spectator membership.

Most spectators either slowly began to identify with the group (most core members started out as spectators) or stopped frequenting the punk events. There were, however, some loyal spectators. They would frequent the club. They knew most of the punks at least slightly. The punks generally liked this sort of spectator because they provided the punks with an audience. Every type of punk thrived on an audience. The punks needed people to shock. The spectator served that function. The attitude that the members had toward the spectators was one of tolerance and indifference. As one core member said of them:

> They're into it for the novelty. It's like going to the circus for them, to be a part of something new and exciting. But that's okay. I like going to the circus, too; I just like being in it better.

Thus though spectators were only peripheral to the scene they provided an alternative set of norms that functioned to delineate the social boundaries of the counterculture. . . .

NOTES

1. I am indebted to David Matza and John Torpey for this analogy.

2. With the exception of the term *softcore,* all of the distinctions between categories came directly from the participants. The members did make a distinction between hardcore and what I am calling softcore punks. However, the softcores were referred to simply as "punks" by the hardcores, in an effort to distinguish the "hardcore" quality they attributed to themselves. I chose to refrain from using the term punk to apply to one specific category so that I can use the term more freely and generally, and to avoid confusion.

3. Very little information is provided in this text about the class, race, and ethnicity of these participants. The community from which these data come is relatively homogeneous. The few references to class are more impressionistic; that is, based upon knowledge of family occupations, school districts, and so on. However, the dearth of this kind of data stems from the fact that I was more interested in the features the members had in common than in the distinctions between them, with the exception of their differing levels of commitment and their styles.

REFERENCES

Adler, P. A. (1985). *Wheeling and Dealing.* New York: Columbia Univ. Press.

Becker, H. S. (1960). "Notes on the concept of commitment." *Amer. J. of Sociology* 66: 32–40.

Berger, B. (1967). "Hippie morality—more old than new." *Transaction* 5: 19–27.

Brake, M. (1980). *The Sociology of Youth Culture and Youth Subcultures.* London: Routledge.

———. (1985). *Comparative Youth Culture: The Sociology of Youth Culture and Subcultures in America, Britain, and Canada.* London: Routledge.

Cohen, A. (1955). *Delinquent Boys.* Glencoe, IL: Free Press.

Cohen, S. (1972). *Folk Devils and Moral Panics.* New York: St. Martin's.

Davis, F. (1970). "Focus on the flower children: Why all of us may be hippies some day." In J. Douglas (ed.), *Observations of Deviance* (pp. 327–340). New York: Random House.

Davis, F., and L. Munoz. (1968). "Heads and freaks: Patterns and meanings of drug use among hippies." *J. of Health and Social Behavior* 9: 156–164.

Douglas, J. (1970). *Youth in Turmoil.* Washington, DC: National Institute of Mental Health.

———. (1976). *Investigative Social Research.* Newbury Park, CA: Sage.

Flacks, R. (1967). "The liberated generation: An exploration of the roots of student protest." *J. of Social Issues* 23: 52–75.

———. (1971). *Youth and Social Change,* Chicago: Markham.

Frith, S. (1982). *Sound Effects.* New York: Pantheon.

Hall, S., and T. Jefferson (eds.). (1975). *Resistance Through Rituals.* London: Hutchinson.

Hebdige, D. (1981). *Subcultures: The Meaning of Style.* New York: Methuen.

Henslin, J. (1972). "Studying deviance in four settings: Research experiences with cabbies, suicides, drug users, and abortionees." In J. Douglas (ed.), *Research on Deviance* (pp. 35–70). New York: Random House.

Kanter, R. M. (1972). *Commitment and Community: Communes and Utopia in Sociological Perspective.* Cambridge, MA: Harvard Univ. Press.

Kinsey, B. A. (1982). "Killum and eatum: Identity consolidation in a middle class poly-drug abuse subculture." *Symbolic Interaction* 5: 311–324.

Lofland, J. and R. Stark. (1965). "Becoming a world saver: A theory of conversion to a deviant perspective." *Amer. Soc. Rev.* 30: 862–875.

Matza, D. (1964). *Delinquency and Drift.* New York: John Wiley.

Miller, W. (1958). "Lower class culture as a generating milieu of gang delinquency." *J. of Social Issues* 14: 5–19.

Mungham, G., and G. Pearson (eds.). (1976). *Working Class Youth Culture.* London: Routledge.

Pfohl, S. (1985). *Images of Deviance and Social Control.* New York: McGraw-Hill.

Rochford, E. B., Jr., (1985). *Hare Krishnas in America.* New Brunswick, NJ: Rutgers Univ. Press.

Snow, D. A., E. B. Rochford, Jr., S. K. Worden, and R. D. Benford. (1986). "Frame alignment and mobilization." *Amer. Soc. Rev.* 51: 464–481.

Snow, D. A., L. Zurcher, Jr., and S. Ekland-Olson. (1980). "Social networks and social movements: A micro-structural approach to differential recruitment." *Amer. Soc. Rev.* 45: 787–801.

Stebbins, R. A. (1971). *Commitment to Deviance.* Westport, CT: Greenwood.

Street, J. (1986). *Rebel Rock: The Politics of Popular Music.* Oxford: Basil Blackwell.

Sykes, G., and D. Matza. (1957). "Techniques of neutralization." *Amer. Soc. Rev.* 22: 664–670.

Taylor, I. (1982). "Moral enterprise, moral panic, and law-and-order campaigns." In M. M. Rosenberg et al. (eds.), *A Sociology of Deviance* (pp. 123–149). New York: St. Martin's.

Yablonsky, L. (1959). "The delinquent gang as a near-group." *Social Problems* 7: 108–117.

———. (1962). *The Violent Gang.* New York: Macmillan.

———. (1968). *The Hippie Trip.* New York: Pegasus.

Yinger, J. M. (1982). *Countercultures: The Promise and Peril of a World Turned Upside Down.* New York: Free Press.

35

Gender and Victimization Risk among Young Women in Gangs

JODY MILLER

*Miller offers a glimpse into the contemporary urban world of street gangs in this
analysis of the role and dangers faced by female gang members. Gang members not
only associate together, they also need each other's participation in the deviant
act to function (no man or woman is a gang unto him- or herself). Once nearly
faded to obscurity, gangs made a rebound in American society in the late 1980s,
fueled by the drug economy and the increasing economic plight of urban areas.
Since that time they have evolved considerably, adding sophisticated nuances and
female members. Miller finds that while women gain status, social life, and some
protection from the hazards of street life in joining gangs, they exchange this for a
new set of dangers. By entering the gang world, they are exposing themselves to
violence, both from rival gang members as well as from their own homeboys. Miller
discusses the particularly gendered status dilemmas and risks for these young
women, and how these vary depending on their activities, stance, and associations
within the group.*

An underdeveloped area in the gang literature is the relationship between gang
participation and victimization risk. There are notable reasons to consider the
issue significant. We now have strong evidence that delinquent lifestyles are associated with increased risk of victimization (Lauritsen, Sampson, and Laub 1991).
Gangs are social groups that are organized around delinquency (see Klein 1995),
and participation in gangs has been shown to escalate youth's involvement in
crime, including violent crime (Esbensen and Huizinga 1993; Esbensen, Huizinga,
and Weiher 1993; Fagan 1989, 1990; Thornberry et al. 1993). Moreover, research
on gang violence indicates that the primary targets of this violence are other gang
members (Block and Block 1993; Decker 1996; Klein and Maxson 1989; Sanders

From "Gender and Victimization Risk among Young Women in Gangs," Jody Miller,
Journal of Research in Crime and Delinquency, Vol. 35, No. 4, pp. 430, 434, 436, 438–439,
440–447, 450–453, copyright © 1998. Reprinted by permission of Sage Publications, Inc.

1993). As such, gang participation can be recognized as a delinquent lifestyle that is likely to involve high risks of victimization (see Huff 1996:97). Although research on female gang involvement has expanded in recent years and includes the examination of issues such as violence and victimization, the oversight regarding the relationship between gang participation and violent victimization extends to this work as well.

The coalescence of attention to the proliferation of gangs and gang violence (Block and Block 1993; Curry, Ball, and Decker 1996; Decker 1996; Klein 1995; Klein and Maxson 1989; Sanders 1993), and a possible disproportionate rise in female participation in violent crimes more generally (Baskin, Sommers, and Fagan 1993; but see Chesney-Lind, Shelden, and Joe 1996), has led to a specific concern with examining female gang members' violent activities. As a result, some recent research on girls in gangs has examined these young women's participation in violence and other crimes as offenders (Bjerregaard and Smith 1993; Brotherton 1996; Fagan 1990; Lauderback, Hansen, and Waldorf 1992; Taylor, 1993). However, an additional question worth investigation is what relationships exist between young women's gang involvement and their experiences and risk of victimization. Based on in-depth interviews with female gang members, this article examines the ways in which gender shapes victimization risk within street gangs. . . .

METHODOLOGY

Data presented in this article come from survey and semistructured in-depth interviews with 20 female members of mixed-gender gangs in Columbus, Ohio. The interviewees ranged in age from 12 to 17; just over three-quarters were African American or multiracial (16 of 20), and the rest (4 of 20) were White. The sample was drawn primarily from several local agencies in Columbus working with at-risk youths, including the county juvenile detention center, a shelter care facility for adolescent girls, a day school within the same institution, and a local community agency.[1] The project was structured as a gang/nongang comparison, and I interviewed a total of 46 girls. Gang membership was determined during the survey interview by self-definition: About one-quarter of the way through the 50+ page interview, young women were asked a series of questions about the friends they spent time with. They then were asked whether these friends were gang involved and whether they themselves were gang members. Of the 46 girls interviewed, 21 reported that they were gang members[2] and an additional 3 reported being gang involved (hanging out primarily with gangs or gang members) but not gang members. The rest reported no gang involvement.

The survey interview was a variation of several instruments currently being used in research in a number of cities across the United States and included a broad range of questions and scales measuring factors that may be related to gang membership.[3] On issues related to violence, it included questions about peer activities and delinquency, individual delinquent involvement, family

violence and abuse, and victimization. When young women responded affirma-
tively to being gang members, I followed with a series of questions about the
nature of their gang, including its size, leadership, activities, symbols, and so
on. Girls who admitted gang involvement during the survey participated in a
follow-up interview to talk in more depth about their gangs and gang activities.
The goal of the in-depth interview was to gain a greater understanding of the
nature and meanings of gang life from the point of view of its female members.
A strength of qualitative interviewing is its ability to shed light on this aspect of
the social world, highlighting the meanings individuals attribute to their experiences (Adler and Adler 1987; Glassner and Loughlin 1987; Miller and Glassner
1997). In addition, using multiple methods, including follow-up interviews, provided me with a means of detecting inconsistencies in young women's accounts
of their experiences. Fortunately, no serious contradictions arose. However, a
limitation of the data is that only young women were interviewed. Thus, I
make inferences about gender dynamics, and young men's behavior, based
only on young women's perspectives.

GENDER, GANGS, AND VIOLENCE

Gangs as Protection and Risk

An irony of gang involvement is that although many members suggest one thing
they get out of the gang is a sense of protection (see also Decker 1996; Joe and
Chesney-Lind 1995; Lauderback et al. 1992), gang membership itself means
exposure to victimization risk and even a willingness to be victimized. These
contradictions are apparent when girls talk about what they get out of the
gang, and what being in the gang means in terms of other members' expectations
of their behavior. In general, a number of girls suggested that being a gang member is a source of protection around the neighborhood. Erica,[4] a 17-year-old
African American, explained, "It's like people look at us and that's exactly
what they think, there's a gang, and they respect us for that. They won't
bother us.... It's like you put that intimidation in somebody." Likewise, Lisa,
a 14-year-old White girl, described being in the gang as empowering: "You
just feel like, oh my God, you know, they got my back. I don't need to
worry about it." Given the violence endemic in many inner-city communities,
these beliefs are understandable, and to a certain extent, accurate.

In addition, some young women articulated a specifically gendered sense of
protection that they felt as a result of being a member of a group that was predominantly male. Gangs operate within larger social milieus that are characterized
by gender inequality and sexual exploitation. Being in a gang with young men
means at least the semblance of protection from, and retaliation against, predatory
men in the social environment. Heather, a 15-year-old White girl, noted, "You
feel more secure when, you know, a guy's around protectin' you, you know,
than you would a girl." She explained that as a gang member, because "you
get protected by guys ... not as many people mess with you." Other young

women concurred and also described that male gang members could retaliate against specific acts of violence against girls in the gang. Nikkie, a 13-year-old African American girl, had a friend who was raped by a rival gang member, and she said, "It was a Crab [Crip] that raped my girl in Miller Ales, and um, they was ready to kill him." Keisha, an African American 14-year-old, explained, "If I got beat up by a guy, all I gotta do is go tell one of the niggers, you know what I'm sayin'? Or one of the guys, they'd take care of it."

At the same time, members recognized that they may be targets of rival gang members and were expected to "be down" for their gang at those times even when it meant being physically hurt. In addition, initiation rites and internal rules were structured in ways that required individuals to submit to, and be exposed to, violence. For example, young women's descriptions of the qualities they valued in members revealed the extent to which exposure to violence was an expected element of gang involvement. Potential members, they explained, should be tough, able to fight and to engage in criminal activities, and also should be loyal to the group and willing to put themselves at risk for it. Erica explained that they didn't want "punks" in her gang: "When you join something like that, you might as well expect that there's gonna be fights. . . . And, if you're a punk, or if you're scared of stuff like that, then don't join." Likewise, the following dialogue with Cathy, a White 16-year-old, reveals similar themes. I asked her what her gang expected out of members and she responded, "to be true to our gang and to have our backs." When I asked her to elaborate, she explained,

CATHY: Like, uh, if you say you're a Blood, you be a Blood. You wear your rag even when you're by yourself. You know, don't let anybody intimidate you and be like, "Take that rag off." You know, "You better get with our set." Or something like that.

JM: Ok. Anything else that being true to the set means?

CATHY: Um. Yeah, I mean, just, just, you know, I mean it's, you got a whole bunch of people comin' up in your face and if you're by yourself they ask you what's your claimin', you tell 'em. Don't say, "Nothin'."

JM: Even if it means getting beat up or something?

CATHY: Mmhmm.

One measure of these qualities came through the initiation process, which involved the individual submitting to victimization at the hands of the gang's members. Typically this entailed either taking a fixed number of "blows" to the head and/or chest or being "beat in" by members for a given duration (e.g., 60 seconds). Heather described the initiation as an important event for determining whether someone would make a good member:

When you get beat in if you don't fight back and if you just like stop and you start cryin' or somethin' or beggin' 'em to stop and stuff like that, then, they ain't gonna, they'll just stop and they'll say that you're not gang material because you gotta be hard, gotta be able to fight, take punches.

In addition to the initiation, and threats from rival gangs, members were expected to adhere to the gang's internal rules (which included such things as not fighting with one another, being "true" to the gang, respecting the leader, not spreading gang business outside the gang, and not dating members of rival gangs). Breaking the rules was grounds for physical punishment, either in the form of a spontaneous assault or a formal "violation," which involved taking a specified number of blows to the head. For example, Keisha reported that she talked back to the leader of her set and "got slapped pretty hard" for doing so. Likewise, Veronica, an African American 15-year-old, described her leader as "crazy, but we gotta listen to 'im. He's just the type that if you don't listen to 'im, he gonna blow your head off. He's just crazy."

It is clear that regardless of members' perceptions of the gang as a form of "protection," being a gang member also involves a willingness to open oneself up to the possibility of victimization. Gang victimization is governed by rules and expectations, however, and thus does not involve the random vulnerability that being out on the streets without a gang might entail in high-crime neighborhoods. Because of its structured nature, this victimization risk may be perceived as more palatable by gang members. For young women in particular, the gendered nature of the streets may make the empowerment available through gang involvement an appealing alternative to the individualized vulnerability they otherwise would face. However, as the next sections highlight, girls' victimization risks continue to be shaped by gender, even within their gangs, because these groups are structured around gender hierarchies as well.

Gender and Status, Crime and Victimization

Status hierarchies within Columbus gangs, like elsewhere, were male dominated (Bowker, Gross, and Klein 1980; Campbell 1990). Again, it is important to highlight that the structure of the gangs these young women belonged to—that is, male-dominated, integrated mixed-gender gangs—likely shaped the particular ways in which gender dynamics played themselves out. Autonomous female gangs, as well as gangs in which girls are in auxiliary subgroups, may be shaped by different gender relations, as well as differences in orientations toward status, and criminal involvement.

All the young women reported having established leaders in their gang, and this leadership was almost exclusively male. While LaShawna, a 17-year-old African American, reported being the leader of her set (which had a membership that is two-thirds girls, many of whom resided in the same residential facility as her), all the other girls in mixed-gender gangs reported that their Original Gangster was male. In fact, a number of young women stated explicitly that only male gang members could be leaders. Leadership qualities, and qualities attributed to high-status members of the gang—being tough, able to fight, and willing to "do dirt" (e.g., commit crime, engage in violence) for the gang—were perceived as characteristically masculine. Keisha noted, "The guys, they just harder." She explained, "Guys is more rougher. We have our G's back but, it ain't gonna be like the guys, they just don't give a fuck. They gonna shoot you in a minute."

For the most part, status in the gang was related to traits such as the willingness to use serious violence and commit dangerous crimes and, though not exclusively, these traits were viewed primarily as qualities more likely and more intensely located among male gang members.

Because these respected traits were characterized specifically as masculine, young women actually may have had greater flexibility in their gang involvement than young men. Young women had fewer expectations placed on them—by both their male and female peers—in regard to involvement in criminal activities such as fighting, using weapons, and committing other crimes. This tended to decrease girls' exposure to victimization risk comparable to male members, because they were able to avoid activities likely to place them in danger. Girls *could* gain status in the gang by being particularly hard and true to the set. Heather, for example, described the most influential girl in her set as "the hardest girl, the one that don't take no crap, will stand up to anybody." Likewise, Diane, a White 15-year-old, described a highly respected female member in her set as follows:

> People look up to Janeen just 'cause she's so crazy. People just look up to her 'cause she don't care about nothin'. She don't even care about makin' money. Her, her thing is, "Oh, you're a Slob [Blood]? You're a Slob? You talkin' to me? You talkin' shit to me?" Pow, pow! And that's it. That's it.

However, young women also had a second route to status that was less available to young men. This came via their connections—as sisters, girlfriends, cousins—to influential, high-status young men.[5] In Veronica's set, for example, the girl with the most power was the OG's "sister or his cousin, one of 'em." His girlfriend also had status, although Veronica noted that "most of us just look up to our OG." Monica, a 16-year-old African American, and Tamika, a 15-year-old African American, both had older brothers in their gangs, and both reported getting respect, recognition, and protection because of this connection. This route to status and the masculinization of high-status traits functioned to maintain gender inequality within gangs, but they also could put young women at less risk of victimization than young men. This was both because young women were perceived as less threatening and thus were less likely to be targeted by rivals, and because they were not expected to prove themselves in the ways that young men were, thus decreasing their participation in those delinquent activities likely to increase exposure to violence. Thus, gender inequality could have a protective edge for young women.

Young men's perceptions of girls as lesser members typically functioned to keep girls from being targets of serious violence at the hands of rival young men, who instead left routine confrontations with rival female gang members to the girls in their own gang. Diane said that young men in her gang "don't wanna waste their time hittin' on some little girls. They're gonna go get their little cats [females] to go get 'em." Lisa remarked,

> Girls don't face much violence as [guys]. They see a girl, they say, "we'll just smack her and send her on." They see a guy—'cause guys are

like a lot more into it than girls are, I've noticed that—and they like, well, "we'll shoot him."

In addition, the girls I interviewed suggested that, in comparison with young men, young women were less likely to resort to serious violence, such as that involving a weapon, when confronting rivals. Thus, when girls' routine confrontations were more likely to be female on female than male on female, girls' risk of serious victimization was lessened further.

Also, because participation in serious and violent crime was defined primarily as a masculine endeavor, young women could use gender as a means of avoiding participation in those aspects of gang life they found risky, threatening, or morally troubling. Of the young women I interviewed, about one-fifth were involved in serious gang violence: A few had been involved in aggravated assaults on rival gang members, and one admitted to having killed a rival gang member, but they were by far the exception. Most girls tended not to be involved in serious gang crime, and some reported that they chose to exclude themselves because they felt ambivalent about this aspect of gang life. Angie, an African American 15-year-old, explained,

> I don't get involved like that, be out there goin' and just beat up
> people like that or go stealin', things like that. That's not me. The boys,
> mostly the boys do all that, the girls we just sit back and chill, you
> know.

Likewise, Diane noted,

> For maybe a drive-by they might wanna have a bunch of dudes. They
> might not put the females in that. Maybe the females might be weak
> inside, not strong enough to do something like that, just on the
> insides. . . .If a female wants to go forward and doin' that, and she wants
> to risk her whole life for doin' that, then she can. But the majority of
> the time, that job is given to a man.

Diane was not just alluding to the idea that young men were stronger than young women. She also inferred that young women were able to get out of committing serious crime, more so than young men, because a girl shouldn't have to "risk her whole life" for the gang. In accepting that young men were more central members of the gang, young women could more easily participate in gangs without putting themselves in jeopardy—they could engage in the more routine, everyday activities of the gang, like hanging out, listening to music, and smoking bud (marijuana). These male-dominated mixed-gender gangs thus appeared to provide young women with flexibility in their involvement in gang activities. As a result, it is likely that their risk of victimization at the hands of rivals was less than that of young men in gangs who were engaged in greater amounts of crime.

Girls' Devaluation and Victimization

In addition to girls choosing not to participate in serious gang crimes, they also faced exclusion at the hands of young men or the gang as a whole (see also

Bowker et al. 1980). In particular, the two types of crime mentioned most frequently as "off-limits" for girls were drug sales and drive-by shootings. LaShawna explained, "We don't really let our females [sell drugs] unless they really wanna and they know how to do it and not to get caught and every thing." Veronica described a drive-by that her gang participated in and said, "They wouldn't let us [females] go. But we wanted to go, but they wouldn't let us." Often, the exclusion was couched in terms of protection. When I asked Veronica why the girls couldn't go, she said, "so we won't go to jail if they was to get caught. Or if one of 'em was to get shot, they wouldn't want it to happen to us." Likewise, Sonita, a 13-year-old African American, noted, "If they gonna do somethin' bad and they think one of the females gonna get hurt they don't let 'em do it with them.... Like if they involved with shooting or whatever, [girls] can't go."

Although girls' exclusion from some gang crime may be framed as protective (and may reduce their victimization risk vis-à-vis rival gangs), it also served to perpetuate the devaluation of female members as less significant to the gang—not as tough, true, or "down" for the gang as male members. When LaShawna said her gang blocked girls' involvement in serious crime, I pointed out that she was actively involved herself. She explained, "Yeah, I do a lot of stuff 'cause I'm tough. I likes, I likes messin' with boys. I fight boys. Girls ain't nothin' to me." Similarly, Tamika said, "girls, they little peons."

Some young women found the perception of them as weak a frustrating one. Brandi, an African American 13-year-old, explained, "Sometimes I dislike that the boys, sometimes, always gotta take charge and they think, sometimes, that the girls don't know how to take charge 'cause we're like girls, we're females, and like that." And Chantell, an African American 14-year-old, noted that rival gang members "think that you're more of a punk." Beliefs that girls were weaker than boys meant that young women had a harder time proving that they were serious about their commitment to the gang. Diane explained,

> A female has to show that she's tough. A guy can just, you can just look at him. But a female, she's gotta show. She's gotta go out and do some dirt. She's gotta go whip some girl's ass, shoot somebody, rob somebody or something. To show that she is tough.

In terms of gender-specific victimization risk, the devaluation of young women suggests several things. It could lead to the mistreatment and victimization of girls by members of their own gang when they didn't have specific male protection (i.e., a brother, boyfriend) in the gang or when they weren't able to stand up for themselves to male members. This was exacerbated by activities that led young women to be viewed as sexually available. In addition, because young women typically were not seen as a threat by young men, when they did pose one, they could be punished even more harshly than young men, not only for having challenged a rival gang or gang member but also for having overstepped "appropriate" gender boundaries.

Monica had status and respect in her gang, both because she had proven herself through fights and criminal activities, and because her older brothers were members of her set. She contrasted her own treatment with that of other young women in the gang:

> They just be puttin' the other girls off. Like Andrea, man. Oh my God, they dog Andrea so bad. They like, "Bitch, go to the store." She like, "All right, I be right back." She will go to the store and go and get them whatever they want and come back with it. If she don't get it right, they be like, "Why you do that bitch?" I mean, and one dude even smacked her. And, I mean, and, I don't, I told my brother once. I was like, "Man, it ain't even like that. If you ever see someone tryin' to disrespect me like that or hit me, if you do not hit them or at least say somethin' to them. . . ." So my brothers, they kinda watch out for me.

However, Monica put the responsibility for Andrea's treatment squarely on the young woman: "I put that on her. They ain't gotta do her like that, but she don't gotta let them do her like that either." Andrea was seen as "weak" because she did not stand up to the male members in the gang; thus, her mistreatment was framed as partially deserved because she did not exhibit the valued traits of toughness and willingness to fight that would allow her to defend herself.

An additional but related problem was when the devaluation of young women within gangs was sexual in nature. Girls, but not boys, could be initiated into the gang by being "sexed in"—having sexual relations with multiple male members of the gang. Other members viewed the young women initiated in this way as sexually available and promiscuous, thus increasing their subsequent mistreatment. In addition, the stigma could extend to female members in general, creating a sexual devaluation that all girls had to contend with.

The dynamics of "sexing in" as a form of gang initiation placed young women in a position that increased their risk of ongoing mistreatment at the hands of their gang peers. According to Keisha, "If you get sexed in, you have no respect. That means you gotta go ho'in' for 'em; when they say you give 'em the pussy, you gotta give it to 'em. If you don't, you gonna get your ass beat. I ain't down for that." One girl in her set was sexed in and Keisha said the girl "just do everything they tell her to do, like a dummy." Nikkie reported that two girls who were sexed into her set eventually quit hanging around with the gang because they were harassed so much. In fact, Veronica said the young men in her set purposely tricked girls into believing they were being sexed into the gang and targeted girls they did not like:

> If some girls wanted to get in, if they don't like the girl they have sex with 'em. They run trains on 'em or either have the girl suck their thang. And then they used to, the girls used to think they was in. So, then the girls used to just come try to hang around us and all this little bull, just 'cause, 'cause they thinkin' they in.

Young women who were sexed into the gang were viewed as sexually promiscuous, weak, and not "true" members. They were subject to revictimization and mistreatment, and were viewed as deserving of abuse by other members, both male and female. Veronica continued, "They [girls who are sexed in] gotta do whatever, whatever the boys tell 'em to do when they want 'em to do it, right then and there, in front of whoever. And, I think, that's just sick. That's nasty, that's dumb." Keisha concurred, "She brought that on herself, by bein' the fact, bein' sexed in." There was evidence, however, that girls could overcome the stigma of having been sexed in through their subsequent behavior, by challenging members that disrespect them and being willing to fight. Tamika described a girl in her set who was sexed in, and stigmatized as a result, but successfully fought to rebuild her reputation:

> Some people, at first, they call her "little ho" and all that. But then, now she startin' to get bold. . . . Like, they be like, "Ooh, look at the little ho. She fucked me and my boy." She be like, "Man, forget y'all. Man, what? What?" She be ready to squat [fight] with 'em. I be like, "Ah, look at her!" Uh huh. . . . At first we looked at her like, "Ooh, man, she a ho, man." But now we look at her like she just our kickin'-it partner. You know, however she got in that's her business.

The fact that there was such an option as "sexing in" served to keep girls disempowered, because they always faced the question of how they got in and of whether they were "true" members. In addition, it contributed to a milieu in which young women's sexuality was seen as exploitable. This may help explain why young women were so harshly judgmental of those girls who were sexed in. Young women who were privy to male gang members' conversations reported that male members routinely disrespect girls in the gang by disparaging them sexually. Monica explained,

> I mean the guys, they have their little comments about 'em [girls in the gang] because, I hear more because my brothers are all up there with the guys and everything and I hear more just sittin' around, just listenin'. And they'll have their little jokes about "Well, ha I had her," and then and everybody else will jump in and say, "Well, I had her, too." And then they'll laugh about it.

In general, because gender constructions defined young women as weaker than young men, young women were often seen as lesser members of the gang. In addition to the mistreatment these perceptions entailed, young women also faced particularly harsh sanctions for crossing gender boundaries—causing harm to rival male members when they had been viewed as nonthreatening. One young woman[6] participated in the assault of a rival female gang member, who had set up a member of the girl's gang. She explained, "The female was supposingly goin' out with one of ours, went back and told a bunch of [rivals] what was goin' on and got the [rivals] to jump my boy. And he ended up in the hospital." The story she told was unique but nonetheless significant for what it indicates about the gendered nature of gang violence and victimization. Several young men in her set saw the girl walking down the street, kidnapped her, then brought her to a member's house. The young woman I interviewed, along with several other girls in her set, viciously beat the girl, then to their

surprise the young men took over the beating, ripped off the girl's clothes, brutally gang-raped her, then dumped her in a park. The interviewee noted, "I don't
know what happened to her. Maybe she died. Maybe, maybe someone came and
helped her. I mean, I don't know." The experience scared the young woman
who told me about it. She explained,

> I don't never want anythin' like that to happen to me. And I pray to
> God that it doesn't. 'Cause God said that whatever you sow you're gonna
> reap. And like, you know, beatin' a girl up and then sittin' there watchin'
> somethin' like that happen, well, Jesus that could come back on me.
> I mean, I felt, I really did feel sorry for her even though my boy was in
> the hospital and was really hurt. I mean, we coulda just shot her. You
> know, and it coulda been just over. We coulda just taken her life.
> But they went farther than that.

This young woman described the gang rape she witnessed as "the most brutal
thing I've ever seen in my life." While the gang rape itself was an unusual event,
it remained a specifically gendered act that could take place precisely because
young women were not perceived as equals. Had the victim been an "equal,"
the attack would have remained a physical one. As the interviewee herself
noted, "we coulda just shot her." Instead, the young men who gang-raped the
girl were not just enacting revenge on a rival but on a *young woman* who had
dared to treat a young man in this way. The issue is not the question of which
is worse—to be shot and killed, or gang-raped and left for dead. Rather, this particular act sheds light on how gender may function to structure victimization risk
within gangs.

DISCUSSION

Gender dynamics in mixed-gender gangs are complex and thus may have multiple and contradictory effects on young women's risk of victimization and repeat
victimization. My findings suggest that participation in the delinquent lifestyles
associated with gangs clearly places young women at risk for victimization.
The act of joining a gang involves the initiate's submission to victimization at
the hands of her gang peers. In addition, the rules governing gang members'
activities place them in situations in which they are vulnerable to assaults that
are specifically gang related. Many acts of violence that girls described would
not have occurred had they not been in gangs.

It seems, though, that young women in gangs believed they have traded
unknown risks for known ones—that victimization at the hands of friends, or
at least under specified conditions, was an alternative preferable to the potential
of random, unknown victimization by strangers. Moreover, the gang offered
both a semblance of protection from others on the streets, especially young
men, and a means of achieving retaliation when victimization did occur. . . .

Girls' gender, as an individual attribute, can function to lessen their exposure
to victimization risk by defining them as inappropriate targets of rival male gang

members' assaults. The young women I interviewed repeatedly commented that young men were typically not as violent in their routine confrontations with rival young women as with rival young men. On the other hand, when young women are targets of serious assault, they may face brutality that is particularly harsh and sexual in nature because they are female—thus, particular types of assault, such as rape, are deemed more appropriate when young women are the victims.

Gender can also function as a state-dependent factor, because constructions of gender and the enactment of gender identities are fluid. On the one hand, young women can call upon gender as a means of avoiding exposure to activities they find risky, threatening, or morally troubling. Doing so does not expose them to the sanctions likely faced by male gang members who attempt to avoid participation in violence. Although these choices may insulate young women from the risk of assault at the hands of rival gang members, perceptions of female gang members—and of women in general—as weak may contribute to more routinized victimization at the hands of the male members of their gangs. Moreover, sexual exploitation in the form of "sexing in" as an initiation ritual may define young women as sexually available, contributing to a likelihood of repeat victimization unless the young woman can stand up for herself and fight to gain other members' respect.

Finally, given constructions of gender that define young women as non-threatening, when young women do pose a threat to male gang members, the sanctions they face may be particularly harsh because they not only have caused harm to rival gang members but also have crossed appropriate gender boundaries in doing so. In sum, my findings suggest that gender may function to insulate young women from some types of physical assault and lessen their exposure to risks from rival gang members, but also to make them vulnerable to particular types of violence, including routine victimization by their male peers, sexual exploitation, and sexual assault.

NOTES

1. I contacted numerous additional agency personnel in an effort to draw the sample from a larger population base, but many efforts remained unsuccessful despite repeated attempts and promises of assistance. These included persons at the probation department, a shelter and outreach agency for runaways, police personnel, a private residential facility for juveniles, and three additional community agencies. None of the agencies I contacted openly denied me permission to interview young women; they simply chose not to follow up. I do not believe that much bias resulted from the nonparticipation of these agencies. Each has a client base of "at-risk" youths, and the young women I interviewed report overlap with some of these same agencies. For example, a number had been or were on probation, and several reported staying at the shelter for runaways.

2. One young woman was a member of an all-female gang. Because the focus of this article is gender

dynamics in mixed-gender gangs, her interview is not included in the analysis.

3. These include the Gang Membership Resistance Surveys in Long Beach and San Diego, the Denver Youth Survey, and the Rochester Youth Development Study.

4. All names are fictitious.

5. This is not to suggest that male members cannot gain status via their connections to high-status men, but that to maintain status, they will have to successfully exhibit masculine traits such as toughness. Young women appear to be held to more flexible standards.

6. Because this excerpt provides a detailed description of a specific serious crime, and because demographic information on respondents is available, I have chosen to conceal both the pseudonym and gang affiliation of the young woman who told me the story.

REFERENCES

Adler, Patricia A. and Peter Adler. 1987. *Membership Roles in Field Research*. Newbury Park, CA: Sage.

Baskin, Deborah, Ira Sommers, and Jeffrey Fagan. 1993. "The Political Economy of Violent Female Street Crime." *Fordham Urban Law Journal* 20: 401–17.

Bjerregaard, Beth and Carolyn Smith. 1993. "Gender Differences in Gang Participation, Delinquency, and Substance Use." *Journal of Quantitative Criminology* 4: 329–55.

Block, Carolyn Rebecca and Richard Block. 1993. "Street Gang Crime in Chicago." Research in Brief, Washington, DC: National Institute of Justice.

Bowker, Lee H., Helen Shimota Gross, and Malcolm W. Klein. 1980. "Female Participation in Delinquent Gang Activities." *Adolescence* 15(59): 509–19.

Brotherton, David C. 1996. "'Smartness,' 'Toughness,' and 'Autonomy': Drug Use in the Context of Gang Female Deliquency." *Journal of Drug Issues* 26(1): 261–77.

Campbell, Anne. 1990. "Female Participation in Gangs." Pp. 163–82 in *Gangs in America*, edited by G. Ronald Huff. Beverly Hills, CA: Sage.

Chesney-Lind, Meda, Randall G. Shelden, and Karen A. Joe. 1996. "Girls, Delinquency, and Gang Membership." Pp. 185–204 in *Gangs in America*, 2d ed., edited by C. Ronald Huff. Thousand Oaks, CA: Sage.

Curry, G. David, Richard A. Ball, and Scott H. Decker. 1996. Estimating the National Scope of Gang Crime from Law Enforcement Data Research in Brief. Washington, DC: National Institute of justice.

Decker, Scott H. 1996. "Collective and Normative Features of Gang Violence." *Justice Quarterly* 13(2): 243–64.

Decker, Scott H. and Barrik Van Winkle. 1996. *Life in the Gang*. Cambridge, UK: Cambridge University Press.

Esbensen, Finn-Aage and David Huizinga. 1993. "Gangs, Drugs, and Delinquency in a Survey of Urban Youth." *Criminology* 31(4): 565–89.

Esbensen, Finn-Aage, David Huizinga, and Anne W. Weiher. 1993. "Gang and Non-Gang Youth: Differences in Explanatory Factors." *Journal of Contemporary Criminal Justice* 9(2): 94–116.

Fagan, Jeffrey. 1989. "The Social Organization of Drug Use and Drug Dealing among Urban Gangs." *Criminology* 27(4): 633–67.

———. 1990. "Social Processes of Delinquency and Drug Use among Urban Gangs." Pp. 183–219 in *Gangs in America*, edited by C. Ronald Huff. Newbury Park, CA: Sage.

Glassner, Barry and Julia Loughlin. 1987. *Drugs in Adolescent Worlds: Burnouts to Straights*. New York: St. Martin's.

Huff, C. Ronald. 1996. "The Criminal Behavior of Gang Members and Nongang At-Risk Youth." Pp. 75–102 in *Gangs in America,* 2d ed., edited by C. Ronald Huff. Thousand Oaks, CA: Sage.

Joe, Karen A. and Meda Chesney-Lind. 1995. "Just Every Mother's Angel: An Analysis of Gender and Ethnic Variations in Youth Gang Membership." *Gender & Society* 9(4): 408–30.

Klein, Malcolm W. 1995. *The American Street Gang: Its Nature, Prevalence and Control*. New York: Oxford University Press.

Klein, Malcolm W. and Cheryl L. Maxson. 1989. "Street Gang Violence." Pp. 198–231 in *Violent Crime, Violent Criminals*, edited by Neil Weiner and Marvin Wolfgang. Newbury Park, CA: Sage.

Lauderback, David, Joy Hansen, and Dan Waldorf. 1992. "'Sisters Are Doin' It for Themselves': A Black Female Gang in San Francisco." *The Gang Journal* 1(1): 57–70.

Lauritsen, Janet L., Robert J. Sampson, and John H. Laub. 1991. "The Link between Offending and Victimization among Adolescents." *Criminology* 29(2): 265–92.

Miller, Jody and Barry Glassner. 1997. "The 'Inside' and the 'Outside': Finding Realities in Interviews." Pp. 99–112 in *Qualitative Research,* edited by David Silverman. London: Sage.

Sanders, William. 1993. *Drive-Bys and Gang Bangs: Gangs and Grounded Culture*. Chicago: Aldine.

Taylor, Carl. 1993. *Girls, Gangs, Women and Drugs*. East Lansing: Michigan State University Press.

Thornberry, Terence P., Marvin D. Krohn, Alan J. Lizotie, and Deborah Chard-Wierschem. 1993. "The Role of Juvenile Gangs in Facilitating Delinquent Behavior." *Journal of Research in Crime and Delinquency* 30(1): 75–85.

36

Russian Organized Crime in America

ROBERT J. RUSH, JR., AND FRANK R. SCARPITTI

Traditional depictions of organized crime in America have focused on Italian-Americans organized into associations that are rather large, ethnically homogeneous, connected by familial or pseudo-familial bonds, committed to each other through a blood oath for lifelong participation, insulated from outsiders, ensconced in a currency of violence, vertically and horizontally differentiated, involved in crime at the highest levels, internationally connected, and able to afford the corruption of law enforcement and judicial officials and politicians on a local, state, national, and international plane. Although various ethnic and national groups have dabbled in large-scale crime at this level, particularly as they immigrated to this country and sought to establish themselves, many have moved in and through these activities on their way to legitimate pursuits.

Rush and Scarpitti examine one such group with a moderately recent entry in growing numbers to the American criminal scene: the Russians. They describe a Russian-American criminal syndicate's involvement in a large-scale and highly profitable form of tax fraud, discussing how they learned this business from their Italian-American associates. The view of this criminal organization that they propose is smaller and more flexible than traditional depictions, showing the integration of white-collar crime with more ordinary street crime. Readers will find the model they discuss of the rise of criminal syndicates from street gangs to top-level formal organizations particularly interesting.

M any contemporary writers and criminal justice institutions have identified the crimes of certain Russian émigrés as Russian organized crime. Yet, other scholars who have studied émigré Russian populations and communities in the United States have not been so quick to label their criminal behavior as organized crime. They stress the necessity for further study before the criminal activity of the émigré crime groups can be correctly categorized as organized crime.

To clarify this issue, we set out to determine if there is a separate and distinct criminal phenomenon that can be labeled Russian Organized Crime (ROC). We wanted to establish whether the criminality of the Russian émigrés was characteristic of organized crime in the traditional or classic sense of that term. That is, does their criminality follow a developmental path similar to the progression observed for the organized crime of other ethnic groups, namely Italian-Americans, and, if it does, which existing theoretical paradigm of organized crime, if any, best explains the criminality of the émigré Russians.

To answer these questions, we initially examined the overall concept of organized crime to develop an acceptable understanding of this condition. As a result of our review of the scholarly literature and government reports, we came up with a generic understanding of organized crime. We determined that organized crime is essentially a culturally defined phenomenon, a condition uniquely distinguished by the social system from which it emerges, and regardless of its social or environmental origin, organized crime will always be characterized by similar fundamental traits. It will be a market-driven, economically motivated phenomenon characterized by patron-client networks wherein illegal goods and services are provided to a client public by a criminal patron. The social environment that supports organized crime will be distinguished by political corruption or ineffective apathetic law enforcement. Power and violence will serve as methods of social control to preserve the criminal enterprise. With this understanding of organized crime as a foundation, we analyzed certain Russian émigré crime with the goal of determining if such criminal activity is consistent with organized crime as it has come to be known, identified, and defined in American culture.

Information contained in transcripts of government hearings, investigative reports, and the scholarly literature documents the fact that a number of Russian émigré groups have become involved in a plethora of criminal activities throughout the United States and Canada. Their crimes include theft, gambling, prostitution, drugs, murder, extortion, jewelry theft, insurance fraud, credit card fraud, counterfeiting, money laundering, and various tax scams including their most ambitious venture to date, motor fuel tax fraud.

It also has been established that the émigré crime groups are actively involved with Russian organized crime groups in the former Soviet Union (FSU) and Eastern Europe. Records indicate further that the financial return from their criminal activities is quite substantial, with Russian émigré crime groups operating in the New York City area in the mid-1980s deriving income approaching 100 million dollars a year from crimes perpetrated against fellow émigrés living in the Brighton Beach neighborhood of Brooklyn alone. While some of their criminal activities are characterized by cunning and viciousness, the real domains of Russian criminals are crimes of fraud and deception. This is consistent with the fact that a large number of the émigré criminals were highly educated and trained in the newest technologies in their homeland. Some held responsible professional positions prior to coming to the United States. They have used their sophisticated knowledge and background to develop and operate complicated fraud schemes associated with telecommunications, consumer credit, insurance, money laundering, and motor fuel tax.

For purposes of our study, motor fuel excise tax fraud was selected as a representative example of Russian émigré crime for the simple reason that it has been singularly identified by most government institutions and the scholarly literature as a typical form of Russian émigré crime. Motor fuel tax fraud exemplifies the high level of sophistication that these crime groups bring to their criminal endeavors. It is our intent to assess the role of Russian émigré crime groups in such frauds to determine if their participation has been characteristic of organized crime in the traditional sense. Recent fuel tax frauds that were carried out in the metropolitan New York City and Los Angeles areas served as case studies from which Russian émigré crime was examined and analyzed.

CASE STUDY: *THE UNITED STATES OF AMERICA V. ANTHONY MORELLI ET AL.*

The case of the *United States of America v. Anthony Morelli et al.* focused upon a joint criminal enterprise involving Italian-American organized crime families and Russian émigré crime groups. The investigation upon which *Morelli* was based featured an undercover wholesale fuel company that was operated in Ewing Township, New Jersey, by the federal government for the purpose of infiltrating the bootleg motor fuel industry. The investigation was called "Operation Red Daisy" and ran from approximately May 1991 to December 1992.

The Fraudulent Method: The Daisy Chain

The method used to implement the fraud was based upon the use of multiple companies, both real and fictitious, engaged in a series of transactions called a "daisy chain." The goal of the fraud was simple: buy the fuel tax-free, sell it tax-paid, and pocket the taxable portion of the transaction before the authorities discover the lost revenue. The common element of any motor fuel tax scam is to disguise the true taxpayer and set up a phony trail to mislead investigators and auditors, that is, move the point of taxation away from the real seller. Initially, gasoline was the fuel of interest in most motor fuel evasion frauds and the daisy chain was the usual method used to avoid the payment of gasoline taxes.

In the typical daisy chain evasion scam, a complicated paper trail of transactions involving the purchase and sale of motor fuel by several insulating[1] paper companies is created. At some point in the chain the gasoline is invoiced as being tax-paid fuel, but the company identified as having paid the tax is a fictitious entity. This nonexistent company, commonly referred to as the burn company,[2] is responsible for collecting the taxes but, because it exists only as a paper entity, its detection will result in a dead end for the investigator. All other participants in the chain claim no knowledge of the fact that they were dealing with a fraudulent divergent operation.

An example of an actual daisy chain orchestrated by the Morelli operation worked in the following way. Anthony Petroleum, a company operated by Jeffrey Pressman in Bellmore, New York, purchased 170,000 gallons of gasoline and had it shipped for storage at GATX in Paulsboro, New Jersey, a legitimate storage terminal. On the same day, Pressman issued an invoice citing the sale of the 170,000 gallons of gasoline to an insulating company, Britt Processing and Refining. Also on the same day, Britt Processing sold the gasoline to another insulating company, Philbo Energy Company, which immediately invoiced the sale of the gasoline as part of a larger lot to AFG Trading, another insulating company. AFG Trading then moved the gasoline to yet another insulating or burn company, Ultimate Ameritrade, which turned the gasoline over by selling it to Amco International. Amco purchased the product as tax-paid fuel and wire transferred $171,000 to the account of Anthony Petroleum for the gasoline. Amco then distributed the gasoline to retail service stations for sale to customers who unknowingly supported the scam by purchasing allegedly tax-paid fuel. The money representing the taxes was skimmed from the total and distributed among the coconspirators.

In the chain of events set forth above, only Amco International, a company owned by a Russian émigré, handled the fuel. The other companies in the chain were insulating companies that existed only on paper. One of the insulating companies in this example, Ultimate Ameritrade, was designated as the burn company and was invoiced as having paid the tax on the gasoline. While it appears that the gasoline was passing through a series of wholesale distributors before reaching the retail pumps, the invoiced sales of the fuel were all fraudulent transactions. The fuel never left the initial storage facility until it was sold to a retail outlet. All of the insulating and burn companies were properly incorporated and had bank accounts. All complied with the appropriate record keeping and reporting procedures required for their business by the federal government and the State of New Jersey, albeit with fraudulent records to create the illusion of an existing business. Most corporate officials were real people. However, in the case of the burn company, many were not, or if they were, they were not currently in residence in the United States. . . .

ENTERPRISE THEORY: AN EXPLANATION
OF RUSSIAN EMIGRE CRIME

It is the enterprise paradigm of organized crime that has been advanced by scholars as providing the best understanding and interpretation of this phenomenon (Albanese 1996; Albini 1971; Block 1983; Haller 1997; Ianni 1972; Potter 1994; Scarpitti 1993; Smith 1975). Enterprise theory emerged from the scholarship that critically challenged the issues proposed by the alien-conspiracy and ethnicity theories of organized crime. Like alien-conspiracy and ethnicity theories, enterprise theory is fundamentally rooted in the conditions of organized crime that have traditionally been associated with the criminal activities of the Italian-American crime families. As this paradigm has developed and been refined through scholarly research, enterprise theory has disassociated itself from an exclusive Italian-American orientation. It is now more general in its descriptive characteristics and is distinguished by defining elements that may be applied to any group of individuals associated to fulfill the market demand for illegal goods and services.

Enterprise theory advocates the proposition that most criminal ventures characterized as organized crime are entrepreneurial and are driven by economic concerns. The organizational component of the criminal enterprise is based upon shared values and interests among the participants. The criminal groups themselves lack any real sense of being a formal bureaucratically structured organization. The participants involved in the criminal enterprise are, for the most part, independent entrepreneurs who exhibit flexibility in their criminal activities. They pursue illegal opportunities that are opportunistic, characterized by a public demand, provide a good financial return and, at the same time, expose them to minimal risk.

The enterprise paradigm emphasizes that criminal enterprises identified as organized crime are usually small, centralized operations with little specialization, short hierarchies, and formalization based on socialization. . . .

While some students of organized crime see a highly formalized, hierarchical, and bureaucratic structure as a defining characteristic of organized crime, the present analysis suggests that such a well-defined structure is not an essential characteristic. The Russian émigré crime groups identified in our case studies have been small (less than 15–20 individuals) and lack any formal hierarchical structure, although they do have a leader. The groups are fluid and dynamic. They are market orientated and economically motivated. Their implementation of the necessary and required tasks to achieve the goal of financial gain defines the operational structure of their criminal enterprise. Those participating in the illegal enterprise are like-minded individuals who usually share a common cultural identity and adhere to the same value system. Most come from the same social system.

In this analysis of the Russian émigré crime groups, it is suggested that the émigré organizations are structured and function in a manner totally compatible with the recognized characteristics of enterprise theory. Enterprise theory explains the criminal activity of the Russian émigrés, as demonstrated by their participation in motor fuel tax fraud cases, just as it explains the participation of the Italian-American crime families in similar enterprises. Consequently, the criminality of the Russian émigré crime groups, as demonstrated by the case studies, can be considered to be characteristic of organized crime in the traditional understanding of the term.

LEVEL OF SOPHISTICATION: CRIMINAL ACTIVITY OF THE ÉMIGRÉ CRIME GROUPS

O'Kane (1992: 79–82) has identified six stages of criminal mobility, which he suggests explain the progression of Irish, Jewish, and Italian gangs in criminal activity from the time of each group's immigration to the United States, through the emergence of such gangs as criminal organizations, continuing with their rise to power, and culminating with their fall from dominance. Utilizing O'Kane's assumptions, this research compared the criminal progression of the Russian émigrés in America with that of the Italian immigrants to establish the current performance level of Russian émigré crime groups. It is the intent of our comparison to determine the extent to which criminals in the Russian émigré populations adhere to O'Kane's thesis when compared with the Italians.

O'Kane's Stage One represents the "Individual Criminal," a loner who has no group or organizational identity. Yet, on occasion this individual may join with other like-minded criminals on an ad hoc basis to engage in a singular type of criminal activity. However, such associations are short-lived. For example, the predominant criminal activities of the solitary Italian immigrant criminals were characterized by "Black Hand" methods of intimidation, extortion, and terrorism, which were directed against their fellow Italians. In like manner, most crimes perpetrated by Russian émigré criminals mimic the activities of the earlier Italian extortionists (Rosner 1986; U.S. Congress, Senate 1994). Examples of the crimes that have been carried out by individual Russian émigrés against their fellow émigrés involve a range of offenses from simple confidence scams, such as jewelry switching, to extortion and murder.

In Stage Two, "Intra-Ethnic Gang Rivalry," the individual criminal joins with other like-minded fellow ethnics in recognition that a group can be more effective in achieving criminal success in terms of financial return and power than the individual. Historically, all groups that have immigrated to America have formed gangs to be more successful in the pursuit of their criminal activities. Moreover, at the early stages of their criminal activities the ethnic gangs have historically preyed upon and victimized their fellow immigrants. Among some of the most common crimes committed by Russian émigré groups against other émigrés from the FSU are confidence schemes, theft, extortion, and murder. O'Kane also offers insight into intraethnic criminal group activity and the escalation of violent crime as a predominant characteristic of such groups. He believes that as the individual immigrant gangs become established in their neighborhoods, competition develops among the separate groups for dominance in a struggle for money and power. He suggests violence will emerge as a hallmark of these intraethnic groups when they compete with each other for territorial control of the criminal activities in the community (1992: 80). Examples of intraethnic conflict within the Russian émigré community include the attempted assassination of Victor Zilber and the assassination of Russian gang leader Evsei Agron over disagreements concerning the motor fuel excise tax frauds.[3]

Stage Three, "Inter-Ethnic Gang Rivalry," is the stage in which the intraethnic groups that developed in Stage Two consolidate resources and combine and work together to eliminate competition from other ethnic groups. While the Russian émigré crime groups are currently working in successful illegal ventures both by themselves and with Italian-American crime families, they have not yet made any known overt effort to consolidate among themselves as distinct criminal organizations with a singular Russian ethnic identity. Unlike the Italians, who demonstrated ruthless violent competition against the Irish gangs of O'Banion in Chicago during Prohibition and against the Jewish mobsters,

Dutch Schultz and Vincent Coll, in the New York area following Prohibition, the Russian émigré crime groups have not displayed any concentrated effort to consolidate their separate gangs to eliminate other ethnic crime groups. It could be that they have not yet reached the point where they are confident enough to challenge established organized crime groups, such as the Italian-Americans or the Colombians, in those criminal endeavors where they have a common interest. Or, perhaps at this phase of their criminal progression they are content to work in cooperative ventures with the Italian-Americans who have powerful contacts in the political and economic infrastructure of American society. Unlike their predecessors from other ethnic crime groups, crime groups from the FSU have not been disposed to seek absolute dominance of a specific criminal enterprise. They are culturally predisposed to survive in the face of adversity, and as survivors would rather work with the other crime groups and accept their share of the profits than risk losing everything in an attempt to have it all. It is suggested that the émigré crime groups have essentially bypassed O'Kane's Stage Three of criminal mobility. They have recognized the practicality of accommodation and cooperation as being the best option in the face of overwhelming competition for power and control.

In Stage Four, "Organized Criminal Accommodations," the competing ethnic groups of Stage Three recognize that it is counterproductive for all involved to work against each other. Consequently, they accommodate each other and work together in furthering the success of the criminal enterprise. In essence, a business decision is made to eliminate conflict in order to insure success of the criminal enterprise, reduce risk, and maximize financial return.

Prohibition is arguably the best known example of interethnic gang cooperation. Prohibition took local neighborhood gangs of different ethnicity and turned them into efficient, albeit illegal, business organizations. It was Prohibition that gave birth to organized crime as we know it today. It created the environment that prompted Italian gangs to work in cooperation with Irish gangs and the more dominant gangs composed of Eastern European Jews from Poland, Germany, and Russia. Furthermore, Prohibition forced cooperation among gang leaders from different geographical regions of the United States and Canada because of the logistical demands associated with the manufacture and distribution of illegal alcohol. Syndication was essential to the success of the venture in the interests of maximizing the financial return to the criminal enterprise. Consequently, criminal activities associated with Prohibition actually became an organized endeavor involving the combined efforts of Irish, Italian, and Jewish ethnic gangs from numerous locations throughout the United States and Canada.

A parallel to the Italian involvement with Eastern European Jews in the manufacture and distribution of illegal alcohol during Prohibition can be drawn to the recent association of the Russian émigré crime groups with the Italian-American organized crime families in motor fuel tax frauds. This association is arguably the quintessential example of organized criminal activity involving the Russian émigrés to date. Our analysis of these frauds confirmed the ability of the émigré criminals to infiltrate legitimate areas of the United States economy and coordinate their activities with established and recognized organized crime groups, the

Italian–American traditional crime families. It is at this level, O'Kane's Stage Four, that the Russian émigré crime groups from the New York City metropolitan area and other East Coast regions are presently positioned in their development as organized crime groups.

O'Kane's Stage Five addresses "Ethnic Gang Criminal Supremacy." As the cooperation and accommodation that developed in Stage Four breaks down, violence between the competing ethnic groups reemerges. This is usually because one of the ethnic factions becomes so powerful that it eliminates its rival, either by persuasion or violence. One group will prevail and emerge as dominant. The other groups will either become legitimate, die out, develop a new niche in the world of crime and move into new and competing criminal activities, or work as subordinates to the dominant group (O'Kane 1992: 75–76). The best example of Ethnic Gang Criminal Supremacy was the curtailing of Irish and Jewish competition by the Italian organizations in New York and Chicago during the latter stages of Prohibition and shortly thereafter. Throughout Prohibition and afterward, Italian and Jewish mobsters continued to cooperate with each other to further their entrepreneurial criminal endeavors. But, as the involvement of Jewish racketeers waned, Italians increasingly developed major racketeering interests in a variety of new areas. By the late 1960s practically all of the prominent Irish and Jewish gangsters were gone. They were either victims of competition, investigative and prosecutorial action of the criminal justice system, or they had turned their illegally gained capital into legitimate business ventures. Italian–American crime families thus assumed dominance of organized crime as both the Irish and Jews moved into more socially acceptable occupations and business ventures. As a result, the concept of organized crime became synonymous with Italian–American organized crime and the terms Mafia and La Cosa Nostra. This perception was validated and reinforced well into the 1980s by Congressional investigative committees, law enforcement agencies, and the mass media. The myth that established Italian–American organized crime as the totality of organized crime endured until the President's Commission on Organized Crime issued its report in 1987. The report specifically recognized 10 other ethnic groups as participants in the world of organized crime. Prominent within this group of 10 were the Russian émigré groups which the report specifically identified as Russian organized crime (Albanese 1996: 133).[4]

In Stage Six, "Decline and Fall of the Ethnic Gang," O'Kane demonstrates how criminal enterprises suffer as power and influence are weakened because of competition or increased scrutiny from law enforcement and the criminal justice community in general. Resources are depleted as organizational leaders become preoccupied with prosecutorial concerns. The power and financial successes of the past are diminished. Young ethnics are not as likely to enter the profession. Many leaders reduce their criminal exposure, some retire altogether , others move into legitimate businesses or pursue less intensive, low profile, minimal risk, high financial return activities, such as white-collar type activities. With the move into these more respectable types of criminal activities, the traditional street crimes receive less support and attention, creating a void that is quickly filled by the less sophisticated, newly emerging ethnic crime groups.

In summarizing the criminal progression of the Russian émigré crime groups as measured by O'Kane's six stage scale of progression for organized crime groups, our study suggests that on the East Coast of the United States, Russian émigrés are presently operating at Stage Four. This is aptly demonstrated by their cooperative ventures with the more dominant Italian-American organized crime families in motor fuel tax frauds. However, it is suggested that the Russian émigré crime groups operating in the Los Angeles area have skipped Stage Four on the O'Kane scale and are presently situated in Stage Five, where they hold a dominant position in the criminal enterprises of motor fuel tax fraud. This situation has developed for the California Russian émigré crime groups because the Italian-American organized crime families have never positioned themselves to exclusively control the traditional criminal enterprises of organized crime there. Consequently, the emerging ROC groups on the West Coast have not had to face serious interethnic competition from other crime groups as they develop their criminal enterprises.

NOTES

1. An insulating company is a front company; it may be an existing functioning business or just a company that exists on paper. It is used in motor fuel tax fraud schemes as a device for generating false invoices and other necessary documents to conceal and otherwise facilitate the successful operation of the fraud.

2. In the daisy chain, this is the company that is ostensibly responsible for the tax. It buys the fuel tax-free and sells it tax-paid. This company is a total fabrication created solely for purposes of facilitating the fraud and impeding any investigation or tax enforcement efforts. It exists as a company only on paper and all identifying information associated with the company, such as corporate officials, business records, and physical plant, are fictitious. It will be, at best, a mail drop or telephone.

This is a company that presents a dead end to investigators.

3. See Friedman's (1993) *Vanity Fair* article for a detailed explanation of the Agron assassination. The attempted assassination of Victor Zilber is discussed as part of the Morelli case study.

4. In its final report, the President's Commission on Organized Crime outlined the operations of organized crime by Italian-American groups, prison gangs, outlaw motorcycle gangs, Japanese, Chinese, Vietnamese, Columbian, Cuban, Irish, Canadian, and Russian criminal groups (President's Commission on Organized Crime 1986; Albanese 1996: 133).

REFERENCES

Albanese, Jay S. 1996. *Organized Crime in America.* 3rd ed. Cincinnati, OH: Anderson.

Albini, Joseph L. 1971. *The American Mafia—Genesis of a Legend.* New York: Appleton-Century-Crofts.

Block, Alan. 1983. *East Side—West Side: Organizing Crime in New York 1930–1950.* New Brunswick, NJ: Transaction.

Friedman, Robert I. 1993. "Brighton Beach Goodfellas." *Vanity Fair,* January, pp. 26–41.

Haller, Mark H. 1997. "Bureaucracy and the Mafia: An Alternative View." Pp. 52–58 in *Understanding Organized Crime in Global Perspective : A Reader,* edited by Patrick J. Ryan and George E. Rush. Thousand Oaks, CA: Sage.

Ianni, Francis, A. J. 1972. "Ideology and Field Research Theory on Organized Crime." Columbia University, New York. Unpublished manuscript.

O'Kane, James M. 1992. *The Crooked Ladder. Gangsters, Ethnicity, and the American Dream.* New Brunswick, NJ: Transaction.

Potter, Gary W. 1994. *Criminal Organizations: Vice, Racketeering, and Politics in an American City.* Prospect Heights, IL: Waveland.

President's Commission on Organized Crime. 1986. *The Impact: Organized Crime Today.* Washington, DC: U.S. Government Printing Office.

Rosner, Lydia S. 1986. *Soviet Way of Crime.* South Hadley, MA: Bergin and Garvey.

Scarpitti, Frank R. 1993. "Organized Crime." In *Social Problems,* edited by Craig Calhoun and George Ritzer. New York: McGraw-Hill.

Smith, Dwight, C., Jr., 1975. *The Mafia Mystique.* New York: Basic.

U.S. Congress, Senate. 1994. *International Organized Crime and Its Impact on the United States.* Hearing Before the Permanent Subcommittee on Investigations of the Committee on Governmental Affairs, United States Senate, 104th Congress, 2nd Sess. S. Hrg. 103–899. Washington, DC: Government Printing Office.

CASES

United States of America v. Anthony Morelli et al. USDC—DNJ (93–210), Newark, NJ.

37

War Profiteering: Iraq and Halliburton

DAWN ROTHE

The complementary interplay between governmental and corporate interests has given rise, especially during the Bush years, to a rash of state-sponsored and state-supported crime. During these years we have seen the ascendance of the neo-conservative philosophy that governmental functions are best contracted out to private industry. At the same time, government agencies have lost funding and employees, with an associated weakening in their regulatory and oversight capacities. The original rationale underpinning this transfer was that private industry would operate more efficiently and profitably than government bureaucracy, as it is driven by competitive market forces. But what has happened, instead, is that corporate and governmental interests have become incestuously joined so that favoritism and no-bid contracts have replaced the free, market economy. This climate has fostered one of the greatest environments for the growth of white-collar crime ever seen. The oil industry, the tobacco industry, and the pharmaceutical industry are three that notably come to mind as enmeshed in governmental connections and decision making.

The twenty-first century has seen one company after another taken down, their employees' pension funds and shareholders' investments robbed by the corporate fraud of top executives. Yet an even higher price has been less visibly paid by members of the ordinary public through state-supported corporate crime. Politicians have fostered this practice by decreasing business regulation and creating a revolving door between business, lobbying, and government.

One of the most dramatic case examples lies in the connection between the Bush administration, through Vice-President Dick Cheney, and Halliburton. In this chapter, Rothe gives us some background on the rise and proliferation of Halliburton and its subsidiaries, charting their history of international, flagrant corporate malfeasance and ethical violations. We see how controls imposed to prevent crony-fueled corporate crime by the Clinton administration were

Michalowski, Raymond J., and Ronald C. Kramer, eds., *State-Corporate Crime: Wrongdoing at the Intersection of Business and Government.* Copyright © Rutgers, the State University. Reprinted by permission of Rutgers University Press.

dismantled by the Bush regime, paving the way for Halliburton to gain an ever-greater share of governmental contracts. Rothe shows us the movement back and forth between top political position and corporate power, with an elite inner circle of those with the right connections amassing great wealth while critical domestic reconstruction and international warfare are shabbily underfunded. At the same time, today's taxpayers and their children will be left facing the burden of financing these state-corporate white-collar crimes.

The intersection of state and corporate interests during times of war is a fundamental part of the war-making process. Every capitalist country must rely on private-sector production to produce the weapons of war. In the United States, for example, major auto manufacturers such as Chrysler, Ford, and Chevrolet retooled to produce tanks, guns, and missiles instead of cars during World War II, while many other companies refocused some or all of their production to serve the war effort. With the introduction of a permanent wartime economy after the end of World War II, amid concerns that the United States was coming to be dominated by a military-industrial complex, major providers of weapons and logistical support such as General Electric, Boeing, Bechtel Group, and Lockheed Martin became regular recipients of government contracts. They were also repeatedly at the center of controversies concerning cost overruns and questionable charges.

The close alignment of corporate and government interests in the production and procurement of the weapons of war is a vivid example of the "revolving door" effect as described by C. Wright Mills (1956) in *The Power Elite*. As executives from major military contractors full elected or appointed government positions, the interests of the state become increasingly entangled with prior corporate loyalties.

In recent years the integration of state interests with those of the private corporation has intensified. This integration began with efforts to adapt to a downsized military through increased reliance on just on time privatized logistic contracts. The move to an active war footing following the attacks of 9/11, including the wars in Afghanistan and Iraq and the permanent "war on terror," further cemented the private-public strategy for war making in the United States.

The controversy surrounding links between Vice President Dick Cheney and Halliburton, the company he formerly headed, provides a demonstration of the potential for state-corporate crime embedded in this new policy of war by subcontract. There have been claims that the association between Cheney and Halliburton resulted in no-bid, cost-plus contractual work without competitive pricing or oversight. According to some, the affiliation between Cheney and Halliburton has established war profiteering as an acceptable and systematic practice within the Bush administration by rewarding "corporations for who they know rather than what they know, and a system in which cronyism is more important than competence" (W. Hartung 2004: 26).

This chapter examines the relationship between Halliburton, the current Bush administration, the "war on terrorism," and war profiteering. In doing so, I incorporate themes from previous state-corporate crime literature, particularly the concepts of state-facilitated and state-initiated forms of state-corporate crime in delineating the relationships between Halliburton and the U.S. government. On the one hand, I suggest that war profiteering in the form of overcharges is state-facilitated crime insofar as the government was aggressive in its refusal to take appropriate and available regulatory action. On the other hand, the Bush administration's repeal of the Clinton administration ruling regulating state contracts can be analyzed as state-initiated crime to the extent that these repeals allowed Halliburton to attain state contracts for which it would have not qualified previously....

BRIEF HISTORY OF HALLIBURTON

Halliburton was first established in 1919. Since that time the company has purchased several subsidiaries that include Brown and Root (including the consortium of Devonport Management Ltd.), Dresser Industries (known as KBR after the purchase of M. W. Kellogg by Dresser), Landmark Graphics Corporation, Wellstream, Well Dynamics, Eventure, and Subsea 7. The success of the Halliburton Corporation is the consequence of a range of strategic practices. Since the mid-1990s, however, many of Halliburton's corporate actions have come under the scrutiny of several U.S. governmental oversight organizations (such as the Securities and Exchange Commission, U.S. General Accounting Office, and congressional leaders), criminologists, and international critics. More specifically, the GAO charged Halliburton with improper billing for questionable expenses associated with its logistical work in the Balkans and the Defense Department charged fraudulent claims were submitted for work at Fort Ord, California. Among these corporate actions are a series of Halliburton's practices that can be classified as corporate crime. Halliburton's actions of systematically overcharging the U.S. government for contracted work, utilizing bribes to attain foreign contracts, and using subsidiaries and foreign joint ventures to bypass U.S. law restricting trade embargos have a long history.

Foreign Trade Barred under International Sanctions

Halliburton's participation in the practices of overcharging, bribing, and utilizing "loopholes" in general to bypass U.S. trade law or international sanctions cannot be traced solely to the 2003 invasion and subsequent occupation of Iraq. The company's business practices of corruption and corporate crime were evident during the 1990s when Dick Cheney was chief executive officer. For example, in 2001 the Treasury Department opened an inquiry into whether Halliburton had used transnational trade loopholes that allowed the company to circumvent sanctions on Iran by doing business with that country through foreign subsidiaries. In another instance, Halliburton's subsidiary, Dresser Inc., did substantial business with Iraq from 1997 through the summer of 2000, closing $73 million in deals with Saddam Hussein at a time when such dealings were prohibited by international trade sanctions. Many of these contracts occurred under the

United Nation's oil-for-food program through joint ventures with the Ingersoll-Rand Company via subsidiaries known as the Dresser-Rand and Ingersoll Dresser Pump companies. Although Cheney claimed that Halliburton divested itself of the subsidiaries in 1998 as soon as it learned of the trading in violation of U.S. legislation prohibiting business ventures while Iraq was under sanctions by the United States and the United Nations, the firms continued trading with Baghdad for over a year past the time of Cheney's initial "awareness." Cheney signed nearly $30 million in contracts before he sold Halliburton's 49 percent stake in Ingersoll Dresser Pump Company in December 1999 and the 51 percent stake in Dresser-Rand in February 2000.

International Bribery Charges

In 2004 Halliburton came under investigation by the French government, the United States Department of Justice, and the Securities and Exchange Commission for international bribery. The U.S. Department of Justice conducted a criminal investigation into an alleged $180 million bribe paid by Halliburton and three other companies to the government of Nigeria. The alleged bribe was paid in exchange for awarding a contract to the companies to build a $4 billion natural gas plant in Nigeria's southern delta region. France is also investigating a former Halliburton executive for his role in the scheme. Investigators said $5 million of the bribes intended for Nigeria were deposited into the Swiss bank account of former KBR chairman Jack Stanley, who retired from the company on December 31, 2003. In 2002, the U.S. Securities and Exchange Commission investigated a second bribery case involving Nigeria. Halliburton admitted that its employees paid a $2.4 million bribe to a government official of Nigeria in order to receive favorable tax treatment, in violation of the U.S. Foreign Corrupt Practices Act and the convention adopted by the Organization for Economic Cooperation and Development prohibiting making bribes in the course of commercial transactions. At this writing the cases remain ongoing.

History of Fraud and Overcharges

In 1997 the U.S. General Accounting Office charged Halliburton with billing the U.S. Army for questionable expenses associated with its logistical work in the Balkans. These charges included billing $85.98 per sheet of plywood that cost Halliburton $14.06 and "cleaning" offices up to four times a day. A 2000 follow-up report by the GAO regarding Halliburton's contract in the Balkans found continuous systematic overcharges via inflated costs submitted in its billing to the army. Halliburton paid $2 million in fines to resolve the fraudulent overcharges.

In a separate inquiry, the Defense Department inspector general and a federal grand jury had investigated allegations that a subsidiary of Halliburton, KBR, defrauded the government of millions of dollars through inflated prices for repairs and maintenance for work at Fort Ord in California. In February 2002 Halliburton paid an additional $2 million in fines to resolve the fraud claims.

As these examples show, Halliburton has a history of corporate criminality and questionable organizational practices. However, it is the recent intermingling of Halliburton and Vice President Cheney that makes its corporate practices a case of state-initiated and/or state-facilitated corporate crime.

HALLIBURTON–CHENEY CONNECTIONS

The connections between Halliburton and Cheney date back to the early 1990s. Dick Cheney, as Ronald Reagan's secretary of defense, assigned Halliburton subsidiary Brown & Root the task of conducting a classified survey detailing how private corporations such as itself could provide logistical support for U.S. military forces scattered around the world. At that time Halliburton prepared a report to implement the privatization of everyday military activities and logistical planning, for which it earned $3.9 million. During the latter part of 1992, Halliburton received an additional $5 million for a follow-up study to outline how private firms could supply the logistical needs of several contingency plans at the same time. This led to "a five-year contract to be the U.S. Army's on-call private logistics arm," what was known as the Logistics Civil Augmentation Program (LOGCAP). Halliburton held this contract until 1997, when Dyncorp, another private firm specializing in providing military logistical support, beat Halliburton's competitive bid.

When Cheney's term as secretary of defense ended in 1993, he and his deputy secretary of defense, David Gribbin, were hired by Halliburton: Cheney as acting CEO and Gribbin as the official go-between to obtain contracts from the U.S. government. Cheney gave Halliburton "a level of access that no one else in the oil sector could duplicate" (Dean 2004: 43). Between 1995 and 2000 intense lobbying efforts paid off with a doubling of the value of Halliburton's government contracts. During 1999 and 2000, Halliburton reported spending $1.2 million on lobbying efforts while obtaining approximately $1 billion in defense contracts. The pattern of paying lobbyists to represent and to obtain future contracts quickly diminished once Cheney took office as vice president. As will be seen, the ratio of money spent on lobbying declined significantly while the value of contracts awarded to Halliburton increased notably. The Cheney–Halliburton marriage, as with all marriages, incorporated distant family members that include some members of the Bush family tree. William H. (Bucky) Bush, the uncle of George W., is a trustee on the board of Lord and Abbott, one of Halliburton's top shareholders. Indeed, this relationship may have played a role in prime contracts recently awarded to Lord Abbott in Iraq. The purchase of Dresser Inc. (1998) by Halliburton further enhanced the family tree of the Halliburton–Cheney marriage. Prescott Bush, father of George H. W. Bush, was the bank representative who financed Dresser and a member of the company's board for several years. The former president of Dresser Industries, Neil Mallon, was a close family friend and mentor to George H. W. Bush, and it was Mallon who was responsible for getting Bush into the oil industry.

Republican Party presidential nominee George W. Bush hired Cheney to oversee the selection process for a vice-presidential candidate. Just as Cheney was finding the right candidate (May–June 2000) to run on Bush's ticket, he sold 100,000 shares of Halliburton stock for 50.97 dollars a share for an approximate value of $5.1 million. The first week of July 2000 the news was official: it would be a Bush-Cheney ticket. Cheney resigned from Halliburton in August 2000 with a retirement package on a future contract worth $45 to $62.2 million, which included stocks, options, and deferred income payments. Shortly after his role as CEO of Halliburton ended and he announced he was running for vice president, Cheney sold another 660,000 shares of stock (worth approximately $36 million). However, he continued to hold 433,000 stock options. . . .

Normally politicians put their stock assets into a blind trust after being elected to high office, so that someone else manages their stock portfolio. However, Cheney continued to hold 140,000 shares of unvested stock that is worth $7.6 million at 2005 stock prices. This stock could not be managed or sold by anyone until 2002, because it was unvested; thus it could not be put into a blind trust. Therefore Cheney had a direct financial interest in the value of Halliburton's stock and the financial profitability of Halliburton the first two years of his vice presidency. Cheney also owns great amounts of Halliburton stock that are safely tucked away in a blind trust for post-vice presidency personal use. . . .

The Cheney–Halliburton relationship had indeed proved to be mutually beneficial while Cheney acted as CEO. Halliburton's revenue from state defense contracts such as LOGCAP nearly doubled (from $1.1 to $2.3 billion) under Cheney's five-year tenure as CEO compared with the five prior years. For example, in 1995 Halliburton jumped from seventy-third to eighteenth on the Pentagon's list of top contractors, benefiting from at least $3.8 billion in federal contracts and taxpayer-insured loans, according to the Center for Public Integrity.

HALLIBURTON AND CHENEY—POST-''RETIREMENT''

Although it may appear to many that the Cheney-Halliburton marriage ended with the selling of stocks and options, the relationship continued. During his tenure as vice president, Cheney drew income both from U.S. taxpayers and from Halliburton. As Bivens (2004: 1) stated, "billions are slyly, secretively and controversially showered on the entity [Halliburton] that paid Dick Cheney 178,437 dollars this year, by the entity [the federal government] that paid him 198,600 dollars." For the previous year, 2002, Cheney received $162,392 from Halliburton and $190,134 as vice-presidential pay.

A Halliburton financial statement of January 2, 2001, showed that $147,579 was paid to Cheney that day as payout of salary from the company's Elective Deferral Plan. This amount was a part of the salary Cheney earned in 1999 but had chosen to receive in five installments spread over five years. Another financial statement, of January 18, 2001, showed an additional $1,451,398 was paid under the company's "Incentive Plan C" for senior executives. This sum was Cheney's incentive compensation (yearly bonus based on the company's performance in 2000). Cheney's personal financial disclosure forms show that Cheney received $398,548 in deferred salary from Halliburton as vice president. Moreover, Cheney received another payment in 2004, and a final payment is to be made in 2005. As previously mentioned, Halliburton drastically reduced its lobbying for governmental contracts once Cheney took office. In the two years prior to Cheney's election as vice president, Halliburton spent $1.2 million lobbying and attained contracts worth approximately $900 million a year. During 2001–2002, Halliburton spent $600,000 dollars lobbying for contracts. In 2003 Halliburton lobbyist Charles Dominy (a retired general with the Army Corps of Engineers, which oversees contracts awarded in Iraq) spent only $300,000 lobbying Congress on behalf of Halliburton. Overall, from 2001 through 2003 Halliburton and its subsidiaries earned $6 billion in U.S. government contracts while spending only $900,000 lobbying, less than they would have spent in one year before their former CEO became vice president. As Thurber (2004: 1) observed, "They are already in; they don't need to lobby any more."

Although the relationship between Cheney and Halliburton appeared to be a major factor in awarding contracts to Halliburton, Cheney denied any involvement in the contracting process. On "Meet the Press" he said: "As Vice President, I have absolutely no influence of, involvement of, knowledge of in any way, shape or form of contracts let by the Army Corps of Engineers or anybody else in the Federal Government." Private memos, however, proved otherwise. An internal Pentagon e-mail (March 5, 2003) sent by an Army Corps of Engineer official claimed that Douglas Feith, the Defense Department's undersecretary for policy, approved arrangements for a multibillion dollar contract for Halliburton "contingent on informing the WH tomorrow. We anticipate no issues since action has been coordinated w/VP office" (*Time* 2004: 1). Within three days Halliburton received one of the first State Department contracts for Iraq worth as much as $7 billion, according to information on the Army Corps of Engineers Web site. Thus state-initiated actions put Halliburton in a favorable position to attain future multibillion-dollar contracts.

Halliburton Contracts

Immediately after September 11, 2001, the neoconservatives and many corporations seized the moment for personal, political, and economic gain. The "war on terrorism" was announced, the long-sought plans to attack Iraq were jump-started, and Halliburton was ready to cash in on its relationship with its past CEO. In November 2001, Halliburton subsidiary Brown & Root received a contract worth $2 million to reinforce the U.S. embassy in Tashkent, Uzbekistan. By December 2001, Halliburton had won back the cost-plus LOGCAP contract to provide facilities, logistic support, and provisions for U.S. troops in places such as Afghanistan, Qatar, Kuwait, Georgia, Jordan, Djibouti, and Uzbekistan. The contract was originally won back through a competitive bid. However, additional no-bid contracts were awarded as the holder of the quick-response LOG-CAP contract was already providing logistical help for the U.S. military and thus was ready at hand. "Halliburton was the obvious choice," said Army Corps of Engineers commander Lt. Gen. Robert Flowers. "To invite other contractors to complete to perform a highly classified requirement that KBR [Halliburton] was already under a competitively awarded contract to perform would have been a wasteful duplication of effort" (Flowers 2003). Unlike the previous LOG-CAP contract, the new contract extended over a ten-year period with an estimated value of $830 million to over $1 billion.

Halliburton (KBR) also received contracts worth $110 million to build prison cells and other facilities at Guantanamo Bay. With the recent addition of Camp 5 (a permanent addition also built by KBR), the prison capacity grew to 1,100 detainees....

During September 2002, 1,800 Halliburton employees were present in Kuwait and Turkey (under a contract for nearly $1 billion) to provide temporary housing (tents) and logistical support for the invasion into Iraq....

Under the competitively bid contract, KBR provides for the support of the Reception, Staging, Onward Movement, and Integration (RSOI) process of U.S. forces as they enter or depart their theater of operation by sea, air, or rail. The post for the Halliburton employees, Camp Arifjan, includes a gymnasium and fast food outlets, including, Burger King, Baskin-Robbins, and Subway, all paid for by U.S. taxpayers. While Halliburton employees are eating fast food, U.S. troops are served mess hall food and live in tents. Excessive U.S. taxpayer money is spent on cost-plus contracts allowing posh conditions for private contractors such as the one in Camp Arifjan while troops are left without the necessary equipment such as bullet-proof vests and armored vehicles. Moreover, Camp Arifjan is the same base in which a Saudi subcontractor hired by KBR billed for 42,042 meals a day on average but served only 14,053 meals a day....

Not only has Halliburton received billions of dollars from the state through competitive and noncompetitive contracts, but most of these were cost-plus contracts. Cost-plus contracts are essentially blank checks that ensure that Halliburton is reimbursed for whatever it bills for its services as well as an additional percentage (between 2 percent and 7 percent) for the company's profits (fees). These types of open-ended contracts are incentives to maximize expenditures to attain an increase in the total value of the contract and profits. Moreover, the larger the contract, the more valuable becomes Halliburton's stock. In October 2002, for example, Halliburton's stock was $12.62 a share. When the KBR Iraq restructure contract was awarded, its stock rose to $23.90 a share (see Halliburton stock portfolios 2001–2004). According to Henry Bunting's testimony to the Democratic Policy Committee, the Halliburton motto in Iraq is "don't worry about it, it's cost plus" (Bunting 2004). In essence, no one questioned pricing. "The comment by both Halliburton buyers and management was 'it's cost plus,' don't waste time finding another supplier.".

Under the Bush-Cheney administration, the profit from the war on terror benefited those corporations that were part of the "family."[1] Few other United States, foreign, or most significantly, Iraqi contractors benefited from U.S. government contracts resulting from the Iraq war. In May 2004, a Pentagon program management office in Baghdad reported that only 24,179 Iraqis were employed in the postwar rebuilding efforts (less than 1 percent of Iraq's total work force of seven million). Meanwhile, just 25 percent of Halliburton's employees in Iraq—6,000 out of 24,000 total—were Iraqis.

The close relationship between the administration and Halliburton constitutes a form of state-initiated war profiteering. While Halliburton may be guilty of inflating total contract values through overcharges and/or charges for services not provided, the opportunities for profiteering were the products of the cozy relationship between the company and a sitting administration whose vice president was Halliburton's former CEO.

STATE FACILITATION OF CORPORATE CRIME

Halliburton and its subsidiaries have engaged in systematic and significant overcharging for services for contracts awarded in Iraq. This is not new behavior for the company. Thus it is not surprising that Halliburton would use "war on terrorism" cost-plus contracts as an opportunity for overcharging. Previously, systemic overcharging by corporations potentially could have resulted in periods of ineligibility for new federal contracts. Specifically, some of the added language required a satisfactory record of integrity and business ethics, including satisfactory compliance with the relevant labor and employment, environmental, antitrust, and consumer protection laws.

The burden of enforcement was on the contracting officers to consider all relevant credible information, with the greatest weight given to offenses adjudicated within the past three years. Thus it relied on the compliance of contracting officers for enforcement.

During the Clinton administration, requirements for contractors bidding on federal contracts were strengthened. New "blacklisting" regulations would have barred contractors from future contracts if they had committed past labor, environmental, or violations of federal trade laws. On April 1, 2001, however, the Bush administration revoked this regulation (no. 65 FR 80255) with no. 66 FR 17754, thus undoing the tightening of regulations put forth by the Clinton administration. This change made it possible for Halliburton and other corporate criminals to obtain contracts regardless of previous or current allegations of illegal practices. Moreover, although payments to corporations under investigation by a federal agency are supposed to be deferred, Halliburton continued to receive its pay as investigations were being carried out, thereby facilitating Halliburton's alleged illegal activities. For example, Halliburton was under investigation by the SEC for accounting fraud that was alleged to have occurred from 1999 to 2001, and the Justice Department was conducting an investigation on charges of bribery connected with the Nigerian official to attain an oil contract in Nigeria. After contracts were awarded to Halliburton in Iraq, the company continued to come under investigations by the Congressional Oversight Committee and the Department of Defense for overcharges, as we shall see.

In 2003 a Pentagon audit found that Halliburton (KBR) had overcharged the U.S. government for approximately fifty-seven million gallons of gasoline delivered to Iraqi citizens under a no-bid contract. These overcharges totaled nearly $61 million from May through September 2003. KBR was charging from $2.27 to $3.09 a gallon to import gasoline from Kuwait, while a different contractor delivered gas from Turkey to Iraq for $1.18 a gallon. Iraq's state oil company (SOMA) was charging 96 cents a gallon for gasoline delivered to the same depots in Iraq, utilizing the same military escorts.

Further exacerbating the problem, part of the money for the KBR gas service contract came from the United Nation's oil-for-food program (now the Development Fund for Iraq). Under the terms of UN Security Resolution 1483, an independent board called the International Advisory and Monitoring Board was to be created to ensure that UN oil-for-food funds were spent for the benefit of Iraqi citizens. The purpose of this agency was to be "the primary vehicle for guaranteeing the transparency of the DFI and for ensuring the DFI funds are used properly" (U.S. House 2003: 6). As of the end of 2003, this body had not been created. Thus the use of these funds to pay inflated prices to Halliburton ($600 million out of $1 billion in funds was transferred to Halliburton) by the U.S. Coalition Provisional Authority went uncontested. In this case the government's failure to provide the necessary oversight agency constitutes a case of state-facilitated wrongdoing.

After it won a ten-year contract worth $3.8 billion to provide food, wash clothes, deliver mail, and other basic services, Halliburton continued to systematically overcharge the government for services rendered (and unrendered). In one case Halliburton charged the government $67 million more for military dining services than the corporation had paid to the actual subcontractors who provided the service. The *Wall Street Journal* reported that Pentagon auditors believed the corporation overcharged $16 million for meals served at Camp Arifjan (subcontracted to Tamimi Global Company). Moreover, the military was already paying Halliburton $28 a day per soldier. In December 2003, Halliburton had estimated it served twenty-one million meals to 110,000 soldiers at forty-five sites in Iraq. Early in 2004 military auditors suspected the corporation was cooking the numbers and overcharging the government millions of dollars. The fact that these overcharges occurred in at least five of Halliburton's facilities suggests they were part of a systematic effort to increase profit margin.

Not only did Halliburton overcharge the government (nearly three times the number of meals than were actually provided to soldiers in Kuwait over a nine-month period), it had been repeatedly warned and audited for its dining service conditions by the Coalition Provisional Authority's inspector general Bowen. The claims were that the food was dirty, blood was consistently found on the kitchen floors, the utensils were dirty, and meats were rotting in four of the military messes the company operates in Iraq. Moreover, Halliburton's promises of improvement in these messes were empty as they "have not been followed through" (NBC News 2003). Regardless of the Pentagon report warning of "serious repercussions," no actions to date have been taken. The government has thus facilitated illegal and unethical profiteering by allowing charges for meals never provided and unsafe food handling and preparations (NBC News 2003). . . .

Beyond overcharging and purchasing unnecessary and/or excessive products, Halliburton systematically billed for labor never performed. For example,

Representative Henry Waxman posted whistleblower testimony on his Web site revealing systematic practices of overpaying employees for hours of labor never performed and billed to the U.S. government. One of the testimonials, from Mike West, stated he was hired as a labor foreman at a salary of $130,000. Moreover, he stated he was paid despite the fact that he had no work; "I only worked one day out of six in Kuwait." During his tenure at Al Asad he claimed to have worked one of every five days, although he was told by his supervisor to bill for twelve hours of labor every day.

The failure of the House of Representatives Government Reform Committee to hear whistleblower testimony suggests that Halliburton's actions clearly fall in the category of state-facilitated corporate crime.

Halliburton's practice of billing for delivering supplies has amounted to systematic overcharges. The U.S. taxpayer has been billed nearly $327 million as of mid-2004 for these runs, and Halliburton was expecting to charge an additional $230 million more. Yet many of these runs have been unnecessary, because at least one in three trucks makes the three-hundred-mile trip empty, while others may carry only one pallet of supplies. Moreover, of the fleet of trucks (Mercedes and Volvos), dozens have not been used. Trailers are left along the roadside when the slightest mechanical problem or flat tire occurs, or when the convoy lacks necessary maintenance items. It was also reported that one Halliburton employee took a video in January 2004 of fifteen empty trailers on the road and stated, "This is just a sample of the empty trailers we're handling called sustainers. And there's more behind me. . . . his is fraud and abuse."

As controversy continues to shroud the Bush-Cheney administration, little to no public outcry has occurred over the blatant misuse of a political office. In part this is due to the ideology of unquestioning patriotism during times of war, and also due to the lack of attention by news media to these issues. This has led to state-facilitated profiteering by Halliburton as an agent of the state. After all, Halliburton is the "biggest contractor to the U.S. government in Iraq earning three times as much as Bechtel, its nearest competitor . . . earning $3.9 billion dollars from the military in 2003, a dizzying 680% increase from 2002 when it earned $483 million" (Chatterjee 2004: 39).

The Bush-Cheney team has achieved a level of secrecy and symbolic power unlike that of any other administration. Criminologists and the public must delve into understanding how and why the Cheney–Halliburton marriage has been able to remain and prosper in the White House. Not doing so increases the probability of future abuses of power, corrupt relationships, and war profiteering. . . .

CONCLUSION

The atrocity of September 11, 2001, clearly was a factor in enabling the Bush administration to gain public support for the war on Iraq. Moreover, the timing of the attacks coincided with the "right" administration being placed in the White House. Halliburton was perfectly positioned to seize the moment to attain additional state contracts. Moreover, the direct connection to Cheney provided further opportunities for Halliburton that would not have been present without the aforementioned conditions.

While war profiteering by Halliburton has been characterized by defense officials as stupid mistakes that can be made right by *nole contendere* payments, the systematic practice of over-costs, overcharges, failure to provide services charged for, and kickback profits are actually a result of the incestuous relationship between Halliburton, Cheney, and Bush. Beyond their existing empowered roles within the state, their political agendas have paved the way for further privatizing areas of the world deemed U.S. interests. More specifically, the Bush administration's imperial agenda, coupled with its private interest in fostering economic expansion through privatization for its "family" companies, has created an environment conducive to war profiteering. Thus the motive for economic gain is both personal and political, serving both administration members' individual financial situations and their neoconservative political agenda. Moreover, Cheney's ties with Halliburton "represent a continuing financial interest in those employers which makes them potential conflicts of interests" (Congressional Research Service, quoted in Chatterjee 2004: 44). Hence the Supreme Court rulings appointing Bush and Cheney into the White House, coupled with the September 11, 2001, attacks, brought an explosion of opportunities for the administration, Halliburton, and war profiteering . . . efforts to control these forms of state-corporate crimes must address all levels of the model: actors; the organizational, structural, and economic culture of hypercapitalism; and the international arena. As state and state-corporate crime literature suggests, there must be internal and external controls. Until measures are taken to provide international regulations for transnational corporations and accountability measures for states utilizing privatization of defense work, no external controls will exist. Currently there are little to no incentive or enforcement mechanisms for the state to adhere to any international laws. Therefore an empowered universal criminal court that is empowered to address states and corporations must be allowed to exist within the international arena.

As long as the current administration's use of secrecy and flagrant violations of checks and balances remains unchallenged by Congress, the public, and the press, little will be done to stop the practices of granting unregulated defense contracts. Internal regulations must be established to separate state from corporate interests. These must include an end to the practice of lobbying and campaign contributions. The revolving door between public and private must be closed. Without these internal and external controls, war profiteering will continue.

NOTE

1. The Bush-Cheney family includes the Carlyle Group (where the senior Bush was a board member and holds significant stock shares); Lockhead Martin (Lynne Cheney was a board member from 1994 to January 2001, as well as a director); Engineered Support Systems (William Bush is on the board of directors); and Lord and Abbott (William Bush is acting trustee).

REFERENCES

Bivens, M. 2004. "Vice President Halliburton." *The Nation*, April 14.

Bunting, H. 2004. In *Testimony Given to Senate Democratic Policy Committee*. Session 12: Democratic Policy Committee on Iraq Contact Abuses. Washington, DC, February 13.

Chatterjee, P. 2004. *Iraq Inc.: A Profitable Occupation*. New York: Seven Stories Press.

Dean, J. 2004. *Worse than Watergate: The Secret Presidency of George W. Bush*. New York: Little, Brown and Company.

Flowers, General Robert, 2003. Memo Response to House of Representative Waxman. http://www.house.gov/waxman/. Also available in testimony of Senate Democratic Policy Committee Hearing, http://democrats.senate.gov/dpc/hearings/hearings24/transcript.pdf.

Hartung, W. 2004. *How Much Are You Making on the War Daddy? A Quick and Dirty Guide to War Profiteering in the Bush Administration*. New York: Nation Books.

Mills, C. W. 1956. *The Power Elite*. New York: Oxford University Press.

NBC News. 2003. Pentagon Report News Coverage. September 14. Archived Transcripts. http://www.nbc.com.

Time. 2004. "The Paper Trail: Did Cheney Okay a Deal?" May 30, 1.

Thurber, J. 2004. The *Progress Report* Archive. *American Progress Report*. http://www.americanprogress.org/site/pp.asp?c=biJRJ8OVF&b=9328 (accessed November 17, 2004).

U.S. House of Representatives. 2003. See Flowers (2003).

Structure of the Deviant Act

Although the structure of deviant associations is revealing, in Part VII we investigate the characteristics of the acts of deviance themselves. Deviant acts involve one or more people aiming to accomplish a particular deviant goal. These vary widely in character from those enduring over a period of months to the more fleeting encounters that last only a few minutes, from those conducted alone to those requiring the participation of several or many people, and from those where the participants are face-to-face to those where they are physically separated. At the same time, acts of deviance can be looked at in terms of what they have in common. All deviant acts consist of purposeful behavior intended to accomplish a gratifying end, require the coordination of participants (if more than one), and depend on individuals reacting flexibly to unexpected events that may arise in this relatively unstructured and unregulated arena. Like the relations among deviants, deviant acts fall along a continuum of sophistication and organizational complexity. Following Best and Luckenbill's (1981) typology, we arrange them here according to the minimum number of their participants and the intricacy of the relations among these participants, moving again from the least to the most organizationally sophisticated.

Some deviant acts can be accomplished by a lone **individual**, without recourse to the assistance or presence of other people. This does not mean that others cannot accompany the deviant, either before or during the deviant act, or even that two deviants cannot commit acts of individual deviance together. Rather, the defining characteristic of individual deviance is that it can be committed by one person, to that person, on that person, for that person. A teenager's suicide, a drug addiction, a skid row transient's alcoholism, and a self-induced abortion are all examples of individual behavioral deviance. Nonbehavioral forms of individual deviance include obesity, minority group status, a physical disability, and a deviant belief system (such as alternative religious or political beliefs). Individual deviance has areas of overlap with loner deviance, discussed in Part VII, but there are also ways in which they diverge. Sexual asphyxia, eating disorders, and suicide all fall within both categories, able to be accomplished alone and usually done without the benefit of associations with other like deviants. But individual deviants, unlike loners, can hang around and participate in subcultures and countercultures with other fellow deviants as long as they can accomplish their deviant act alone. Individual deviants, such as illicit drug users, stutterers, transvestites, the depressed, the obese, and the homeless, are colleagues. On the other hand, loners, unlike individual deviants, need not rely exclusively on themselves to accomplish their acts of deviance; they may have victims. Rapists, murderers, embezzlers, physician drug addicts, and obscene telephone callers are loners, but not individual deviants. Turvey's chapter on sexual asphyxia (Chapter 38) discusses such a solitary deviant practice, where people diminish the flow of oxygen to their brains during autoeroticism in order to enhance their sexual climaxes.

A second type of deviant act involves the **cooperation** of at least two voluntary participants. This cooperation usually involves the transfer of illicit goods, such as pornography, arms, or drugs, or the provision of deviant services, such as those in the sexual or medical realm. Cooperative deviant acts may involve the exchange of money. Where this is absent, participants usually trade reciprocal acts. They both come to the interaction wanting to give and get something. In deviant sales, one participant supplies an illicit good or service in exchange for money. One or more of the participants in such acts may be earning a living through this means.

Bullock's chapter on lesbian cruising (Chapter 39) illustrates the way gay women navigate roles, displays of availability, and seductions that stand apart from any financial exchange. Pasko's chapter on strippers (Chapter 40) shows how these voluntary transactions are different when stage dancers lead their customers through a simulation of intimacy designed to entice them into tipping heavily for erotic performances.

The final type of deviant act is one of **conflict** between the involved parties. One or more perpetrators force the interaction on the unwilling other(s), or an act seemingly entered into through cooperation turns out with one party "setting up" the other. In either case, the core relationship between the interactants is one of hostility, with one person getting the more favorable outcome. Conflictual acts may be carried out through secrecy, trickery, or physical force, but they end up with one person giving up goods or services to the other, involuntarily and without adequate compensation. Conflictual acts may be highly volatile in character, as victims may complain to the authorities or enlist the aid of outside parties if they have the chance. To be successful, therefore, perpetrators must control not only their victims' activities, but also victims' perception of what is going on. Such acts can range from kidnapping and blackmail, to theft, fraud, arson, pickpocketing, trespassing, and assault.

Various types of conflictual and exploitive deviance abound and are thriving, both domestically and internationally. We have witnessed the re-popularization of kidnapping abroad for political, military, and financial purposes. Domestic rates of rape, committed by strangers, friends, and family members, have never been higher. The prevalence of fraud is also rising precipitously, aided by the Internet, through fake stock tips, travel scams, identity theft, "phishing" (where victims disclose account passwords and other data online in response to e-mails that seem to come from legitimate business), "pharming" (where experienced hackers are able to redirect people from a legitimate site to a bogus site without users even knowing it), mail-order bride schemes, and "advance fee" 419 scams (where you are contacted by a solicitor from abroad offering fabulous riches if you will help them recover some lost fortune).

The reading by Armstrong, Hamilton, and Sweeney (Chapter 41), on college rape deals with simple coercion. The article by Trahan, Marquart, and Mullings (Chapter 42) on fraud victimization considers one of the forms of white-collar crime described in the Introduction to Part VII: Ponzi schemes.

38

Autoerotic Sexual Asphyxia

BRENT TURVEY

Turvey debunks previous images of sexual asphyxia, otherwise known as the "ultimate orgasm," offering rare insights into this highly secretive and hidden form of deviance. Primarily a solitary, loner practice, sexual asphyxia is pursued by individuals who desire to raise their level of orgasmic pleasure through self-strangulation. So deviant that practitioners rarely reveal their interest in this act even to friends and lovers, it is accomplished outside of any deviant support group or subculture. Turvey outlines some of the features of this highly secretive form of individual deviance, the characteristics of individuals who engage in it, and the driving force behind their involvement with an emphasis on the myths, criteria, and issues faced by a medical examiner or coroner dealing with autoerotic fatalities.

Autosexuality, the same thing as autoeroticism, is defined as "perversions performed on oneself, including masturbation." (Holmes [6]) It must be assumed that shared sexuality, by inference, would be defined by Holmes as perversions performed on others. Holmes, and others who engage in the autoerotic debate, often suffer from such failures to resolve conflicts between public and private views on sexuality. This tendency often brings poorly defined morality to the debate, and detracts from the purposes of the Forensic Sciences. Autoeroticism may be defined as masturbation and other forms of self-gratification. Autoerotic fatality may be defined as any death that occurs as a result of behavior that is performed as a form of self-gratification.

Example: "The victim, a twenty-six-year-old white male, died while suspended by leather wrist restraints from a hook in the ceiling. When found, he was wearing a commercially produced 'discipline Mask' and a had a bit in his mouth. A length of rope was attached to each end of the bit and ran over his shoulders, going through an eyelet at the back of a specially designed belt he was wearing. The pieces of rope ran to eyelet's on both sides of his body and were connected to wooden dowels

Reprinted by permission of the author.

that extended the length of his legs. The ropes were attached to two plastic water bottles, one on each ankle. The bottles were filled with water and each weighed 7 pounds. The victim's ankles had leather restraints about them. A clothespin was affixed to each of the victim's nipples. The victim's belt had a leather device that ran between his buttocks and was attached to the rear in front of his belt. This belt device included a dildo that was inserted in his anus and an aperture through which his penis protruded. His penis was encased in a piece of pantyhose and a toilet-paper cylinder. A small red ribbon was tied in a bow at the base of his penis."(Hazelwood, Burgess, & Dietz[5])

This case demonstrates a blatant, undeniable autoerotic fatality that is complete with props. But not every case will look like it, or even be as overt. Each autoerotic fatality is as different or as similar in its engagement as the personality of the individual who orchestrates it. Every autoerotic death has its own personal script. That is due in part to the fact that masturbation is a deeply personal, sexual act. This is a critical notion that is often dismissed by even the most seasoned death investigators. Failure to accept this notion will assist in the pathologist's failure to recognize autoerotic fatalities when they are so confronted. Again, autoerotic behavior is a kind of masturbation, and masturbation is a personal and dynamic behavior.

Specific circumstantial elements that can help a forensic pathologist distinguish an autoerotic fatality will be discussed later on.

WHY?

The general appeal of autoerotic behavior is fairly straightforward: sexual gratification.

There are dangerous autoerotic behaviors that individuals find similarly appealing. Of these, perhaps the most prevalent is autoerotic asphyxia.

An important question surrounding autoerotic asphyxia, for those who are not familiar with the term, is why. Why is engagement in autoerotic asphyxia pleasurable? Where does the pleasure come from?

The pleasure can be generally understood as arriving from two distinct sources (distinct but not exclusive). The first source of pleasure is physical, and the second is psychological. Both are equally potent pleasure sources on their own. Put them together and the combination can easily provide the foundation for a dangerous habit.

The physical pleasure from autoerotic asphyxia comes with the reduction of oxygen to the brain, or Hypoxia, which just means partial oxygen deprivation (DiMaio & DiMaio[3]). This is simple euphoric asphyxia. Less oxygen to the brain equals a semi-hallucinogenic, lucid state. It is pleasurable enough on its own to be engaged in without bondage or genital manipulation. It can be performed while entirely clothed.

Dietz[2] gives us an example of a death by sexual asphyxia (classified as an autoerotic fatality), with all of the classic earmarks, yet little sexual obviousness at the scene:

> A 42-year-old Asian man was found hanging by the neck, suspended by a rope attached to the raised shovel of a John Deere Model JD410, diesel powered, backhoe tractor.... The decedent was suspended in a semi-sitting position by a cloth safety harness strap wrapped around his neck and clipped to a rope that was hooked to the raised shovel of the backhoe tractor. A towel was between the loose fitting strap and the victim's neck. A long piece of plastic pipe was connected on one end by conduit tape to the hydraulic control lever of the shovel in the operator's compartment of the tractor. A broom stick was taped to the other end of the pipe and was partially under the decedent's buttocks. The hydraulic shovel could be easily raised or lowered by slight pressure applied to the broomstick. The decedent was fully clothed, and his genitals were not exposed. No pornographic materials, women's clothing items, or mirrors were at the scene.... He had no known psychiatric illness.

Determination of autoerotic death was made from decedent history and circumstantial indicators. The victim kept a journal of love poetry dedicated to his tractor that he had named "Stone," outlining his desire for them to "soar high" together. The victim was unmarried and lived with his parents on their farm. He also had a reasonable expectation of privacy for an extended period, as he engaged in this behavior in the late evening down by the barn. Cause of death was determined to be accidental autoerotic asphyxiation with carbon monoxide intoxication as a contributor.

The psychological source of pleasure from autoerotic asphyxia is personal, and therefore difficult to generalize. The pleasure is best understood as residing in the fantasy. It is widely accepted that the fantasy is fueled by the masochistic/cordophilic aspect of the behavior. The sexual pleasure associated with binding oneself up in a restrictive and/or complex fashion can have a twofold effect: (1) Self-restriction, and (2) Pain/pleasure. The psychological pleasure derived from either of those fantasy behaviors is sufficient reason for many individuals to engage in that form of sexual activity. Bondage and masochism are widely practiced without the element of hypoxia.

An individual involved in dangerous autoerotic behavior of any kind generally has a rich and intense fantasy life. The most accurate way to understand that form of fantasy and pleasure, on an individual basis, is to have the individual explain it to you themselves, in their own words, or to read about it in a journal of love poetry they wrote in honor of their backhoe tractor.

What can be seen and understood of the fantasy, by the pathologist, is the object. Every fantasy requires an object to fulfill it. In the above case it was the backhoe tractor. In other cases it may be complex ligature, pornographic paraphernalia, women's undergarments, or perhaps some other masochistic element.

All of these objects assist the fantasy and accelerate sexual arousal. If the pathologist or death investigator can identify the object of the decedent's fantasy, the autoerotic behavior will begin to explain itself.

Additional explanations include masochistic sensations when approaching death and backing off at the last possible second. Prevailing over that brush with mortality is often perceived as empowering. On that note, many individuals who engage in autoerotic asphyxia may do so in concert with genital masturbation and insertion of objects into their body's orifices for further sexual stimulation, although not necessarily.

Combining all of these elements together, it becomes clearer as to why some extreme types of autoerotic behavior (including autoerotic asphyxia), although dangerous, are a behavior of choice for certain individuals. Individuals of all ages and social strata find behaviors akin to autoerotic asphyxia pleasurable. This is not disputed.

MYTHS

There are many common myths surrounding autoerotic fatalities. This is largely the product of social fear and social ignorance which girdles the issue. It is widely documented that parents or loved ones alter the scene of victims in such cases[5] in order to mislead investigators. They wish to avoid embarrassment. They wish to hide the truth of their loved one's demise and put it away where no one can see it. In addition, investigators and pathologists have their own problems with the issue. Some have problems because they consider the behavior perverse, others because they simply do not recognize it for what it is.

Autoerotic asphyxia is, as discussed earlier, a form of masturbation. Masturbation is one of the few cultural taboos that older society pretends to hold on to. There is also a great deal of cultural evidence to suggest that we are still very publicly ashamed of our sexuality. It is therefore no great wonder that bad and often judgmental information can lead the way to investigating or even discussing autoerotic fatalities.

In discussions on this topic with colleagues, the author has noted that several individuals labor under the impression that the deceased in these cases should be regarded as criminal. Holmes[6] lists Erotic Asphyxiation under the heading "Dangerous Sex Crimes," along with others including Lust Killers, Pyromaniacs, and Necromaniacs. That classification is an egregious error for too many reasons to list here.

It is important to express that individuals who die while engaging in autoeroticism are not guilty of a crime, unless perhaps they live in Missouri or Mississippi. There is nothing criminal about exercising careless masturbation unless someone else is killed or property is damaged. In autoerotic asphyxia and other types of autoerotic fatalities, this is rarely the case. Once a death is ruled to be an autoerotic fatality, law enforcement is generally out of the picture.

Moreover, the author perceives four principal myths surrounding autoerotic fatalities. To effectively engage in the death investigation of a possible autoerotic fatality, the pathologist or death investigator must reconcile these myths with the documented facts. They are as follows:

Myth # 1—Age

"Decedents tend to be young boys"

Adelson[1], in his book, *The Pathology of Homicide,* published in 1973, was a proponent of this theory. He stated, in reference to autoerotic deaths, that they were "a unique group of accidental hangings usually involving young boys ages nine and fourteen or fifteen." And he described it as the termination of a dangerous game played by boys. He was later cited by Geberth[4] and Holmes[6] concerning this specific issue.

This statement is a myth because it is only partly accurate. According to the study of autoerotic asphyxial deaths published by Hazelwood, Burgess, and Dietz[5], operating with the largest subject population of any study to date (N = 132), 37 of the subjects were white teenage males. However, 42 of the subjects were white males (plus one black) between the ages of 20 and 29. An additional 28 subjects were white males (plus one Hispanic and two black) between the ages of 30 and 39. This clearly shows that each of those three age clusters are strongly represented by the data. In light of this sampling, the age generalization cannot practically be applied to the occurrence of autoerotic fatalities. When applied to a death investigation, the need for reconciliation of this myth is compounded.

> *Example* [5]: "A 47-year-old divorced dentist was found on the floor of his office by the janitor. He was lying face down on the floor with an anesthesia mask held in place over his mouth and nose by the weight of his head. His shirt was open and his trousers unzipped. He was alleged to have taken sexual advantage of his patients while they were under anesthesia in his office."

Failure to terminate inhalation of the anesthetic agent, nitrous oxide, resulted in the death of this individual. He was an older man; a professional with an education. A man with some intimate understanding of anatomy, physiology, and the potency of the hypoxia caused by nitrous oxide. He was not ignorant to the potential danger of his activities and he was not a teenager engaging in sexual self, discovery–type behavior. His age group is not to be ignored for the possibility of autoerotic fatalities.

Myth # 2—Gender

"Autoerotic fatalities occur only in men"

This is by far the most concerning myth. It harkens back to a misguided male mind set that females do not have the same sexual desires as males, and therefore do not masturbate. Not only is it supported by Adelson[1], but Geberth[4].

Similarly, DiMaio and DiMaio[3] minimize female occurrences by saying of sexual asphyxia: "Such deaths are rare, with the victim virtually always a male. Only a few cases involving a females have been reported."

The inferences made by the above authors are entirely misleading. Although female occurrences are not as frequent as male occurrences, they do exist in the data. If the Pathologist or death investigator does not think to look at female cases with the possibility of autoerotic fatality in mind, then something may be overlooked in the examination of the body, the scene or the history.

In a lecture that the author attended in the winter of this year given by S.A. Max Thiel (ret.)[8] of the FBI on the subject of autoerotic fatalities, there was a marked show of public embarrassment and disbelief in the audience. One female attendee asked how any female could possibly perform an autoerotic act. She seemed unbelieving of the answer that she received.

Female data are presented by Hazelwood, Burgess, and Dietz[5], Sass[7], and Holmes[6].

> *Example* [5]: "A thirty-year-old single woman was found dead in her locked apartment. She was nude and lying supine on a blanket on the bedroom floor. A pillow beneath her buttocks elevated them. Her legs were slightly spread, and her arms were by her sides. A blouse was lodged in her mouth and covered her face. Next to the body was a dental plate belonging to the victim. Near her left foot were an empty beer can, an ashtray, and a drinking glass. Neither the body nor the scene exhibited signs of a struggle. The victim's clothes and purse (containing her keys) were on her bed. A vibrator and leather bondage materials were found in her closet. The door was locked with a spring bolt. The autopsy report indicated that she had choked to death."

> *Example* [5]: "A black woman in her early twenties was discovered in a severely decomposed condition. She was found on a bed in her locked apartment where she had resided alone. She was nude and lay face down with a pillow under her abdomen and her buttocks in the air. Her right hand was beneath her, near her vagina. Her face was turned to one side, and a knife was beneath her cheek. On the bed immediately below her vagina lay a long sausage, which in all probability, fell from her vagina after death. On the kitchen counter a package of similar sausages, once frozen, had since thawed. The apartment door had been locked from within, and no other persons were known to posses a key."

Hazelwood, Burgess, and Dietz[5] note that the cause of death in this second example remained undetermined due to severe decomposition, but included it because of a similar Japanese case that presented a woman who died while masturbating with a carrot (Tomita & Uchida[9]). The cause of death was determined, in that case, to have been a subarachnoid hemorrhage, which is a form of stroke.

> *Example* [7]: "This case occurred in a midwestern city . . . and involved a woman 35 years of age, a divorcee with a nine-year-old daughter.

The mother was found deceased by the daughter in the morning after the child arose from a night's sleep in an adjoining room.

The child had gone to bed at 10:00 p.m. . . . upon awakening noticed a strange humming noise coming from her mother's room. After entering the room, she found her mother hanging deceased in a small closet off the bedroom.

The victim was found completely nude lying on a small shelved space at the rear of the closet. Her feet were against the wall and her body was extended in a prone position, head downward . . . There was a folded quilt placed on the front portion of the shelf that was immediately under he abdomen and upper thighs . . .

. . . She died of strangulation. An electric vibrator connected to an extension cord was found running. The vibrator was positioned between her thighs and the hard rubber massage head in contact with the victim's vulva. There was a string-type clothespin on the nipple of her right breast, compressing the nipple, and another clothespin of the same type was found immediately below her left breast.

. . . Over the place where the body was lying on this shelved area was a small narrow shelf 66 in. above the floor. This was attached to the wall by two steel brackets and the one closest to the shelved area had a nylon hose tied around it which formed a long loop. The victim has placed her head in the loop and placed a hand towel between her neck and the nylon hose. Her face was turned toward the wall and laying against it."

Women masturbate. Women engage in autoerotic behavior. Women who engage in dangerous forms of autoerotic behavior can die from it. Asphyxia and trauma are not sexually discriminate in their effects on human anatomy. If that tenet of autoerotic fatalities is ignored by a death investigator, or a pathologist, then autoerotic fatalities among women may be overlooked. The belief that it doesn't happen among women may create classification errors in future reporting despite any previous documentation to the contrary.

Myth # 3—Nudity

"Nudity, or partial nudity, is always a feature
in autoerotic death scenes."

As Thiel[8] reminds us, nudity must be explained at any death scene. Nudity can be a feature of an autoerotic fatality. However, nudity is not a necessary feature of an autoerotic death, as was demonstrated by the example in the section entitled "Why". . . . Many serious researchers make the statement that nudity or partial nudity is an element of an autoerotic fatality (DiMaio & DiMaio[3], Holmes[6], and Thiel[8]). This is misleading, again, because it is only mostly true.

Nudity or partial nudity can help with a pathologist's determination as to whether or not the decedent was engaged in autoerotic behavior. However there are many documented cases of autoerotic fatalities that do not have a feature of nudity ([2]; [5]). If cases that do not have a feature of nudity are excluded as possible autoerotic fatalities on that basis alone, a classification error may occur.

Myth # 4—Transvestitism

"Most autoerotic deaths are characterized by transvestitism."

Transvestitism[5]: Recurrent and persistent cross-dressing by a heterosexual male for whom there is no explanation other than sexual excitement for the cross-dressing behavior.

There are widespread anecdotal reports of male decedents being discovered wearing female undergarments or clothing. A sampling of the literature yields these results: Holmes[6] notes cross-dressing as a typical element in an autoerotic hanging; DiMaio and DiMaio[3], on the other hand, note a less common tendency in decedents to wear female articles of clothing. Adelson[1] takes his conjecture further, saying, "...one aspect of sexual psychopathy is observed occasionally in...fatal hangings involving boys and men who are transvestites.... When discovered dead by hanging, they are wearing female undergarments and occasionally, female outer clothing." This demonstrates the diversity of thought on the matter among an array of published experts. This also likely demonstrates that what is typical to these experts is probably a function of personal experience, and not necessarily typical for the entire population of autoerotic asphyxial decedents.

Hazelwood, Burgess, and Dietz[5] and Thiel[8] are more careful in their reporting of the data. They note that sometimes the decedent is attired in one or more articles of female clothing. 20.5% of their sample population were cross-dressed at death. But Hazelwood, Burgess, and Dietz[5] also note that this is probably not transvestitism. They hypothesize that individuals found cross-dressed are probably utilizing the female articles for their masochistic value. They diagnosed only 3.9% of their sample population as transvestites.

Geberth[4] includes their comments in his chapter on the phenomenon of autoerotic fatalities. Not surprisingly, his criterion for making a determination of autoerotic death does not include transvestitism or cross-dressing. Geberth[4] also cites DiMaio and DiMaio[3].

The best data does not support the claim that most autoerotic deaths are characterized by transvestitism. Transvestitism occurs in only a slight portion of autoerotic death cases. Cross-dressing is a little more common, but only one-fifth of the time. In either case, the idea that tranvestitism is prevalent in either form is untrue.

The trained pathologist or death investigator has an obligation to the decedent to resolve these myths and conduct an objective investigation. The failure to resolve any one of these myths is enough to mislead even the most seasoned death investigator. The importance of reconciling personal experience with overall data is equally important.

THE OBJECTIVE CRITERIA

There are many laundry lists available that characterize typical elements of an autoerotic death. Unfortunately, they are generalizations. While they are good compasses for a bewildered investigation, they do not apply to every case.

The following is a list of the objective criteria that should be used to assist the pathologist or death investigator to determine autoerotic death. Recall the definition:

Autoerotic fatality: Any death that occurs as a result of behavior that is performed as a form of self-gratification.

Criteria[4]

1. Evidence of a physiological mechanism for obtaining or enhancing sexual arousal that provides a self-rescue mechanism or allows the victim to voluntarily discontinue its effect
2. Reasonable expectation of privacy
3. Evidence of solo sexual activity
4. Evidence of sexual fantasy aids
5. Evidence of prior dangerous autoerotic practice
6. No apparent suicidal intent

> *Example*[5]: "A twenty-two year old single woman was discovered dead by her sister, who had been staying with the victim temporarily. The sister had been away for two days and returned on Sunday at 9:00 p.m. . . . . She went directly to her bedroom and did not discover the victim until the following morning.
>
> The deceased was found in an arched position with an electrical cord attached to her neck by a slip knot, passing over a doorknob, and wrapped around her ankles. Her abdomen, thighs, and forearms rested on the floor, and her feet were pulled back toward her head. The right side of her head was against the door's edge, and her hair was entangled in the slip knot. She was clothed in a blouse that she normally wore for sleeping. Commercial lubrication cream was found in the victim's vagina, and a battery-operated vibrator was found 4 feet from her body. The only trauma was a 1 1/2-inch contusion above and behind her right ear. The scene was not disturbed, and there was no sign of a struggle. On her bed were a series of drafted letters she had written in response to an advertisement seeking a possible sexual liaison.
>
> Autopsy revealed no evidence of recent intercourse, and no alcohol or other drugs were detected in the body. The cause of death was determined to be asphyxia due to laryngeal compression.
>
> The victim had been in excellent physical condition, had made plans for a canoe trip on the day following her death, and had recently been in good spirits. She was sexually active but was reportedly disappointed

in her sexual relationships as she had difficulty attaining orgasm. She used contraceptive cream and a diaphragm to prevent pregnancy, and these items were located in her car."

This case demonstrates all of the above mentioned criteria. Her hands were free to push herself up at any time and interrupt the laryngeal compression, but she clearly passed out before she could do so. She was dating, and concerned about pregnancy, but those items associated with blocking potential pregnancy during sexual intercourse were located in her vehicle. If she were not alone while engaging in her sexual behavior, it would stand to reason that those articles would have been found to be in use at the scene of her death.

The presence of the vibrator and the letters on her bed demonstrate the presence of the sexual fantasy aids. The complexity of the bondage involved demonstrates complex fantasy behavior, and is indicative of prior fantasy and therefore represents an escalation of fantasy behavior. Rarely will an individual hog-tie themselves to a doorknob with a slip knot on their first time out. The planning of the canoe trip and the reports of her being in good spirits indicate that she was not suicidal. Also, there was no suicide note. By itself that fact has little meaning as suicide notes are not always left by those who commit suicide. Circumstantially, however, the absence of a suicide note does contribute to the picture of an autoerotic death. If there had been a note present, it would have been helpful in dismissing the possibility of an autoerotic death.

> *Example*[5]: "An eighteen-year-old white male was discovered by his mother in what she believed to be an unconscious state. She drove him to a local hospital where he was pronounced dead. Upon arrival at the hospital, he wore only a pair of red women's panties. Markings around his chest led to further questioning of his mother who then stated that when she found her son, there had been a plastic bag around his head that was secured around his neck with a rubber band, and he had been wearing a brassiere and a slip. She had removed all of these things before taking him to the hospital. A search of his room at home disclosed an extensive collection of women's clothing and several other plastic bags. The plastic bag he had used during the fatal episode had holes in it that had been repaired with masking tape."

This case is a good example of suffocation to induce euphoric hypoxia, rather than the more common compression/hanging techniques. His hands were not bound, so he could have punctured the bag at any time to discontinue its effect (as he had obviously done in the past). From all indications, he was alone at the time of death. The female clothing represents a significant fantasy aid. Holes in the one bag, the presence of other bags, and the presence of a collection of women's clothing suggests a great deal of previous engagement in this dangerous autoerotic behavior. The existence of prior incidents using the same repaired bag, in tandem with sexual fantasy material, would suggest that his intention was pleasure, not death.

The above criteria for determining autoerotic fatalities are fairly objective. They do not exclude on the basis of gender, skin color, age or lifestyle. They also depend heavily on victim history. Adherence to these criteria could assist greatly in keeping each of the four main myths previously mentioned from misleading a death investigation. It would also assist in barricading against other classic personal and political considerations that are prone to interfere with investigations of a private, sexual, and potentially socially embarrassing nature.

MANNER OF DEATH

Determining the manner of death is sometimes very problematic for a medical examiner or coroner. A definition is in order.

> **Manner of Death** [3]: The manner of death explains how the cause of death came about. Manners of death are generally considered to be natural, homicide, suicide, accident, and undetermined. . . . The manner of death as determined by the forensic pathologist is an opinion based on the known facts concerning the circumstances leading up to and surrounding the death in conjunction with the autopsy findings and the laboratory tests.

Autoerotic fatalities are, by definition, accidental deaths. They are not suicides. Suicide is the intentional taking of one's own life. Accidents are unforeseen and unintentional.

Last month, a colleague wrote the author of this work and stated that he "would consider it [autoerotic death] (assuming that the pathologist has identified the death as autoerotic asphyxia) as suicide, simply because the action is as voluntary as suicide." This is not an uncommon argument. However, the argument is only sustainable on a superficial level. By that same reasoning, traffic fatalities would be determined as suicides, because everyone knows that being in a car or walking along the side of the road can kill you with more frequency than any dangerous autoerotic behavior.

DiMaio and DiMaio[3] opine on the case of a 21-year-old white male in their discussion of autoerotic asphyxiation and end their section on the subject with the following:

> The danger of this[transitory anoxia] is that a few seconds miscalculation can cause loss of consciousness and the individual will hang himself. The key to the diagnosis of this entity is the presence of a towel or some other article of clothing between the noose and the neck. . . . The intention is not to die, but to produce sexual gratification.

Those who engage in autoerotic behavior are doing so for pleasure. They have no expectation or intention of death. They have made provisions for their future and for future autoerotic behavior. Failure to surmise the intent of sexual arousal in autoerotic fatalities should be avoided in determining manner of death. There is little disagreement among the experts ([1], [2], [3], [4], [5], [6], [7], and [8]) on this issue.

CONCLUSION

Dr. Park Dietz[5] begins the second chapter of his work with a statement as to the nature of autoerotic asphyxia that pathologists and other death investigators cannot afford to ignore:

> For certain individuals, the preferred or exclusive mode of producing sexual excitement is to be mechanically or chemically asphyxiated to or beyond the point at which consciousness or perception is altered by cerebral Hypoxia (diminished availability of oxygen to the brain).

There is a diverse population of individuals that find autoerotic asphyxia and other dangerous autoerotic behavior sexually gratifying. Their autoerotic behavior is not criminal. It is a dangerous form of masturbation. Their subsequent death is accidental. For whatever social or personal reasons, death investigators and pathologists can fall into the ignorance of some alluring myths surrounding autoerotic death. This is a failing that cannot be afforded in light of the pathologist's responsibility to the decedent.

There is a great deal of literature that works against reconciling popular myths in favor of good data. That literature must be reconciled with objective data, and objective criteria for determining autoerotic fatalities. If death investigators and pathologists are to maintain any scientific credibility at all in this area, the need for the more universal, more objective criteria must not be ignored.

BIBLIOGRAPHY

1. Adelson, Lester. *The Pathology of Homicide* (Springfield, Ill: Charles C. Thomas, Pub., 1973), pp. 552–553.

2. Dietz, P. E., & O'Halloran, Ronald. "Autoerotic Fatalities with Power Hydraulics," *Journal of Forensic Sciences,* No. 2, March 1993, pp. 359–364.

3. DiMaio, D., & DiMaio, V. *Forensic Pathology,* (Boca Raton: CRC Press, 1993), pp. 246–247.

4. Geberth, Vernon. *Practical Homicide Investigation,* 2nd Ed., (Boca Raton: CRC Press, 1993).

5. Hazelwood, Roy, & Burgess, Ann, & Dietz, Park. *Autoerotic Fatalities* (Lexington, Mass: D.C. Heath & Co., 1983).

6. Holmes, Ronald. *Sex Crimes,* (London: Sage Publications, 1991), pp. 62–65.

7. Sass, F. A., "Sexual Asphyxia in the Female" *Journal of Forensic Sciences,* No. 20, 1973, pp. 181–185.

8. Thiel, Max, S. A. (ret.). Lecture: "Autoerotic Death," University of New Haven, Feb. 2, 1995.

9. Tomita, K., & Uchida, M. "On a Case of Sudden Death While Masturbating." *Japanese Journal of Legal Medicine,* Vol. 26, 1972, pp.42–45.

39

Lesbian Cruising

DENISE BULLOCK

Once again tackling the issue of gender deviance, this chapter examines what was once thought be to an exclusively male domain: homosexual cruising by women. Traditionally socialized to take a passive role in flirting and dating, women who love women find that in order to hook up, they must take more active stances. This behavior involves a trade of services between cooperating members. Drawing on her lesbian participation in bar scenes and several years of focused data gathering, Bullock illuminates the behaviors of this hidden population. She outlines lesbians' goals in looking for other women, framing them within a continuum ranging from one-night stands to committed relationships. Further, she offers a typology of cruising styles incorporating the method, approach, intent, and investment of each. Bullock's chapter highlights the diversity within the lesbian community that explodes the simple typology of butch-femme styles and discusses the difficulty women have in breaking out of their traditionally passive feminine gender roles.

The lesbian community, like other subcultural groups, is often viewed by the larger society as a homogeneous entity. Lesbians are perceived to have singularly common lives; however, the lesbian subculture is a diverse and dynamic group. Studies regarding lesbian lifestyle issues, therefore, must encompass this diversity and range of behavior while, at the same time, examine the common themes that exist within the group. This research attempts to broaden our scope and understanding of one aspect of the lesbian social world, lesbian bars, by ascertaining a sense of this special social space as it is experienced by its participants. . . .

Sexual awareness and what is perceived as acceptable sexual behavior have changed in both the general society and the lesbian/gay community. While both sexual and gender norms have either changed or softened in recent years, my experience and observations indicate that women in general remain more reluctant than males to make the initial approach to a prospective partner.

Both academic and popular writers have noted the difficulty lesbians have initiating interaction. The dilemma is so widely known and shared among the lesbian community that it has become comedic material.

In this society, the single heterosexual woman in a bar setting generally conforms to the larger societal rules and sexual scripts of interaction. If she wishes to meet a man she will make herself "presentable" and receptive to approach. While she may seem a passive participant, she is, in fact, actively engaged in the interaction. In a lesbian setting, little or nothing would happen if all of the women waited to be approached. Kitaka, who owned and operated a sex club called Ecstasy Lounge,[1] found that women were hesitant to approach other women, particularly to initiate a sexual encounter (1999). Realizing that women were hesitant to initiate interaction, volunteers, who acted as hostesses, introduced women or engaged in sex when desired.

> At the Ecstasy Lounge, they (the hostesses) did not have the social
> barriers that usually kept them from pursuing women at bars or other
> public places. They had a reason to approach women for sex, it was their
> "job" as a hostess! (Kitaka: 181–82)

Kitaka also devised games to ease interactional tensions. A lesbian club designed for public sexual encounters is clearly a different interactional space than the typical lesbian bar. What Kitaka's experiences illustrate, however, is that interaction rituals that fall outside of normative scripts are problematic in a lesbian social environment. Lesbians, constituting a sororal group, have had to adapt, create, and maintain their own environment and rules of interaction.

Previous definitions of cruising have narrowly focused on the impersonal sexual pursuits of gay men. This limited perspective privileges stereotypical male, aggressive roles and reinforces negative stereotypical images of a promiscuous gay lifestyle. When I first observed men cruising in gay bars, I quickly noted their overt acts of cruising. One of the first interactions I noted was a man who approached a second man. The approaching man grabbed the other man's groin area, gave a quick head nod toward a back room and both men left the bar area for that back room. I later was told (I was not permitted to enter) the room was an area where sexual exchanges occurred. This type of behavior fits the stereotypical image of cruising. If I had stopped observing at that point or if I simply looked no closer I would have come to the conclusion that this style was all there was to cruising. I had not observed women acting as aggressively or overtly and did not see special rooms[2] for sexual encounters in the women's clubs. I would have come to the conclusion, based upon these observations, that lesbians do not cruise. However, I did look closer over a period of many years at the interactional patterns occurring in both men's and women's bar settings and discovered a range of cruising styles which led to my expanding the concept of cruising.

I have broadened the definition of cruising to include a range of behavior as evidenced in both the lesbian and gay community. Cruising will be defined as the purposeful search for a socio-sexual partner, in some cases for a limited relationship (one-night stand) and in others for an indeterminate period. There are numerous styles and methods of cruising, and I have developed a typology of cruising, highlighting seven styles based on method, whether the individual approaches or not, intent, and investment. The primary and most accessible cruising ground for the lesbian community is the lesbian/gay bar or club.

RESEARCH METHODS

This research, using an ethnographic approach, focuses narrowly on the behavior of women as observed in lesbian clubs. Entry into the settings was not a problem for me as I was a known member of the community. I have frequented many lesbian clubs in the large southern, metropolitan city I examined and around the country for both social and observational purposes.

My personal exposure to lesbian and gay bars and to cruising began in 1979 in a large southern city. I entered the clubs as a naive, single woman looking for friendship, companionship, and a good time. I joined the student gay organization associated with the local university and met quite a few people who were more than willing to show me the town. Within the first few months I had visited all of the lesbian and gay clubs in the city (numbering approximately 20). I then began venturing out on my own, finding my niche within the community.

In my young adulthood I was outgoing and presented myself as very confident and self-assured. These personal characteristics, along with my desire to expand my sexual experiences and remain unattached, impacted the cruising method I chose to use in the bar settings. I frequented the clubs often and actively participated in cruising (a style which will later be identified as "Home Base"). This exposure and experience gave me a context in which to place my observations and respondent accounts of cruising. The ability to contextualize and personalize my findings provided me an insider's view into the interactional dramas unfolding in the lesbian and gay bar settings.

I examined lesbian cruising between 1979 and 1992, conducting interviews and focused observations during three separate time periods: 1982, 1986, and 1992. These data provide a historical perspective of the lesbian community, bar participation, and of cruising activity and attitudes. The observations and interviews were conducted concurrently, using an open interview guide with a base set of questions. The focus of the interviews was on the respondent's past and present bar participation, how they meet potential partners and their perceptions, attitudes and actions relating to the factors affecting cruising. All interviews were recorded on cassette tape and then later transcribed. Field notes were recorded primarily during the three indicated time periods in the two or three most popular clubs of the time. So as to be less conspicuous, I generally took notes in the privacy of a bathroom stall or my car. Although I continued to patronize the lesbian clubs during the ten-year span, the frequency of structured observations was the highest during the three specified time periods. I generally visited one or two clubs during the weekend and one club on the weekdays, attempting to assess the flow and characteristics of clientele. In addition, I altered my focus over the course of an observation period to differing aspects of cruising interaction. At first, I thought there were only two distinct styles based on the desire for a short- or long-term relationship. As my research progressed I soon realized the complexity of various factors, including intent. The seven styles of cruising emerged over the course of the study as the complexity and distinct elements unfolded.

My respondent sample was composed of 46 women who identified themselves as lesbian. The women ranged in age from 18 to 47 and were predominately White (38 White, 6 Black, 2 Hispanic). In all three studies notices asking for participants were posted in clubs. The remaining participants were gathered by "snowball" sampling technique. . . .

METHODS OF CRUISING

Styles of Cruising

A woman's cruising style and the methods she uses will be a reflection of her personal characteristics and experiences, socialization, social context, social ecology, and intentions for cruising. If a woman wishes to meet a prospective partner she must decide if she will approach or be receptive to approach. If she decides to approach an unacquainted woman she may use an opening line, not dissimilar from a heterosexual setting (i.e., "Would you like to dance?"; "Can I buy you a drink?"; "Have you seen this video before?"), in order to initiate interaction. My research uncovered many variations of cruising styles which I have distilled into seven basic categories, based on method, intent, and investment (see Table 39.1). These seven types are intended to provide a general view of cruising styles.

> Katherine (1992, 31-year-old, White, accountant):
>
> There's also different styles of approaching people, but you got the sense that they circulated and they watched and they paid attention and they looked for the right face. And when they saw it, they zeroed in on them and depending on what kind of a person they were, maybe they would watch for weeks or maybe they would watch for ten minutes, depending on how much self-confidence they had . . . everybody had their sort of style of making contact, and some of it was through friends, you know, do you know that person, do you know that person, do you know that person, go up and introduce me . . . some people were very bold and just went up to people they didn't know and they would say things or ask to dance or offer to buy a drink or whatever.

The cruising styles are not, however, fixed or inflexible. A particular woman may overlap into one or more categories and may even use different styles during the same night or evening at the club.

Strutter   A Strutter circulates throughout the club, occasionally pausing to speak with acquaintances or friends. She projects self-confidence and determination in

TABLE 39.1 Styles of Cruising

STYLE	METHOD	APPROACH	INTENT	INVESTMENT
Strutter	Circulates through club, occasionally stopping to speak with friends	Yes or very receptive	Find a partner, any term	High
Home Base	Stands in one visible location, surveying area	Yes	Find a partner, short term	Medium
Mingler	Comes in alone or with a group and stands or sits with a group	Yes with coercion or support of friends	Have fun with friends, maybe find a partner	Low
Ego Booster	Circulates through club, attempting to draw attention to herself	Yes	Impress herself or make a lover jealous	Low
Sojourner	Circulates through club, periodically stopping to survey areas and be seen. She will distance herself from others.	Unlikely	Hopes to find a long-term partner	High
Woman in Waiting	Stakes out a territory to be visible, surveying area	Yes after long observation	Find a partner, long-term	Medium
Game Player	Uses a game, such as pool as method of approach	Yes through the game	Find a partner, any term	Low

her search for a partner, but she is not as visibly overt as a Home Base Cruiser. Her intent is to find a partner. A Strutter is outgoing and usually knows many women in the club. She does not, however, stop and chat with people just for the sake of mingling—which would take time away from her search for a partner. If she sees a prospective partner among a group of people she knows, she can easily step into the group to be introduced. A Strutter will approach or make herself very receptive to approach. The length of involvement with the relationship is flexible and not pre-determined. However, like the Ego Booster, a Strutter needs to stroke her ego and likes to know that she can "strut her stuff" and not necessarily be tied into a relationship. Her investment is high, but she will rationalize her failure if she does not find a partner.

Chris (1986, 26-year-old, White, teacher):

They strut around. They talk to as many people as they can. They don't stand in one place like we normally do. They'll cruise around the bar and you'll see them talking to one group and then another group. They're presenting themselves.

Clubs with a circuit path (an aisle or walking path that encircles the dance floor or club) work best for a Strutter; however, it is not a necessity. The circuit path is the Strutters domain. She can both present herself as available while at the same time peruse for prospective partners.

Home Base Cruiser   A Home Base Cruiser is what most people envision when someone is described or labeled as a "cruiser." If there is an area in a club that has a prominent "lookout point" (a place where she can see most or all of the club and be highly visible) she will make this her "home base." She will remain in her home base for an extended period of time (20–40 minutes or more), continually surveying the area and targeting prospective partners. A Home Base Cruiser is very conspicuous and self-confident and may circulate occasionally but will always return to her chosen lookout point. A Home Base Cruiser is seeking a short-term relationship and will continue to pursue a woman as long as she determines that contact will successfully lead to a sexual encounter for the evening. She is a regular patron to the club, and her home base often becomes known to other patrons as her regular designated spot in the club. If she sees someone she is interested in she will approach them, using a variety of opening lines and methods for first contact.

Megan (1992, 24-year-old, White, X-ray technician):

... there is one girl that I know that is the cruiser from hell. She's this little Vietnamese girl and I see her quite often in different bars. She thinks she is like the biggest stud and it just cracks us up. That is a classic example.... I mean her sole purpose is to go out and pick up a woman ... and it's not a pretty sight. Well, she goes up and she'll light their cigarettes, you know ... she'll stalk her prey, you know and then they'll get a cigarette out and she'll be out there [she mimics lighting a cigarette] ... you know, hey baby can I buy you a drink, you know. And she does this ... Hey ... Frank Sinatra head shake (laugh). I mean, it's a classic. I mean if you saw it you would totally think cruiser, you know.

I observed in 1986, on a regular basis, three women whose actions typify what would be identified as Home Base cruising behavior. The women were well acquainted with each other and appeared to be friends. Typically they would enter the club alone, get a drink and take up position in their usual place, standing at the top of the stairs in the most prominent observation point in the club. As each of these women arrived they would greet the other Home Base Cruisers then quickly establish individual distance from them (typically two to three feet) to visually show that they were each alone. Once one of the women had identified a prospective partner for the evening she would approach the woman. As soon as a Home Base Cruiser was successful she would leave the club immediately with her catch, only to return again the next evening, weekend, or even later that evening to repeat the same process. In all of my observations of these three women I never saw them leave the club alone.

As soon as a Home Base Cruiser determines she will not be successful with a prospective woman she breaks off contact and returns to her home base to prospect for another candidate. She may lower her standards as an unsuccessful evening progresses, and approach women whom at the beginning of the evening she would not have approached.

Bobbi (1986, 23-year-old, White):

> My observation, they pick out... they don't necessarily have to be too terribly cute ... In fact they can be quite homely. All they're looking for is a yes. Like when I think of someone who likes to just cruise people. [A particular woman] she'll walk into the bar she looks constantly. And you know, just anybody that shows her any attention she tries to cruise ... just for the night.

The Home Base Cruiser's investment is medium because her confidence level allows her to feel that she can find a partner when she "really" wants one. She will rationalize her lack of success as perhaps a night with poor prospects or her disinterest in "actually" wanting a partner for the evening.

Mingler The most popular and frequently used style of cruising, the Mingler, comes in alone or with a group and will stand or sit with a group. She will only circulate with a specific verbalized or rationalized intent, such as going to get a drink from the bar, going to the bathroom, meeting with another group or person she already knows. A Mingler may, however, take the longest route to her "destination," possibly choosing a path that will take her near someone she is interested in meeting. A woman going to a club with friends or to meet with friends has a valid and safe reason (sociability) for being at the club. It may be that she intends to find a partner for a short-term or long-term relationship, but her friends provide a buffer against possible rejection. Friends also provide support to her in approaching or being receptive to a prospective partner. The following is an excerpt from the interview with Judi (1982, 20-year-old, White, emergency medical technician) which is an example of peer support:

J: Probably just sit back and tell my friends that I would like to meet her (laughter).

I: What if she made eye contact with you?

J: I'd probably still leave it up to her. I don't know, I would continue the eye contact.

I: What conditions would have to be met before you would approach a woman?

J: I suppose it would just depend on my state of mind. How I felt. If I were seeing anyone else at the time. If I had been dared into it maybe (laughter).

The more women a Mingler knows in the club the greater her social exposure and her subsequent extended number of potential contacts for partners. Also, women tend to look among their known group of friends for potential partners.

This is one reason why in lesbian social gatherings there are typically many women who have been in relationships within their social network. Several women interviewed said that the only time they would consider approaching a woman they were interested in meeting was if the woman was standing or sitting with someone she knew. Then they could join the group and arrange to be introduced. A Mingler will generally not approach another woman directly unless coerced or with approval from her friends. The investment for a Mingler is low. She is patient and will often observe a potential prospect over an extended period. If she doesn't find a prospective partner during an evening out she leaves the bar content that she had a good time with her friends anyway. This group of friends, interviewed together in 1986, describe how this type of cruiser and they themselves as a group observe prospective partners over a period of time:

BETH: They take their time. They pick out the feature in the other person that they want and then watch. They study them for a while.

JESSE: Before we even approach someone we've had our eye on them for a while.

BETH: We watch the kind of friends they have. And what they do when they're at the club.

BOBBI: The more I like a person the longer I wait and check them out.

Minglers, comprising the largest group of the cruising syles, have the added advantage of being able to utilize a variety of settings. Open areas allow individuals more freedom of movement and interaction. A Mingler would naturally be found in these areas, providing there are other individuals known to them in the area.

Ego Booster Ego Boosters circulate throughout the club, often trying to draw the attention of everyone in the area. The Ego Booster is cruising to impress herself or someone else and may already be involved in a relationship with another woman. Her primary objective of cruising is to build her own self-confidence or make someone else jealous. She is very overt in her cruising style and may appear to be a Strutter; however, she is not looking for a relationship. An Ego Booster will approach, but, if she intends to pick up a woman only to show that she can be successful in this venture. Her investment is low as she is not really looking for a partner.

Dawn (1986, 23-year-old, White, student):

It's almost like they let down the barriers to present themselves to a lot of different people. To me it's an act of showing off. You're trying to show people . . . Maybe you feel good about yourself that night. You look good and you just go around and let everybody see you. It's kind of like a modeling show.

Like a Strutter, clubs that have a circuit path allow the Ego Booster to be visible to other patrons.

Sojourner   These are the ever-hopeful women seen standing around the fringes of the dance floor, bar, or pool table area. A Sojourner circulates around a club, stopping periodically to observe an area and be seen, remaining there for as long there are potential prospects. When the potential decreases she will float around the club to try different areas. She wants others to notice that she is available and, therefore, will distance herself from other women and present herself as interested; utilizing a different method of proceptivity. For example, if she is near the dance floor she may move her body with the music to show that she is receptive to dancing. Additionally, she may situate herself in close proximity to a potential prospect and try to establish eye contact with her.

Katherine (1992, 31-year-old, White, accountant):

Those are the people that would take two weeks or, you know, they would watch for six weekends in a row. They would come to the bar with hopes of seeing somebody that they thought was cute and they would gather information for weeks about this person and daydream about them and fantasize about them and do all this stuff. And then, finally, by some either highly orchestrated or totally unplanned event they would meet this person. And, proceed from there.

The confidence level for this type of cruiser is usually very low. Some Sojourners appear sad, projecting a defeatist attitude and quite often they receive just what they project. Other Sojourners may possess a happier disposition but still lack enough confidence to approach. A Sojourner goes to a club hoping someone will find them for a long-term relationship. Their investment is high because success is directly connected to their self-esteem.

GC (1992, 31-year-old, White, unemployed):

Well you have your little cruiser, one who just walks back and forth through the bar all night long. She's always by herself and it's like, you can tell when she spots someone she thinks is attractive because she'll stand ten to fifteen feet from her and watch and then hope for that eye contact.

The Sojourner style of cruising was illustrated in an observation made one evening at the most popular lesbian bar complex in the city examined:

I went to the Promenade with my partner and we stayed in NXS for most of the evening. It was a slow night for all three clubs. At the peak, NXS only had between 50 and 75 women. I only noted a few singles. There were three in particular that I watched. The first single woman was Hispanic and she came in, got a drink and stood at the dance floor rail. The second single woman was White and she came in and sat on the raised middle seating platform, on the top, at the end closest to the dance floor. The third single woman was White and she came in, got a drink and sat in a chair at a table between the two platform seating areas. After about ten minutes, from the time the third single woman arrived, the third single woman approached the first single woman, still standing at the rail, and asked her to dance. The first single woman, who had appeared to either be bored or sad, immediately smiled and her whole body image changed. The two women danced for quite a long time, talking periodically. They then went to the table the third woman had been sitting at. The second single woman was still sitting up on the platform, which was only a few feet away from the table where the other two women were now sitting. After a short time the two women decided to dance again and the third single woman asked if the second single woman wanted to join them. All three danced for a couple of long songs and all three went to the table. Single woman one and three were sitting fairly close to each other and were having a good time talking with each other. The second woman was sitting on the lower platform at the table across from the two women. She was not involved in the conversation and was restless. She started out trying to be involved in the conversation. She sat up to the table and was leaning in. After around five minutes she leaned back and started looking around. A short time later she moved up to the upper platform above the table. The two other women didn't acknowledge her departure. They were involved in their conversation. A little while longer, the second woman moved down to the end of the upper platform, her original position. Single woman one and three dance periodically throughout the evening and were sitting at their table when my partner and I left NXS to go to Ms. T's around 12:45 am. When we returned to NXS at 1:25 am all three single women were gone. (Friday, 11-13-92, 9:00 pm to 1:50 am)

In this observation, I would characterize single women #1 and #2 as sojourners. Their lack of self-confidence was evident. They chose locales to stand or sit in which other singles would take note of them and waited to be approached. I would tentatively classify single woman #3 as a Home Base Cruiser. Since I did not see single #1 and #3 leave or later interview either woman, it is difficult for me to determine intent. However, single woman #3 quickly identified a prospective partner and made an approach.

Woman in Waiting   A Woman in Waiting stakes out a territory, arriving at a club early to claim a table for herself. She will choose a prominent location to see and be seen, circulating around the club periodically. A Woman in Waiting wants others to know that she is available; therefore, she distances herself from others like a Sojourner, with usually one location in the club that she claims for herself. She usually comes in alone and will present herself as available and interested. As a regular patron in the club, she studies potential partners over an extended period of time in order to be more certain of her choice in a potential partner before she considers approaching. If eye contact is continually maintained or there are other positive signs this often will speed up the process. She may approach another Woman in Waiting if there is interest in each other and time has passed and neither has made an approach to someone else. A Woman in Waiting is looking for a long-term, committed relationship. Her investment level is medium. She is patient and will watch and wait for many weeks.

Gloria (1992, 39-year-old, White, production coordinator):

I was more the person who waited and was approached. But I was real friendly with everybody and I would try and be friendly with everybody and it seemed like almost every time someone would approach me and want to get to know me a little better. And, you know, they had that sparkle in their eyes . . . (laughter) . . . and they wanted to be more than friends. And . . . but I was usually not the initiator.

A Woman in Waiting needs a prominent location from which to observe. Unlike a Home Base Cruiser, she is less overt in her method and manner.

Game Player   This style of cruiser uses a gaming area, such as a pool table as a means to meet a prospective partner. If, for example, a woman sees an interesting prospect playing pool, she can put her name on a play board or place coins on the table to challenge her.

Angela (1986, 24-year-old, Hispanic, student):

I always go to the pool tables. . . . I find it so easy to meet people playing pool; you put your quarters up, you challenge somebody, you play them. You got to get their name at least.

Pool etiquette requires that introductions are made prior to the start of the game. Casual conversation is as much a part of the game as the billiards themselves (except during tournament play). A Game Player may approach a prospective partner in relation to the game. Her intent is to find a partner for a relationship of any term. Her investment is low because contact is maintained throughout the game and her public intent is to have fun and socialize. If the interaction does not progress beyond the game there is no visible sign of failure, thereby saving her being rejected publicly.

Almost all lesbian clubs have at least one pool table and most have additional games available; therefore, a Game Player can utilize each of these clubs for cruising. In addition to pool, there may be other games which allow women to interact as part of the game. . . .

CONCLUSION

As is indicated by the research, cruising is a complex issue reflecting the diversity within the lesbian and gay community. My research has shown that previous definitions of cruising have privileged the behavior of males and has been limited to an overt style of cruising for the intent of short-term sexual encounters. This position has reinforced negative stereotypical images of gay men and the gay lifestyle in general and has devalued women's more subtle styles and longer-term relationships. I have broadened the definition of cruising to encompass a range of behaviors and intents which more closely coincide with the variety of meanings attached to this concept within the homosexual community.

Since 1979, I have observed many changes in the surrounding society, lesbian community, and lesbian clubs themselves. As a subculture, lesbians are becoming more visible; therefore, the society at large is increasingly aware of the diverse nature of the lesbian community. The lesbian community reflects the changing values and life patterns of the surrounding society, assimilating those values and patterns as well as creating their own. While I did find consistent interactional patterns in the form of the styles of cruising over the course of the study, I also noted an increased openness in public displays of affection and sexual activity in lesbian bars. As the fear of police raids in the clubs has decreased, young lesbians in the clubs seem to be bursting out of the closet—unapologetically—creating their own versions of what it means to be a lesbian, bisexual, or queer.

The method of cruising chosen by a woman directly relates to her own individual characteristics, the learned experiences she brings to the setting, the accessibility of and her intent for cruising. Specifically, the style chosen will be determined by (1) whether she is comfortable approaching an unacquainted woman; (2) whether she is looking for a short, long, or any term relationship; and (3) the amount of risk she is willing to place herself in in order to achieve her goal. The ease or degree of comfortableness a woman feels regarding approaching an unacquainted woman will be, in part, an outgoing, confident personality; a reflection of her ability to disregard learned gender socialization; a feeling of necessity, desire, or desperation; and/or the adoption of an assertive role.

A woman who is cruising will generally have an intended goal in mind prior to entering the cruising setting. My research indicates that the majority of women seek a long-term, committed relationship. This response not only coincides with female socio-sexual socialization but also politically correct lesbian feminist beliefs and behavior. For some, promiscuous behavior was simply a phase that some young lesbians go through in their process of lesbian identity formation prior to settling down into a stable relationship. Lesbian promiscuous behavior was viewed negatively by most respondents (except in the cases where it was seen as a phase), and it was not seen as a legitimate socio-sexual choice.

The amount of risk a woman is willing to place herself in to achieve her goal will be a reflection of her desire to meet a particular woman, her confidence level, her intended goal, and the amount of investment she places in the meet/ date/mate process. My research indicates that a woman who is looking for a long-term relationship is willing to wait for the "right" woman to come along and, therefore, will engage in a protracted cruising style. If her confidence level permits her to handle rejections in a constructive manner then she will engage in more assertive styles of cruising. Those women seeking more immediate sexual gratification may risk more in order to achieve that goal.

The most frequently used styles are the Mingler and the Game Player, which require the least amount of investment and risk of public failure while allowing flexible motives. A high percentage of women also fit the Sojourner and Woman in Waiting styles of cruising which I believe is a reflection of the difficulty women feel in the meet/date/mate process. A smaller percentage of women will engage in the Strutter style, due to the outgoing, gregarious nature required. The most infrequently used styles are the Home Base Cruiser and the Ego Booster styles, used by women who are looking specifically for a one-night stand (and are willing to be overt and "labeled" as such) or are cruising solely for narcissistic gratification.

American society, in particular, and the global society, in general, have throughout history restricted and impinged upon the sexual freedom of women. These restrictions have in large part stemmed from conservative social, moral, and religious movements. The lesbian feminist movement, while trying to create a healthier environment for women, has added to the conservative dialogue which has served to restrict sexual behavior and what is perceived as acceptable lesbian sexuality. These restrictions continue to create feelings of guilt for behavior (or thoughts of behavior) which fall outside the accepted limits of approved lesbian sexuality. I believe the time is long overdue for lesbians and women in general to open their minds, lift restrictions, and accept and embrace a broader range of socio-sexual behavior. My research will continue to evaluate and examine the dynamic lesbian social environment and the practice and politics of cruising.

NOTES

1. The Ecstasy Lounge was designed and created as a public sex club for lesbians. It was in operation in San Francisco from 1991 to 1996.

2. While I have never personally seen a room officially specified for sexual encounters in women's clubs, I have seen areas where romantic and sexual encounters take place. Two clubs that I frequented in the early 1980s had loft areas with sofas and virtually no lighting. It was widely known in each club that the lofts were for "making out." Home Base Cruisers, in particular, might use the sofas for quick sexual encounters and then return to their "base" in search of their next sexual partner. By the 1990s it was not uncommon in certain clubs to see sexual encounters-discretely or not so discretely-performed on the dance floor or primary bar area.

REFERENCE

Kitaka. 1999. "Kitaka's Experiment; or, Why I Started the Ecstasy Lounge," in *The Lesbian Polyamory Reader: Open Relationships, Non-Monogamy, and Casual Sex.* Edited by Marcia Munson and Judith P. Stelboum. New York: Haworth.

40

Naked Power

Stripping as a Confidence Game

LISA PASKO

Pasko fascinatingly applies the framework of the con game to this deviant sales exchange, where strippers ply their bodies seductively in order to earn tips from bar patrons. Using qualitative interviews and observations, Pasko discusses the way customers at a strip bar move through their evening's entertainment, focusing on how strippers engage them and earn a living. Differential levels of experience, skill, knowledge, and authority between strippers and their patrons characterize this transaction, a classic deviant sale. Pasko describes the entrance to the club and the way bouncers and waiters prepare customers for their experience, the way strippers approach their potential marks and socialize them to the norms of tipping, and the way they disengage from customers after the interaction is complete without delivering the actual sex their bodies had suggested might be forthcoming.

On the unassuming streets of most of urban America is the inviting neon glimmer of an old and established sight—the strip club. The strip club has survived the decades as an accessible and acceptable place for men to realize and visually experience their sexual fantasies. From the flirtatious pull of her garter to the counterfeiting of emotional intimacy, the stripper, or exotic dancer, entices and entertains customers through a complex negotiation of power. Using a structure that resembles a "confidence game" (Goffman, 1952) strippers manipulate symbolic communication and create emotional control over their patrons. All the while, however, their customers still possess a pervasive power: the sex–object role dancers must assume and perform is defined and managed by men and their desires.

... This study explores the false intimacy and sexual objectification created by strippers and their customers. It explains how, in Meadian (1934) terms, strippers employ significant symbols in order to assess their situation, take the role of the other and adjust their behavior accordingly. Unlike previous sex work studies, this study analyzes exotic dance from the theoretical framework of a confidence game. Borrowing from Hochschild's work (1983) on emotional labor, it pays particular attention to strippers' "feelings" management and describes how strippers keep customers controlled.

From *Sexualities*, 5(1), pp. 49–66, copyright © 2002 by Sage Publications. Reprinted by permission of Sage Publications Inc.

METHODOLOGY

Because the stripper–customer interaction is an intricate exchange of sex, power and inequality that influences experiences both inside and outside the club, I employed several techniques in order to collect data. I gathered the information for this study through three main methods: (1) the six-month participant observation of one of Hawaii's strip clubs; (2) unstructured interviews of exotic dancers; and (3) observation of dancers' behavior and interactions outside the strip club setting. I attended the club varying nights of the week and recorded my observations qualitatively. Because the club is a public bar occasionally attended by women, my presence as a customer was inconspicuous. As with much field research, information was collected through informal conversations with customers and other employees of the club.

Altogether, thirteen interviews were completed with three respondents; each informant varied in age, race and marital status. Two of the dancers were white, single and in their late 20s and one was a Japanese-Hawaiian divorced mother of two in her early 30s. Each of the informants in the study reported that a desperate financial situation preceded their first exotic dance experience; they needed to earn a lot of money quickly. Afterwards, they could not find other opportunities that pay as well as exotic dance or that allow them as much freedom with their time. The informants reported that they could not afford to quit.

THE CONFIDENCE GAME

While Western culture has often venerated the image of the confidence man in theatre and literature, only a limited selection of academic research has recognized the con game as a mode of organizing and manipulating social interactions (Blum, 1972; Goffman, 1952; Leff, 1976; Leo, 1996; Prus and Sharper, 1977; Schur, 1958). The confidence game is an act of trust development, fake pretenses and duplicity in order to acquire some kind of gain, usually monetary (Goffman, 1952; Leo, 1996; Schur, 1958). The confidence game, like many acts of deception, is an assumption of power; power over the victim is necessary in obtaining the desired item (Schur, 1958: 299). The confidence person, or swindler, often enjoys the development and exercise of power over his or her victims, repetitively proving his or her cleverness and superiority (Blum, 1972: 13).

The confidence person manufactures a false social relationship and develops rapport with the prospective victims, or "marks," for the purpose of exploiting them. Swindlers need to be likeable individuals. They are skillful actors who are unafraid of risks and well-versed in human nature. Finally, they can and must change masks easily and frequently (Leo, 1996; Schur, 1958). Using their "grift sense" they are able to vary their personalities, in order to entice the mark and keep him or her interested (Leo, 1996: 264). They understand their victims well, appearing to be similar and empathetic to them in many ways. Preying on the psychological vulnerabilities of their victims, con people "may rehearse and make up different roles, draw on a repertoire of tricks, and adapt their techniques to fit a particular event" (Leo, 1996: 264). They continually assess the situation and the behavior of others and readjust their actions accordingly. The fundamental message that all con people must effectively deliver to their victims is this: do the transaction (Leff, 1976: 9). Swindlers must convince their victims that the deal is not only believable and desirable, but also a guaranteed bargain. The cost to the mark must seem particularly low for the commodity or service they will receive.

Since the marks often believe they are shrewd individuals who could not be taken by a con game (Goffman, 1952: 451), the confidence person must be able to pacify and "cool out" their victims. They keep their victims from feeling conned or abused. As a cooler, the con person uses safeguards and strategies to console marks and allow them careful and smooth adjustment to the con. Swindlers are careful that their deceit is not discovered, at least until after the con is completed. They ensure they are never caught lying. The cooler provides status for the mark—a redefinition of the self along defensible lines that allow the mark to feel he or she is still a smart person (Goffman, 1952). Once the con is over (and especially if it is realized), the mark employs several techniques to deal with the exit from the false relationship. The mark may joke or maintain an unserious nature to the involvement. He or she may use the process of "hedging," an assertion that he or she was not completely taken by the con. Victims may also choose to keep the involvement secret (Goffman, 1952: 460).

THE STRIP CLUB

The moment the customer nears the entrance of the strip club, the confidence game begins. Outside the entrance, strippers walk around in their lingerie or gowns, in order to entice potential marks to go into the club. Closer to the entrance, customers are immediately greeted by a bouncer who asks for their cover charge. The dancers frequently approach the bouncer and offer affectionate exchanges. The customers see he is well-liked by the dancers and that he is equipped to maintain order in the club. While these contacts seem to signal to patrons that he is someone designated to enforce club rules, these actions are, for the most part, only symbolic. The dancers and other wait staff must socialize the patrons to the conventions of the club and ultimately they who are responsible for maintaining control.

Once customers enter the club, they are immediately greeted by a male waiter who asks their drink preference (one drink minimum); the waiter then finds seating for the customer. Customers have the option of sitting at the bar, in any of the booths that line the perimeter of the club or in the tables and seats that line the stages. The club has three stages that, at any given time, have eight to ten dancers on each of them. Legally, customers are not allowed to touch the dancers, and, likewise, it is illegal for strippers to touch customers. Lap dancing is illegal as well. As with most strip clubs in Hawaii, this club has full nudity, and it makes its profits through the sale of alcohol. The club's clientele are mostly male tourists and military, although female customers and local residents sometimes come to the club.

QUALIFYING THE MARK

After the customer encounters the bouncer and enters the club, the customer becomes a mark for the stripper. While she is the primary confidence person in the game, male waiters also participate in the con for a share of the tips. They will encourage and pressure customers to experience a dance from the stripper. A customer may decline close interaction with a stripper; they then will remain seated at the bar that is in the back of the club and away from the stages. Often customers who sit at the bar are returning customers who understand the game and choose not to play it. If a customer sits at the booths that line the club or in a chair along the stage, he becomes a mark.

In sizing up their marks, strippers decide which customers are likely to be good tippers and try to charm them into a dance. Strippers try to entice those customers they believe will be desirable victims and respond to their advances. Looking uncertain of the obligations of the club or, conversely, wandering into the club with bills in hand, are signs of easy marks. The stripper knows that these customer types can be good marks. They are likely ignorant of the club's structure and can be easily conned into tipping. Conversely, they may know the game and be willing to play it. Some customers may be more difficult to qualify and, subsequently, con. To avoid difficult swindles, strippers will frequently ask for a tip before showing any nudity for customers who do not present cash in front of them. The stripper does this in order to teach the customer that he must tip continuously during the performance. One dancer explains both how this works and how it can go wrong:

> I had a tourist one time who sat down in front of me and didn't have a stack of money in front of him so I knew this was going to be tough. He motioned for me to spread my legs. I pulled my garter and he tipped. So I continued my act and removed my bikini bottoms. He continually tried to kiss my inner thigh. I kept pushing his head away but he kept trying. Finally he ... grabbed me and kissed my crotch. I punched him right in his ... nose and knocked him off his chair. [Security] came and took him away. He was back the next night but, man, he came nowhere near me.

Using her presentation of self, her dance, eye contact and snapshots of nudity, the stripper will entice the customer to sit in front of her. For those customers who appear hesitant, a male waiter will approach them and promote a dance, asking "isn't she pretty ... wouldn't you like a closer look?" Once the customer sits in front of a dancer, the con begins.

CULTIVATING THE MARK

The stripper plays two primary roles in order to cultivate the mark: (1) as sex object for the customer; and (2) an impersonator of "counterfeit intimacy." When a customer takes a seat in front of a stripper on stage and makes eye contact with her, the stripper begins her dance and signals her expectations. She uses flattery and other emotional appeals in order to negate any refusals or unwillingness the customer may express. She also does this in order to develop "trust" between her and the mark. The dancer fosters a trust that she will provide him intimacy and sexual gratification at an insignificant cost. A stripper may remove a small piece of clothing and offer a quick peak at her breast. If a preview of her nudity is given, the stripper insists on a tip.

The tip is the primary significant gesture in the strip club. Whereas the average tip from a customer is usually only a dollar, it is the frequency of tipping the dancers aim to increase. The stripper lets the customer know that for the sexual pleasure to continue, he must be willing to give a monetary reward. The dancer, having a sixth sense or "grift sense" about men's sex fantasies, tries different sexual poses, pulling the garter as she finishes each movement. She analyzes the customer's movements, responses and demeanor in order to detect his preferences and weaknesses. Finding a position and type of dance the customer likes, the stripper will disguise fatigue and/or irritation and repeat those movements which reaped her the biggest tip. Likewise, if the customer discontinues tipping, the stripper will discontinue her act.

For the stripper and the customer, the principal significant symbol which aids in this social interaction is the garter. When a stripper pulls her garter, the customer must be trained that this is a signal that he is to tip her. For the dance to continue, the customer must understand the significance of the gesture. If the customer ignores the garter pull, the stripper may offer an affectionate touch and pull her garter again. Or, she may stand up, put her clothes back on, and lure another customer. One informant illustrates the importance of the garter pull:

> Knowing when to pull the garter is truly a talent. If you pull it too much or too soon, you could turn a customer off. If you don't pull it enough, you look easy, and then you have to work that much harder to get your money. The key is to show a little, you know, entice, and then pull it every time you touch yourself, or push something in their face, or reveal something. The best customers are those who have the stack of bills ready to go. You know they are ready and eager to tip. Those are the good ones.

CONNING THE MARK

After the stripper attracts the customer to her and cultivates him to the expectations of tipping, the con continues. She cons the mark into surrendering significant amounts of tip money. She assures the customer that, in order to receive sexual satisfaction, it is in his best interest to tip her; the satisfaction is indeed worth its price. Language is of key importance at this stage of the strip act. The stripper will flirt, tease and joke with the customer to see what level of emotional intimacy is required to con this particular customer: does she merely need to display her nude body or does this customer want a more intimate and personal experience, one that includes emotion and conversation?

Strippers quickly decide which role will furnish them the greatest tips—to be a sex object, devoid of facial expression or emotion, or to be the intimate partner. The stripper reads the behavior of the customer and acts accordingly. Because of past emotion memories (memories that recall feelings), strippers are able to respond appropriately to customers' stimuli and are able to understand a customer's desires. Knowing when to act seductive, when to remain still and simply be nude, when to shock a shy customer with a sudden revelation of nudity and when to be talkative, come from a dancer's ability to decode men's sexual wishes. If a customer presents a stack of dollar bills, sits back in his seat and gives no verbal response to the stripper's flirtations, the stripper knows this attitude and decides he would like a less intimate strip act to begin. In this situation, the stripper would react to the customer's response by performing the act with less conversation and more concentration on the provocative removal of her clothes. One informant illustrates:

> Men come here for many reasons—some want a sexy girlfriend for the night; some want to see a fantasy, a sex object; some want a therapist; and others just want to stare [at your body]—they couldn't care less what my face or anything else looks like. They couldn't care less if you dance ... nothing exotic about it. They just want to see my [nudity]. So ... I'll be the girlfriend, the counselor, the playmate, the object—whatever. I don't care. Just keep it [the money] coming.

... Strippers may touch the customer, press their breasts or other body parts against his face, or place their feet on his shoulders, as they lie on the stage and simulate sexual intercourse. They may allow the customer to touch them or kiss their personal body parts. They may perform pelvic thrusts within inches of the customer's face and touch themselves while doing so. Strippers carefully use physical space, their appearance, cigarettes and sex acts—sexual touching and masturbation—to deceive the customer into accepting the pseudo closeness and staged attention of the strip act.

The confidence game varies, according to customer's preferences and vulnerabilities. Unsure of the cost, some come to see the nude bodies of women and to be sexually aroused by the dancers' performance of the strip act. Some men come to acquire the attention of a beautiful woman and to have that woman listen to his problems and his life stories. For those customers who desire a more intimate interaction with the stripper, the stripper creates an atmosphere of physical as well as emotional connectedness to the customer. The increased amount of interaction, the range of interaction and the intensity of emotion ensure that the interaction reflects counterfeit intimacy, an inauthentic relation (see Foote, 1954). In order to create this environment, the stripper compliments the customer and whispers comments in his ear which make the customer feel special and important. All the while she physically touches the customer and maintains close proximity. The stripper will perform sexual poses as she complements her sexual movements with hugs and light kisses. Once again, for the act to continue, the customer has to tip.

For men who are looking for conversation, the stripper creates intimacy in a different way. Instead of a strip act, the stripper will sit with the customer and

have a drink with him. Often the customer will furnish the stripper with free cigarettes or an occasional token of affection, such as a cheap bracelet. The stripper
will listen to his stories and become his companion and sometimes his confessor
for the night. She will often create a fictitious background for herself, which she
tells all the customers. One respondent demonstrates:

> I often tell people I am a student, which I am. I will tell them I am a med
> student or a law student, though. The guys who want a "girlfriend"—
> they want you to be more than just a stripper. They want you to be
> a sex object but to have an interesting personal life, too. So I tell them
> something and this makes them feel special and they keep buying me
> drinks.

Drawing on emotional memories, the stripper will create a persona which
will fulfill for the customer the desired feeling he has come to the club to acquire.
The stripper will pretend that she could potentially become his girlfriend and
carry out the customer's desired image, such as college student, professional
dancer, dominatrix or an innocent. The stripper continues to create this level
of intimacy as long as the customer continuously and repeatedly buys her drinks.
Each drink a customer has with a stripper costs 20 dollars. Often the male waiters
serve as "pseudopimps" and arrange the drinks between strippers and customers.
For their part, the waiters receive half the cost of the drink as a tip. They are also
the ones responsible for collecting the cost and cooling the customer out.

COOLING OUT

Cooling out the mark becomes necessary when tipping becomes less frequent.
Strippers want the customer to believe she had genuine interest in him, that
he is worthy of her sexual display and advances. Strippers must shy away from
their dance and yet make the mark unmindful of the con: he must stay to buy
alcohol and other dances. Positive reinforcement is often used by strippers in
their exit from the con. The dancer may take a well-paying mark to a booth
for a drink and require the waiter to collect the fee. She may also end her performance by replacing her clothing while offering a hug and smile. She may also
cool the mark out by referring him to another dancer who will then begin
attracting him to her sexual mystique. Occasionally, well-paying customers are
offered a back-room table dance for an established fee. Some dancers are reluctant
to exit the con this way. The back room is a private place and a source of vulnerability for the dancer. She could lose the power negotiation and be victimized.

If a customer wishes to see the dancer socially outside the club, the dancer
most often refuses and the trust is violated. The confidence game becomes obvious to the customer: the dancer is not sexually or socially interested in him,
despite the high cost paid. To cool these customers out, the stripper pacifies
the mark by stating that she does not date customers or that it is against the
law for her to go home with him. Appealing to a legal authority or universal
rule keeps the customers from feeling abused.

Other exits from the dance require a quick ending. Strippers have freedom to cool out their marks with force. The primary "feeling" offense to a dancer is a customer's initiation of hostility. If customers who violate the emotional rules through a display of unfriendliness towards the stripper, the dancer has the latitude to respond to the customer; she may respond with anger and aggressive actions. Any indication of hostility or rejection by the customer, either through degrading language or through inappropriate touching (such as repeated touches which the dancer has warned the customer to stop), will result in the stripper terminating her act. In some cases, the stripper may react violently or ask security to remove the patron. Because the club is continuously busy, these actions are performed without consequence for the stripper. Some customers may actually be enticed by a display of "toughness."

CONCLUSION

Dancers are experienced suppliers of sexuality and utmost feminine attention; they direct the social, sexual act for economic gain. They have freedom and power in their individual performances and presentations. Consistent with previous research (Ronai and Ellis, 1989; Sijuwade, 1995), this research has found that the two primary roles enacted by the exotic dancer are the pseudo-girlfriend and the sex object. Strippers must nurture pseudo-relationships and false personas in order to satisfy customers' needs for attention. Through their precise control over emotions and symbolic communication, strippers continually read the meaning of customers' conduct—"the glances of an eye and the attitude of the body"—in order to complete the confidence game.

While the confidence game gives them power in their individual interactions with customers, this power does not extend into the outside world. Strippers are cognizant of the consequences of their choice of profession: the victimization, stigmatization and isolation. Because they continually enact the role of sex object, strippers often lose facility with other alternative, nonsexual identities, and they begin to access their sex object personas in nonwork-related relationships. They become estranged from genuine emotions and real feelings of pleasure and sexuality. As a consequence, the stripping experience is not a liberating experience. The sexual persona the dancers must construct is of someone who can be easily dominated—a young, available, naked "girl." This role reflects and reinforces the inequality in gender power relations that exist beyond the club. It responds to patriarchal understandings of female sexuality and femininity and does not challenge mainstream conceptions of masculinity (see Segal, 1997; Nead, 1997).

For the male customers, the stripping experience initially seems to provide opportunities to enact masculinity by using economic power to achieve emotional and sexual intimacy. The setting of the club reinforces the general social construction of women as ubiquitously available sex objects, despite the fact that dancers actually maintain a certain form of power in their particularized interaction. It is power, though, that does not really challenge male authority, and ultimately embeds even the con woman in stigmatizing, exhausting and dangerous work. The money may be good, but the social consequences for the women are profound.

The impact of participating in stripping cons would also be worth exploring. As men are immersed with carefully scripted contexts and images of seemingly available but realistically unattainable beautiful women, their performance in other personal relationships, especially with their female partners, may become affected. The mixture and confusion of the genuine and the counterfeit could influence men's understandings of, not only their own sexuality and of femininity, but also those attributes of their partners. The sexual objectification and commodification of women reinforces notions of possession, authority and aggression in masculinity. How these masculine elements are negotiated, contested and internalized by men as they visit sites of constructed sexuality requires further exploration.

REFERENCES

Blum, R. (1972) *Deceivers and Deceived: Observations on Confidence Men and Their Victims, Informants and Their Quarry, Political and Industrial Spies and Ordinary Citizens.* Springfield, IL: Charles Thomas.

Foote, N. (1954) "Sex as Play," *Social Problems* 1: 159–63.

Goffman, E. (1952) "On Cooling the Mark Out: Some Aspects of Adaptation to Failure," *Psychiatry* 15: 451–63.

Hochschild, A. (1983) *The Managed Heart: Commercialization of Human Feeling.* Berkeley: University of California Press.

Leff, A. (1976) *Swindling and Selling.* New York: Free Press.

Leo, R. (1996) "Miranda's Revenge: Police Interrogation as a Confidence Game," *Law and Society Review* 30: 259–88.

Mead, G. (1934 [1962]) *Mind, Self, and Society: From the Standpoint of a Social Behaviorist.* Chicago, IL: University of Chicago Press.

Nead, L. (1997) "The Female Nude: Pornography, Art, and Sexuality," in Laura O'Toole and Jessica Schiffman (eds) *Gender Violence,* pp. 374–82. New York: New York University Press.

Prus, R. and Sharper, C. R. D. (1977) *Road Hustler: The Career Contingencies of Professional Card and Dice Hustlers.* Lexington, MA: Lexington Books.

Ronai, C. and Ellis, C. (1989) "Turn-Ons for Money," *Journal of Contemporary Ethnography* 18: 271–98.

Schur, E. (1958) "Sociological Analysis of Swindling," *Journal of Criminal Law, Criminology and Political Science* 48: 296–304.

Segal, L. (1997) "Pornography and Violence: What Experts Really Say," in Laura O'Toole and Jessica Schiffman (eds) *Gender Violence,* pp. 400–13. New York: New York University Press.

Sijuwade, P. (1995) "Counterfeit Intimacy: A Dramaturgical Analysis of an Erotic Performance," *Social Behavior and Personality* 23: 369–76.

41

Sexual Assault on Campus

ELIZABETH A. ARMSTRONG, LAURA HAMILTON,
AND BRIAN SWEENEY

In the twenty-first century, American college and universities have been rocked by scandals involving alcohol-related deaths and sexual abuse. School administrators have tried to clamp down on students with punitive measures such as "strikes," probations, and expulsions. Students living in university housing have been particularly vulnerable, as they are under the watchful eye of resident advisors. These restrictions have pushed student partying, increasingly into the fraternity scene, which, Armstrong, Hamilton, and Sweeney find, is dominated by white, middle-class populations. There, they find that women with these demographics are most at risk of rape.

We have known for more than a decade that college women's risk of sexual assault by people they knew far outweighed that of stranger rape. Armstrong, Hamilton, and Sweeney explore some of the popular explanations for the pervasiveness of fraternity rape, from individual bad boys, to fraternity culture, to dangerous environments, and integrate all these by going beyond them. They locate the problem for women in a structural situation where they are forced out of their dorms to party by harsh sanctions against alcohol in the residence halls. The party scene they find is located in private environments where men control access to alcohol, transportation, and, more importantly, social status through their attentions to women. Here, traditional gender roles end up contributing to the victimization of women as men and women voluntarily engage in a dance where women seek flirtation to gain status and men offer flirtation and status to gain sex. Most women, being disempowered in these male-dominated settings, put themselves into situations where they are at high risk of victimization. When this occurs, they are likely to attribute blame to their girlfriends individually (remember Reinarman's [Chapter 15] comments about the vocabulary of

From "Sexual Assault on Campus: A Multilevel, Integrative Approach to Party Rape," by Elizabeth A. Armstrong, Laura Hamilton, and Brian Sweeney. *Social Problems*, Vol. 53, No. 4, 2006. All rights reserved. Reprinted by permission of University of California Press Journals.

attribution being individualistic rather than structural) rather then banding together to fight the unequal power structure in which men have the control and advantage. Older and higher-statused women look down on their less fortunate female schoolmates as "stupid" or "asking for it," despite the likelihood that they may have experienced these same troubles themselves at an earlier age.

Armstrong, Hamilton, and Sweeney offer disturbing insight into how women fight to protect their access to the party scene despite its risk for them to be taken advantage of and disempowered. We are reminded that fraternity culture, despite its deviant aspects, represents an enclave of the dominant culture where women are complicit in their own victimization and men use their gender advantage to callously foster their own ends.

A 1997 National Institute of Justice study estimated that between one-fifth and one-quarter of women are the victims of completed or attempted rape while in college (Fisher, Cullen, and Turner 2000). College women "are at greater risk for rape and other forms of sexual assault than women in the general population or in a comparable age group" (Fisher et al. 2000: iii). 2 At least half and perhaps as many as three-quarters of the sexual assaults that occur on college campuses involve alcohol consumption on the part of the victim, the perpetrator, or both (Abbey et al. 1996; Sampson 2002). The tight link between alcohol and sexual assault suggests that many sexual assaults that occur on college campuses are "party rapes." A recent report by the U.S. Department of Justice defines party rape as a distinct form of rape, one that "occurs at an off-campus house or on- or off-campus fraternity and involves ... plying a woman with alcohol or targeting an intoxicated woman" (Sampson 2002: 6).[1] While party rape is classified as a form of acquaintance rape, it is not uncommon for the woman to have had no prior interaction with the assailant, that is, for the assailant to be an in-network stranger (Abbey et al. 1996).

Colleges and universities have been aware of the problem of sexual assault for at least 20 years, directing resources toward prevention and providing services to students who have been sexually assaulted. Programming has included education of various kinds, support for *Take Back the Night* events, distribution of rape whistles, development and staffing of hotlines, training of police and administrators, and other efforts. Rates of sexual assault, however, have not declined over the last five decades (Adams-Curtis and Forbes 2004: 95; Bachar and Koss 2001; Marine 2004; Sampson 2002: 1).

Why do colleges and universities remain dangerous places for women in spite of active efforts to prevent sexual assault? While some argue that "we know what the problems are and we know how to change them" (Adams-Curtis and Forbes 2004: 115), it is our contention that we do not have a complete explanation of the problem. To address this issue we use data from a study of college life at a large midwestern university and draw on theoretical developments in the sociology of gender. Continued high rates of sexual assault can be viewed as a case of the reproduction of gender inequality—a phenomenon of central concern in gender theory.

We demonstrate that sexual assault is a predictable outcome of a synergistic intersection of both gendered and seemingly gender neutral processes operating at individual, organizational, and interactional levels. The concentration of homogenous students with expectations of partying fosters the development of sexualized peer cultures organized around status. Residential arrangements intensify students' desires to party in male-controlled fraternities. Cultural expectations that partygoers drink heavily and trust party-mates become problematic when combined with expectations that women be nice and defer to men. Fulfilling the role of the partier produces vulnerability on the part of women, which some men exploit to extract non-consensual sex. The party scene also produces fun, generating student investment in it. Rather than criticizing the party scene or men's behavior, students blame victims. By revealing mechanisms that lead to the persistence of sexual assault and outlining implications for policy, we hope to encourage colleges and universities to develop fresh approaches to sexual assault prevention.

APPROACHES TO COLLEGE SEXUAL ASSAULT

Explanations of high rates of sexual assault on college campuses fall into three broad categories. The first tradition, a psychological approach that we label the "individual determinants" approach, views college sexual assault as primarily a consequence of perpetrator or victim characteristics such as gender role attitudes, personality, family background, or sexual history. While "situational variables" are considered, the focus is on individual characteristics. For example, Antonia Abbey and associates (2001) find that hostility toward women, acceptance of verbal pressure as a way to obtain sex, and having many consensual sexual partners distinguish men who sexually assault from men who do not. Research suggests that victims appear quite similar to other college women (Kalof 2000), except that white women, prior victims, first-year college students, and more sexually active women are more vulnerable to sexual assault (Adams-Curtis and Forbes 2004; Humphrey and White 2000).

The second perspective, the "rape culture" approach, grew out of second wave feminism. In this perspective, sexual assault is seen as a consequence of widespread belief in "rape myths," or ideas about the nature of men, women, sexuality, and consent that create an environment conducive to rape. For example, men's disrespectful treatment of women is normalized by the idea that men are naturally sexually aggressive. Similarly, the belief that women "ask for it" shifts responsibility from predators to victims. This perspective initiated an important shift away from individual beliefs toward the broader context. However, rape supportive beliefs alone cannot explain the prevalence of sexual assault, which requires not only an inclination on the part of assailants but also physical proximity to victims.

A third approach moves beyond rape culture by identifying particular contexts—fraternities and bars—as sexually dangerous. Ayres Boswell and Joan Spade (1996) suggest that sexual assault is supported not only by "a generic culture surrounding and promoting rape," but also by characteristics of the "specific settings" in which men and women interact (p. 133). Mindy Stombler and

Patricia Yancey Martin (1994) illustrate that gender inequality is institutionalized on campus by "formal structure" that supports and intensifies an already "high-pressure heterosexual peer group" (p. 180). This perspective grounds sexual assault in organizations that provide opportunities and resources.

We extend this third approach by linking it to recent theoretical scholarship in the sociology of gender. Martin (2004), Barbara Risman (1998; 2004), Judith Lorber (1994) and others argue that gender is not only embedded in individual selves, but also in cultural rules, social interaction, and organizational arrangements. This integrative perspective identifies mechanisms at each level that contribute to the reproduction of gender inequality (Risman 2004). Socialization processes influence gendered selves, while cultural expectations reproduce gender inequality in interaction. At the institutional level, organizational practices, rules, resource distributions, and ideologies reproduce gender inequality. Applying this integrative perspective enabled us to identify gendered processes at individual, interactional, and organizational levels that contribute to college sexual assault. . . .

METHOD

Data are from group and individual interviews, ethnographic observation, and publicly available information collected at a large midwestern research university. Located in a small city, the school has strong academic and sports programs, a large Greek system, and is sought after by students seeking a quintessential college experience. Like other schools, this school has had legal problems as a result of deaths associated with drinking. In the last few years, students have attended a sexual assault workshop during first-year orientation. Health and sexuality educators conduct frequent workshops, student volunteers conduct rape awareness programs, and *Take Back the Night* marches occur annually.

The bulk of the data presented in this paper were collected as part of ethnographic observation during the 2004–05 academic year in a residence hall identified by students and residence hall staff as a "party dorm." While little partying actually occurs in the hall, many students view this residence hall as one of several places to live in order to participate in the party scene on campus. This made it a good place to study the social worlds of students at high risk of sexual assault—women attending fraternity parties in their first year of college. The authors and a research team were assigned to a room on a floor occupied by 55 women students (51 first-year, 2 second-year, 1 senior, and 1 resident assistant [RA]). We observed on evenings and weekends throughout the entire academic school year. We collected in-depth background information via a detailed nine-page survey that 23 women completed and conducted interviews with 42 of the women (ranging from 1 ¼ to 2 ½ hours). All but seven of the women on the floor completed either a survey or an interview.

With at least one-third of first-year students on campus residing in "party dorms" and one-quarter of all undergraduates belonging to fraternities or sororities, this social world is the most visible on campus. As the most visible scene on campus, it also attracts students living in other residence halls and those not in the

Greek system. Dense pre-college ties among the many in-state students, class and race homogeneity, and a small city location also contribute to the dominance of this scene. Of course, not all students on this floor or at this university participate in the party scene. To participate, one must typically be heterosexual, at least middle class, white, American-born, unmarried, childless, traditional college age, politically and socially mainstream, and interested in drinking. Over three-quarters of the women on the floor we observed fit this description.

There were no non-white students among the first and second year students on the floor we studied. This is a result of the homogeneity of this campus and racial segregation in social and residential life. African Americans (who make up 3% to 5% of undergraduates) generally live in living-learning communities in other residence halls and typically do not participate in the white Greek party scene. We argue that the party scene's homogeneity contributes to sexual risk for white women. We lack the space and the data to compare white and African American party scenes on this campus, but in the discussion we offer ideas about what such a comparison might reveal....

Selves and Peer Culture in the Transition from High School to College

Student characteristics shape not only individual participation in dangerous party scenes and sexual risk within them but the development of these party scenes. We identify individual characteristics (other than gender) that generate interest in college partying and discuss the ways in which gendered sexual agendas generate a peer culture characterized by high-stakes competition over erotic status.

Non-Gendered Characteristics Motivate Participation in Party Scenes
Without individuals available for partying, the party scene would not exist. All the women on our floor were single and childless, as are the vast majority of undergraduates at this university; many, being upper-middle class, had few responsibilities other than their schoolwork. Abundant leisure time, however, is not enough to fuel the party scene. Media, siblings, peers, and parents all serve as sources of anticipatory socialization (Merton 1957). Both partiers and non-partiers agreed that one was "supposed" to party in college. This orientation was reflected in the popularity of a poster titled "What I Really Learned in School" that pictured mixed drinks with names associated with academic disciplines. As one focus group participant explained:

> You see these images of college that you're supposed to go out and have fun and drink, drink lots, party and meet guys. [You are] supposed to hook up with guys, and both men and women try to live up to that. I think a lot of it is girls want to be accepted into their groups and guys want to be accepted into their groups.

Partying is seen as a way to feel a part of college life. Many of the women we observed participated in middle and high school peer cultures organized around status, belonging, and popularity (Eder 1985; Eder, Evans, and Parker 1995; Milner 2004). Assuming that college would be similar, they told us that they wanted to fit in, be popular, and have friends. Even on move-in day, they were supposed to already have friends. When we asked one of the out-siders, Ruth, about her first impression of her roommate, she replied that she found her:

> Extremely intimidating. Bethany already knew hundreds of people here. Her cell phone was going off from day one, like all the time. And I was too shy to ask anyone to go to dinner with me or lunch with me or anything. I ate while I did homework.

Peer Culture as Gendered and Sexualized   Partying was also the primary way to meet men on campus. The floor was locked to non-residents, and even men living in the same residence hall had to be escorted on the floor. The women found it difficult to get to know men in their classes, which were mostly mass lectures. They explained to us that people "don't talk" in class. Some complained they lacked casual friendly contact with men, particularly compared to the mixed-gender friendship groups they reported experiencing in high school.

Meeting men at parties was important to most of the women on our floor. The women found men's sexual interest at parties to be a source of self-esteem and status. They enjoyed dancing and kissing at parties, explaining to us that it proved men "liked" them. This attention was not automatic, but required the skillful deployment of physical and cultural assets. Most of the party-oriented women on the floor arrived with appropriate gender presentations and the money and know-how to preserve and refine them. While some more closely resembled the "ideal" college party girl (white, even features, thin but busty, tan, long straight hair, skillfully made-up, and well-dressed in the latest youth styles), most worked hard to attain this presentation. They regularly straightened their hair, tanned, exercised, dieted, and purchased new clothes.

Women found that achieving high erotic status in the party scene required looking "hot" but not "slutty," a difficult and ongoing challenge. Mastering these distinctions allowed them to establish themselves as "classy" in contrast to other women. Although women judged other women's appearance, men were the most important audience. A "hot" outfit could earn attention from desirable men in the party scene. A failed outfit, as some of our women learned, could earn scorn from men. One woman reported showing up to a party dressed in a knee length skirt and blouse only to find that she needed to show more skin. A male guest sarcastically told her "nice outfit," accompanied by a thumbs-up gesture. . . .

Men also sought proof of their erotic appeal. As a woman complained, "Every man I have met here has wanted to have sex with me!" Another inter-viewee reported that: this guy that I was talking to for like ten/fifteen minutes says, "Could you, um, come to the bathroom with me and jerk me off?" And I'm like, "What!" I'm like, "Okay, like, I've known you for like, fifteen minutes, but no." The women found that men were more interested than they were in having sex. These clashes in sexual expectations are not surprising: men derived status from securing sex (from high-status women), while women derived status from getting attention (from high-status men). These agendas are both complementary and adversarial: men give attention to women en route to getting sex, and women are unlikely to become interested in sex without getting attention first.

University and Greek Rules, Resources, and Procedures

Simply by congregating similar individuals, universities make possible heterosexual peer cultures. The university, the Greek system, and other related organizations structure student life through rules, distribution of resources, and procedures.

Sexual danger is an unintended consequence of many university practices intended to be gender neutral. The clustering of homogeneous students intensifies the dynamics of student peer cultures and heightens motivations to party. Characteristics of residence halls and how they are regulated push student partying into bars, off-campus residences, and fraternities. While factors that increase the risk of party rape are present in varying degrees in all party venues (Boswell and Spade 1996), we focus on fraternity parties because they were the typical party venue for the women we observed and have been identified as particularly unsafe (see also Martin and Hummer 1989; Sanday 1990). Fraternities offer the most reliable and private source of alcohol for first-year students excluded from bars and house parties because of age and social networks.

University Practices as Push Factors   The university has latitude in how it enforces state drinking laws. Enforcement is particularly rigorous in residence halls. We observed RAs and police officers (including gun-carrying peer police) patrolling the halls for alcohol violations. Women on our floor were "documented" within the first week of school for infractions they felt were minor. Sanctions are severe—a $300 fine, an 8-hour alcohol class, and probation for a year. As a consequence, students engaged in only minimal, clandestine alcohol consumption in their rooms. In comparison, alcohol flows freely at fraternities.

The lack of comfortable public space for informal socializing in the residence hall also serves as a push factor. A large central bathroom divided our floor. A sterile lounge was rarely used for socializing. There was no cafeteria, only a convenience store and a snack bar in a cavernous room furnished with big-screen televisions. Residence life sponsored alternatives to the party scene such as "movie night" and special dinners, but these typically occurred early in the evening. Students defined the few activities sponsored during party hours (e.g., a midnight trip to Wal-Mart) as uncool. . . .

Male Control of Fraternity Parties   The campus Greek system cannot operate without university consent. The university lists Greek organizations as student clubs, devotes professional staff to Greek-oriented programming, and disbands fraternities that violate university policy. Nonetheless, the university lacks full authority over fraternities; Greek houses are privately owned and chapters answer to national organizations and the Interfraternity Council (IFC) (i.e., a body governing the more than 20 predominantly white fraternities).

Fraternities control every aspect of parties at their houses: themes, music, transportation, admission, access to alcohol, and movement of guests. Party themes usually require women to wear scant, sexy clothing and place women in subordinate positions to men. During our observation period, women attended parties such as "Pimps and Hos," "Victoria's Secret," and "Playboy Mansion"—the last of which required fraternity members to escort two scantily clad dates. Other recent themes included: "CEO/Secretary Ho," "School Teacher/Sexy Student," and "Golf Pro/Tennis Ho."

Some fraternities require pledges to transport first-year students, primarily women, from the residence halls to the fraternity houses. From about 9 to 11 P.M. on weekend nights early in the year, the drive in front of the residence hall resembled a rowdy taxi-stand, as dressed-to-impress women waited to be carpooled to parties in expensive late-model vehicles. By allowing party-oriented first-year women to cluster in particular residence halls, the university made them easy to find. One fraternity member told us this practice was referred to as "dorm-storming."

Transportation home was an uncertainty. Women sometimes called cabs, caught the "drunk bus," or trudged home in stilettos. Two women indignantly described a situation where fraternity men "wouldn't give us a ride home." The women said, "Well, let us call a cab." The men discouraged them from calling the cab and eventually found a designated driver. The women described the men as "just dicks" and as "rude."

Fraternities police the door of their parties, allowing in desirable guests (first-year women) and turning away others (unaffiliated men). Women told us of abandoning parties when male friends were not admitted. They explained that fraternity men also controlled the quality and quantity of alcohol. Brothers served themselves first, then personal guests, and then other women. Non-affiliated and unfamiliar men were served last, and generally had access to only the least desirable beverages. The promise of more or better alcohol was often used to lure women into private spaces of the fraternities.

Fraternities are constrained, though, by the necessity of attracting women to their parties. Fraternities with reputations for sexual disrespect have more success recruiting women to parties early in the year. One visit was enough for some of the women. A roommate duo told of a house they "liked at first" until they discovered that the men there were "really not nice."

The Production of Fun and Sexual Assault in Interaction

Peer culture and organizational arrangements set up risky partying conditions, but do not explain *how* student interactions at parties generate sexual assault. At the interactional level we see the mechanisms through which sexual assault is produced. As interactions necessarily involve individuals with particular characteristics and occur in specific organizational settings, all three levels meet when interactions take place. Here, gendered and gender neutral expectations and routines are intricately woven together to create party rape. Party rape is the result of fun situations that shift—either gradually or quite suddenly—into coercive situations. Demonstrating how the production of fun is connected with sexual assault requires describing the interactional routines and expectations that enable men to employ coercive sexual strategies with little risk of consequence.

College partying involves predictable activities in a predictable order (e.g., getting ready, pre-gaming, getting to the party, getting drunk, flirtation or sexual interaction, getting home, and sharing stories). It is characterized by "shared assumptions about what constitutes good or adequate participation"—what Nina Eliasoph and Paul Lichterman (2003) call "group style" (p. 737). A fun partier throws him or herself into the event, drinks, displays an upbeat mood, and evokes revelry in others. Partiers are expected to like and trust party-mates. Norms of civil interaction curtail displays of unhappiness or tension among party-goers. Michael Schwalbe and associates (2000) observed that groups engage in scripted events of this sort "to bring about an intended emotional result" (p. 438). Drinking assists people in transitioning from everyday life to a state of euphoria.

Cultural expectations of partying are gendered. Women are supposed to wear revealing outfits, while men typically are not. As guests, women cede control of turf, transportation, and liquor. Women are also expected to be grateful for men's hospitality, and as others have noted, to generally be "nice" in ways that men are not. The pressure to be deferential and gracious may be intensified by men's older age and fraternity membership. The quandary for women, however, is that fulfilling the gendered role of partier makes them vulnerable to sexual assault.

Women's vulnerability produces sexual assault only if men exploit it. Too many men are willing to do so. Many college men attend parties looking for casual sex. A student in one of our classes explained that "guys are willing to do damn near anything to get a piece of ass." A male student wrote the following description of parties at his (non-fraternity) house:

> Girls are continually fed drinks of alcohol. It's mainly to party but my roomies are also aware of the inhibition-lowering effects. I've seen an old roomie block doors when girls want to leave his room; and other times I've driven women home who can't remember much of an evening yet sex did occur. Rarely if ever has a night of drinking for my roommate ended without sex. I know it isn't necessarily and assuredly sexual assault, but with the amount of liquor in the house I question the amount of consent a lot.

Another student—after deactivating—wrote about a fraternity brother "telling us all at the chapter meeting about how he took this girl home and she was obviously too drunk to function and he took her inside and had sex with her." Getting women drunk, blocking doors, and controlling transportation are common ways men try to prevent women from leaving sexual situations. Rape culture beliefs, such as the belief that men are "naturally" sexually aggressive, normalize these coercive strategies. Assigning women the role of sexual "gatekeeper" relieves men from responsibility for obtaining authentic consent, and enables them to view sex obtained by undermining women's ability to resist it as "consensual" (e.g., by getting women so drunk that they pass out).[2]

In a focus group with her sorority sisters, a junior sorority woman provided an example of a partying situation that devolved into a likely sexual assault.

ANNA: It kind of happened to me freshman year. I'm not positive about what happened, that's the worst part about it. I drank too much at a frat one night, I blacked out and I woke up the next morning with nothing on in their cold dorms, so I don't really know what happened and the guy wasn't in the bed anymore, I don't even think I could tell you who the hell he was, no I couldn't.

SARAH: Did you go to the hospital?

ANNA: No, I didn't know what happened. I was scared and wanted to get the hell out of there. I didn't know who it was, so how am I supposed to go to the hospital and say someone might've raped me? It could have been any one of the hundred guys that lived in the house.

SARAH: It happens to so many people, it would shock you. Three of my best friends in the whole world, people that you like would think it would never happen to, it happened to. It's just so hard because you don't know how to deal with it because you don't want to turn in a frat because all hundred of those brothers...

ANNA: I was also thinking like, you know, I just got to school, I don't want to start off on a bad note with anyone, and now it happened so long ago, it's just one of those things that I kind of have to live with.

This woman's confusion demonstrates the usefulness of alcohol as a weapon: her intoxication undermined her ability to resist sex, her clarity about what happened, and her feelings of entitlement to report it. We collected other narratives in which sexual assault or probable sexual assault occurred when the woman was asleep, comatose, drugged, or otherwise incapacitated.

Amanda, a woman on our hall, provides insight into how men take advantage of women's niceness, gender deference, and unequal control of party resources. Amanda reported meeting a "cute" older guy, Mike, also a student, at a local student bar. She explained that, "At the bar we were kind of making out a little bit and I told him just cause I'm sitting here making out doesn't mean that I want to go home with you, you know?" After Amanda found herself stranded by friends with no cell phone or cab fare, Mike promised that a sober friend of his would drive her home. Once they got in the car Mike's friend refused to take her home and instead dropped her at Mike's place. Amanda's concerns were heightened by the driver's disrespect. "He was like, so are you into ménage à trois?" Amanda reported staying awake all night. She woke Mike early in the morning to take her home. Despite her ordeal, she argued that Mike was "a really nice guy" and exchanged telephone numbers with him. These men took advantage of Amanda's unwillingness to make a scene. Amanda was one of the most assertive women on our floor. Indeed, her refusal to participate fully in the culture of feminine niceness led her to suffer in the social hierarchy of the floor and on campus. It is unlikely that other women we observed could have been more assertive in this situation. That she was nice to her captor in the morning suggests how much she wanted him to like her and what she was willing to tolerate in order to keep his interest.[3] This case also shows that it is not only fraternity parties that are dangerous; men can control party resources and work together to constrain women's behavior while partying in bars and at house parties. What distinguishes fraternity parties is that male dominance of partying there is organized, resourced, and implicitly endorsed by the university. Other party venues are also organized in ways that advantage men.

We heard many stories of negative experiences in the party scene, including at least one account of a sexual assault in every focus group that included heterosexual women. Most women who partied complained about men's efforts to control their movements or pressure them to drink. Two of the women on our floor were sexually assaulted at a fraternity party in the first week of school—one was raped. Later in the semester, another woman on the floor was raped by a friend. A fourth woman on the floor suspects she was drugged; she became disoriented at a fraternity party and was very ill for the next week.

Party rape is accomplished without the use of guns, knives, or fists. It is carried out through the combination of low level forms of coercion—a lot of liquor and persuasion, manipulation of situations so that women cannot leave, and sometimes force (e.g., by blocking a door, or using body weight to make it difficult for a woman to get up). These forms of coercion are made more effective by organizational arrangements that provide men with control over how partying happens and by expectations that women let loose and trust their party-mates. This systematic and effective method of extracting non-consensual sex is largely invisible, which makes it difficult for victims to convince anyone—even themselves—that a crime occurred. Men engage in this behavior with little risk of consequences.

Student Responses and the Resiliency of the Party Scene

The frequency of women's negative experiences in the party scene poses a problem for those students most invested in it. Finding fault with the party scene potentially threatens meaningful identities and lifestyles. The vast majority of heterosexual encounters at parties are fun and consensual. Partying provides a chance to meet new people, experience and display belonging, and to enhance social position. Women on our floor told us that they loved to flirt and be admired, and they displayed pictures on walls, doors, and websites commemorating their fun nights out.

The most common way that students—both women and men—account for the harm that befalls women in the party scene is by blaming victims. By attributing bad experiences to women's "mistakes," students avoid criticizing the party scene or men's behavior within it. Such victim-blaming also allows women to feel that they can control what happens to them. The logic of victim-blaming suggests that sophisticated, smart, careful women are safe from sexual assault. Only "immature," "naïve," or "stupid" women get in trouble. When discussing the sexual assault of a friend, a floor resident explained that:

She somehow got like sexually assaulted ... by one of our friends' old roommates. All I know is that kid was like bad news to start off with. So, I feel sorry for her but it wasn't much of a surprise for us. He's a shady character.

Another floor resident relayed a sympathetic account of a woman raped at knife point by a stranger in the bushes, but later dismissed party rape as nothing to worry about "'cause I'm not stupid when I'm drunk." Even a feminist focus group participant explained that her friend who was raped "made every single mistake and almost all of them had to with alcohol. . . . She got ridiculed when she came out and said she was raped." These women contrast "true victims" who are deserving of support with "stupid" women who forfeit sympathy (Phillips 2000). Not only is this response devoid of empathy for other women, but it also leads women to blame themselves when they are victimized (Phillips 2000).

Sexual assault prevention strategies can perpetuate victim-blaming. Instructing women to watch their drinks, stay with friends, and limit alcohol consumption implies that it is women's responsibility to avoid "mistakes" and their fault if they fail. Emphasis on the precautions women should take—particularly if not accompanied by education about how men should change their behavior— may also suggest that it is natural for men to drug women and take advantage of them. Additionally, suggesting that women should watch what they drink, trust party-mates, or spend time alone with men asks them to forgo full engagement in the pleasures of the college party scene.

Victim-blaming also serves as a way for women to construct a sense of status within campus erotic hierarchies. As discussed earlier, women and men acquire erotic status based on how "hot" they are perceived to be. Another aspect of erotic status concerns the amount of sexual respect one receives from men. Women can tell themselves that they are safe from sexual assault not only because they are savvy, but because men will recognize that they, unlike other women, are worthy of sexual respect. For example, a focus group of senior women explained that at a small fraternity gathering their friend Amy came out of the bathroom. She was crying and said that a guy "had her by her neck, holding her up, feeling her up from her crotch up to her neck and saying that I should rape you, you are a fucking whore." The woman's friends were appalled, saying, "no one deserves that." On other hand, they explained that: "Amy flaunts herself. She is a whore so, I mean . . . " They implied that if one is a whore, one gets treated like one.[4]

Men accord women varying levels of sexual respect, with lower status women seen as "fair game." On campus the youngest and most anonymous women are most vulnerable. High-status women (i.e., girlfriends of fraternity members) may be less likely victims of party rape.[5] Sorority women explained that fraternities discourage members from approaching the girlfriends (and ex-girlfriends) of other men in the house. Partiers on our floor learned that it was safer to party with men they knew as boyfriends, friends, or brothers of friends. One roommate pair partied exclusively at a fraternity where one of the women knew many men from high school. She explained that "we usually don't party with people we don't know that well." Over the course of the year, women on the floor winnowed their party venues to those fraternity houses where they "knew the guys" and could expect to be treated respectfully.

Opting Out   While many students find the party scene fun, others are more ambivalent. Some attend a few fraternity parties to feel like they have participated in this college tradition. Others opt out of it altogether. On our floor, 44 out of the 51 first-year students (almost 90%) participated in the party scene. Those on the floor who opted out worried about sexual safety and the consequences of engaging in illegal behavior. For example, an interviewee who did not drink was appalled by the fraternity party transport system. She explained that:

> All those girls would stand out there and just like, no joke, get into these big black Suburbans driven by frat guys, wearing like seriously no clothes, piled on top of each other. This could be some kidnapper taking you all away to the woods and chopping you up and leaving you there. How dumb can you be?

In her view, drinking around fraternity men was "scary" rather than "fun."

Her position was unpopular. She, like others who did not party, was an outsider on the floor. Partiers came home loudly in the middle of the night, threw up in the bathrooms, and rollerbladed around the floor. Socially, the others simply did not exist. A few of our "misfits" successfully created social lives outside the floor. The most assertive of the "misfits" figured out the dynamics of the floor in the first weeks and transferred to other residence halls.

However, most students on our floor lacked the identities or network connections necessary for entry into alternative worlds. Life on a large university campus can be overwhelming for first-year students. Those who most needed an alternative to the social world of the party dorm were often ill-equipped to actively seek it out. They either integrated themselves into partying or found themselves alone in their rooms, microwaving frozen dinners and watching television. A Christian focus group participant described life in this residence hall: "When everyone is going out on a Thursday and you are in the room by yourself and there are only two or three other people on the floor, that's not fun, it's not the college life that you want.". . .

DISCUSSION AND IMPLICATIONS

Individual characteristics and institutional practices provide the actors and contexts in which interactional processes occur. We have to turn to the interactional level, however, to understand *how* sexual assault is generated. Gender neutral expectations to "have fun," lose control, and trust one's party-mates become problematic when combined with gendered interactional expectations. Women are expected to be "nice" and to defer to men in interaction. This expectation is intensified by men's position as hosts and women's as grateful guests. The heterosexual script, which directs men to pursue sex and women to play the role of gatekeeper, further disadvantages women, particularly when virtually *all* men's methods of extracting sex are defined as legitimate.

The mechanisms identified should help explain intra-campus, cross-campus, and overtime variation in the prevalence of sexual assault. Campuses with similar students and social organization are predicted to have similar rates of sexual assault. We would expect to see lower rates of sexual assault on campuses characterized by more aesthetically appealing public space, lower alcohol use, and the absence of a gender-adversarial party scene. Campuses with more racial diversity and more racial integration would also be expected to have lower rates of sexual assault because of the dilution of upper-middle class white peer groups. Researchers are beginning to conduct comparative research on the impact of university organization on aggregate rates of sexual assault. For example, Meichun Mohler-Kuo and associates (2004) found that women who attended schools with medium or high levels of heavy episodic drinking were more at risk of being raped while intoxicated than women attending other schools, even while controlling for individual-level characteristics. More comparative research is needed.

This perspective may also help explain why white college women are at higher risk of sexual assault than other racial groups. Existing research suggests that African American college social scenes are more gender egalitarian (Stombler and Padavic 1997). African American fraternities typically do not have houses, depriving men of a party resource. The missions, goals, and recruitment practices of African American fraternities and sororities discourage joining for exclusively social reasons (Berkowitz and Padavic 1999), and rates of alcohol consumption are lower among African American students (Journal of Blacks in Higher Education 2000; Weschsler and Kuo 2003). The role of party rape in the lives of white college women is substantiated by recent research that found that "white women were more likely [than non-white women] to have experienced rape while intoxicated and less likely to experience other rape" (Mohler-Kuo et al. 2004: 41). White women's overall higher rates of rape are accounted for by their high rates of rape while intoxicated. Studies of racial differences in the culture and organization of college partying and its consequences for sexual assault are needed.

Our analysis also provides a framework for analyzing the sources of sexual risk in non-university partying situations. Situations where men have a home turf advantage, know each other better than the women present know each other, see the women as anonymous, and control desired resources (such as alcohol or drugs) are likely to be particularly dangerous. Social pressures to "have fun," prove one's social competency, or adhere to traditional gender expectations are also predicted to increase rates of sexual assault within a social scene.

This research has implications for policy. The interdependence of levels means that it is difficult to enact change at one level when the other levels remain unchanged. Programs to combat sexual assault currently focus primarily or even exclusively on education. But as Ann Swidler (2001) argued, culture develops in response to institutional arrangements. Without change in institutional arrangements, efforts to change cultural beliefs are undermined by the cultural commonsense generated by encounters with institutions. Efforts to educate about sexual assault will not succeed if the university continues to support organizational arrangements that facilitate and even legitimate men's coercive sexual strategies. Thus, our research implies that efforts to combat sexual assault on campus should target all levels, constituencies, and processes simultaneously. Efforts to educate both men and women should indeed be intensified, but they should be reinforced by changes in the social organization of student life.

Researchers focused on problem drinking on campus have found that reduction efforts focused on the social environment are successful (Berkowitz 2003: 21). Student body diversity has been found to decrease binge drinking on campus (Weschsler and Kuo 2003); it might also reduce rates of sexual assault. Existing student heterogeneity can be exploited by eliminating self-selection into age-segregated, white, upper-middle class, heterosexual enclaves and by working to make residence halls more appealing to upper-division students. Building more aesthetically appealing housing might allow students to interact outside of alcohol-fueled party scenes. Less expensive plans might involve creating more living-learning communities, coffee shops, and other student-run community spaces.

While heavy alcohol use is associated with sexual assault, not all efforts to regulate student alcohol use contribute to sexual safety. Punitive approaches sometimes heighten the symbolic significance of drinking, lead students to drink more hard liquor, and push alcohol consumption to more private and thus more dangerous spaces. Regulation inconsistently applied—e.g., heavy policing of residence halls and light policing of fraternities—increases the power of those who can secure alcohol and host parties. More consistent regulation could decrease the value of alcohol as a commodity by equalizing access to it.

Sexual assault education should shift in emphasis from educating women on preventative measures to educating both men and women about the coercive behavior of men and the sources of victim-blaming. Mohler-Kuo and associates (2004) suggest, and we endorse, a focus on the role of alcohol in sexual assault. Education should begin before students arrive on campus and continue throughout college. It may also be most effective if high-status peers are involved in disseminating knowledge and experience to younger college students....

NOTES

1. On party rape as a distinct type of sexual assault, see also Ward and associates (1991). Ehrhart and Sandler (1987) use the term to refer to group rape. We use the term to refer to one-on-one assaults. We encountered no reports of group sexual assault.

2. In ongoing research on college men and sexuality, Sweeney (2004) and Rosow and Ray (2006) have found wide variation in beliefs about acceptable ways to obtain sex even among men who belong to the same fraternities. Rosow and Ray found that fraternity

men in the most elite houses view sex with intoxicated women as low status and claim to avoid it.

3. Holland and Eisenhart (1990) and Stombler (1994) found that male attention is of such high value to some women that they are willing to suffer indignities to receive it.

4. Schwalbe and associates (2000) suggest that there are several psychological mechanisms that explain this behavior. Trading power for patronage occurs when a subordinate group accepts their status in exchange for

compensatory benefits from the dominant group. Defensive othering is a process by which some members of a subordinated group seek to maintain status by deflecting stigma to others. Maneuvering to protect or improve individual position within hierarchical classification systems is common; however, these responses support the subordination that makes them necessary.

5. While "knowing" one's male party-mates may offer some protection, this protection is not comprehensive.

Sorority women, who typically have the closest ties with fraternity men, experience more sexual assault than other college women (Mohler-Kuo et al. 2004). Not only do sorority women typically spend more time in high-risk social situations than other women, but arriving at a high-status position on campus may require one to begin their college social career as one of the anonymous young women who are frequently victimized.

REFERENCES

Abbey, Antonia, Pam McAuslan, Tina Zawacki, A. Monique Clinton, and Philip Buck. 2001. "Attitudinal, Experiential, and Situational Predictors of Sexual Assault Perpetration." *Journal of Interpersonal Violence* 16: 784–807.

Abbey, Antonia, Lisa Thomson Ross, Donna McDuffie, and Pam McAuslan. 1996. "Alcohol and Dating Risk Factors for Sexual Assault among College Women." *Psychology of Women Quarterly* 20: 147–69.

Adams-Curtis, Leah and Gordon Forbes. 2004. "College Women's Experiences of Sexual Coercion: A Review of Cultural, Perpetrator, Victim, and Situational Variables." *Trauma, Violence, and Abuse: A Review Journal* 5: 91–122.

Bachar, Karen and Mary Koss. 2001. "From Prevalence to Prevention: Closing the Gap between What We Know about Rape and What We Do." Pp. 117–42 in *Sourcebook on Violence against Women,* edited by C. Renzetti, J. Edleson, and R. K. Bergen. Thousand Oaks, CA: Sage.

Berkowitz, Alan. 2003. "How Should We Talk about Student Drinking—And What Should We Do about It?" *About Campus* May/June: 16–22.

Berkowitz, Alexandra and Irene Padavic. 1999. "Getting a Man or Getting Ahead: A Comparison of White and Black Sororities." *Journal of Contemporary Ethnography* 27: 530–57.

Boswell, A. Ayres and Joan Z. Spade. 1996. "Fraternities and Collegiate Rape Culture: Why Are Some Fraternities More Dangerous Places for Women?" *Gender & Society* 10: 133–47.

Eder, Donna. 1985. "The Cycle of Popularity: Interpersonal Relations among Female Adolescents." *Sociology of Education* 58: 154–65.

Eder, Donna, Catherine Evans, and Stephen Parker. 1995. *School Talk: Gender and Adolescent Culture.* New Brunswick, NJ: Rutgers University Press.

Ehrhart, Julie and Bernice Sandler. 1987. "Party Rape." *Response* 9: 205.

Eliasoph, Nina and Paul Lichterman. 2003. "Culture in Interaction." *American Journal of Sociology* 108: 735–94.

Fisher, Bonnie, Francis Cullen, and Michael Turner. 2000. "The Sexual Victimization of College Women." Washington, DC: National Institute of Justice and the Bureau of Justice Statistics.

Holland, Dorothy and Margaret Eisenhart. 1990. *Educated in Romance: Women, Achievement, and College Culture.* Chicago: University of Chicago Press.

Humphrey, John and Jacquelyn White. 2000. "Women's Vulnerability to Sexual Assault from Adolescence to Young Adulthood." *Journal of Adolescent Health* 27: 419–24.

Journal of Blacks in Higher Education. 2000. "News and Views: Alcohol Abuse Remains High on College Campus, But Black Students Drink to Excess Far Less Often Than Whites." *The Journal of Blacks in Higher Education.* 28: 19–20.

Kalof, Linda. 2000. "Vulnerability to Sexual Coercion among College Women: A Longitudinal Study." *Gender Issues* 18: 47–58.

Lorber, Judith. 1994. *Paradoxes of Gender.* New Haven, CT: Yale University Press.

Marine, Susan. 2004. "Waking Up from the Nightmare of Rape." *The Chronicle of Higher Education.* November 26, p. B5.

Martin, Patricia Yancey. 2004. "Gender as a Social Institution." *Social Forces* 82: 1249–73.

Martin, Patricia Yancey and Robert A. Hummer. 1989. "Fraternities and Rape on Campus." *Gender & Society* 3: 457–73.

Merton, Robert. 1957. *Social Theory and Social Structure.* New York: Free Press.

Milner, Murray. 2004. *Freaks, Geeks, and Cool Kids: American Teenagers, Schools, and the Culture of Consumption.* New York: Routledge.

Mohler-Kuo, Meichun, George W. Dowdall, Mary P. Koss, and Henry Weschler. 2004. "Correlates of Rape While Intoxicated in a National Sample of College Women." *Journal of Studies on Alcohol* 65: 37–45.

Phillips, Lynn. 2000. *Flirting with Danger: Young Women's Reflections on Sexuality and Domination.* New York: New York University.

Risman, Barbara. 1998. *Gender Vertigo: American Families in Transition.* New Haven, CT: Yale University Press.

———. 2004. "Gender as a Social Structure: Theory Wrestling with Activism." *Gender & Society* 18: 429–50.

Rosow, Jason and Rashawn Ray. 2006. "Getting Off and Showing Off: The Romantic and Sexual Lives of High-Status Black and White Status Men." Department of Sociology, Indiana University, Bloomington, IN. Unpublished manuscript.

Sampson, Rana. 2002. "Acquaintance Rape of College Students." Problem-Oriented Guides for Police Series, No. 17. Washington, DC: U.S. Department of Justice, Office of Community Oriented Policing Services.

Sanday, Peggy. 1990. *Fraternity Gang Rape: Sex, Brotherhood, and Privilege on Campus.* New York: New York University Press.

Schwalbe, Michael, Sandra Godwin, Daphne Holden, Douglas Schrock, Shealy Thompson, and Michele Wolkomir. 2000. "Generic Processes in the Reproduction of Inequality: An Interactionist Analysis." *Social Forces* 79: 419–52.

Stombler, Mindy. 1994. "'Buddies' or 'Slutties': The Collective Reputation of Fraternity Little Sisters." *Gender & Society* 8: 297–323.

Stombler, Mindy and Patricia Yancey Martin. 1994. "Bringing Women In, Keeping Women Down: Fraternity 'Little Sister' Organizations." *Journal of Contemporary Ethnography* 23: 150–84.

Stombler, Mindy and Irene Padavic. 1997. "Sister Acts: Resisting Men's Domination in Black and White Fraternity Little Sister Programs." *Social Problems* 44: 257–75.

Sweeney, Brian. 2004. "Good Guy on Campus: Gender, Peer Groups, and Sexuality among College Men." Presented at the American Sociological Association Annual Meetings, August 17, Philadelphia, PA.

Swidler, Ann. 2001. *Talk of Love: How Culture Matters.* Chicago: University of Chicago Press.

Ward, Sally, Kathy Chapman, Ellen Cohn, Susan White, and Kirk Williams. 1991. "Acquantance Rape and The College Social Scene." *Family Relations* 40: 65–71.

Weschsler, Henry and Meichun Kuo. 2003. "Watering Down the Drinks: The Moderating Effect of College Demographics on Alcohol Use of High-Risk Groups." *American Journal of Public Health.* 93: 1929–33.

42

Fraud and the American Dream

The Ponzi Scheme

ADAM TRAHAN, JAMES W. MARQUART,
AND JANET MULLINGS

*Returning once again to the crimes of the powerful, we see a smaller-scale example
of fraud and its relation to American culture than the Halliburton fiasco. Not just
the domain of the rich and well connected, fraud is alive and flourishing in our
capitalist system from top to bottom. Trahan, Marquart, and Mullings lay out a
pyramid scheme fashioned in a very basic style by a Texas financier. Jack Barnes
created a fictitious investment opportunity that attracted mainstream Americans
from children to retirees, preying on all who would take advantage of the American
dream to make their savings grow into profits for their future security and wealth.
While Barnes was taking in money from new investors, he was using it to pay off
early clients, hoping that word-of-mouth would make his business grow exponentially. Taking his ill-gotten gains to support a lavish lifestyle, he stole money
from all of his clients. One of the ultimate ironies of this tale is that even after his
schemes tumbled down and his victims recognized their loss and betrayal, the
victims were not as critical of his behavior as they might have been of other forms of
deviance because they regarded it as merely an overzealous application of accepted
American values.*

Consumer fraud and most other forms of white-collar crime have been virtually
ignored compared to "violent" street crimes. This is illustrated in both the
Federal Bureau of Investigation's Uniform Crime Report and the Justice Department's National Crime Victimization Survey that provide data on property and violent crime, but report no information concerning fraud victimization (Titus et al.
1995). Studies have shown that fraud often causes more harm to victims and society
than most other forms of crime (Glassner 1999; Lynch and Michalowski 2000).

In 1992, the U.S. Department of Justice estimated the total cost of street
crime to be approximately $17.6 billion (Klaus 1994). This figure is only a frac-
tion of the cost incurred annually by fraud crimes. The U.S. General Accounting
Office (GAO) estimated the cost of fraud alone to be $100 billion in 1995 (Davis

1990; Thompson 1992). Total costs for all white-collar crimes are approximately $400 billion—"22 times higher than the total cost of serious street crime" (Lynch and Michalowski 2000: 66). According to the GAO, fraud costs the U.S. Government approximately 10% of its total national budget, or $164 billion annually (Lynch and Michalowski 2000). Many studies also have found that white-collar crime causes more deaths annually than "violent" crimes (Mizell 1997; Lynch and Michalowski 2000). Such findings have led to the conclusion that "corporate crime represents the most widespread and costly form of crime in America" (Michalowski 1985: 325).

The relevant scholarly literature base on fraud also is sketchy. Except for the work of Titus et al. (1995), Mizell (1997), and Doherty and Smith (1981), we know very little about the perpetrators of fraud. Many anecdotal accounts exist along with first-person narratives by offenders about fraud. We know virtually nothing about the victims of fraud. Little systematic inquiry has examined the general characteristics of fraud victims or their motivations to invest in fraudulent schemes. We contribute to the research literature by examining a Ponzi scheme. . . .

THE PONZI SCHEME

Fraud and white-collar crime take many forms and are often difficult to classify. American society has complicated the classification of fraud. The economic system's reliance on free market structures allows entrepreneurial ventures to closely resemble fraud crimes. As such, categorizing enterprises in a legal/illegal dichotomy often becomes blurred. Alan Greenspan (2002) stated, "It is not that humans have become any more greedy than in generations past. It is that the avenues to express greed have grown so enormously."

Ponzi schemes occur when an offender(s) uses fictitious investment opportunities to bilk funds from victims by promising high returns compounded over and above typical investments. This elaborate form of deception is named after its creator, Charles "Carlo" Ponzi (Kitchens 1993).

These schemes are unique in that the offenders maintain their con over relatively long periods of time by paying existing investor/victims their promised returns using the investment capital of new investor/victims. This circular nature requires that the amount of new money coming in always outweighs the demands of established investors. Further, the demands of established investors will eventually be larger than that which can be satiated by new investment capital. Ponzi schemes are doomed to collapse as a result of their own structure.

It is important to note that although there are some levels of victim culpability in other forms of fraud (e.g., pyramid schemes), victims of Ponzi schemes are true victims, not knowing that they are involved in any illegal activity. Aspects of fraud victimization that will be explored include motivations to invest, why victim/investors invest in a seemingly "hasty" manner (i.e., not rigorously exploring the validity of the investment and/or offender), the cultural and social states that foster such desires, and why, when offenders are apprehended, investigated, and prosecuted, their victims often continue to support them.

Fraud and the American Dream

Messner and Rosenfeld (2001) . . . concluded that America is organized for crime. They contended that the American Dream itself and the social conditions it creates are criminogenic. The American Dream fosters an overwhelmingly strong cultural ethos where the pursuit of material gain is valued over all other things and the means to attain such "success" are devalued. This creates an environment that promotes the attainment of material success by any means necessary. The American Dream also promotes such ambition on an individual basis, whereby monetary gain is to be sought by each person regardless of the impacts on others. This mentality has created a social reality where open and free competition is celebrated. As such, any one person's material success is contingent upon another's failure and is highly prized (Messner and Rosenfeld 2001). . . .

Monetary success has no limit. There are always possibilities to acquire more. When money has inherent value as it does in America, and a person's "success" is measured in financial terms, there is also no limit to a person's status. American culture perpetuates these assumptions because to do so is productive to its advancement as a corporate nation. If American citizens become satiated with wealth at a certain level, American industry can move no further than this limit. Therefore, as Messner and Rosenfeld (2001) stated, the American Dream requires constant and never-ending acquisition of money by any means necessary. Their theory attempts to explain only perpetrator behavior. This paper shows that these same theoretical contentions are the underlying motivations for fraud victims as well.

DATA SOURCES

Data were gathered from the investor files confiscated by the Consumer Fraud Division of the Harris County District Attorney's Office (HCDA) from the offender and a survey administered by the Division to the victim population. We were given access to these sources by the HCDA with the agreement that the confidentiality of the victims would be upheld. A third source of information came from a single interview conducted with Russell Turbeville, the Chief of the Consumer Fraud Division and supervisor of the case. This interview was relatively informal as its primary purpose was simply to construct a timeline of the offense and a general profile of Ponzi schemes. . . .

THE STATE OF TEXAS V. JACK BARNES—
A CASE HISTORY

This research examines a "classic" Ponzi scheme that unfolded in the late 1990s in Houston, Texas. The defendant, Jack Barnes, was a young and highly successful financier with many contacts in the business world. He established a small but unlicensed business, One West Financial Services, with offices in Houston and several other large cities. He began advertising an elaborate, but entirely fictitious "investment opportunity."

He claimed to have a contact overseas that sold and manufactured inexpensive raw titanium. He also claimed that an Admiral in the U.S. Navy was interested in buying this titanium to use in American nuclear submarines. Barnes alleged that by acting as the "middle-man" (i.e., buying this titanium overseas and selling it to the U.S. Navy) profits were immense, unlimited, and assured. Barnes had no trouble enticing people to turn over large sums of money as investments. He required a minimum of $1,000 to participate and promised an average 10% return compounded monthly. Barnes was able to pay these returns in full and on time using the investment capital of other victims. As a result, the number and amount of investments grew as satisfied "investors" informed others of a "great investment opportunity."

Barnes used most of the victims' monies to fund his lavish lifestyle. He frequently traveled to Europe, bought expensive cars, paid cash for several multi-million dollar houses, and purchased extravagant jewelry. One victim testified that Barnes never wore the same piece of jewelry twice and often changed watches and rings throughout the day. This scheme saw its first investor/victim in 1997 and continued to acquire new investors until March of 2003. Throughout this time period there were no known complaints by victims or employees of Barnes. In Mid-March of 2003, a man contacted the Harris County District Attorney's Office and stated that he had met with Jack Barnes and was considering investing with him until his research of the investment led him to suspect that Barnes had been engaged in illegal activities. Investigators with the Consumer Fraud Division ultimately determined Barnes's "investment opportunity" to be a Ponzi scheme. This case involved 434 victims from 19 states in the U.S., Canada, and the West Indies. Illegally acquired funds totaled over $19 million. . . .

What Happened?

The data revealed that Barnes's scheme experienced very little growth from 1997 to 2000 but rapidly expanded by the end of 2001 and grew exponentially in 2002. Initially, Barnes drew victims from Texas and California only. By the year 2003, his victims represented 19 different states in the U.S., Canada, and the West Indies.

The Ponzi scheme also grew rapidly with respect to the frequency of new investors. The number of new investors each year did not begin any significant growth until the end of 2001. However, the frequency of new investors greatly expanded from 2002 until the scheme collapsed in March of 2003.

Barnes claimed to be returning an average of 10% to investors each month. Because of the fraudulent nature of this investment, these returns were entirely fictitious (i.e., no "real" growth was occurring because no investment was actually made). However, the monthly statements submitted by Barnes to his victims led them to believe they were accruing interest on their investment. Each victim's amount of perceived gain was derived from the monthly statements included in the investor files. Perceived gain ranged from $0 to $2,377,584 with a median value of $40,960. When compared to the average investment amount, these data revealed that investors believed they had slightly more than doubled the amount of their investment.

By the time Barnes's Ponzi scheme collapsed he had acquired a total of $19,141,866.81 in investor funds. Experts appointed by the Texas Attorney General's Office to handle the redistribution of funds estimated that the victims would get back only 20% of their investment...

DISCUSSION AND CONCLUSION

There is a very important difference between the victims of fraud (and many other white-collar crime subcategories) and the victims of violent street crime—choice. Consumer fraud victims choose to "invest." The criminal nature of consumer fraud dictates that each scheme will fall apart under the least bit of scrutiny (i.e., potential investors researching the legitimacy of the investment opportunity). Therefore, consumer fraud victims make uninformed decisions with unpredictable outcomes. This study suggests that the American Dream is the catalyst that makes people vulnerable to fraud victimization. As such, the victims' apparent motivations and behavior are "the very same values and behaviors that are conventionally viewed as part of the American success story" (Messner and Rosenfeld 2001: 5)....

Why do only some people participate in fraudulent investment schemes?.... The data analyzed here suggest that fraud victimization occurs when victims are provided an outlet to express the values ingrained in them by the culture of the economic structure. Thus, two catalysts are necessary—opportunity and the successful socialization of capitalistic values. From this realization, several inductions can be made that should guide subsequent research.

First, fraudulent investment schemes obviously must be present for victimization to occur. Although fraud occurs much more often than popular sentiment suggests, there are spatial and temporal variations in its prevalence. Victimization rates may be a function of access to fraudulent investment opportunities. Also, the data suggest that an overwhelming majority of the victims behaved in accordance with the values of the American Dream. American culture, not solicitation by offenders, instills the behavioral tendencies conducive to fraud victimization. Therefore, most Americans share the ingrained motivations that lead to fraud victimization.

Second, the socialization process of capitalist culture must be relatively successful to induce fraud victimization. Most all people are subjected to relatively equivalent socialization attempts. However, the success enjoyed by these attempts may vary across subcultures and/or individuals. Thus, the likelihood of investing in a fraudulent scheme may be a function of the strength with which these values are indoctrinated. People who support the utility and benefit of notions such as an inherent value of money may be at greater risk for fraud victimization. This assertion is not meant to suggest that the victims of fraud are exceptionally greedy. Rather, the sociopolitical context in which fraud occurs is highly conducive to the necessities of these crimes. The deification of money and the status it provides are two of these necessities. The theories employed here suggest that the structure and scope of the American economy create these cultural values. As such, the "greed" upon which fraud thrives is not a function of the individual attributes of its victims. Instead, it is a characteristic of American economic culture.

REFERENCES

Davis, M. 1990. City of Quartz: *Excavating the Future in Los Angeles*. London: Verso.

Doherty, V. P., and M. E. Smith. 1981. "Ponzi Schemes and Laundering—How Illicit Funds are Acquired and Concealed." *FBI Law Enforcement Bulletin* 50(11): 5–11.

Glassner, B. 1999. *The Culture of Fear*. New York: Basic Books.

Greenspan, A. 2002, July 16. Testimony before the U.S. Senate Banking Committee. Retrieved November 23, 2004, from http://www.federalreserve.gov/boarddocs/hh/2002/july/testimony.htm

Klaus, P. 1994. *The Costs of Crime to Victims*. Washington, DC: Bureau of Justice Statistics.

Kitchens, T. L. 1993. "Cash Flow Analysis Method: Following the Paper Trail in Ponzi Schemes." *FBI Law Enforcement Bulletin* 62(8): 10–13.

Lynch, M. J. and R. Michalowski. 2000. *The New Primer in Radical Criminology: Critical Perspectives on Crime,* *Power, and Identity,* 3rd ed. Monsey, NY: Criminal Justice Press.

Messner, S. F., and R. Rosenfeld. 2001. *Crime and the American Dream,* 3rd ed. Belmont, CA: Wadsworth.

Michalowski, R. 1985. *Law, Order, and Power*. New York: Random House.

Mizell, L. R. 1997. *Masters of Deception*. New York: Wiley.

Thompson, J. 1992, May. Health Insurance: Vulnerable Payers Lose Billions to Fraud and Abuse. Report to Chairman, Subcommittee on Human Resources and Intergovernmental Operations. United States General Accounting Office, Washington, DC.

Titus, R. M., F. Heinzelmann, and J. M. Boyle. 1995. "Victimization of Persons by Fraud." *Crime and Delinquency* 41(1): 54–72.

PART VIII

Deviant Careers

One of the fascinating things about people's involvement in deviance is that it evolves, yielding a shifting and changing experience. Doing something for the first time is very different from doing it for the hundredth time. It is fruitful, then, to consider involvement in deviance from a career perspective, to see what the nature of deviance is, and how it develops over the course of people's involvement with it. Sociologists have documented various stages of people's participation in such things as drug use, drug dealing, fencing, carrying out a professional hit, engaging in prostitution, and shoplifting. Although these activities are very different in character, they have structural similarities in the way people experience them according to the stage of people's involvement. In fact, the career analogy has been applied fruitfully to the study of deviance because people go through many of the same cycles of entry, upward mobility, achieving career peaks, aging in the career, burning out, and getting out of deviance as they do in legitimate work.

Six themes tend to be most commonly addressed in the literature on deviant careers. **Entering deviance** attracts the greatest bulk of scholarly attention for two reasons: Policy makers have great interest in finding out how and why people enter deviance so that they can prevent it, and it represents fairly easy data for most scholars to gather since every individual or group of deviants can tell their story of how they got into the scene. Over the last couple of decades sociologists have worked to discover and disseminate information that has been adopted by the public in making decisions aimed at influencing and deterring potential deviants. First, they have developed the concept of "at risk" populations and articulated a range of risk factors associated with various forms of deviance such as gang membership, dropping out of school, unwed pregnancy, suicide, depression, eating disorders, detainment, arrest, incarceration, and so on. (Loeber and Farrington 1998; Werner and Smith 1992). Complementing these, they have identified protective factors that, in the face of exposure to multiple risks, help prevent individuals from becoming involved in these forms of deviance (Werner and Smith 1982).

Although some people venture into deviance on their own, the vast majority do it with the encouragement and assistance of others, often joining cooperative deviant enterprises. The turning points that mark significant phases in their transitions have been explored, as well as their changing self-identities. Most commonly, people who become involved in deviance do so through a process of shifting their circle of friends. They drift into new peer groups as they are drifting into deviance, or their whole peer group drifts into deviance together as the members enter a new phase of the lifecycle.

Second, the career perspective incorporates an interest with the **training** and **socialization** of new deviants. Relatively little has been written about this area for several reasons. First, most deviants, although they might be socialized to the norms and values of their activity through their contacts with fellow deviants, get relatively little explicit training in how to perform their deviance, how to avoid detection, how to deal with the police, and other important concerns. Second, real training generally occurs when deviants are working together, side by side, as a team in their enterprise. That leaves only crews and deviant formal organizations as the likely places where training is apt to occur, and these forms of deviance organization are relatively rare. In their analysis of the career of a professional burglar and fence, Steffensmeier and Ulmer (2005) have suggested that the kind of professional criminals engaged in crew operations has declined.

Third, focusing on careers in deviance enables people to study how individuals' involvement in their deviant worlds and with their deviant associates and activities **change over time.** This type of processual analysis represents a highly nuanced understanding of deviance that cannot easily be captured by the frozen-in-time snapshot of most survey research. Longitudinal studies of deviant careers are rare but valuable assets in this literature, as they distinguish for us some of the different motivations, rewards, conflicts, and problems that deviants encounter over the course of their participation in deviance. These are enormously helpful to both people who struggle to understand and to help themselves, friends, and family members caught in the lure of deviance as well as policy makers, as they illuminate motivations and deterrents that are more or less effective at different career stages.

Over the course of their deviant careers, participants must face the various challenges involved in managing their deviance. They must navigate the changing dimensions of available opportunities to commit their deviance, evolving technologies that can enable and catch them, their relationships within deviant communities, their safety from agents of social control, and must evolve a personal style for their deviance. They must balance their deviance with the nondeviant aspects of their lives, such as their relationships with family members, people in the community, and those on whom they rely to meet their legitimate needs. Yip's (1997) work on gay male Christian couples, for example, illustrates some of the creative ways that homosexuals find to maintain their relational commitment in a social environment where their union is sanctioned by neither church nor state.

Wanting out or **exiting deviance** represents the fourth major area in this literature. As in entering deviance there is a high political interest in this topic, as policy makers are looking for ways to induce people to quit their deviance. Information on longer-term deviants and their attitudes toward the scene, the people in them, their satisfactions and dissatisfactions, and their hopes or dreams for the future is somewhat hard to get. People tend to feel most comfortable talking to others like themselves, and the most active researchers are young. Yet there are valuable studies of aging deviants.

A variety of factors "push" people out of the deviant life and "pull" them back into the conventional world. Individuals with whom they interact, such as their victims or spouses, may push them out of the deviant world, as Pryor (1996) found in his study of child molesters. This is particularly likely in extended exploitive transactions, such as incest perpetrators have with their victims. In deviant exchanges such as the drug traffickers we studied, participants may hold more satisfying relationships with co-participants but burn out from extended drug use and their increasing risk of arrest and death. The drug traffickers show us also how people change during their involvement with deviance, becoming transformed by the easy money they earned dealing and smuggling. Yet returning to legitimate jobs where their earning potential is reduced may involve restricting their spending patterns, something that people find difficult. They also become accustomed to the freewheeling lifestyle and open value system associated with a deviant community, where conventional norms are disdained.

Reentering the straight world with its morality may chafe. Finding legitimate work may be difficult, especially if the deviants were involved in occupational deviance where they were making money through illicit means. Former deviants often have difficulty putting together a resume that accounts for their gap in legitimate employment and finding someone who will hire them. They may find adhering to the structure of the 9-to-5 straight world overly constraining. Yet most people do not want to spend their whole lives engaged in deviance.

Very little information is available on the **postdeviant** features of individuals' lives. These are the hardest data to get because just as developing a deviant identity, as we described in Part V, moves people out of their conventional friend-ships and social worlds into those populated by deviants, quitting deviance usually requires exiting from these same relationships and scenes. Once people decide to get out and actually make that move, they disperse and leave no forwarding address.

Finally, there is a literature in this field on crime and deviance as work, which compares occupational deviance with legitimate jobs and examines **deviant versus legitimate careers.** Working in deviant fields holds many similarities to the skills, professionalism, connections, and attitudes needed for conventional jobs (Letkemann 1973). Goods and services may be bought and sold, credit arranged and extended, costs and profits calculated, and business associates, supp-liers, and customers assessed. Contracts cannot be legally enforced in deviant occupations, however, nor are associates as reliable or durable. Due to the high turnover of personnel and the greater likelihood of drug use involved in all facets of deviant work, people are less likely to have expertise in their trade or to deliver on promises made.

There are also limitations to the career analogy, since while legitimate work may have several structures (the compressed career, the bureaucratic career, the entrepreneurial career), the patterns for deviant careers are more generally entre-preneurial. Entry may take many shapes and lengths of time. Behavioral shifts once in deviance may be lateral and downward as well as upward, precipitous as well as gradual and controlled, repetitive as well as dissimilar, and involve con-tinuity or complete shift into other venues (Luckenbill and Best 1981). Exits are problematic, varying in their degree of self- or other-directed degree of initiation, being temporary or lasting, and involve anything from going out on top to slink-ing away in debt and disgrace. Perhaps the biggest contrast between deviant and legitimate careers (taking the bureaucratic, organizational form for the latter) lies in the legitimate career's slower ascent at the beginning and the greater stability and security toward the end, compared to the deviant career's rapid upward mobility and earlier burnout (as we see in Chapter 45 on the pimp-controlled prostitute's career).

43

Deciding to Commit a Burglary

RICHARD T. WRIGHT AND SCOTT H. DECKER

In this classical occupational study of deviance, Wright and Decker share with us their insights into the motivations and behavior of residential burglars. Simply written and filled with rich quotes, this chapter affirms the view that most burglaries are committed spontaneously by semiskilled criminals. Crimes of opportunity, these burglaries are fueled by perpetrators' desire for money. Although most attempt to diminish the stigma of their crimes by rationalizing that they steal to support their basic living expenses, Wright and Decker undercut these accounts as impression management, citing subjects' behavior and alternative explanations that they steal to gain money for drugs, partying, impressing women, and to sustain an overall high lifestyle with the trappings of material success. Enmeshed in the world of instant gratification, these burglars give lip service to their desire for legitimate jobs but have neither the patience nor the interest in developing the skills of legitimate work or working their way up the ladder of legitimate career success. For most of Wright and Decker's subjects, burglary is their "main line," although not the only line of deviant income. In addition to the financial yield, they are attracted to residential burglary by the excitement, the freedom, the adventure, the spontaneity, the identity, and the instant gratification. These people enact a criminal lifestyle that is reinforced by the norms and values of the deviant subculture in which they are ensconced.

The demographic characteristics of residential burglars have been well documented. As Shover (1991) has observed, such offenders are, among other things, disproportionately young, male, and poor. These characteristics serve to identify a segment of the population more prone than others to resort to breaking in to dwellings, but they offer little insight into the actual causes of residential burglary. Many poor, young males, after all, never commit any sort of serious offense, let alone a burglary. And even those who carry out such crimes are not offending

Richard T. Wright and Scott H. Decker, *Burglars on the Job: Streetlife and Residential Break-ins*, pp. 35–61. © 1994 by Richard T. Wright and Scott H. Decker. Reprinted by permission of University Press of New England, Hanover, NH.

most of the time. This is not, by and large, a continually motivated group of criminals; the motivation for them to offend is closely tied to their assessment of current circumstances and prospects. The direct cause of residential burglary is a perceptual process through which the offense comes to be seen as a means of meeting an immediate need, that is, through which a motive for the crime is formed. Walker (1984: viii) has pointed out that, in order to develop a convincing explanation for criminal behavior, we must begin by "distinguishing the states of mind in which offenders commit, or contemplate the commission of, their offenses." Similarly, Katz (1988: 4), arguing for increased research into what he calls the foreground of criminality, has noted that all of the demographic information on criminals in the world cannot answer the following questions: "Why are people who were not determined to commit a crime one moment determined to do so the next?" This is the question to which the present chapter is addressed. The aim is to explore the extent to which the decision to commit a residential burglary is the result of a process of careful calculation and deliberation.

In the overwhelming majority of cases, the decision to commit a residential burglary arises in the face of what offenders perceive to be a pressing need for cash. Previous research consistently has shown this to be so and the results of the present study bear out this point. More than nine out of ten of the offenders in our sample—95 of 102—reported that they broke into dwellings primarily when they needed money.

> Well, it's like, the way it clicks into your head is like, you'll be thinking about something and, you know, it's a problem. Then it, like, all relates. "Hey, I need some money! Then how am I going to get money? Well, how do you know how to get money quick and easy?" Then there it is. Next thing you know, you are watching [a house] or calling to see if [the occupants] are home. (Wild Will—No. 010) . . .

These offenders were not motivated by a desire for money for its own sake. By and large, they were not accumulating the capital needed to achieve a long-range goal. Rather, they regarded money as providing them with the means to solve an immediate problem. In their view, burglary was matter of day-to-day survival.

> I didn't have the luxury of laying back in on damn pinstriped [suit]. I'm poor and I'm raggedy and I need some food and I need some shoes . . . So I got to have some money some kind of way. If it's got to be the wrong way, then so be it. (Mark Smith—No. 030) . . .

Given this view, it is unsurprising that the frequency with which the offenders committed burglaries was governed largely by the amount of money in their pockets. Many of them would not offend so long as they had sufficient cash to meet current expenses.

> Usually what I'll do is a burglary, maybe two or three if I have to, and then this will help me get over the rough spot until I can get my shit straightened out. Once I get it straightened out, I just go with the

flow until I hit that rough spot where I need the money again. And then I hit it . . . the only time I would go and commit a burglary is if I needed the money at that point in time. That would be strictly to pay light bill, gas bill, rent. (Dan Whiting—No. 102)

Long as I got some money, I'm cool. If I ain't got no money and I want to get high, then I go for it. (Janet Wilson—No. 060)

You know how they say stretch a dollar? I'll stretch it from here to the parking lot. But I can only stretch it so far and then it breaks. Then I say, "Well, I guess I got to go put on my black clothes. Go on out there like a thief in the night." (Ralph Jones—No. 018)

A few of the offenders sometimes committed a burglary even though they had sufficient cash for their immediate needs. These subjects were not purposely saving money, but they were unwilling to pass up opportunities to make more. They attributed their behavior to having become "greedy" or "addicted" to money.

I have done it out of greed, per se. Just to be doing it and to have more money, you know? Say, for instance, I have two hundred dollars in my pocket now. If I had two more hundreds, then that's four hundred dollars. Go out there and do a burglary. Then I say, "If I have four hundred dollars, then I can have a thousand." Go out there and do a burglary. (No. 018) . . .

Typically, the offenders did not save the money that they derived through burglary. Instead, they spent it for one or more of the following purposes: (1) to "keep the party going"; (2) to keep up appearances; or (3) to keep themselves and their families fed, clothed, and sheltered.

KEEPING THE PARTY GOING

Although the offenders often stated that they committed residential burglaries to "survive," there is a danger in taking this claim at face value. When asked how they spent the proceeds of their burglaries, nearly three-quarters of them—68 of 95—said they used the money for various forms of (for want of a better term) high-living. Most commonly, this involved the use of illicit drugs. Fifty-nine of the 68 offenders who spent the money obtained from burglary on pleasure-seeking pursuits specifically mentioned the purchase of drugs. For many of these respondents, the decision to break into a dwelling often arose as a result of a heavy session of drug use. The objective was to get the money to keep the party going.

The drug most frequently implicated in these situations was "crack" cocaine.

[Y]ou ever had an urge before? Maybe a cigarette urge or a food urge, where you eat that and you got to have more and more? That's how that crack is. You smoke it and it hits you [in the back of the throat] and you got to have more. I'll smoke that sixteenth up and get through, it's like I never had none. I got to have more. Therefore, I gots to go do another burglary and gets some more money. (Richard Jackson—No. 009) . . .

Lemert (1953: 304) has labelled situations like these "dialectical, self-enclosed systems of behavior" in that they have an internal logic or "false structure" which calls for more of the same. Once locked into such events, he asserts, participants experience considerable pressure to continue, even if this involves breaking the law.

> A man away from home who falls in with a group of persons who have embarked upon a two or three-day or even a week's period of drinking and carousing . . . tends to have the impetus to continue the pattern which gets mutually reinforced by [the] interaction of the participants, and [the pattern] tends to have an accelerated beginning, a climax and a terminus. If midway through a spree a participant runs out of money, the pressures immediately become critical to take such measures as are necessary to preserve the behavior sequence. A similar behavior sequence is [evident] in that of the alcoholic who reaches a "high point" in his drinking and runs out of money. He might go home and get clothes to pawn or go and borrow money from a friend or even apply for public relief, but these alternatives become irrelevant because of the immediacy of his need for alcohol. (Lemert, 1953: 303)

Implicit in this explanation is an image of actors who become involved in offending without significant calculation; having embarked voluntarily on one course of action (e.g., crack smoking), they suddenly find themselves being drawn into an unanticipated activity (e.g., residential burglary) as a means of sustaining that action. Their offending is not the result of a thoughtful, carefully reasoned process. Instead, it emerges as part of the natural flow of events, seemingly coming out of nowhere. In other words, it is not so much that these actors consciously choose to commit crimes as that they elect to get involved in situations that drive them toward lawbreaking.

Beyond the purchase of illicit drugs and, to a lesser extent, alcohol, 10 of the 68 offenders—15 percent—also used the proceeds from their residential burglaries to pursue sexual conquests. All of these offenders were male. Some liked to flash money about, believing that this was the way to attract women. . . .

> [I commit burglaries to] splurge money with the women, you know, that's they kick, that's what they like to do. (Jon Monroe—No. 011) . . .

Like getting high, sexual conquest was a much-prized symbol of hipness through which the male subjects in our sample could accrue status among their peers on the street. The greatest prestige was accorded to those who were granted sexual favors solely on the basis of smooth talk and careful impression management. Nevertheless, a few of the offenders took a more direct approach to obtaining sex by paying a streetcorner prostitute (sometimes referred to as a "duck") for it. While this was regarded as less hip than the more subtle approach described above, it had the advantage of being easy and uncomplicated. As such, it appealed to offenders who were wrapped up in partying and therefore reluctant to devote more effort than was necessary to satisfy their immediate sexual desires.

> I spend [the money] on something to drink, ... then get me some [marijuana]. Then I'm gonna find me a duck. (Ricky Davis—No. 015)

It would be misleading to suggest that any of the offenders we spoke to committed burglaries *specifically* to get money for sex, but a number of them often directed a portion of their earnings toward this goal.

In short, among the major purposes for which the offenders used the money derived from burglary was the maintenance of a lifestyle centered on illicit drugs, but frequently incorporating alcohol and sexual conquests as well. This lifestyle reflects the values of the street culture, a culture characterized by an openness to "illicit action" (Katz, 1988: 209–15), to which most of our subjects were strongly committed. Viewed from the perspective of the offenders, then, the oft-heard claim that they broke into dwellings to survive does not seem quite so farfetched. The majority of them saw their fate as inextricably linked to an ability to fulfill the imperatives of life on the street.

KEEPING UP APPEARANCES

Of the 95 offenders who committed residential burglaries primarily for financial reasons, 43 reported that they used the cash to purchase various "status" items. The most popular item was clothing; 39 of the 43 said that they bought clothes with the proceeds of their crimes. At one level, of course, clothing must be regarded as necessary for survival. The responses of most of the offenders, however, left little doubt that they were not buying clothes to protect themselves from the elements, but rather to project a certain image; they were drawn to styles and brand names regarded as chic on the streets.

> See, I go steal money and go buy me some clothes. See, I likes to look good. I likes to dress. All I wear is Stacy Adams, that's all I wear.
> [I own] only one pair of blue jeans cause I likes to dress. (No. 011) ...

After clothes, cars and car accessories were the next most popular status items bought by the offenders. Seven of the 43 reported spending at least some of the money they got from burglaries on their cars.

> I spent [the money] on stuff for my car. Like I said, I put a lot of money into my car ... I had a '79 Grand Prix, you know, a nice car. (Matt Detteman—No. 072)

The attributes of a high-status vehicle varied. Not all of these offenders, for example, would have regarded a 1979 Grand Prix as conferring much prestige on its owner. Nevertheless, they were agreed that driving a fancy or customized car, like wearing fashionable clothing, was an effective way of enhancing one's street status. ...

> I don't know if you've ever thought about it, but I think every crook likes the life of thieving and then going and being somebody better. Really, you are deceiving people; letting them think that you are well off . . . You've got a nice car, you can go about and do this and do that. It takes money to buy that kind of life.

Shover and Honaker (1990: 11) have suggested that the concern of offenders with outward appearances, as with their notorious high-living, grows out of what is typically a strong attachment to the values of street culture; values which place great emphasis on the "ostentatious enjoyment and display of luxury items." In a related vein, Katz (1988) has argued that for those who are committed to streetlife, the reckless spending of cash on luxury goods is an end in itself, demonstrating their disdain for the ordinary citizen's pursuit of financial security. Seen through the eyes of the offenders, therefore, money spent on such goods has not been "blown," but rather represents a cost of raising or maintaining one's status on the street.

KEEPING THINGS TOGETHER

While most of the offenders spent much of the money they earned through residential burglary on drugs and clothes, a substantial number also used some of it for daily living expenses. Of the 95 who committed burglaries to raise money, 50 claimed that they needed the cash for subsistence.

> I do [burglaries] to keep myself together, keep myself up.
> (James Brown—No. 025) . . .

Quite a few of the offenders—13 of 50—said that they paid bills with the money derived from burglary. Here again, however, there is a danger of being misled by such claims. To be sure, these offenders did use some of their burglary money to take care of bills. Often, though, the bills were badly delinquent because the offenders avoided paying them for as long as possible—even when they had the cash—in favor of buying, most typically, drugs. It was not until the threat of serious repercussions created unbearable pressure for the offenders that they relented and settled their accounts.

> [Sometimes I commit burglaries when] things pressuring me, you know? I got to do somethin' about these bills. Bills. I might let it pass that mornin'. Then I start trippin' on it at night and, next thing you know, it's wakin' me up. Yeah, that's when I got to get out and go do a burglary. I *got* to pay this electric bill off, this gas bill, you know? (No. 009) . . .

Spontaneity is a prominent feature of street culture (Shover and Honaker, 1992); it is not surprising that many of the offenders displayed a marked tendency to live for the moment. Often they would give every indication of intending to take care of their obligations, financial or otherwise, only to be distracted by more immediate temptations. For instance, a woman in our sample, after being paid for an interview, asked us to drive her to a place where she could buy a pizza for her children's lunch. On the way to the restaurant, however, she changed her mind and asked to be dropped off at a crack house instead. . . .

Katz (1988: 216) has suggested that, through irresponsible spending, persistent offenders seek to construct "an environment of pressures that guide[s] them back toward crime." Whether offenders spend money in a conscious attempt to create such pressures is arguable; the subjects in our sample gave no indication of doing so, appearing simply to be financially irresponsible. One offender, for example, told us that he never hesitated to spend money, adding, "Why should I? I can always get some more." However, the inclination of offenders to freespending leaves them with few alternatives but to continue committing crimes. Their next financial crisis is never far around the corner.

The high-living of the offenders, thus, calls into question the extent to which they are driven to crime by genuine financial hardship. At the same time, though, their spendthrift ways ensure that the crimes they commit will be economically motivated (Katz, 1988). The offenders perceive themselves as needing money, and their offenses typically are a response to this perception. Objectively, however, few are doing burglaries to escape impoverishment.

WHY BURGLARY?

The decision to commit a residential burglary, then, is usually prompted by a perceived need for cash. Burglary, however, is not the only means by which offenders could get some money. Why do they choose burglary over legitimate work? Why do they elect to carry out a burglary rather than borrow the money from a friend or relative? Additionally, why do they select burglary rather than some other crime?

Given the streetcorner context in which most burglary decisions were made, legitimate work did not represent a viable solution for most of the offenders in our sample. These subjects, with few exceptions, wanted money there and then and, in such circumstances, even day labor was irrelevant because it did not respond to the immediacy of their desire for cash (Lemert, 1953). Moreover, the jobs available to most of the offenders were poorly-paid and could not sustain their desired lifestyles. It is notable that 17 of the 95 offenders who did burglaries primarily to raise money *were* legitimately employed.

> [I have a job, but] I got tired of waiting on that money. I can get money like that. I got talent, I can do me a burg, man, and get me five or six hundred dollars in less than a hour. Working eight hours a day and waiting for a whole week for a check and it ain't even about shit. (No. 022) . . .

Beyond this, a few of the offenders expressed a strong aversion to legitimate employment, saying that a job would impinge upon their way of life.

> I ain't workin' and too lazy to work and just all that. I like it to where I can just run around. I don't got to get up at no certain time, just whenever I wake up. I ain't gotta go to bed a certain time to get up at a certain time. Go to bed around one o'clock or when I want, get up

when I want. Ain't got to go to work and work eight hours. Just go
in and do a five minute job, get that money, that's just basically it.
(Tony Scott—No. 085) . . .

Nevertheless, a majority of the offenders reported that they wanted lawful
employment; 43 of the 78 unemployed subjects who said that they did burglaries
mostly for the money claimed they would stop committing offenses if someone
gave them a "good" job.

I'm definitely going to give it up as soon as I get me a good job. I don't
mean making fifteen dollars an hour. Give me a job making five-fifty
and I'm happy with it. I don't got to burglarize no more. I'm not doing
it because I like doing it, I'm doing it because I need some [drugs].
(No. 079) . . .

When faced with an immediate need for cash, then, the offenders in our
sample perceived themselves as having little hope of getting money both quickly
and legally. Many of the most efficient solutions to financial troubles are against
the law (Lofland, 1969). However, this does not explain why the subjects decided
specifically on the crime of residential burglary. After all, most of them admitted
committing other sorts of offenses in the past, and some still were doing so. Why
should they choose burglary?

For some subjects, this question held little relevance because they regarded
residential burglary as their "main line" and alternative offenses were seldom
considered when the need for money arose. . . .

[I do burglary] because it's easy and because I know it. It's kind of getting
a speciality or a career. If you're in one line, or one field, and you
know it real well, then you don't have any qualms about doing it. But
if you try something new, you could really mess up . . . At this point,
I've gotten away with so much [that] I just don't want to risk it—it's too
much to risk at this point. I feel like I have a good pattern, clean; go
in the house, come back out, under two minutes every time. (Darlene
White—No. 100) . . .

When these subjects did commit another kind of offense, it typically was triggered by the chance discovery of a vulnerable target . . . most of the burglars
we interviewed identified themselves as hustlers, people who were always looking to "get over" by making some fast cash; it would have been out of character
for them to pass up any kind of presented opportunity to do so.

If I see another hustle, then I'll do it, but burglary is my pet.
(Larry Smith—No. 065) . . .

THE SEDUCTIONS OF RESIDENTIAL BURGLARY

For some offenders, the perceived benefits of residential burglary may transcend
the amelioration of financial need. A few of the subjects we interviewed—7 of
102—said that they did not typically commit burglaries as much for the

money as for the psychic rewards. These offenders reported breaking into dwellings primarily because they enjoyed doing so. Most of them did not enjoy burglary per se, but rather the risks and challenges inherent in the crime.

> [I]t's really because I like [burglaries]. I know that if I get caught I'm a do more time than the average person, but still, it's the risk. I like doin' them. (No. 013)

> I think [burglary is] fun. It's a challenge. You don't know whether you're getting caught or not and I like challenges. If I can get a challenging [burglary, I] like that. It's more of the risk that you got to take, you know, to see how good you can really be. (No. 103)

These subjects seemingly viewed the successful completion of an offense as "a thrilling demonstration of personal competence" (Katz, 1988: 9). Given this, it is not surprising that the catalyst for their crimes often was a mixture of boredom and an acute sense of frustration born of failure at legitimate activities such as work or school.

> [Burglary] just be something to do. I might not be workin' or not going to school—not doing anything. So I just decide to do a burglary. (No. 017)

The offense provided these offenders with more than something exciting to do; it also offered them the chance to "be somebody" by successfully completing a dangerous act. Similarly, Shover and Honaker (1992: 288) have noted that, through crime, offenders seek to demonstrate a sense of control or mastery over their lives and thereby to gain "a measure of respect, if not from others, at least from [themselves]." . . .

While only a small number of the subjects in our sample said that they were motivated *primarily* by the psychic rewards of burglary, many of them perceived such rewards as a secondary benefit of the offense. Sixteen of the 95 offenders who did burglaries to raise cash also said that they found the crime to be "exciting" or "thrilling." . . .

> [Beyond money], it's the thrill. If you get out [of the house], you smile and stand on it, breathe out. (No. 045)

> It's just a thrill going in undetected and walking out with all they shit. Man, that shit fucks me up. (No. 022) . . .

Finally, one of the offenders who did burglaries chiefly for monetary reasons alluded to the fact that the crime also provided him with a valued identity. . . .

SUMMARY

Offenders typically decided to commit a residential burglary in response to a perceived need. In most cases, this need was financial, calling for the immediate acquisition of money. However, it sometimes involved what was interpreted as a need to repel an attack on the status, identity, or self-esteem of the offenders.

Whatever its character, the need almost invariably was regarded by the offenders as pressing, that is, as something that had to be dealt with immediately. Lofland (1969: 50) has observed that most people, when under pressure, have a tendency to become fixated on removing the perceived cause of that pressure "as quickly as possible." Those in our sample were no exception. In such a state, the offenders were not predisposed to consider unfamiliar, complicated, or long-term solutions (see Lofland, 1969: 50–54) and instead fell back on residential burglary, which they knew well. This often seemed to happen almost automatically, the crime occurring with minimal calculation as part of a more general path of action (e.g., partying). To the extent that the offense ameliorated their distress, it nurtured a tendency for them to view burglary as a reliable means of dealing with similar pressures in the future. In this way, a foundation was laid for the continuation of their present lifestyle which, by and large, revolved around the street culture. The self-indulgent activities supported by this culture, in turn, precipitated new pressures; and thus a vicious cycle developed.

That the offenders, at the time of actually contemplating offenses, typically perceived themselves to be in a situation of immediate need has at least two important implications. First, it suggests a mind-set in which they were seeking less to maximize their gains than to deal with a present crisis. Second, it indicates an element of desperation which might have weakened the influence of threatened sanctions and neutralized any misgivings about the morality of breaking into dwellings. . . .

REFERENCES

Katz, J. (1988). *Seductions of Crime: Moral and Sensual Attractions in Doing Evil*. New York: Basic Books.

Lemert, E. (1953). "An Isolation and Closure Theory of Naive Check Forgery." *Journal of Criminal Law, Criminology, and Police Science* 44: 296–307.

Lofland, J. (1969). *Deviance and Identity*, Englewood Cliffs, NJ: Prentice Hall.

Shover, N. (1991). "Burglary." In Tonry, M., *Crime and Justice: A Review of Research*, vol. 14, pp. 73–113, Chicago: University of Chicago Press.

Shover, N., and Honaker, D. (1990). "The Criminal Calculus of Persistent Property Offenders:

A Review of Evidence." Paper presented at the Forty-second Annual Meeting of the American Society of Criminology, Baltimore, November.

———. (1992). "The Socially Bounded Decision Making of Persistent Property Offenders." *Howard Journal of Criminal Justice* 31, no. 4: 276–93.

Walker, N. (1984). "Foreword." In Bennett, T., and Wright, R., *Burglars on Burglary: Prevention and the Offender*, pp. viii–ix, Aldershot: Gower.

44

Hard Drugs in a Soft Context

Managing Crack Use on a College Campus

CURTIS JACKSON-JACOBS

Jackson-Jacobs offers us a fascinating glimpse into a drug subculture that runs counter to the prevailing wisdom about hard drugs. He presents an intimate and longitudinal portrait of four white, middle-class crack users in a college setting. Here, he juxtaposes conventional attitudes about the behavior of crack users with the social standing of these participants and shows how this hybrid clash of class, culture, and crack unfolds. Jackson-Jacobs traces the lives of his crack-smoking friends as they manage their deviant careers, hiding their stigmatized drug use from their fellow students, navigating the often-dangerous venues where they must buy their drugs, and negotiating the conflicting values implicit in their deferred-gratification college lives and their immediate-gratification drug experiences. Their lives, he notes, consist of a series of managed interactions and relationships with groups of different others. He concludes that while their environment and social capital protects them from the kinds of economic, legal, and physical danger associated with inner-city crack users, when compared to their white, middle-class college peers, they have not fared so well. Overall, Jackson-Jacobs's study offers us insight into the importance of set and setting in drug-using environments.

C rack cocaine inspired fear in many Americans in the 1980s. As for many "drug scares" before it and since, the drug of the moment was felt to be the most dangerous ever.... Crack started to become relatively available in several cities across the country in 1984 ... and was quickly blamed for widespread addiction,

From Curtis Jackson-Jacobs, "Hard Drugs in a Soft Context: Managing Crack Use on a College Campus," *The Sociological Quarterly.* Copyright © 2004 by Blackwell Publishers. Reprinted by permission of the publisher.

urban street crime, and rampant gang violence. The first sources of information about the new drug were lurid news reports (e.g., four articles printed in *Newsweek* June 16, 1986) and frenzied political speeches (e.g., Reagan and Reagan [1986] 1989), introducing crack as an addictive menace of "epidemic" proportion. Americans overwhelmingly came to believe drugs were the most pressing threat to their society by 1989 (Reinarman and Levine 1997).

Crack, or "rock," was a marketing innovation, quickly adopted by urban street-corner sellers.... It is a smokeable form of cocaine, often sold in pea-sized units costing $20 or less. Unlike powder cocaine, traditionally popular among the rich, crack produces a considerably more powerful high in smaller doses, and appears more conducive to patterns of binge use.... Crime and addiction were attributed to the more powerful pharmacological effects of the rock form of cocaine.

Strict laws and sentencing guidelines were imposed in response to the perceived tide of lawlessness, disorder, and inhumane violence thought to be riding a wave of crack addiction.... American jails and prisons filled with unprecedented numbers of young minority males.... A pervasive cultural stereotype of the users, colloquially dubbed "crack-heads," quickly spread, conjuring images of crazed criminal men with superhuman strength and wasting women prostituting in abject degradation.

Journalistic stories in the 1980s and sociological studies in the 1990s depicted images of distinctively "ghetto" and "street" users. Crime, violence, and exploitation were at the core of these accounts of the worlds of crack use. Poverty, minority status, and desperation characterized the users.

In this paper I present an ethnographic case study of a network of four regular crack users on a major university campus. The environment in which I did my fieldwork is perhaps as socially distant from the sites of previous crack research as one can get in America. As young, upper-middle-class, mostly white college students, the subjects of this research were at the other end of the social-inequality spectrum, commanding economic, social, and cultural resources.

The result was that I found users whose experiences of crack use were equally distant from those reported in other research. Unlike the users described elsewhere, those described here did not commit crime, were not afraid of victimization by criminals or the law, and did not face the same risks of suffering negative social esteem....

The focus of the paper is purposely comparative with data from other sources. In reviewing previous findings and then in presenting my own, I highlight variations in how users experience trouble. Two conditions increase the probability of successfully avoiding crack-related trouble: (1) using in secure contexts and (2) bounding crack use from conventional life. By comparing users' experiences across settings, I discuss how these conditions are differentially distributed across social worlds, especially with regard to economic resources and "conventional ties" (Becker 1955). The emphasis here is on making the link between class and crack-related trouble by examining particularly consequential qualities of daily life.

SOCIOLOGICAL UNDERSTANDINGS
OF CRACK USE

The best known and most powerful sociological studies of marijuana and heroin use have countered popular beliefs that the effects of drug use are due to properties of the drugs alone.... The role of social rituals, patterns of interaction, and users' understandings of drugs have often been developed by researchers who found that there are users who "get away with it."...

Sociologists have studied crack users in a strikingly narrow range of contexts. Researchers have focused nearly universally on the poorest, most desperate users they can find, especially those involved in dealing, prostitution, and violence. As Katz points out, although ethnographers often "debunk" prevailing stereotypes, ethnographers of crack users have portrayed the inside of crack houses as places where "humanity is tortured and degraded," even more so than posed by popular claims (1997, p. 395). Indeed, such places do exist. By portraying only the most shocking settings, though, researchers have done little to convey the diversity of experiences people have using rock cocaine. Such selective attention risks reinforcing simple stereotypes at the expense of understanding the wide variation in how crack is used and how it affects its users....

There are ... reasons why it is important to continue to study users who do not get in trouble.... Although we would expect, given past experience, that any new drug will be used without destroying all of its users, there is a powerful, recurrent, countervailing argument. Historically, each time a new drug inspires "moral panic" (Goode and Ben-Yehuda 1994), the panic is sustained by the belief that the new drug is "the worst ever," representing a level of inevitable destructiveness never previously known.... Thus, popular understandings continually seek to abolish, for the present drug, the relevance of previous sociological explanation. As new drugs emerge, so do new demands for critical investigation and tests to see if the limits of social explanations have been reached....

"Street Predators" and "Street Victims"

The biographies of "street predators" and "street victims," and their corresponding modes of use, illustrate the ways in which local troubles common to many residents become intensified in the course of crack use....

"Campus Users"

In this paper I offer a ... description of the social world of a new variant, ... "campus users...." The characterization is not meant to suggest that the type is strictly limited to campuses. More research into the diversity of crack use may well reveal a similar mode of using in other contexts. Alternatively, research on other college students may reveal their crack use to be organized differently. What is most important is not that these people were "college students" but that:

they used crack in a social world comparatively quite safe from violence and financial or legal trouble; important spheres of their lives such as housing, work, leisure, and family were loosely coupled relative to other social worlds; they used crack only in highly ritualized ways; and they understood crack use as something that should be subordinate to conventional activities. The logic of this comparative analysis suggests that under similar circumstances—whether on campus or not—we should find other users who do not get into trouble.

METHODS

All of the observation for this research was conducted in a Midwestern American college town with a population around 200,000. Around the time of my research, the town was labeled one of the most crime-free and best cities in which to live and raise a family by national magazines like *Money* and *Parenting*.

After moving out of their dormitory, the primary participants moved into and around a 15-square block student neighborhood that I call "Midtown." In neither the town nor the Midtown neighborhood did I note media or other attention to drug problems, much less concern about crack.

I met the primary participants in this study in the fall of 1996. All were living on the same dormitory floor on the campus of a well-respected public university. Jon, Martin, and Mark were white, and Casey was Korean American. They all came from upper-middle-class families, their parents working in professional positions, some requiring graduate or professional degrees. Over the years I knew them, they all went to school and worked jobs at least part time, even as they used crack and other drugs. Before that year all had used drugs of some kind, including marijuana, hallucinogens, and narcotics, but only Jon had used crack. Jon had used crack throughout high school, traveling with suburban friends to buy it in the inner city of the adjacent large urban center.

Jon and Casey were out-of-state students and did not come to school with any friends. Mark and Martin had moved from other cities in the state. Mark was in his second year of college, having already lived on the floor the previous year, making new friends during that time. During the fall I met them, they were in the process of forming new friendship groups, largely organized by their common residence.

Between fall 1996 and fall 1997, I hung out with the group and a wider circle of dozens of friends and acquaintances on campus. During the fall of 1997, I asked if I could watch them whenever they smoked, which they permitted. I began to take field notes on their activities and interview them about their experiences with crack and family and social life generally. During the binges lasting several hours, I would sit outside their circle and take jottings and do homework alternately. By doing homework (sometimes only pretending to do homework), I was able to write without making them visibly uncomfortable.

I kept up taking notes for the duration of a one-semester fieldwork course and for a short time the following semester. In total my systematic observation continued for six months, although I had known and interacted with the users for a year prior to that time. My friendships with and observations of the group continued until I moved away from Midtown at the end of 1998. After that I kept speaking with Casey by phone for 2 years and have continued to speak with Jon.

Around the time I left Midtown, Mark had started to "lose control" of his drug use. He reportedly took out and squandered loans, lost his job, dropped out of school, and wrote bad checks. His mother arrived there and took him back to his hometown. Since that time I have been unable to locate him, though Jon had heard he relocated to another university in the same state. Casey transferred to a college in his home state at the beginning of 1999 and continued attending until I lost track of him in 2000. By May 2001 both Jon and Martin had graduated from the university with "B" average grades or better. Martin continued his studies toward a Ph.D. Jon has since worked in various service jobs while moving around the United States and then back to his hometown. In the past four years we have gotten together several times. He developed what he and I both considered to be a serious heroin problem during this time, though he is working to keep it under control through treatment. Now seven years since we first met, he tells me he continues to use rock occasionally without any detriment to his life that he can identify.

SAMPLE-SIZE CONSIDERATIONS

By the end of my observation, I had seen at least 20 male and female college students smoking crack. Yet my data focus only on Jon, Martin, Casey, and Mark, except incidentally. A brief justification for the study of only four participants in this case is in order.

First, my reasons for focusing on these four: The primary participants in this study were the only ones who smoked crack regularly for very long. Many of the others only smoked once or only once every few months. While I can attest to their presence in the setting, I did not gather the data to provide a detailed analysis of the numerous occasional users.

The core group developed a particular culture, body of local knowledge, and set of practices together not shared to the same degree by the nonregular smokers. By focusing in detail over an extended period of time, I was able to gather varied and in-depth data on four persistent users. The methodological decision was to seek deep understanding of a few participants rather than thinner data on a larger sample. My goal was to use a portrait of these students' experience to contribute to a developing understanding of varieties of using crack....

FEATURES OF CAMPUS LIFE:
SOCIAL-ORGANIZATIONAL DIMENSIONS

Two social-organizational features of social life among the campus users are brought into relief when compared with reports from other contexts. First, they used crack in a context that afforded them economic, social, and physical security. They had grown up with the typical economic, social, and cultural rewards of upper-middle-class suburban life in America. Neither crime nor victimization had been especially relevant to their youthful cultures, nor did they become relevant to life on campus. There was no ghetto near Midtown. Street crime was not a visible feature of Midtown life.

Second, they were able to bound their lives as crack users from their other roles and relationships. As I describe below, they were also geographically removed from their families and experienced little pressure to show up at school regularly and sober. They successfully maintained the boundaries of their identities as crack users to such a degree that they could often even deceive certain friends and roommates about their use.

How they understood crack also helped them to maintain their conventional lives. First, they oriented to crack as a social object to be used in leisure rituals. Before ever using crack together they had used drugs only in similar groups, especially among circles of marijuana-smoking friends. Social and ritual drug use were widespread among students, while the kinds of paranoid and predatory relationships found in poor urban neighborhoods were completely absent.

Second, the perceptual horizons of crack use stopped at the edge of the circle of users. In contrast to users in other reports, the place of crack in these users' lives grew out of the context of developing friendships as a shared interest and social ritual (see Jackson-Jacobs 2001). Crack was, for them, something to be controlled and placed secondary to other concerns, namely their conventional relationships and activities. They overwhelmingly used it in their "spare time," organizing consumption not to interfere with conventional pursuits, not the other way around, as observed among hard-core users.

THE SOCIAL WORLDS OF THE CRACK USERS

The relationship between resources and trouble is not as simple as might be assumed. Campus life—not simply upper-middle-class status—at once provided social resources to these men and a locus for bounding identity as crack users, and it provided them with the understandings of crack, trouble, and social life that motivated them to actively maintain conventional identities. On the one hand, their conventional identities were central concerns, things to be managed vigilantly. They wanted to avoid being identified as crack users by their roommates, at school, at work, or by their families. On the other hand, related aspects of these same relationships served as resources, allowing them to hide their drug use. They were able to move residences often to manage their use, were able to miss school without drawing attention, were largely free from parental supervision, and almost never interacted with street criminals.

Residential Mobility

Like many students, the small group of crack users and their larger group of friends moved often, affording them control over their crack use. In the town there was a virtually uniform term for apartment leases. All Midtown landlords required one-year leases beginning and ending on August 15. Despite the one-year imposition on apartment leases, many of these men moved more often than once per year. They did this by subleasing apartments to and from conventional acquaintances and friends.

As many college students in areas like Midtown do, they organized groups of friends to rent large apartments. For the crack smokers the ease of mobility allowed them to shape their drug use. At times they moved in with roommates who did not use crack in order to cut back. At other times they moved in with other users in order to get high more often.

At one point Jon and Mark moved into an apartment together, where they frequently had company over to use various drugs. Attracted to the steady source of clientele, a cocaine dealer soon began spending several hours each day in the house. When Jon decided that he could not stand to live where a cocaine dealer spent so much time, he looked for another place to live. It was too crazy, he told me once; he had to get out of there.

Martin, who had been spending a lot of time at this house, also hoped to cut down on his use. The two moved into another house together. They had five women roommates who did not use crack but smoked marijuana from time to time and drank alcohol on weekends. Jon signed his lease over to their regular dealer, Paul, while Martin simply stopped paying rent at his old apartment.

Moving residences to get away from drugs was, compared to the residential troubles described in other studies of crack users, quite simple. As nice-looking white college students looking for housing in a student neighborhood, they did not face the type of discrimination frequently experienced by black renters, especially poor black renters. Also, they had access to financial resources from parents, providing them with the cash to put down damage deposits and rent at any time. The combination of race, financial resources, and type of neighborhood is especially important in light of findings that African Americans are less likely to move out of neighborhoods that they describe as "undesirable" than are whites. Furthermore, this type of movement would be very difficult and costly for many members of the middle-class, especially homeowners, who cannot move on a moment's notice. Additionally, moving around middle-class neighborhoods frequently, among middle-aged adults, would be suspicious and damaging to financial credit.

Residential mobility among the group provided some control of the crack-using experience. Migrating from apartment to apartment might have raised suspicion in typical suburban and city environments. And it is often unmanageable for many ghetto residents, especially unemployed women. For the men I knew, though, moving in order to control crack use was seen as nothing more than typical migration through friendship circles, at the same time depending on student status and access to social networks.

Roommates

For several months during my observation, Martin lived with Casey and two other male roommates. A coworker of the two roommates, Ray, who coincidentally was also a crack user, slept on the couch without paying rent. The other two never used crack, although one of them, Lonnie, did use powder cocaine a few times. Martin and his three roommates signed a joint lease, which he defaulted on by moving out. When Martin moved out to live with Jon, he quit paying his share of the rent.

His roommates called his mother to complain. Lonnie told me, "I called Martin's mom and told her that he had money to spend on crack but not on rent. I said he was worthless and the product of worthless parents." Martin confirmed that his roommates had called his mother and been confrontational, but he did not know exactly what they said. He said, though, that he wished he hadn't told them about his crack use and that they were unfairly trying to use this against him. Especially important, Martin said, was the possibility that they would use this information against him when suing for unpaid rent—the only serious reference to legal trouble related to crack I heard in Midtown.

Having moved out of this house, Martin and Jon began to share a room in a house with five female college-student roommates. Although Martin and Jon were friends with their roommates, they did not at first tell any of the five that they smoked crack in their room. Two of them eventually did find out, but Jon and Martin continued to try to deceive them while high and to keep the others in the dark completely.

They kept their roommates from learning of their crack use by assigning a "door man." His job was specifically to watch out for roommates. The door man was instructed in how to deal with a knock.

Jon said on one occasion, "Hey, Martin, you be the door man, okay?"

"What do you want me to do?" Martin asked.

"Tell them we're naked."

This role was often crucial in delaying entry when a roommate did knock; during the delay everyone else had time to hide the paraphernalia.

Inciardi, Lockwood, and Pottieger (1993) also described the role of door man, but in the context of "crack houses." He was someone "who lets people in, checks for weapons, and watches for police" (Inciardi, Lockwood, and Pottieger 1993, p. 154). Looking at the door man as a security guard gives us a way to see what the users feel threatened by in a specific context. In inner-city "crack houses" violence and arrest are pressing threats from the outside world.

For the Midtown smokers, however, violence and arrest were not real concerns. Their environment posed little threat to their life or liberty. Instead, they feared losing their social standing and relationships. Roommates' gazes, not predators' violence, were the most prescient sources of trouble. We see in these interactions not only the physical security of life on campus but the success these men had at bounding crack use from relationships: even among friends and close acquaintances, only a select group were let in on the secret.

Parents

All of the men claimed to keep close ties with their parents. The special problems of concealing crack use from them were primarily dealt with on the phone.

I observed all of the central members speaking on the telephone with their parents on occasion. All of their parents lived in different cities, so they never feared a surprise visit.

Mark would almost never answer his phone when he was high on cocaine or crack. When it rang, he asked someone else to pick it up and ask who was calling. If no one would do this, he was so afraid it might be his mother and she would know he was on drugs that he allowed the phone to ring until the caller hung up. When someone did answer for him, even before the caller was announced, he would wave his arms, shake his head, and mouth the words, "I don't want to talk to my mom." Martin and Casey used similar strategies.

While the other three went to the extreme of not answering potential calls from parents, Jon said, "when they call, actually, I become instantly sober." He would talk to them, but, he continued, "I try to answer in short, direct statements and don't try to keep them on the phone for too long."

Their dealings with parents show how fearful these men could be in some situations. Although they hid their drugs from roommates, they would often "pass" as sober just minutes after smoking. Over the phone with their parents, though, they would not take this risk.

It seemed to me that, in fact, these men could speak just fine on the telephone when they were high on crack. The invasion of the family world into the crack circle inspired an immediate shame. These brief collisions of the worlds of parental supervision and campus freedom opened a window into a tension these men experienced in their drug lives. While they were, for the most part, free to use drugs in their campus life, they were at rare times compelled to bear the constraints of conventional family life.

Work

Being identified as a crack user at work poses some of the same problems as detection by parents and roommates: "It's the stigma of it. That people will do anything to get it," Jon explained. But, in addition to negative social esteem, he expected that the stigma generally associated with crack users would present practical concerns for his employer: "It's kind of implicit that I would be stealing to buy crack." Their strategies for avoiding roommates and parents were inapplicable to work. Hiding from employers may lead employees to be fired for missing work. Crack smokers cannot just say to their employers, as they might to their roommates or parents, "I can't deal with you right now."

Jon and Martin were able to maintain both their jobs and school activities despite using crack several times a month, often very soon before going to work. Jon explained that they would decide to "stop [using crack] at least three hours early for work, which might mean to finish as quickly as possible." Usually smoking together, they had a pact that they would smoke whatever was left well before going to work. Doing this was hazardous, though, as they were sometimes "coming down" and "strung out" at work; symptoms include diarrhea, talking nonsense, "fiending"[1] for another hit, minor hallucinations, and various other acute health problems. This, they believed, could result in a noticeable change from their sober behavior. On the few occasions that he was noticed, Jon claimed to be hung over or sick, more legitimate excuses for withdrawn and sickly behavior than claiming to be fiending for crack. The group orientation to maintaining conventional identities, in this case as "good employees," was a basis for concerted and individual efforts to bound their crack-using identity.

School

Whereas the main responsibility of conventional American adult worlds, besides family, is to maintain employment, for college students it is to go to school. This responsibility was fairly easy for my sample to maintain despite their crack use. All that these crack users really felt obligated to do was pass their courses in order to make progress toward graduation and to satisfy their parents.

Oftentimes after a night or more of heavy smoking, these men would miss class. And, unlike a full-time job, no one cared much. At worst they lost some credit for class participation. Furthermore, when they felt motivated, it was no problem for them to sit quietly through classes while under the influence of crack without being detected, especially in large lectures.

The schedule of "work" required of college students also made it easy for them to manage crack use and homework. Many college students spend days in a row binge drinking, ... playing video games, seeking out sex partners, ... or wasting time without doing homework. Like more conventionally idle students, these crack smokers could simply cram schoolwork in large doses during their days-long hiatuses from crack use. . . .

Strangers

One window into the kind of trouble most relevant to their life on campus, their intense anxiety about being stigmatized, came when these users purchased legal paraphernalia at stores. In order to "cook up" crack from powder cocaine and smoke it, they had to buy certain households items before a binge. Although they interacted only with strangers in the store, they showed reluctance to buy the necessary items.

All the supplies needed to produce and smoke crack, other than the cocaine, are legal to buy: spoons, candles, baking soda, copper scouring pads, metal sockets, lighters, etc. Many of the crack smokers claimed that although legal, some items are known by the general public to be primarily used for crack. Baking soda and Chore-Boy brand scouring pads (specifically the copper ones that come in bright orange boxes), according to several of the smokers, are easily recognized by "most people" as crack paraphernalia. Jon spoke once of a convenience store in his hometown that "put a sign up next to the Chore Boy that said 'crack pipe screens,' because that's all people buy it for." Martin, Mark, and Casey also agreed that making crack screens "is the only reason people buy Chore-Boy."

Several times Jon gave money to one of the others and asked him to purchase the Chore-Boy. I pointed out that there was nothing illegal about it. "Yeah," Jon said and laughed one evening. "But I don't want people to see me buy it. I mean, cause they'll know what it's for."

Jake, another student who smoked rock with them on several occasions, recounted a story to me and to the group members that seemed to confirm their fears. Jake was buying Chore Boy and baking soda, he explained, when a woman behind him in line made a chilling statement. Out loud she said, "I know what that's for. You're going to smoke crack with that." The only negative consequence that came of this episode was that Jake was verbally "outed." Nonetheless, this was enough to mortify him and was accepted as a serious, noteworthy event in itself by the rest of the group. Being verbally identified was terrifying, regardless of any secondary consequences that could imaginably follow.

Jake's story illustrated both the product and the process of fear. His story confirmed to the group that their anxieties were well founded. His statement, and Jon's, show just how deeply these men feared being "outed," even by strangers. That this fear was felt so intensely illustrates how bounded their identities as crack users were and how strictly they wanted to maintain that boundary. Not only people who could "do something" about it ... but people who might simply interpret their behavior as deviant were felt to seriously threaten their identities.

Neither Jon nor any of the others ever showed any apprehension about buying the cocaine. The transaction was always conducted in a private home with a well-known and trusted dealer. When buying the cocaine, because of the privacy of the transaction, they were confident that no one outside the cocaine community would know what they were doing. In buying the screen material, though, they perceived that they were at risk of detection by people in the store. Although they couldn't be arrested, their concern was that some anonymous other in the community would recognize them as crack users. . . .

Experiences of Trouble

None of the men became seriously involved in street crime or were criminally victimized in Midtown, except in the rare instances I describe below. They were not around nor did they interact with anyone who they thought might victimize them. Nor did they feel they were likely to be arrested. Unlike users described elsewhere, these men bought and used cocaine privately in the company of friends. They did not buy on street corners, in crack houses, or from strangers, especially likely ways to get caught.

So far I have described how Jon, Martin, Casey, and Mark stayed out of trouble and, in fact, hardly ever thought of any trouble besides stigma. Yet these data offer two suggestive, but limited, contrasts within. First is Mark's experience of "losing control" of both his drug use and social life. Second are Jon and his hometown friends' descriptions of concerns and experiences using crack in an urban ghetto when Jon was not living in Midtown. Each show that the troubles associated with crack both are shaped by and shape contexts of use.

After I left the field, Mark experienced what might be considered "serious" loss of social standing. After using for two years, he dropped out of school and lost his job. He inherited $3,000, took out a loan of the same size, and sold much of his property, spending all in under a year. Finally, after exhausting his supply of legal economic resources, he began writing bad checks. After a year out of school, his mother realized something was wrong and took him back home.

Mark did not commit serious crime but instead turned to a form of misdemeanor, low-level white-collar crime, one that in its illegality was identical to a legal counterpart behavior: writing bad checks. Mark's experience shows that this group was not immune from trouble or addiction. Nonetheless, in the midst of "losing control" of his use, Mark's troubles were shaped by his social standing. The most serious consequences turned out to be those that he had feared most. He was thought to be a "crack-head," was considered a failure in social life by his friends, and was "found out" by his mother. Mark's trouble coincided precisely, and was itself constituted by, a progressive failure to maintain a boundary between his crack use and his work, school, and family life. The process of getting into social trouble, Mark's case illustrates, is a process of a breakdown of the boundaries between crack use and conventional activities.

Jon and his high school friends' experience, using in urban slums before he moved to Midtown and on his trips home, points to the critical importance of the immediate contexts of use. They suffered many of the same troubles as the users who lived in these places. Back home, they told me, they would drive from their affluent suburban neighborhood to the most impoverished ghettos of the city to buy rock. Part of the lore of this circle of friends was the violent experiences they had when buying crack in the ghetto.

Jon or his friends reported that on separate occasions they were: punched in the face; dragged in an alley and choked in a robbery attempt; confronted with racial threats; arrested while buying on a public street; beaten by police; forced to make an undercover buy to avoid arrest; and pursued in a high-speed car chase by a dealer who thought they were buying with a fake twenty-dollar bill. Such stories were of exotic interest to the rest of the Midtown group who had no such experiences.

While Mark's case shows that troubles emerge when the boundaries between crack using and conventional identities break down, Jon's experience shows that troubles depend also on the environment of use. Although the drug, the user, and the user's understandings were held constant, Jon experienced serious trouble only in the dangerous context of the ghetto. Furthermore, he experienced the kinds of trouble endemic to interacting with the street economy. Though victimization and arrest were pressing concerns, becoming stigmatized, the fear on campus, was not relevant.

CONCLUSION

The users Americans feared in the 1980s are precisely the ones documented in sociological field research: poor minority addicts from urban ghettos. Among these users, in these places, we see not just the effects of crack but, more specifically, the effects of crack on people who live in some of the most impoverished conditions in America. The "social pathologies" sociologists find in the ghetto, including violence, crime, and addiction, are largely due to the concentration of face-to-face interactions among those who experience the worst of America's structural inequalities (Wilson 1987; Massey and Denton 1993). These structurally distributed troubles infuse the lives of crack users, blurring the results of research on the effects of crack with the effects of marginality.

In the ghetto there are users who do not use compulsively (Ratner 1993, p. viii) and who may not get into serious trouble. Research has yet to document their lives. That no one has recognized or tried to account for nondestructive crack use should trouble sociologists, especially given the tremendous social costs associated with the societal reaction it has received. Professional sociology appears to have contributed to what Best calls the "Iron Quadrangle" that institutionalizes perspectives on social problems (1999, pp. 63–69): mass media, activists, government, and experts (including, in this case, sociologists and anthropologists) all perceived some advantage in treating crack as universally devastating.

The experiences of the men I observed were only colored by a faint shadow of the stigma and other troubles that met the "street" users widely reported elsewhere. Nestled in the world of postadolescence and preadulthood, this group of crack smokers shared an experience of both social security and the resources to bound spheres of social life. These cushy, lax conditions made it possible to manage crack use alongside conventional pursuits.

The practical necessities and moral constraints placed on crack users have a great deal to do with whether and how they experience negative outcomes. Variation in these constraints across context, not simply conventionally defined class dimensions, should be a topic of sociological research. For those in urban slums it may be necessary to buy crack in public, thus risking arrest or the dreaded label "crack-head." Poor smokers, if they lose an apartment or a job, may not have the financial resources to regain a stable lifestyle. Middle-class adults are in constant contact with their families and their colleagues. If they are not careful about the hours or company they keep, their use may be discovered.

The causal arguments developed here also suggest lines of inquiry in the study of drug use, stigma, and trouble more generally. We can locate "hardness" and "softness" in particular social worlds and biographies, not only in particular kinds of drugs and personalities. Socially, economically, and physically insecure communities present endemic threats. Harder or softer contexts present variable future implications for the relationship between one trouble and potential future troubles in other spheres of social life. Where individuals and social groups cannot or do not keep typical, local problems from spreading across these boundaries, their suffering becomes intensified and diversified.

Moreover, different people understand their potentially discrediting activities, conventional life, and the prospects of trouble differently. Not only are some individuals more vulnerable, but they orient to threats differently. More or less troubled trajectories in life shape how people experience their criminal or deviant activities. For individuals successfully navigating conventional institutions, including university life, serious trouble is likely to be seen as both realistically avoidable and to be avoided. In social worlds permeated by serious suffering, in contrast, members appear more often to see trouble as unavoidable and to adopt perspectives that embrace or disregard it.

NOTE

1. "Fiending" is the feeling of intense craving for another hit. They describe it as almost overwhelming; all one can think about is another hit. Like low-status "fiends," the users would sometimes scour the carpet looking for a rock that might have fallen.

REFERENCES

Becker, Howard. 1955. "Marihuana Use and Social Control." *Social Problems*. 3: 35–44.

Best, Joel. 1999. *Random Violence: How We Talk about New Crimes and New Victims*. Berkeley and Los Angeles: University of California Press.

Goode, Erich, and Nachman Ben-Yehuda. 1994. *Moral Panics: The Social Construction of Deviance*. Cambridge, MA: Blackwell.

Inciardi, James, Dorothy Lockwood, and Anne E. Pottieger. 1993. *Women and Crack-Cocaine*. New York: Macmillan Publishing Company.

Jackson-Jacobs, Curtis. 2001. "Refining Rock: Practical and Social Features of Self-Control among a Group of College-Student Crack Users." *Contemporary Drug Problems* 28: 597–624.

Katz, Jack. 1997. "Ethnography's Warrants." *Sociological Methods and Research* 25: 391–423.

Massey, Douglas, and Nancy Denton. 1993. *American Apartheid: Segregation and the Making of the Underclass* Cambridge, MA: Harvard University Press.

Ratner, Mitchell, ed. 1993. *Crack Pipe as Pimp: An Ethnographic Investigation of Sex-for-Crack Exchanges*. Lexington, MA: Lexington Books.

Reagan, Ronald, and Nancy Reagan. [September 14, 1986] 1989. "Address to the Nation on the Campaign Against Drug Abuse." Pp. 1178–1182 in *Public Papers of the Presidents of the United States 1986*. Washington, DC: United States Government Printing Office.

Reinarman, Craig, and Harry Levine. 1997. "Crack in Context: America's Latest Demon Drug." Pp. 1–17 in *Crack in America*, edited by Craig Reinarman and Harry Levine. Berkeley and Los Angeles: University of California Press.

Wilson, William Julius. 1987. *The Truly Disadvantaged: The Inner City, the Underclass, and Public Policy*. Chicago: University of Chicago Press.

45

Pimp-Controlled Prostitution

CELIA WILLIAMSON AND TERRY CLUSE-TOLAR

Although underrepresented in the sociological literature on prostitution, the relationship between pimps and prostitutes is legendary, yet often misunderstood, in popular culture. Based on qualitative interviews, Williamson and Cluse-Tolar's study offers an insightful analysis of pimp-controlled prostitutes' careers. Pimps often "turn prostitutes out," introducing them to the sex work industry. But while these women are seduced into accepting a pimp as their manager by the impression he gives that he is romantically interested in them, this behavior often turns out to be part of his "game," where he does whatever it takes to lure them into an obligatory relationship that supports him in flashy style. The authors show how the power relationship between these prostitutes and their pimps shifts dramatically in a predictable fashion over the course of their deviant careers. At the beginning the women have the greatest cachet, as the men must woo them into their service by "gaming" them with their charm and enticements of money and control over the customers. For a brief "honeymoon" period the women have the freedom to move around from one pimp to another, but even here the men dominate these exchanges ("bros before ho's"). The further they get into the relationship, the less the pimps need to woo them or treat them well, and they eventually become subjected to severe forms of emotional and physical manipulation that is often violent in nature. Their careers in pimp-controlled prostitution peak early and go through a gradual but early decline, landing them worse off than before they entered the relationship. Only once they have hit bottom do they get the courage to exit, often fleeing with nothing more than their lives.

A pimp is one who controls the actions and lives off the proceeds of one or more women who work the streets. Pimps call themselves "players" and call their profession "the game." The context in which this subculture exists is

From Celia Williamson and Terry Cluse-Tolar, "Pimp-Controlled Prostitution," *Violence Against Women*, Vol. 8, No. 9, Copyright © 2002. Reprinted by permission of Sage Publications, Inc.

called "the life" (Milner & Milner, 1972). Social scientists of the 1960s and 1970s devoted a significant amount of research energies toward exposing and understanding pimp-controlled prostitution within street-level prostitution (Goines, 1972; Heard, 1968; Milner & Milner, 1972; Slim, 1967, 1969). Street-level prostitution entails sexual acts for money or for barter that occur on and off the streets and include sexual activities in cars and motels, as dancers in gentlemen's clubs, massage parlor work, truck stops, and crack house work (Williamson, 2000). It represents that segment of the prostitution industry where there is the most violence.

... This study aims to examine pimp-related violence toward women involved in street-level prostitution within the context of pimp-controlled prostitution. To understand contemporary pimp-controlled prostitution and, more specifically, pimp-related violence, it is necessary to examine the type of relationships between pimps and prostitutes, the roles that each play in the business, and the social rules that accompany the lifestyle.

METHOD

... Information regarding the traditional pimp-prostitute phenomenon was obtained from a larger study that included both independent and pimp-controlled women. Criteria for inclusion in the study were women 18 years of age and older who were no longer involved in prostitution activities. Participants were selected through a process of purposive, or snowball, sampling by word of mouth. In total, 21 former street prostitutes from the Midwest were interviewed. Respondents ranged in age from 18 to 35. Of the total sample, 13 were Appalachian White women, 7 were African American women, and 1 was a Hispanic woman. The time spend in prostitution ranged from 3 months to 13 years. From this total sample, 6 of the women had pimps and 13 women worked independently. The small number of women found by the researcher who worked for pimps and were willing to be interviewed may underscore the limited access researchers have to this population and hence the importance of research in this area.

Of the six women who are the focus for this report, five were Appalachian White and one was Hispanic. They ranged in age from 18 to 28. For this subgroup, the time spent in prostitution ranged from four to eight years.

The researcher spent six months on the streets, three days per week, learning the culture, language, and geographic layout of the streets. The researcher learned where the dope houses were, who the pimps were, and how to identify a customer from a typical passerby.

Subsequently, in-depth, face-to-face interviews were conducted with six participants who were involved in pimp-controlled prostitution, and one interview was conducted with a pimp. Each interview lasted approximately two hours. Data were analyzed line by line. Codes were developed from the raw data. Codes were collapsed into larger themes. By connecting relevant themes, the researcher was able to develop the theoretical propositions that supported the subsequent theory of the lived experience of pimp-controlled prostitution.

In addition to these interviews, added interviews were conducted with some of the participants for the purpose of member checking, a process of clarification for qualitative methods, and to gather any additional missing data. Interviews were taped and transcribed verbatim. In addition to the member-checking techniques, the researcher engaged a group of social work experts in the area of street prostitution to critique the methodology and to provide guidance toward accurate interpretation of the data. This gathering is known as peer debriefing for the purpose of challenging the researcher's interpretations to increase the accuracy of the study findings. Both member checks and peer debriefing were used to enhance the credibility of the study.

FINDINGS

Pimping: Rules of the Game

Pimps involved in prostitution activities refer to this sector of the underground economy as "the game." Players, pimps, and macks are those at the top of the pimping game. To these men in power, it is a game in which they control and manipulate the actions of others subordinate to them. Monica, a prostituted woman for three years, explains,

> It's all about the game. Nothing in the game changes, but the name. It's all about getting that money. Some women have pimps that they give the money to, some are just out there on their own. (Monica)

A player or pimp has a particular manner or style of playing the game. The pimping game requires strict adherence to the rules. The idea of a game parallels the formal economy in that one can be said to be in a game; for example, he is in the real estate game. Pimps are also said to "have" game. To have game is to possess a certain amount of charisma and smooth-talking, persuasive conversation toward women.

There are several rules that one must be willing to follow to be a successful and professional pimp. Massi, a "bottom bitch"[1] to a pimp who boasts having six women in his stable,[2] outlines the rules for pimping. The most paramount rule in the pimping game is "the pimp must get paid" (Massi). This means there cannot be any "shame in your game" (Massi); one must require and, if necessary, demand the money without shame. Second, any successful pimp will remember that the game is "sold and not told." This means that pimps are expected to sell it to a prospective prostitute that he wants to occupy his stable without revealing his entire game plan. To do this, he has to develop his game or "his rap." These consist of a series of persuasive conversations similar to poetic and rhythmic scats that are philosophical in nature and ideological about life and making money. For Sonya, it was the combination of his rap and her need to feel loved by someone.

> For me, it was wanting to be loved and liked the words that was said. And you know, the nice things you got. I have two beautiful children

that I wanted to take care of, and I guess that's the kind of hold
they have on you. (Sonya)

... The third and final ingredient for successful pimping is that a pimp must
have a woman or women that want to see him on top. He is looking for dedication. He is looking for someone who wants to see her man in fine clothes and
driving fine cars. His success or lack of success is a reflection on her. If her man is
not looking his best, then she is not a very successful ho, and this will make for an
embarrassing impression. As a prostituted woman, she must work very hard to
earn his respect and his love and to keep him achieving the best in material possessions. He in return invites her into his underground social network with the
sense of belonging it brings and the promise of material possessions it provides.

You just, you just take control of the tricks. You know what you gotta
do to make your man happy.... Some prostitutes are out there for a
man, for a pimp. We're out there bustin' our ass to get our money for
a man. (Sonya)

I worked for months to get my man into a new Cadillac. (Sandra)

The most well-respected pimps are called "macks." They are at the top of
their game and employ many hardworking and successful prostitutes. Dominating the pimp scene are "players" who have an average stable of women, are well
respected, and make a good living. Lowest in the hierarchy of pimping are tennis
shoe pimps. These pimps may have one or possibly two prostitutes on the street.
They are seen as least successful in the game, and unlike more successful pimps,
they may do drugs and allow their women to do drugs. In this study, six of the
women previously had pimps ranging from tennis shoe pimps to players.

Turning a Woman Out

A pimp's chance at gaining a woman's attention is by looking good, smelling
good, flashing his possessions, and presenting himself as someone who can
counter boredom with both adventure and excitement. This is rarely enough,
however, to get a woman to prostitute herself. A pimp must be skilled at assessing
a woman's needs and vulnerabilities. Understanding how to exploit those vulnerabilities and fulfill those unmet needs will enable him to prostitute her. Reese, a
player in Toledo, Ohio, explains how he appeals to what women need:

I tell her "Now, you need to leave them drugs alone" and I get her
cleaned up. She may come here, on drugs or not on drugs, with nothing.
I mean nothing. Dirty, strung out. Some of them don't even have a social
security card or state ID, nothing. I ask 'em if they want something
better, you know, you can make some money. I'll set you up right.
Let you have a few things in your life. You wanna have nice clothes,
some good jewelry, be able to have your own place, maybe a little car to
drive around in? (Reese)

A pimp offers hope for the future, and women see this as an opportunity to
be financially successful. During the time a prostitute is entering the profession of

street-level prostitution, the pimp is said to be "turning her out" or has "turned her out" on the streets to make a profit.

> I knew this guy, and he brought me here and turned me out on the streets. He was a pimp.... The first day, I was scared, but I got the money. And once I seen the money, I mean, my first day I made $600 in a three-hour period. (Sonya)

Women involved with a pimp in this study were typically not engaged in drug abuse. Pimps realize that crack is the competition and frown upon any drug abuse in their stable. However, two women involved with tennis shoe pimps indulged in drug use along with their pimp.

Pimp-controlled women in the study were told they were beautiful and that men wanted them—that they desired them so much, they would pay hard-earned money for them. In the words of Massi, the message is conveyed early in the relationship that women are literally "sitting on a gold mine. If they could work the game good enough, the game would work for them."

Although pimps never guaranteed emotional or financial security, the potential for success inspired women to test the waters in this new life. There was a sense of belonging that women longed for, a sense of exciting hope for the future, an adventure that would take them from their meager existence into a life with a man who told them they had special skills, intelligence, and beauty. In return for his attention, protection, and love, she would be required to work to bring their dream into reality. Reese speaks in terms of "goals" as he works his women:

> I have them set little goals for themselves. Say they want to buy something they want; well, we would set a little goal. Say you make this certain amount of money: Keep working and I'll set aside a little at a time and you'll have what you want. Say if she wants a little piece a car, she can work and I'll make sure that she gets that car. So I try to have them set goals for themselves, something they're working for.

...Over time, as women learn the game and have become proficient in playing, they are known as thoroughbreds. Thoroughbreds are professionals in the prostitution industry and are responsible for maintaining the market rates in the profession. A thoroughbred is able to handle customers, command money, and conduct business effectively and efficiently to maximize profits.

> When you're turned out, you're just out there. You don't know what you're doing. You're just being turned out for a new job. You're being trained for it. And then once you get down the steps, you know, you become a thoroughbred. You don't let the guy take control of you, you take control of it. (Sonya)

Free Enterprise and Choosing Up

Pimps understand the meaning of business over personal ventures, that is, marketing a product and investing in your product first so your product can return profits. Thus, there is a honeymoon period or courting time between pimps and prostitutes. This is the time in which the pimp "runs his game." This may last one day or several months.

> He progressively led up to the fact that that's what he wanted. You know, he didn't come out that night when I met him and tell me, "This is what I am. This is what you need to do." . . . I think they really feel like they have to gain your trust before they can dump something like that on you. We spent a lot of time together. I mean . . . we would go out to eat, go to the movies, and we did, you know, normal couple things. But . . . in my head I'm just thinking it's just normal couple things, but he's thinking that he's winning . . . that he's gonna win and I'm gonna end up doing what he wanted. And he was right. (Tracey)

Pimps understand the meaning of capitalism in that it is a pimp's prerogative to entice any woman away from another pimp. It is viewed as a component of free enterprise. Therefore, other pimps are free to attempt to seduce a woman away from her current pimp and into his stable for his financial gain. He may do this without retaliation from the current pimp, as the street rule is "bros before hoes" (Massi). The woman being approached is instructed not to respond to the seducing pimp's advances. She is never to make eye contact with another pimp. If she does, then she is "out of pockets," a term referring to a woman who puts a pimp's money at risk, and she is subject to "being broke," meaning physically reprimanded. On the other hand, the seducing pimp may also choose to "break her" and take all her money. These rules vary in situations where a woman is prohibited from making eye contact with a pimp to situations in which she is not allowed to make eye contact with any African American male.

> I mean, most of the time, if they're a true pimp, they're not gonna play like that. You know, they'll harass you and you mainly just turn away and look in the other direction or whatever and try not to come in contact with them, because if you do, then they do what you call "break you," they take your money. . . . He's allowed to harass you as much as he wants. But if I don't talk back to him, then I'm cool. But if I'm "out of pockets" that means you're doing something that you ain't suppose to be doing. You know, some pimps will beat you or you go through a lot of stuff. . . . They're in control. You do what they say. (Sonya)

In the event that a woman is dissatisfied with her current pimp, the appropriate way to switch pimps is to make a definite decision and "choose up."

> You choose up. And if you're with a pimp and you want to go with another pimp, you have to put the money in the other pimp's hand and let your man know, you know, you're leaving and going with somebody else. . . . I've been with three. (Elsie)

Reese explains the transaction between pimps when a woman chooses up:

> If he comes to me like a man and tells me "That's my ho now," and she done gave him the money, then that's cool. Leave your ho clothes

here and go. You can take your regular clothes, the clothes I bought you to go see your family now and then, but you leave the ho clothes. But if she leaves here and is gone and then don't come back with my money and she been out there making money and giving it to him, "Nigga, that's *my* money you got."

Pimp and Prostitute Relationships in the Game

Each woman in the study had a pimp who set the rules, controlled her actions, and took her earnings. Most reported they were infatuated with their pimp, but not always. Women involved with a tennis shoe pimp, a man who had only one or two women, were more likely to consider themselves in love and defined the involvement with their pimp as a relationship. The more corporate the pimp, for instance, a player possessing three or more women, the less likely it was for women to describe their feelings as love or to define their interaction as a relationship. Women's feelings were instead described as ones of infatuation, admiration, and loyalty. The more women involved with a pimp, the less probable it became for each woman to achieve a status that allowed for the comforts of his affections, time, and attention. It was more likely that each became a part of his pimp family or stable that was made up of many women. This type of arrangement between women is known as a "wife-in-law" situation, in which each prostituted woman is a member of the family that works for the benefit of the same pimp. They are known as wife-in-laws to each other. However, some women did not tolerate such arrangements and moved on, whereas others welcomed the prestige of being with a successful pimp and willingly took on the challenge and responsibilities as a prostituted woman under his direction.

> It's just like when a pimp goes out and gets another girl and she's in the family. She's whoring for him like you are. Like a wife-in-law is what they call it. Sometimes you just getting tired of it and you know cheating on me and you know the wife-in-law stuff where you know the wife-in-law is another girl that is working for him, and I just couldn't handle all that. (Tonya)

Wife-in-laws may be responsible for ongoing training of recent inductees. However, the availability of wife-in-law training depends on how large the stable and how corporate the pimp. A bottom bitch, or number one lady, may also be required to work but may only use her mouth or hands when working and to save intercourse for her pimp. She may live with her pimp and may be required to train the new women joining his stable. Women may even drop off money to her after work in the event that their pimp is otherwise occupied.

> I know about the game because I was [his] bottom bitch. I knew everything about hoeing, tricking, or whatever. I was with [him] for eight years. He had women out here working their asses off. Wouldn't even ask him for money or nothing, not even $5, thinking that's making him respect them more. (Massi)

The true talents of a pimp, however, are in his ability to keep his women happy, command money, and portray a deep mysterious and somewhat mean demeanor about him, one that conveys the message that he is not to be crossed. He is then said to be "cold-blooded" or "icy," able to turn off any warm feelings and loving affection in exchange for certain emotional cruelty and physical harm. Two famous and successful pimps, Iceberg Slim and Ice Tea, were said to be so cold blooded they called themselves "Ice" to let everyone know their capacity for heartlessness.

> He would just snap. Like his whole expression would change. One day, he came to my motel room to beat my ass. And made it clear that he came over to beat on me. He said he had some extra time on his hands, that he didn't have anything to do, so he wanted me to know that he knew I was thinking about doing something stupid. And I was too. I was thinking about leaving him again. The last time I left him, I ended up in Cleveland. . . . He beat me until I blacked out. . . . But he was like that. He could be so much fun one time, silly and playing around, and the next minute, he could be something else, somebody you don't want to fuck with. (Massi)

A pimp's approach is never to cow down to his woman at any time. He cannot let love cloud his judgments concerning business. If he lets these weaknesses show, he will be left vulnerable and runs the risk of being less successful. Although pimps appear to be in control, in a sense every pimp becomes a whore to his prostitutes. The pimp rule is "purse first, ass last" (Massi). He may treat his hoes in loving ways in return for the amount of money he requested she bring him. She must pay for his love with her sheer tenacity to work and bring him the money. She must in turn request little emotionally and financially. Because of his generosity, he gives her what he thinks she needs.

Pimp-Related Violence: Physical Control of Women

The extent to which women felt threatened by a pimp was, in part, a function of her evaluation of the likelihood that he will become violent. This threat had been realized by all of the women in the study. Pimp-related violence was sometimes unpredictable and took on many forms. However, the most revealing form of pimp-related violence was immediate attack following a violation of the rules. One such violation is leaving the "ho stroll" or designated work area early without making one's daily quota.

> Different pimps have different rules. I mean some of 'em set quotas with the girls, you have to make a certain amount of money before you can go home. . . . When I first started, I was bringing in $1,000 a day. (Tonya)

After it was found out that Debbie was holding back some money from her pimp, without hesitation she was quickly and brutally assaulted.

> He ended up getting mad at me one day and punched me in my chest and cracked my rib. That was cracked, and all I could remember is

that I couldn't breathe. I mean, I passed out. I was knocked out all day. I was unconscious. (Debbie)

Some pimps use violence as a means of discouraging freelance work and coercing money from women who work within pimp territory but not for the benefit of any pimp.

> There was this guy that kept beatin' me in my head, telling me I was gonna pay him, and every time he'd see me, if I didn't give him some money, he'd punch me in my forehead. (Cara)

In the instance that a woman is found to be "out of pockets," she is subject to being broke. Carol told of an incident when she took an unauthorized leave without the permission of her pimp.

> When he caught me, he was like "I got you now," and he jumped out. We was in the projects. We were high as fuck off crack, me and Tony was. We were like "Oh, fuck." He was like "I got you now." He had a baseball bat, and Tony ran and left me. So, yes, I got the baseball bat, he beat me in my legs and told me "If you fall, bitch, I'm a hit you in your head and kill you." So I didn't fall, I just stood there and screamed and took it. The police came . . . and they asked if I was going to press charges, and I said, "No." My face was swelling, I looked liked a cabbage patch, I was horrible. (Carol)

Pimp-Related Violence: Emotional Control of Women

A pimp's success is dependent on arousing love and fear in his women. By giving his attention to more than one woman at the same time, he heightens both the love each woman desires all to herself and the fear that she may lose any part of it. However, the negative consequences of such arrangements may be jealousy and rivalry for his affections.

> He had got back with the girl that he had kids by, which she was already a seasoned hooker. She hated me from the jump start. Since they were in that life, he made her deal with me. When she found out that I wanted him for me, she wanted to fight me every time she would see me and we did. . . . She hit me in the head with a beer glass, and I had stitches in my head. (Chris)

Relationships require a level of trust and a degree of vulnerability, and pimp–prostitute relationships are no different. Trust determines how vulnerable the person is willing to be. Without some degree of trust, interactions are limited to explicit contracts (Holmes, 1991), which is what prostitutes have with customers. "Trust involves coming to terms with the negative aspects of a partner, accepting or perhaps tolerating issues by buffering them in the broader context of the lifestyle" (Holmes, 1991, p. 79). Women take abuses from pimps in stride. They learn to cope with this relationship by not focusing on the abusive aspects for what they are but by instead encapsulating those aspects of their pimp that serve their needs for security and protection. Therefore, a pimp–prostitute relationship often lacks cognitive and behavioral consistency. What is believed and desired on the part of the prostitute and what actually happens in the relationship do not correspond and often require repeated leaps of faith on the part of the prostitute.

Leaving Pimp-Controlled Prostitution

Many factors prevent women from pursuing legal assistance. Often, women are fearful, intimidated by what may happen as a result of reporting, and may love their pimp despite his abuses. It is likely that she may buy into the rules of the pimping game and blame herself for violating them. Women who have previously experienced the reluctance of law enforcement to take their claims about customer-related violence seriously are reluctant to take such risks where pimps are concerned.

> I had this guy pull a gun on me, and he made me do things that I didn't want to do.... I ran to a gas station and called the police. The guy who worked at the gas station gave me his jacket to wear. All I had on was a shirt. I called the police and told them. They came. I described the guy to the police and showed them the spot. I told them what kind of car he had, what we was wearing, what he said, what he did, everything. They never even wrote anything down. They ran me for warrants, and when I didn't have any, they left. (Jerri)

Women who are fed up may choose to leave prostitution. The primary means for leaving pimp-controlled prostitution was escape.

> I had gotten like four calls that day. And I hadn't seen him, so I had all the money on me and I just took it. I mean, none of my clothes, none of my nothing. I just took a cab to the bus station, and I went in to Amtrak police and told them what was going on. He had his own driver, and he knew that when I didn't get in the driver's car, that something was up. I took a regular cab, and I went to Amtrak police and I knew that he was coming, so I told 'em what was going on.... And, um, then the Amtrak police ... paid the cab driver to take me to the airport, and I caught a plane home. (Tracey)

DISCUSSION

Using the definition of pimping as controlling and living off the proceeds of one or more women, the findings suggest that pimp-controlled prostitution is still an integral part of street-level prostitution for some women and girls. Just how many is difficult to determine, because pimp-controlled women and girls would be those most unlikely to be able to respond to requests for interviews. Pimp-controlled women in the study were reportedly subject to following the rules of the game. The old adage that "nothing in the game changes but the name" may be truer than not when viewing the dynamics of pimp-controlled prostitution. It is clear that many of the themes identified in this study have appeared in varied form in earlier studies of the 1960s and 1970s (Goines, 1972; Heard, 1968; Milner & Milner, 1972; Slim, 1967, 1969). The important point is that although pimps and prostitutes may differ with the extent to which they apply and adhere to the rules, even allowing for the wide range of situational differences, this code of conduct, termed *pimpology,* is a common practice in this underground society and still exists today (Hughes, Hughes, & Messick, 1999; Milner & Milner, 1972; Owens & Shepard, 1998; Williamson, 2000).

On an interpersonal level, the power and control pimps maintain over women in their stable is akin to that used in abusive relationships. Just as pimps resemble batterers in intimate relationships (Giobbe, 1993), women working in pimp-controlled prostitution seem to be similar to those who are survivors of domestic violence. They often express feelings of love and admiration for the pimp, have their freedom and finances controlled, and may feel they somehow deserve the violence they are dealt. However, there are differences in terms of the cycle of violence. Domestic violence survivors will often express that they knew when the violence was about to occur as evidenced by the building up of tension in their mate before an explosive episode. Beatings and other forms of violence occurring among pimp-controlled women may not follow a familiar pattern and may instead occur by surprise.

NOTES

1. In a more corporate pimp family, the term *bottom bitch* refers to a woman who is the closest in rank to her pimp.

2. A *stable* is what a pimp calls a group of women that prostitute for him.

REFERENCES

Giobbe, E. (1993). A Comparison of Pimps and Batterers. *Michigan Journal of Gender and Law,* 1(1), 33–57.

Goines, D. (1972). *Whoreson: The Story of a Ghetto Pimp.* Los Angeles: Holloway House.

Heard, N. C. (1968). *Howard Street.* New York: Signet.

Holmes, J. G. (1991). Trust and the Appraisal Process in Close Relationships. In W. H. Jones & D. Perlman (Eds.), *Advances in Personal Relationships* (pp. 57–104). London: Jessica Kingsley.

Hughes, A., Hughes, A., & Messick, K. (1999). *American Pimp* [Film documentary]. United States: Metro Goldwyn Mayer Pictures.

Milner, C. A., & Milner, R. B. (1972). *Black Players.* Boston, MA: Little, Brown.

National Center for Missing and Exploited Children. (1992). *Female Juvenile Prostitution: Problem and Response.* Arlington, VA: Author.

Owens, B. (Producer/Director), & Shepard, B. (Coproducer). (1998). *Pimp Up Ho Down* [Film documentary]. United States: Home Box Office.

Slim, I. (1967). *Trick Baby.* Los Angeles: Holloway House.

Slim, I. (1969). *Pimp: The Story of My Life.* Los Angeles: Holloway House.

Williamson, C. (2000). Entrance, Maintenance, and Exit: The Socioeconomic Influences and Cumulative Burdens of Female Street Prostitution. *Dissertation Abstracts International, 61*(02). (UMI No. 9962789)

46

Shifts and Oscillations in Upper-Level Drug Traffickers' Careers

PATRICIA A. ADLER AND PETER ADLER

Adler and Adler discuss the process by which people burn out of deviance in this selection on exiting drug trafficking. After spending several years in the upper echelons of the drug trade, many marijuana dealers and smugglers, who were initially attracted to drug trafficking by the same lure of the excitement, high life, and spontaneity that drew Wright and Deckers' burglars to their deviance, eventually find that the drawbacks of the lifestyle exceed the rewards. Their initial challenges and thrills turn to paranoia, people whom they know get busted all around them, and their risk of arrest grows. Years of excessive drug use take its toll on them physically, and they come to reevaluate the straight life they formerly rejected as boring. Yet they cannot easily quit dealing; they have developed a high-spending lifestyle that they are loath to abandon. One thing that they do is to shift around in the drug world, making changes in their involvement. When this doesn't bring them satisfaction, or when the factors pushing them out continue to grow, they try to retire from trafficking. Commonly, however, they quickly spend all their money and are drawn back into the business. Their patterns of exiting, thus, often resemble a series of oscillations, or quittings and restartings, as they move out of deviance with great difficulty. It is interesting to note that this mode of oscillating in and out of deviance as a means of finally making an exit is often found in other forms of deviant careers such as quitting cigarette smoking and leaving an abusive relationship.

The upper echelons of the marijuana and cocaine trade constitute a world which has never before been researched and analyzed by sociologists. Importing and distributing tons of marijuana and kilos of cocaine at a time, successful operators

From "Shifts and Oscillations in Deviant Careers: The Case of Upper-Level Drug Dealers and Smugglers," Patricia A. Adler and Peter Adler, *Social Problems,* Vol. 31, No. 2, 1983. © Society for the Study of Social Problems. Reprinted by permission of University of California Press Journals.

can earn upwards of a half million dollars per year. Their traffic in these so-called "soft"[1] drugs constitutes a potentially lucrative occupation, yet few participants manage to accumulate any substantial sums of money, and most people envision their involvement in drug trafficking as only temporary. In this study we focus on the career paths followed by members of one upper-level drug dealing and smuggling community. We discuss the various modes of entry into trafficking at these upper levels, contrasting these with entry into middle- and low-level trafficking. We then describe the pattern of shifts and oscillations these dealers and smugglers experience. Once they reach the top rungs of their occupation, they begin periodically quitting and reentering the field, often changing their degree and type of involvement upon their return. Their careers, therefore, offer insights into the problems involved in leaving deviance.

Previous research on soft drug trafficking has only addressed the low and middle levels of this occupation, portraying people who purchase no more than 100 kilos of marijuana or single ounces of cocaine at a time (Anonymous, 1969; Atkyns and Hanneman, 1974; Blum, 1972; Carey, 1968; Goode, 1970; Langer, 1977; Lieb and Olson, 1976; Mouledoux, 1972; Waldorf et al., 1977). Of these, only Lieb and Olson (1976) have examined dealing and/or smuggling as an occupation, investigating participants' career developments. But their work, like that of several of the others, focuses on a population of student dealers who may have been too young to strive for and attain the upper levels of drug trafficking. Our study fills this gap at the top by describing and analyzing an elite community of upper-level dealers and smugglers and their careers.

We begin by describing where our research took place, the people and activities we studied, and the methods we used. Second, we outline the process of becoming a drug trafficker, from initial recruitment through learning the trade. Third, we look at the different types of upward mobility displayed by dealers and smugglers. Fourth, we examine the career shifts and oscillations which veteran dealers and smugglers display, outlining the multiple, conflicting forces which lure them both into and out of drug trafficking. We conclude by suggesting a variety of paths which dealers and smugglers pursue out of drug trafficking and discuss the problems inherent in leaving this deviant world.

SETTING AND METHOD

We based our study in "Southwest County," one section of a large metropolitan area in southwestern California near the Mexican border. Southwest County consisted of a handful of beach towns dotting the Pacific Ocean, a location offering a strategic advantage for wholesale drug trafficking.

Southwest County smugglers obtained their marijuana in Mexico by the ton and their cocaine in Colombia, Bolivia, and Peru, purchasing between 10 and 40 kilos at a time. These drugs were imported into the United States along a variety of land, sea, and air routes by organized smuggling crews. Southwest County dealers then purchased these products and either "middled" them directly to another buyer for a small but immediate profit of approximately $2 to $5 per kilo of marijuana and $5,000 per kilo of cocaine, or engaged in "straight deal-ing." As opposed to middling, straight dealing usually entailed adulterating the cocaine with such "cuts" as manitol, procaine, or inositol, and then dividing the marijuana and cocaine into smaller quantities to sell them to the next-lower level of dealers. Although dealers frequently varied the amounts they bought and sold, a hierarchy of transacting levels could be roughly discerned. "Wholesale" marijuana dealers bought directly from the smugglers, purchasing anywhere from 300 to 1,000 "bricks" (averaging a kilo in weight) at a time and selling in lots of 100 to 300 bricks. "Multi-kilo" dealers, while not the smug-glers' first connections, also engaged in upper-level trafficking, buying between 100 to 300 bricks and selling them in 25- to 100-brick quantities. These were then purchased by middle-level dealers who filtered the marijuana through low-level and "ounce" dealers before it reached the ultimate consumer. Each time the marijuana changed hands its price increase was dependent on a number of factors: purchase cost; the distance it was transported (including such transpor-tation costs as packaging, transportation equipment, and payments to employees); the amount of risk assumed; the quality of the marijuana; and the prevailing pri-ces in each local drug market. Prices in the cocaine trade were much more pre-dictable. After purchasing kilos of cocaine in South America for $10,000 each, smugglers sold them to Southwest County "pound" dealers in quantities of one to 10 kilos for $60,000 per kilo. These pound dealers usually cut the cocaine and sold pounds ($30,000) and half-pounds ($15,000) to "ounce" dealers, who in turn cut it again and sold ounces for $2,000 each to middle-level cocaine dealers known as "cut-ounce" dealers. In this fashion the drug was middled, dealt, div-ided and cut—sometimes as many as five or six times—until it was finally pur-chased by consumers as grams or half-grams.

Unlike low-level operators, the upper-level dealers and smugglers we studied pursued drug trafficking as a full-time occupation. If they were involved in other businesses, these were usually maintained to provide them with a legitimate front for security purposes. The profits to be made at the upper levels depended on an individual's style of operation, reliability, security, and the amount of product he or she consumed. About half of the 65 smugglers and dealers we observed were suc-cessful, some earning up to three-quarters of a million dollars per year.[2] The other half continually struggled in the business, either breaking even or losing money.

Although dealers' and smugglers' business activities varied, they clustered together for business and social relations, forming a moderately well-integrated community whose members pursued a "fast" lifestyle, which emphasized inten-sive partying, casual sex, extensive travel, abundant drug consumption, and lavish spending on consumer goods. The exact size of Southwest County's upper-level dealing and smuggling community was impossible to estimate due to the secrecy of its members. At these levels, the drug world was quite homogeneous. Partic-ipants were predominantly white, came from middle-class backgrounds, and had little previous criminal involvement. While the dealers' and smugglers' social world contained both men and women, most of the serious business was con-ducted by the men, ranging in age from 25 to 40 years old.

We gained entry to Southwest County's upper-level drug community largely by accident. We had become friendly with a group of our neighbors who turned out be heavily involved in smuggling marijuana. Opportunistically (Riemer, 1977), we seized the chance to gather data on this unexplored activity. Using key informants who helped us gain the trust of other members of the community, we drew upon snowball sampling techniques (Biernacki and Waldorf, 1981) and a combination of overt and covert roles to widen our network of contacts. We supplemented intensive participant-observation, between 1974 and 1980,[3] with unstructured, taped interviews. Throughout, we employed extensive measures to cross-check the reliability of our data, whenever possible (Douglas, 1976). In all, we were able to closely observe 65 dealers and smugglers as well as numerous other drug world members, including dealers' "old ladies" (girlfriends or wives), friends, and family members. . . .

SHIFTS AND OSCILLATIONS

. . . Despite the gratifications which dealers and smugglers originally derived from the easy money, material comfort, freedom, prestige, and power associated with their careers, 90 percent of those we observed decided, at some point, to quit the business. This stemmed, in part, from their initial perceptions of the career as temporary ("Hell, nobody wants to be a drug dealer all their life"). Adding to these early intentions was a process of rapid aging in the career: dealers and smugglers became increasingly aware of the restrictions and sacrifices their occupations required and tired of living the fugitive life. They thought about, talked about, and in many cases took steps toward getting out of the drug business. But as with entering, disengaging from drug trafficking was rarely an abrupt act (Lieb and Olson, 1976: 364). Instead, it more often resembled a series of transitions, or oscillations,[4] out of and back into the business. For once out of the drug world, dealers and smugglers were rarely successful in making it in the legitimate world because they failed to cut down on their extravagant lifestyle and drug consumption. Many abandoned their efforts to reform and returned to deviance, sometimes picking up where they left off and other times shifting to a new mode of operating. For example, some shifted from dealing cocaine to dealing marijuana, some dropped to a lower level of dealing, and others shifted their role within the same group of traffickers. This series of phase-outs and reentries, combined with career shifts, endured for years, dominating the pattern of their remaining involvement with the business. But it also represented the method by which many eventually broke away from drug trafficking, for each phase-out had the potential to be an individual's final departure.

Aging in the Career

Once recruited and established in the drug world, dealers and smugglers entered into a middle phase of aging in the career. This phase was characterized by a progressive loss of enchantment with their occupation. While novice dealers and smugglers found that participation in the drug world brought them thrills and status, the novelty gradually faded. Initial feelings of exhilaration and awe began to dull as individuals became increasingly jaded. This was the result of both an extended exposure to the mundane, everyday business aspects of drug trafficking and to an exorbitant consumption of drugs (especially cocaine). One smuggler described how he eventually came to feel:

> It was fun, those three or four years. I never worried about money or anything. But after awhile it got real boring. There was no feeling or emotion or anything about it. I wasn't even hardly relating to my old lady anymore. Everything was just one big rush.

This frenzy of overstimulation and resulting exhaustion hastened the process of "burnout" which nearly all individuals experienced. As dealers and smugglers aged in the career they became more sensitized to the extreme risks they faced. Cases of friends and associates who were arrested, imprisoned, or killed began to mount. Many individuals became convinced that continued drug trafficking would inevitably lead to arrest ("It's only a matter of time before you get caught"). While dealers and smugglers generally repressed their awareness of danger, treating it as a taken-for-granted part of their daily existence, periodic crises shattered their casual attitudes, evoking strong feelings of fear. They temporarily intensified security precautions and retreated into near-isolation until they felt the "heat" was off.

As a result of these accumulating "scares," dealers and smugglers increasingly integrated feelings of "paranoia"[5] into their everyday lives. One dealer talked about his feelings of paranoia:

> You're always on the line. You don't lead a normal life. You're always looking over your shoulder, wondering who's at the door, having to hide everything. You learn to look behind you so well you could probably bend over and look up your ass. That's paranoia. It's a really scary, hard feeling. That's what makes you get out.

Drug world members also grew progressively weary of their exclusion from the legitimate world and the deceptions they had to manage to sustain that separation. Initially, this separation was surrounded by an alluring mystique. But as they aged in the career, this mystique became replaced by the reality of everyday boundary maintenance and the feeling of being an "expatriated citizen within one's own country." One smuggler who was contemplating quitting described the effects of this separation:

> I'm so sick of looking over my shoulder, having to sit in my house and worry about one of my non–drug world friends stopping in when I'm doing business. Do you know how awful that is? It's like leading a double life. It's ridiculous. That's what makes it not worth it. It'll be a lot less money [to quit], but a lot less pressure.

Thus, while the drug world was somewhat restricted, it was not an encapsulated community, and dealers' and smugglers' continuous involvement with the straight world made the temptation to adhere to normative standards and "go straight" omnipresent. With the occupation's novelty worn off and the "fast life" taken for granted, most dealers and smugglers felt that the occupation no longer resembled their early impressions of it. Once they reached the upper levels of the occupation, their experience began to change. Eventually, the rewards of trafficking no longer seemed to justify the strain and risk involved. It was at this point that the straight world's formerly dull ambiance became transformed (at least in theory) into a potential haven.

Phasing Out

Three factors inhibited dealers and smugglers from leaving the drug world. Primary among these factors were the hedonistic and materialistic satisfactions the drug world provided. Once accustomed to earning vast quantities of money quickly and easily, individuals found it exceedingly difficult to return to the income scale of the straight world. They also were reluctant to abandon the pleasure of the "fast life" and its accompanying drugs, casual sex, and power. Second, dealers and smugglers identified with, and developed a commitment to, the occupation of drug trafficking (Adler and Adler, 1982). Their self-images were tied to that role and could not be easily disengaged. The years invested in their careers (learning the trade, forming connections, building reputations) strengthened their involvement with both the occupation and the drug community. And since their relationships were social as well as business, friendship ties bound individuals to dealing. As one dealer in the midst of struggling to phase-out explained:

> The biggest threat to me is to get caught up sitting around the house with friends that are into dealing. I'm trying to stay away from them, change my habits.

Third, dealers and smugglers hesitated to voluntarily quit the field because of the difficulty involved in finding another way to earn a living. Their years spent in illicit activity made it unlikely for any legitimate organizations to hire them. This narrowed their occupational choices considerably, leaving self-employment as one of the few remaining avenues open.

Dealers and smugglers who tried to leave the drug world generally fell into one of four patterns.[6] The first and most frequent pattern was to postpone quitting until after they could execute one last "big deal." While the intention was sincere, individuals who chose this route rarely succeeded; the "big deal" too often remained elusive. One marijuana smuggler offered a variation of this theme:

> My plan is to make a quarter of a million dollars in four months during the prime smuggling season and get the hell out of the business.

A second pattern we observed was individuals who planned to change immediately, but never did. They announced they were quitting, yet their outward actions never varied. One dealer described his involvement with this syndrome:

> When I wake up I'll say, "Hey, I'm going to quit this cycle and just run my other business." But when you're dealing you constantly have people dropping by ounces and asking, "Can you move this?" What's your first response? Always, "Sure, for a toot."

In the third pattern of phasing-out, individuals actually suspended their dealing and smuggling activities, but did not replace them with an alternative source
of income. Such withdrawals were usually spontaneous and prompted by exhaustion, the influence of a person from outside the drug world, or problems with the
police or other associates. These kinds of phase-outs usually lasted only until the
individual's money ran out, as one dealer explained:

> I got into legal trouble with the FBI a while back and I was forced
> to quit dealing. Everybody just cut me off completely, and I saw the
> danger in continuing, myself. But my high-class tastes never dwindled.
> Before I knew it I was in hock over $30,000. Even though I was hot,
> I was forced to get back into dealing to relieve some of my debts.

In the fourth pattern of phasing out, dealers and smugglers tried to move into
another line of work. Alternative occupations included: (1) those they had previously pursued; (2) front businesses maintained on the side while dealing or
smuggling; and (3) new occupations altogether. While some people accomplished this transition successfully, there were problems inherent in all three alternatives.

1. Most people who tried resuming their former occupations found that these
    had changed too much while they were away. In addition, they themselves
    had changed: they enjoyed the self-directed freedom and spontaneity asso-
    ciated with dealing and smuggling, and were unwilling to relinquish it.

2. Those who turned to their legitimate front business often found that these
    businesses were unable to support them. Designed to launder rather than
    earn money, most of these ventures were retail outlets with a heavy cash flow
    (restaurants, movie theaters, automobile dealerships, small stores) that had
    become accustomed to operating under a continuous subsidy from illegal
    funds. Once their drug funding was cut off they could not survive for long.

3. Many dealers and smugglers utilized the skills and connections they had
    developed in the drug business to create a new occupation. They ex-changed
    their illegal commodity for a legal one and went into import/export, man-
    ufacturing, wholesaling, or retailing other merchandise. For some, the
    decision to prepare a legitimate career for their future retirement from the
    drug world followed an unsuccessful attempt to phase-out into a "front"
    business. One husband-and-wife dealing team explained how these legiti-
    mate side businesses differed from front businesses:

    > We always had a little legitimate "scam" [scheme] going, like mail-order
    > shirts, wallets, jewelry, and the kids were always involved in that. We made
    > a little bit of money on them. Their main purpose was for a cover. But
    > [this business] was different; right from the start this was going to be a legal
    > thing to push us out of the drug business.

About 10 percent of the dealers and smugglers we observed began tapering off
their drug world involvement gradually, transferring their time and money

into a selected legitimate endeavor. They did not try to quit drug trafficking altogether until they felt confident that their legitimate business could support them. Like spontaneous phase-outs, many of these planned withdrawals into legitimate endeavors failed to generate enough money to keep individuals from being lured into the drug world.

In addition to voluntary phase-outs caused by burnout, about 40 percent of the Southwest County dealers and smugglers we observed experienced a "bustout" at some point in their careers.[7] Forced withdrawals from dealing or smuggling were usually sudden and motivated by external factors, either financial, legal, or reputational. Financial bustouts generally occurred when dealers or smugglers were either "burned" or "ripped-off" by others, leaving them in too much debt to rebuild their base of operation. Legal bustouts followed arrest and possibly incarceration: arrested individuals were so "hot" that few of their former associates would deal with them. Reputational bustouts occurred when individuals "burned" or "ripped-off" others (regardless of whether they intended to do so) and were banned from business by their former circle of associates. One smuggler gave his opinion on the pervasive nature of forced phase-outs:

> Some people are smart enough to get out of it because they realize, physically, they have to. Others realize, monetarily, that they want to get out of this world before this world gets them. Those are the lucky ones. Then there are the ones who have to get out because they're hot or someone else close to them is so hot that they'd better get out. But in the end when you get out of it, nobody gets out of it out of free choice; you do it because you have to.

Death, of course, was the ultimate bustout. Some pilots met this fate because of the dangerous routes they navigated (hugging mountains, treetops, other aircrafts) and the sometimes ill-maintained and overloaded planes they flew. However, despite much talk of violence, few Southwest County drug traffickers died at the hands of fellow dealers.

Reentry

Phasing out of the drug world was more often than not temporary. For many dealers and smugglers, it represented but another stage of their drug careers (although this may not have been their original intention), to be followed by a period of reinvolvement. Depending on the individual's perspective, reentry into the drug world could be viewed as either a comeback (from a forced withdrawal) or a relapse (from a voluntary withdrawal).

Most people forced out of drug trafficking were anxious to return. The decision to phase out was never theirs, and the desire to get back into dealing or smuggling was based on many of the same reasons which drew them into the field originally. Coming back from financial, legal, and reputational bustouts was possible but difficult and was not always successfully accomplished. They had to reestablish contacts, rebuild their organization and fronting arrangements, and raise the operating capital to resume dealing. More difficult was the problem of overcoming the circumstances surrounding their departure. Once smugglers and dealers resumed operating, they often found their former colleagues suspicious of them. One frustrated dealer described the effects of his prison experience:

> When I first got out of the joint [jail], none of my old friends would have anything to do with me. Finally, one guy who had been my partner told me it was because everyone was suspicious of my getting out early and thought I made a deal [with police to inform on his colleagues].

Dealers and smugglers who returned from bustouts were thus informally subjected to a trial period in which they had to reestablish their trustworthiness and reliability before they could once again move in the drug world with ease.

Reentry from voluntary withdrawal involved a more difficult decision-making process, but was easier to implement. The factors enticing individuals to reenter the drug world were not the same as those which motivated their original entry. As we noted above, experienced dealers and smugglers often privately weighed their reasons for wanting to quit and wanting to stay in. Once they left, their images of and hopes for the straight world failed to materialize. They could not make the shift to the norms, values, and lifestyle of the straight society and could not earn a living within it. Thus, dealers and smugglers decided to reenter the drug business for basic reasons: the material perquisites, the hedonistic gratifications, the social ties, and the fact that they had nowhere else to go.

Once this decision was made, the actual process of reentry was relatively easy. One dealer described how the door back into dealing remained open for those who left voluntarily:

> I still see my dealer friends, I can still buy grams from them when I want to. It's the respect they have for me because I stepped out of it without being busted or burning someone. I'm coming out with a good reputation, and even though the scene is a whirlwind—people moving up, moving down, in, out—if I didn't see anybody for a year I could call them up and get right back in that day.

People who relapsed thus had little problem obtaining fronts, reestablishing their reputations, or readjusting to the scene.

Career Shifts

Dealers and smugglers who reentered the drug world, whether from a voluntary or forced phase-out, did not always return to the same level of transacting or commodity which characterized their previous style of operation. Many individuals underwent a "career shift" (Luckenbill and Best, 1981) and became involved in some new segment of the drug world. These shifts were sometimes lateral, as when a member of a smuggling crew took on a new specialization, switching from piloting to operating a stash house, for example. One dealer described how he utilized friendship networks upon his reentry to shift from cocaine to marijuana trafficking:

> Before, when I was dealing cocaine, I was too caught up in using the drug and people around me were starting to go under from getting into "base" [another form of cocaine]. That's why I got out. But now I think I've got myself together and even though I'm dealing again I'm staying away from coke. I've switched over to dealing grass. It's a whole different circle of people. I got into it through a close friend I used to know before, but I never did business with him because he did grass and I did coke.

Vertical shifts moved operators to different levels. For example, one former smuggler returned and began dealing; another top-level marijuana dealer came back to find that the smugglers he knew had disappeared and he was forced to buy in smaller quantities from other dealers.

Another type of shift relocated drug traffickers in different styles of operation. One dealer described how, after being arrested, he tightened his security measures:

> I just had to cut back after I went through those changes. Hell, I'm not getting any younger and the idea of going to prison bothers me a lot more than it did 10 years ago. The risks are no longer worth it when I can have a comfortable income with less risk. So I only sell to four people now. I don't care if they buy a pound or a gram.

A former smuggler who sold his operation and lost all his money during phase-out returned as a consultant to the industry, selling his expertise to those with new money and fresh manpower:

> What I've been doing lately is setting up deals for people. I've got fool-proof plans for smuggling cocaine up here from Colombia; I tell them how to modify their airplanes to add on extra fuel tanks and to fit in more weed, coke, or whatever they bring up. Then I set them up with refueling points all up and down Central America, tell them how to bring it up here, what points to come in at, and what kind of receiving unit to use. Then they do it all and I get 10 percent of what they make.

Reentry did not always involve a shift to a new niche, however. Some dealers and smugglers returned to the same circle of associates, trafficking activity, and commodity they worked with prior to their departure. Thus, drug dealers' careers often peaked early and then displayed a variety of shifts, from lateral mobility, to decline, to holding fairly steady.

A final alternative involved neither completely leaving nor remaining within the deviant world. Many individuals straddled the deviant and respectable worlds forever by continuing to dabble in drug trafficking. As a result of their experiences in the drug world they developed a deviant self-identity and a deviant *modus operandi*. They might not have wanted to bear the social and legal burden of full-time deviant work but neither were they willing to assume the perceived confines and limitations of the straight world. They therefore moved into the entrepreneurial realm, where their daily activities involved some kind of hustling or "wheeling and dealing" in an assortment of legitimate, quasi-legitimate, and deviant ventures, and where they could be their own boss. This enabled them to retain certain elements of the deviant lifestyle, and to socialize on the fringes of the drug community. For these individuals, drug dealing shifted from a primary occupation to a sideline, though they never abandoned it altogether.

LEAVING DRUG TRAFFICKING

This career pattern of oscillation into and out of active drug trafficking makes it difficult to speak of leaving drug trafficking in the sense of final retirement. Clearly, some people succeeded in voluntarily retiring. Of these, a few managed to prepare a post-deviant career for themselves by transferring their drug money into a legitimate enterprise. A larger group was forced out of dealing and either didn't or couldn't return; the bustouts were sufficiently damaging that they never attempted reentry, or they abandoned efforts after a series of unsuccessful attempts. But there was no way of structurally determining in advance whether an exit from the business would be temporary or permanent. The vacillations in dealers' intentions were compounded by the complexity of operating successfully in the drug world. For many, then, no phase-out could ever be definitely assessed as permanent. As long as individuals had skills, knowledge, and connections to deal they retained the potential to reenter the occupation at any time. Leaving drug trafficking may thus be a relative phenomenon, characterized by a trailing-off process where spurts of involvement appear with decreasing frequency and intensity.

SUMMARY

Drug dealing and smuggling careers are temporary and fraught with multiple attempts at retirement. Veteran drug traffickers quit their occupation because of the ambivalent feelings they develop toward their deviant life. As they age in the career their experience changes, shifting from a work life that is exhilarating and free to one that becomes increasingly dangerous and confining. But just as their deviant careers are temporary, so too are their retirements. Potential recruits are lured into the drug business by materialism, hedonism, glamor, and excitement. Established dealers are lured away from the deviant life and back into the mainstream by the attractions of security and social ease. Retired dealers and smugglers are lured back in by their expertise, and by their ability to make money quickly and easily. People who have been exposed to the upper levels of drug trafficking therefore find it extremely difficult to quit their deviant occupation permanently. This stems, in part, from their difficulty in moving from the illegitimate to the legitimate business sector. Even more significant is the affinity they form for their deviant values and lifestyle. Thus few, if any, of our subjects were successful in leaving deviance entirely. What dealers and smugglers intend, at the time, to be a permanent withdrawal from drug trafficking can be seen in retrospect as a pervasive occupational pattern of mid-career shifts and oscillations. More research is needed into the complex process of how people get out of deviance and enter the world of legitimate work.

NOTES

1. The term "soft" drugs generally refers to marijuana, cocaine, and such psychedelics as LSD and mescaline (Carey, 1968). In this paper we do not address trafficking in psychedelics because, since they are manufactured in the United States, they are neither imported nor distributed by the group we studied.

2. This is an idealized figure representing the profit a dealer or smuggler could potentially earn and does not include deductions for such miscellaneous and hard-to-calculate costs: time or money spent in arranging deals (some of which never materialize); lost, stolen, or unrepaid money or drugs; and the personal drug consumption of a drug trafficker and his or her entourage. Of these, the single largest expense is the last one, accounting for the bulk of most Southwest County dealers' and smugglers' earnings.

3. We continued to conduct follow-up interviews with key informants through 1983.

4. While other studies of drug dealing have also noted that participants did not maintain an uninterrupted stream of career involvement (Blum, 1972; Carey, 1968; Lieb and Olson, 1976; Waldorf et al., 1977), none have isolated or described the oscillating nature of this pattern.

5. In the dealers' vernacular, this term is not used in the clinical sense of an individual psychopathology rooted in early childhood traumas. Instead, it resembles Lemert's (1962) more sociological definition which focuses on such behavioral dynamics as suspicion, hostility, aggressiveness, and even delusion. Not only Lemert, but also Waldorf et al. (1977) and Wedow (1979) assert that feelings of paranoia can have a sound basis in reality, and are therefore readily comprehended and even empathized with others.

6. At this point, a limitation to our data must be noted. Many of the dealers and smugglers we observed simply "disappeared" from the scene and were never heard from again. We therefore have no way of knowing if they phased-out (voluntarily or involuntarily), shifted to another scene, or were killed in some remote place. We cannot, therefore, estimate the numbers of people who left the Southwest County drug scene via each of the routes discussed here.

7. It is impossible to determine the exact percentage of people falling into the different phase-out categories: due to oscillation, people could experience several types and thus appear in multiple categories.

REFERENCES

Adler, Patricia A., and Peter Adler. 1982. "Criminal commitment among drug dealers." *Deviant Behavior* 3: 117–135.

Anonymous. 1969. "On selling marijuana." In Erich Goode (ed.), *Marijuana* (pp. 92–102). New York: Atherton.

Atkyns, Robert L., and Gerhard J. Hanneman. 1974. "Illicit drug distribution and dealer communication behavior." *Journal of Health and Social Behavior* 15(March): 36–43.

Biernacki, Patrick, and Dan Waldorf. 1981. "Snowball sampling." *Sociological Methods and Research* 10(2): 141–163.

Blum, Richard H. 1972. *The Dream Sellers*. San Francisco: Jossey-Bass.

Carey, James T. 1968. *The College Drug Scene*. Englewood Cliffs, NJ: Prentice Hall.

Douglas, Jack D. 1976. *Investigative Social Research*. Beverly Hills, CA: Sage.

Goode, Erich. 1970. *The Marijuana Smokers*. New York: Basic.

Langer, John. 1977. "Drug entrepreneurs and dealing culture." *Social Problems* 24(3): 377–385.

Lemert, Edwin. 1962. "Paranoia and the dynamics of exclusion." *Sociometry* 25(March): 2–20.

Lieb, John, and Sheldon Olson. 1976. "Prestige, paranoia, and profit: On becoming a dealer of illicit drugs in a university community." *Journal of Drug Issues* 6(Fall): 356–369.

Luckenbill, David F., and Joel Best. 1981. "Careers in deviance and respectability: The analogy's limitations." *Social Problems* 29(2): 197–206.

Mouledoux, James. 1972. "Ideological aspects of drug dealership." In Ken Westhues (ed.), *Society's Shadow: Studies in the Sociology of Countercultures* (pp. 110–122). Toronto: McGraw-Hill, Ryerson.

Riemer, Jeffrey W. 1977. "Varieties of opportunistic research." *Urban Life* 5(4): 467–477.

Waldorf, Dan, Sheila Murphy, Craig Reinarman, and Bridget Joyce. 1977. *Doing Coke: An Ethnography of Cocaine Users and Sellers*. Washington, DC: Drug Abuse Council.

Wedow, Suzanne. 1979. "Feeling paranoid: The organization of an ideology." *Urban Life* 8(1): 72–93.

Epilogue

I t has been over a decade since Sumner (1994) rang the death knell for the sociology of deviance. Two notable scholars in particular, Goode and Best, have
debated the vibrancy of the field, from the liveliness of its empirical and theoretical
contributions to its interest to students, scholars, and publishers. We believe in the
continuing vitality of the intellectual and empirical contributions of the field of
deviance in the early twenty-first century. There are many reasons to make such
a claim.

Here we sit, more than a dozen years after Sumner's proclamations, those
same textbooks and readers now in their tenth (and beyond) editions, but also
new ones cropping up on the horizon. Our courses are filled, and a seemingly
endless number of empirical cases have arisen that solidify our strong stance
about how social power creates new categories of deviants. Goode reminds us
that interest in deviance is high, although Best argues that enrollments are not
an accurate measure of vitality. When we began teaching the course (to over
1,000 students a year) in the 1980s, students were mostly attracted to the deviant
aspects of the curriculum. During the 1990s student purpose became geared
more toward pre-law enforcement. The 2000s, however, have witnessed a healthy
mix of the two groups. As Erikson (1966) noted, ever since the earliest years of
American settlement, the discovery, apprehension, and punishment of deviants
have held a central place in the public interest. We *do* think that the continued
popularity of the courses and the research to support them are signs of deviance's
continued contributions to sociology.

We are no more theoretically bereft than any of our counterparts. Theoretical
and conceptual advancements come in increments. Heckert and Heckert's
(Chapter 2) recent analysis of the relationship between positive and negative

typologies of deviance, their connection to how the deviance is socially received as well as to 10 major middle-class norms represent an exciting innovation in the conceptualization of deviance. Their matrix integrates normative expectations (whether deviance "flips out" into nonconformism or "flips in" into overconformism) with society's collective evaluation of that deviation, enabling them to offer conceptual insight into why some instances of overconformism are negatively received and some types of underconformism are positively received.

Although it is rare that we witness the kind of revolutions that paradigmatically change our disciplines, one new metanarrative, still in its infancy, is the cultural studies or postmodern theory of deviance, which focuses on the social creation and historical context for generating meaning (Dotter 2004; Foucault 1979). Building on the interactionist perspective, Dotter focuses on the creation and contestation of stigma by importing concepts from the metaphor of film. He argues that stigmas are conferred in screenplay scenarios in three interactive layers. The first layer is the deviant event, involving acts, actors, normative definitions, and societal reaction. The second layer involves media reconstruction, where media, law enforcement, and other audiences offer interpretations of the deviant event. The third layer is the stigma movie, where mediated reconstructions become ideologized as social control narratives. Deviance, defined and applied, becomes a commodified cultural representation that is consumed through the celebrity-drenched popular culture and mediated through modern power structures. At the levels of individual concepts, process and structural models, and broad theories, conceptualization about deviance remains strong. Studies abound, grants are received, dissertations are produced, and the area remains one of the core foundations of sociology.

Deviance is all around us. It is ubiquitous. Now, more than ever, we see a barrage of case studies that stretch our imaginations of how far deviance can go, how far beyond the evolving limits of human (in)capacity this technologically advanced, warp speed society can take us. Our own current research on self-injury (see Adler and Adler 2007) is a case in point. The more involved with this group we get, the more "normalized" their behavior seems. When we first began the study, we were asked, "Who are these kooks? Why would people do that to themselves?" Many thought that these cutters, burners, and branders were presuicidal, but our research shows that this behavior is an increasingly common coping mechanism for dealing with typical feelings of teen angst, with situations of powerlessness, and with anger or fear coupled with frustration. Whether it be tattoos, cigarettes, new drugs, creative forms of sex, or multibillion-dollar fraud widely perpetrated, people incorporate these new forms of behavior into their repertoire and accept (or reject) the creativity of the human soul for expanding the boundaries of normative behavior. As Frank Zappa, a cultural icon from the 1960s used to say, "Without deviation from the norm, progress is not possible." This is heady stuff and speaks to the heart of the sociological enterprise. Deviance is not marginal; it is central to what we do.

47

The Relevance of the Sociology of Deviance

ERICH GOODE

In this final chapter, Goode introduces us to a debate that has been raging in the field of deviance since the last decade of the twentieth century, ever since its obituary was announced in 1994 by Sumner. Goode reviews some of the relevant arguments and traces two investigations he conducted to assess the field's vibrancy and appeal. It is interesting to note this controversy, but Goode's assessment reassures us that deviance, despite the proclamation of its demise by some scholars, seems to be thriving.

For more than 30 years I've been struggling with what seems to me to be a fundamental paradox. On the one hand, I see evidence everywhere of the central importance of the deviance concept. Now, as an aside, I admit this probably illustrates the principle that, to a herring merchant, everything is fish, or, as C. Wright Mills once said, to a shoemaker, everything is leather. Still, to me, it seems everything is deviance. But, in spite of the ubiquity of manifestations of deviance I see all around me, the very legitimacy of the concept has had more than its share of critics. And for more than a decade, sociological pundits have been proclaiming the "death" of the sociology of deviance (Sumner, 1994); or, even more grandly, "the death of deviance" (Miller, Wright, and Dannels, 2001), as if such a thing were possible. From time to time I've put in my two cents, writing articles explaining why this development has not and cannot come to pass (2002, 2003, 2004). But still, the critics pay no heed to me, and continue to issue their pronunciamentos.

THE POLITICS OF DEVIANCE VERSUS
THE POLITICS OF DEVIANCE

Not long ago, a sociologist named Anne Hendershott published a book entitled *The Politics of Deviance* (2002). When I stumbled across the title, I became very excited. As I said, I've been obsessed with the subject of deviance for decades, and for me the central issue has *always* been "the *politics* of deviance." I'm interested in how definitions of right and wrong are established and maintained, how collectivities in every society struggle, over notions of what is to be demarcated as acceptable and unacceptable behavior, beliefs, and even physical traits; what and who will be stigmatized; what and who will be honored and respected, what and who will be ignored, accepted, tolerated, and condoned. What and who will be regarded as *emblematic* of the society as a whole, and what and who will be relegated to the margins, emblematic only of the society's periphery—not entirely respectable, exemplary, or reputable. What views win out in this struggle that Edwin Schur—in his book of the same title, *The Politics of Deviance* (1980)—refers to as "stigma contests." In the words of Stephen Pfohl: "The story of deviance and social control is a battle story. It is a story of the battle to control the ways people think, feel, and behave. It is a story of winners and losers and of the strategies people use in struggles with one another. Winners in the battle to control 'deviant acts' are crowned with a halo of goodness, acceptability, normality. Losers are viewed as living outside the boundaries of social life as it ought to be, outside the 'common sense' of society itself" (1994: 3).

As I see it, the central idea in the sociology of deviance is that definitions of right and wrong do not drop from the skies. They are not preordained. They are humanly *produced, constructed* as a result of clashes of ideologies, interests—economic, social, cultural, political—the outcome of struggles between and among categories in the society, each vying for dominance, or at least acceptance, of the views and behaviors that characterize them as a social entity. So Hendershott's book, I thought, would offer a full, detailed, and systematic exposition of a perspective I had been thinking, talking, writing, and teaching about for some time.

I was in for quite a shock. Hendershott's book had *nothing* to do with what I thought it was about. In fact, it argued precisely the *opposite* position from my own: that deviance *is* preordained, *does* drop down from the skies, should *not* be thought of as constructed or relative to time and place—but *absolute*. I read the book in horror. I was hurled back into the nineteenth century, into a Platonic or Manichean absolutistic world of light and darkness.

As an aside, one of her claims was easily dispensed with: that no one wants to teach courses in deviance any more, because the field "died" a generation ago. So I checked the enrollments in 20 more or less representative sociology departments, and found, when I compared the 1970s with the 2000s, more departments today are offering a course in deviance during any given semester than was true 30 years ago, and about the same number of students are taking the course per semester for a given department. I also checked the sociology curricula at 25 major universities across the country, and found that the majority, two-thirds (or 16), *do* currently offer a course in deviance (2003). Clearly, *that* aspect of her claim, that the field had "died," at least with respect to course offerings and enrollments, was *completely* false, simply a fantasy on her part.

Hendershott's argument, in a nutshell, begins with the fact that sociology's founding fathers—Durkheim being a clear example—were guided by a firm moral compass. They recognized that some behaviors *are* harmful and *should* be condemned—that is, *should* be regarded *as* deviant. But the 1960s, Hendershott argues, marked a radical break with this traditional idea. Beginning with Becker and his ilk, she asserts, practitioners of the sociology of deviance argued for a kind of moral relativism that recognized no intrinsically evil deeds, only a marketplace of competing claims, each jostling for acceptance. Deviants *should* be condemned and stigmatized, Hendershott believes. Along these lines, then, she rails: *against* medicalizing the deviance of drug use and abuse; *against* removing the stigma from mental illness; *against* the "postmodern" normalization of pedophilia; *against* the removal of stigma against flamboyant and militant gays (and, by implication, against gays generally); *against* "celebrating the sexually adventurous [that is to say, "promiscuous"] adolescent"; and *against* downgrading the deviance of assisted suicide. As a result of the efforts of brave souls such as herself, Hendershott asserts, the pendulum is beginning to swing back once again. The concept of deviance, she says: "is being rediscovered by ordinary people who have suffered the real-world consequences of the academic elite's rejection of the concept. Those whose communities have been broken apart by failed welfare policies, or whose families have fallen apart as a result of teenage pregnancy or divorce, are now speaking out about the moral chaos that is destroying their neighborhoods, their schools and their families" (p.10).

The idea of deviance, she claims, *before* Howard Becker and the labeling theorists came along, was that it was tragically, perniciously harmful behavior, behavior that tore at the fabric of the society, undermined the social order, wreaked havoc on the community. Becker and his colleagues (1963, 1964) have warped, distorted, and *poisoned* the deviance notion, Hendershott argues—took the moral taint out of it, removed it from the negative valuation it so richly deserves—relativized it, made it impossible for us to say that deviance really means bad, evil, morally wrong. Sociologists have the *right*—indeed, the *obligation*—to condemn deviance, she says, and encourage their students and the general public to do so, too.

It was Becker's relativity, Hendershott claims, that killed off the concept that had previously led ordinary people to see that it was immoral, harmful, contrary to natural law and common sense. In opposition to the thesis of the moral relativists, she says, we must "draw from nature, reason and common sense to define what is deviant and reaffirm the moral ties that bind us together" (p. 11). In other words, Hendershott wants to *de-relativize, essentialize,* and *absolutize* deviance. In print, I said a few unkind things about her book (2003: 523–530), and then filed it away in some bin in my mind somewhere as the ravings of a right-wing ideologue, and went about my business.

THE ELECTION OF 2004

Then along came the election of 2004.

In the days following November 3rd, in a communication with Nachman Ben-Yehuda, an Israeli sociologist, I tried to explain the presidential election to him. He in turn told me the Israeli newspapers pointed out that the decisive equation in the election was that a substantial percentage of the population was brought into the polling booth because of their fear of and hatred for gays, especially for gay marriage, which helped Bush win. Then he said: "Maybe those Republicans will help bring deviance back into the center of things."

Anne Hendershott, I realized, was right! (At least about the importance of conventional morality.) She precisely captured the mood of a substantial segment of the ultraconservative fringe of the Republican Party who seek to use traditional values to condemn unconventional behavior. In other words, the 2004 election transformed Anne Hendershott in my mind from a crackbrained lunatic, spouting silly ideas I felt I was forced to critique, to a *harbinger,* a flagbearer of a particular perspective, that, while not widespread in academic sociology, *does* reflect the mood and thinking of a substantial segment of American society. And in this formulation, opposition to homosexuality, in the guise of hostility toward gay marriage, became one fulcrum of the affirmation of traditional institutions and practices, while opposition to abortion became the other. All encapsulated in the term "moral values." Or, even a shade more coded, "family values." Both of which in turn became code words for condemning *sexual activity outside of heterosexual marriage.*

According to a *Washington Post* poll, among voters for whom the issue of "moral values" was important, 80% voted for Bush. More than 40% of Bush's supporters were white, born-again Christians, as opposed to only 22% of the electorate as a whole (Finkel, 2004). I realize that pundits supporting Bush argue against the position I'm spelling out because they don't want it to seem as if the president was reelected by a bunch of yahoos from the boonies (Krauthammer, 2004). But to me, the evidence seems quite convincing. It was Bush who thanked the crowds for their prayers. While Kerry talked about a woman's right to choose, it was Bush who talked about "family values" and "moral values," again, code for conventional sexual practices, i.e., sex blessed by the sanctity of marriage. During the campaign, Bush even used the term the "culture of life," implying that Kerry favored a "culture of death."

It was specifically those states and those social categories where anti-abortion and anti-gay sentiment was the strongest which Bush carried. Though obviously, the "moral values" issue wasn't the only one in this election, it was to a substantial segment of the electorate, and it was the vote of that segment that carried the president to victory in several key states. While the issue of homosexuality did not need to be explicitly spelled out in the presidential races—except for Kerry mentioning the fact that Dick Cheney's daughter is a lesbian—it *was* fought over in state elections. Some form of ban on gay marriage was on the ballot in 11 states, and in all 11 states, the legislation passed. Tom Coburn, the winning Republican senatorial candidate from Oklahoma, claimed that "lesbianism is so rampant in some of the schools in southeastern Oklahoma that they'll only let one girl go to the bathroom [at a time]." "Now think about that issue," Coburn added. "How is it that that's happened to us?" That phrase, "that's happened to us," encapsulates the fear of a substantial proportion of some corners of this society in the early years of the twenty-first century, the fear that deviant behavior is becoming more common and socially acceptable. Coburn also advocated the death penalty for abortion doctors, a position not widely shared even among the electorate of Oklahoma, but one, if expressed by a winning senatorial candidate, nonetheless should give us thought about the mood of the public in some areas. Let me say, so-called "moral values" did *not* constitute fulcrum issues in the so-called "blue" or "metro" states of the West Coast, the upper Great Lakes, or the Northeast—they had very little resonance there—but they *were* hugely important, perhaps even pivotal issues, in several battleground "red" or "retro" states of the South, Midwest, and Rocky Mountain regions.

Here we see on the national canvas the importance of definitions of deviance—of notions of conventional morality—being fought out in the presidential race. And they are likely to play themselves out over the course of the next four years, or more, in those states in which these issues are very much alive, in the form of legislation and court decisions and very possibly as a consequence, personal stigma for gays and women who seek an abortion. And with the election of George Bush, *Roe v. Wade* will be under attack over the next four years. And in a variety of ways, in much of the country, homosexuality, which prior to 2004 had been "departing from deviance" (Minton, 2002; Goode, 2005: 238–246), may very well come to be regarded as *more* deviant when Bush leaves office than when he was reelected.

Without becoming self-righteous or preachy about it, for many of our fellow citizens, the outcome of the election may in the near future make a difference between being *denied* legal rights and *having* legal rights, between deviance and conventionality, between stigma and respectability. And the election forced me to ask, in a somewhat belligerent fashion:

Who says the topic of deviance is irrelevant, dead, or unimportant? A defensive position, I agree, but one born of years of thinking about such issues.

THE TRAJECTORY OF THE DEVIANCE CONCEPT

How had it come to this? I wondered.

The dramatic relevance of the deviance concept in the 2004 campaign, combined with the continued drumbeat of claims that the field of the sociology of deviance is "dead" or at least ailing, forced me to rethink the foundation and the history of the field.

As a field of study, and as I define it, the sociology of deviance is, as sociological subfields go, fairly young. As conceived of as a normative violation that tends to generate negative reactions, the contemporary notion of deviance is only a bit more than half a century old, born, as it was, from a clearly articulated definition in the work of Edwin Lemert (1948, 1951). But in spite of the publication of a major, mainstream textbook in 1957 (Clinard, 1957), the field remained a stodgy, not especially exciting younger brother or sister of the larger field of social problems until 1963 and 1964, when Howard S. Becker published his collection of essays, *Outsiders,* and his anthology, *The Other Side.* These books jet-propelled the field of the sociology of deviance into the limelight of academic prominence, made it seem exciting, a fresh, novel, almost revolutionary way of looking at a newly-carved out subject. But practically from the beginning, the sociology of deviance was the target of criticism, some of it savage and denunciatory. As we know, the price of academic prominence is attracting critics, and beginning in the 1960s, and increasingly throughout the 1970s, the field attracted a lively and spirited host of critiques.

Probably the first of such critiques, by Alvin Gouldner, appeared in 1968; it attacked Howard Becker and his school for portraying the deviant as sneaky rather than defiant, and for not taking sides in defending the underdog against the corporate elite. Four years later, Alexander Liazos condemned the practitioners of the field for concentrating on "nuts, sluts, and deviated preverts," and for ignoring the corporate and high crimes and misdemeanors of the rich and powerful. In short, in the full flush of its initial influence and popularity, the sociology of deviance was under assault. These opening salvos argued that the field was biased and the concept on which it was based, illegitimate, in effect, premised on fatally flawed assumptions. But the most recent attacks have argued that both field and concept have already met their demise, that they are in fact "dead."

Anne Hendershott's screed is far from unique. In 1994, Colin Sumner, approaching the issue from a radical perspective, wrote a volume-length "obituary" for the sociology of deviance. The field arose, Sumner argues, as a mechanism of social control, to defend the interests of the rich and powerful. Given the diversity that prevails today, he says, the field no longer serves the purpose of keeping wrongdoing in check. Polished off by the critiques of radicals, "critical" theorists, and the "new" criminologists, the sociology of deviance is now a corpse.

These recent statements proclaiming the "death" of the sociology of deviance cry out for an evaluation of the claim. Miller, Wright, and Dannels (2001) test the "death" proposition with the use of citations, which indicate that roughly half those in the deviance literature are to works by criminologists (Miller, Wright, and Dannels, 2001). And that among the most frequently cited works in the deviance literature, relatively few are recent (Miller, Wright, and Dannels, 2001). These two findings, the authors conclude, indicate that the field of deviance is less theoretically innovative than it was in the past. The field, they say, while not yet "dead," is less intellectually vital than it was in its heyday in the 1970s. Joel Best agrees; in a slim, book-length study on the "trajectory" of the "career" of the deviance concept (2004), he concludes the field that studies it is, again, not yet dead—but it is not *thriving* either. Fewer articles bearing the word "deviance" in their titles, he found, were published

T A B L E 47.1 Works with "Deviance" or "Deviant" in the Title, by Decade

	ARTICLES (listed in SSCI)			BOOKS (in UM library)	
Decade	Total Number	Number per Year	Decade	Total Number	Number per Year
1950s	3	0.3	1950s	0	0
1960s	129	12.0	1960s	12	1.2
1970s	404	40.4	1970s	73	7.3
1980s	528	52.8	1980s	55	5.5
1990s	423	42.3	1990s	44	4.4
2000s	141	35.2	2000s	18	4.5
2000	37	37	*SOURCE: University of Maryland Library.*		
2001	40	40			
2002	31	31			
2003	33	33			
Total N:	1628	202			

SOURCE: Social Science Citation Index.

Notes: All works with a non-sociological meaning have been eliminated.

in the 1990s in the field's leading journals, the *American Sociological Review,* the *American Journal of Sociology,* and *Social Forces,* than was true 10 or 20 years prior (Best, 2004: ix–xi). On the other hand, in *Creating Deviance,* what he describes as a postmodernist text on the sociology of deviance (2004), Daniel Dotter refers to the "death" claim as an "obituary absent a demise" (pp. 277–278) a brilliant phrase, in my estimation.

So, is the sociology of deviance "dead"—or not? Is it "thriving"—or not? Is it still relevant to the sociologist—or not?

What I've done is a little counting exercise to get some sense of the popularity of the deviance concept. I obtained a print-out of the titles of the 200-plus books in the University of Maryland library with the word "deviance" or "deviant" in their titles, and the 1,600-plus articles published in academic (and a few non-academic) journals indexed by the *Social Science Citation Index,* again, with the terms "deviance" or "deviant" in their titles. Obviously, I eliminated those titles that used the term in a non-sociological way, for instance, if it referred to a statistical or engineering concept, I deleted it.

Articles and books are published in a given area for many reasons. Obviously, the publication in a given year of an article or book reflects research activities that were ongoing two to five years before the year of a publication. Publications and citations are not a perfect reflection of a field's intellectual activity. But when a field becomes irrelevant, nobody talks or writes about it. Even when a field is under attack, if it's a phenomenon of note, it will be talked and written about.

The University of Maryland library has no books with the word "deviance" in the title published before 1960, but 12 that were published in the 1960s, including two designed as textbooks, and three anthologies, also indicated for classroom use. In addition, the journal literature manifests a similarly dramatic leap in interest in the field of deviance: Among the titles of the articles published in the academic journals indexed by the *Social Science Citation Index* that include the words "deviance" or "deviant," only three were published in the 1950s, but 129 were published in the 1960s. So, clearly, by the 1960s, the field had arrived as an academic discipline. In other words, interest in the field grew *substantially* between the 1950s and the 1960s, disproportionate, I'd guess, to the growth of academia generally, and even to the growth of sociology specifically. In any case, the number of book titles grew from 12 in the 1960s to 73 in the 1970s, and the number of articles with deviance in the title indexed in the *Social Science Citation Index,* jumped again from 12 per year in the 1960s to 40 per year in the 1970s. Clearly, by the 1970s, the subject had become *hugely* important in sociology. It was talked about, written about, researched, critiqued, attacked; it had become an intellectual phenomenon of note.

Now, Sumner dates the demise of the sociology of deviance at 1975, that is, after Gouldner's 1968 critique, Liazos' 1972 critique, and the appearance of two books by British criminologists Taylor, Walton, and Young in 1973 and 1975, which attacked the field from a radical or Marxist perspective. But it's not clear what Sumner means by "death," since he admits that studies are still being conducted and books are still being written under the field's umbrella. And what I found was the decade *after* the decade he selects as the era of the field's demise—that is, in the 1980s—the number of articles on deviance grew enormously. With respect to articles, the *1980s* turns out to have been the heyday of the sociology of deviance, with over 50 published each year on the subject. Even more embarrassing for Sumner's argument, there were more articles published in the 1990s bearing the word "deviance" in the title than there were in the 1970s, indicating that Sumner's "obituary" for the field for 1975 was a bit premature. Even in the 2000s, the field seems to be going strong, with respect to both books and articles. Clearly, the field of the sociology of deviance is not "dead".

REFERENCES

Becker, Howard S. 1963. *Outsiders: Studies in the Sociology of Deviance.* New York: Free Press.

———. (ed.). 1964. *The Other Side: Perspectives on Deviance.* New York: Free Press.

Best, Joel. 2004. *Deviance: Career of a Concept.* Belmont, CA: Wadsworth.

Clinard, Marshall B. 1957. *Sociology of Deviant Behavior.* New York: Reinhart.

Dotter, Daniel. 2004. *Creating Deviance: An Interactionist Approach.* Walnut Creek, CA: Altamira Press.

Goode, Erich. 2002. "Does the Death of the Sociology of Deviance Make Sense?" *The American Sociologist,* 33 (Fall): 107–118.

———. 2003. "The MacGuffin that Refuses to Die: An Investigation into the Condition of the Sociology of Deviance." *Deviant Behavior,* 24 (November–December): 507–533.

———. 2004. "The 'Death' MacGuffin Redux: Comments on Best," *Deviant Behavior,* 25 (September–October): 493–509.

———. 2005. *Deviant Behavior* (7th ed.). Upper Saddle River, NJ: Prentice Hall.

Gouldner, Alvin W. 1968. "The Sociologist as Partisan: Sociology and the Welfare State." *The American Sociologist,* 3 (May): 103–116.

Henderschott, Anne. 2002. *The Politics of Deviance.* San Francisco: Encounter Books.

Krauthammer, Charles. 2004. "'Moral Values' Myth." *The Washington Post,* November 12, p. A25.

Lemert, Edwin M. 1948. "Some Aspects of a General Theory of Sociopathic Behavior." *Proceedings of the Pacific Sociological Society,* 16 (1): 23–29.

———. 1951. *Social Pathology.* New York: McGraw-Hill.

Liazos, Alexander. 1972. "The Poverty of the Sociology of Deviance: Nuts, Sluts, and Deviated Perverts." *Social Problems,* 20 (Summer): 103–120.

Miller, J. Mitchell, Richard A. Wright, and David Dannels. 2001. "Is Deviance 'Dead'? The Decline of a Sociological Research Specialization," *The American Sociologist,* 32 (Fall): 43–59.

Minton, Henry L. 2002. *Departing from Deviance: A History of Homosexual Rights and Emancipatory Science in America.* Chicago: University of Chicago Press.

Pfohl, Stephen. 1994. *Images of Deviance and Social Control: A Sociological History* (2nd ed.). New York: McGraw-Hill.

Schur, Edwin M. 1980. *The Politics of Deviance: Stigma Contests and the Uses of Power.* Englewood Cliffs, NJ: Prentice-Hall/Spectrum.

Sumner, Colin. 1994. *The Sociology of Deviance: An Obituary.* Buckingham, UK: Open University Press.

Taylor, Ian, Paul Walton, and Jock Young. 1973. *The New Criminology: For a Social Theory of Deviance.* London: Routledge & Kegan Paul.

Taylor, Ian, Paul Walton, and Jock Young (eds.). 1975. *Critical Criminology.* London: Routledge & Kegan Paul.

References for the General and Part Introductions

Adler, Patricia A., and Peter Adler. 1987. *Membership Roles in Field Research*. Newbury Park, CA: Sage.

———. 1995. "The Demography of Ethnography." *Journal of Contemporary Ethnography* 24: 3–29.

———. 2006. "Deviant Identity." In George Ritzer, *Encyclopedia of Sociology*, Malden, MA: Blackwell.

———. 2007. "The Demedicalization of Self-Injury: From Psychopathology to Sociological Deviance." *Journal of Contemporary Ethnography*, forthcoming.

Becker, Howard S. 1953. "Becoming a Marihuana User." *American Journal of Sociology* 59: 235–43.

———. 1963. *Outsiders: Studies in the Sociology of Deviance*. New York: Free Press.

———. 1973. "Labeling Theory Reconsidered." Pp. 177–212 in *Outsiders*. New York: Free Press.

Best, Joel, and David F. Luckenbill. 1980. "The Social Organization of Deviants." *Social Problems* 28: 14–31.

———. 1981. "The Social Organization of Deviance." *Deviant Behavior* 2: 231–59.

Cloward, Richard, and Lloyd Ohlin. 1960. *Delinquency and Opportunity*. Glencoe, IL: Free Press.

Cohen, Albert. 1955. *Delinquent Boys*. Glencoe, IL: Free Press.

Davis, Fred. 1961. "Deviance Disavowal: The Management of Strained Interaction by the Visibly Handicapped." *Social Problems* 9:120–32.

Dotter, Daniel. 2004. *Creating Deviance*. Walnut Creek, CA: Alta Mira.

Erikson, Kai T. 1966. *Wayward Puritans*. New York: Wiley.

Foucault, Michel. 1979. *Discipline and Punish: The Birth of the Prison*. New York: Pantheon Books.

Galliher, John. 1995. "Chicago's Two Worlds of Deviance Research: Whose Side Are They On?" Pp. 164–187 in *A Second Chicago School?* edited by Gary Alan Fine. Chicago: University of Chicago Press.

Godson, Roy, and William J. Olson. 1995. "International Organized Crime." *Society* 32 (January/February): 18–29.

Goffman, Erving. 1961. *Asylums*. Gardencity, NY: Doubleday.

———. 1963. *Stigma*. Englewood Cliffs, NJ: Prentice Hall.

Gusfield, Joseph. 1967. "Moral Passage: The Symbolic Process in Public Designations of Deviance." *Social Problems* 15: 175–88.

Henry, Jules. 1964. *Jungle People*. New York: Vintage.

Hewitt, John P., and Randall Stokes. 1975. "Disclaimers." *American Journal of Sociology* 38(3): 367–374.

Hilgartner, Stephen, and Charles L. Bosk. 1988. "The Rise and Fall of Social Problems: A Public Arenas Model." *American Journal of Sociology* 94: 53–78.

Hughes, Everett. 1945. "Dilemmas and Contradictions of Status." *American Journal of Sociology* (March): 353–59.

Kitsuse, John. 1980. "Coming Out All Over: Deviants and the Politics of Social Problems." *Social Problems* 28: 1–13.

Krauthammer, Charles. 1993. "Defining Deviancy Up." *The New Republic*, November 22: 20–25.

Lemert, Edwin. 1951. *Social Pathology*. New York: McGraw-Hill.

———. 1967. *Human Deviance, Social Problems, and Social Control*. New York: Prentice Hall.

Letkemann, Peter. 1973. *Crime as Work*. Englewood Cliffs, N.J.: Prentice Hall.

Loeber, Rolf, and David P. Farrington (eds.) 1998. *Serious and Violent Juvenile Offenders: Risk Factors and Successful Interventions*. Thousand Oaks, CA: Sage.

Luckenbill, David F., and Joel Best. 1981. "Careers in Deviance and Respectability: The Analogy's Limitations." *Social Problems* 29: 197–206.

Lyman, Stanford M. 1970. *The Asian in the West*. Reno/Las Vegas, NV: Western Studies Center, Desert Research Institute.

Matza, David. 1964. *Delinquency and Drift*. New York: Wiley.

Merton, Robert. 1938. "Social Structure and Anomie." *American Sociological Review* 3: 672–82.

Miller, Walter. 1958. "Lower Class Culture as a Generating Milieu of Gang Delinquency." *Journal of Social Issues* 14: 5–19.

Mills, C. Wright. 1940. "Situated Actions and Vocabularies of Motive." *American Sociological Review* 5(6): 904–913.

Moynihan, Daniel Patrick. 1993. "Defining Deviancy Down." *The American Scholar* 62: 17–30.

Polsky, Ned. 1967. *Hustlers, Beats, and Others*. Chicago: Aldine.

Pryor, Douglas. 1996. *Unspeakable Acts: Why Men Sexually Abuse Children*. New York: New York University Press.

Quinney, Richard. 1970. *The Social Reality of Crime*. Boston: Little, Brown.

Schur, Edwin M. 1971. *Labeling Deviant Behavior*. New York: Harper and Row.

————. 1979. *Interpreting Deviance*. New York: Harper and Row.

Scott, Marvin, and Stanford Lyman. 1968. "Accounts." *American Sociological Review* 33: 46–62.

Sellin, Thorsten. 1938. "Culture Conflict and Crime." A Report of the Subcommittee on Delinquency of the Committee on Personality and Culture, *Social Science Research Council Bulletin* 41. New York.

Smith, Alexander B., and Harriet Pollack. 1976. "Deviance as a Method of Coping." *Crime and Delinquency* 22: 3–16.

Spector, Malcolm, and John I. Kitsuse. 1977. *Constructing Social Problems*. Menlo Park, CA: Cummings.

Steffensmeier, Darryl, and Jeffery Ulmer. 2005. *Confessions of a Dying Thief: Understanding Criminal Careers and Illegal Enterprise*. New Brunswick: Transaction.

Sumner, Colin. 1994. *The Sociology of Deviance: An Obituary*. New York: Continuum.

Sumner, William. 1906. *Folkways*. New York: Vintage.

Sutherland, Edwin. 1934. *Principles of Criminology*. Philadelphia: J.B. Lippincott.

Sykes, Gresham, and David Matza. 1957. "Techniques of Neutralization: A Theory of Delinquency." *American Sociological Review* 22: 664–70.

Tannenbaum, Frank. 1938. *Crime and the Community*. Boston: Ginn.

Tittle, Charles R., and Raymond Paternoster. 2000. *Social Deviance and Crime*. Los Angeles: Roxbury.

Turner, Ralph H. 1972. "Deviance Avowal as Neutralization of Commitment." *Social Problems* 19 (Winter): 308–21.

Weatherford, Jack. 1986. *Porn Row*. New York: Arbor House.

Werner, Emmy E., and Ruth S. Smith. 1982. *Vulnerable but Invincible: A Longitudinal Study of Resilient Children*. New York: McGraw-Hill.

————. 1992. *Overcoming the Odds: High Risk Children from Birth to Adulthood*. Ithaca, NY: Cornell University Press.

Whitt, Hugh P. 2006. "Where Did the Bodies Go? The Social Construction of Suicide Data, New York City, 1976–1992." *Sociological Inquiry* 76: 166–87.

Yip, Andrew K. T. 1997. "Gay Male Christian Couples and Sexual Exclusivity." *Sociology* 31: 289–306.